Physical Activity and Health

Claude Bouchard, PhD
Pennington Biomedical Research Center

Steven N. Blair, PED
The Cooper Institute

William L. Haskell, PhD
Stanford University School of Medicine

Editors

Human Kinetics

Library of Congress Cataloging-in-Publication Data

Physical activity and health / Claude Bouchard, Steven N. Blair, William Haskell, editors.

    p. cm.
  Includes bibliographical references and index.
  ISBN-13: 978-0-7360-5092-0 (hard cover)
  ISBN-10: 0-7360-5092-2 (hard cover)
  1. Physical fitness--Health aspects. 2. Exercise--Physiological aspects. 3. Health. I. Bouchard, Claude. II. Blair, Steven N. III. Haskell, William L.
  RA781.P59 2006
  613.7'1--dc22

                                      2006011537

ISBN-10: 0-7360-5092-2
ISBN-13: 978-0-7360-5092-0

Copyright © 2007 by Human Kinetics, Inc.

The Web addresses cited in this text were current as of *March 22, 2006,* unless otherwise noted.

Acquisitions Editor: Michael S. Bahrke, PhD; **Developmental Editor:** Elaine H. Mustain; **Assistant Editor:** Lee Alexander; **Copyeditor:** Julie Anderson; **Proofreader:** Kathy Bennett; **Indexer:** Sharon Duffy; **Permission Manager:** Carly Breeding; **Graphic Designer:** Fred Starbird; **Graphic Artists:** Angela K. Snyder, Yvonne Griffith, Dawn Sills, and Kathleen Boudreau-Fuoss; **Photo Manager:** Sarah Ritz; **Cover Designer:** Keith Blomberg; **Photographer (cover):** © Royalty Free/Corbis; **Photographer (interior):** Human Kinetics, unless otherwise noted; **Art Manager:** Kelly Hendren; **Illustrators:** Denise Lowry and Keri Evans; **Printer:** Sheridan Books

We thank the University of Illinois Willard Airport in Savoy, Illinois, for assistance in providing equipment and locations for photographs taken for use in this text.

Printed in the United States of America   10 9 8 7 6 5 4 3 2 1

Human Kinetics
Web site: www.HumanKinetics.com

United States: Human Kinetics
P.O. Box 5076
Champaign, IL 61825-5076
800-747-4457
e-mail: humank@hkusa.com

Canada: Human Kinetics
475 Devonshire Road Unit 100
Windsor, ON N8Y 2L5
800-465-7301 (in Canada only)
e-mail: orders@hkcanada.com

Europe: Human Kinetics
107 Bradford Road
Stanningley
Leeds LS28 6AT, United Kingdom
+44 (0) 113 255 5665
e-mail: hk@hkeurope.com

Australia: Human Kinetics
57A Price Avenue
Lower Mitcham, South Australia 5062
08 8277 1555
e-mail: liaw@hkaustralia.com

New Zealand: Human Kinetics
Division of Sports Distributors NZ Ltd.
P.O. Box 300 226 Albany
North Shore City
Auckland
0064 9 448 1207
e-mail: info@humankinetics.co.nz

Contents

PART II

Effects of Physical Activity on the Human Organism   49

CHAPTER 16 Exercise and Its Effects on Mental Health 247

■ John S. Raglin, PhD ■ Gregory S. Wilson, PED ■ Dan Galper, PhD

CHAPTER 17 Physical Activity, Fitness, and Children. 259

■ Thomas W. Rowland, MD

CHAPTER 18 Physical Activity, Fitness, and Aging. 271

■ Loretta DiPietro, PhD, MPH

CHAPTER 19 Risks of Physical Activity 287

■ Evert A.L.M. Verhagen, PhD ■ Esther M.F. van Sluijs, PhD
■ Willem van Mechelen, MD, PhD

Preface

Physical Activity and Health is designed for upper-level undergraduate and graduate students studying the health benefits associated with a physically active lifestyle and the potential consequences of physical activity. The book is intended for students in kinesiology, exercise science, physical education, public health, health promotion, preventive medicine, and human biology programs.

This textbook provides an integrated treatise of the relationship between physical activity or sedentarism and health outcomes. It also provides a conceptual framework to help the student relate results from single studies or collections of studies to the overall paradigm linking physical activity and physical fitness to health. The book focuses on the prevention of diseases and enhancement of quality of life and well-being. It does not deal with the role of physical activity in the treatment of diseases and rehabilitation and is not intended to be an encyclopedic treatment of the field. Rather, each chapter provides an overview of the key concepts and the most important findings emerging from a collection of studies, discusses the limitations of the current knowledge base, and identifies research needs.

The need for this book became evident to us when reviewing this literature in the context of a variety of consensus conferences. There has not been a single book available to students that brings together the results of key studies and presents the relevant concepts in a detailed yet concise manner. The book was written with the collaboration of the finest scientists in the field from the United States, Canada, Europe, and Australia.

The book is organized into five parts containing 23 chapters. In the first part, three chapters define basic concepts, trace the history of the field, and summarize evidence accumulated on different levels of physical activity and fitness and their variations with age, between women and men, and among ethnic groups. Part II includes five chapters describing the advances made in understanding the effects of acute and chronic exposures to physical activity. Part III follows with 11 chapters that review the relationship between regular physical activity and health outcomes as well as between the level of fitness and the same health outcomes. These health outcomes range from cardiovascular morbidities to mental health to all-cause mortality. Part IV deals, in two chapters, with dose–response issues and types of exercise programs. Finally, Part V explores the challenges posed by advances in genetics in understanding the complex relationships among physical activity, fitness, and health produced by interindividual variability. Finally, the last chapter provides an integrated view of the field and discusses new opportunities for research and implementation of these concepts in a public health perspective.

The student will notice that the chapters are concise and are not referenced extensively. This was done so that the emphasis would be on the key concepts and the most important studies. Many of these references are review papers that the interested student can examine for more in-depth discussion of selected issues and for further references. The student will also recognize common features throughout the chapters, such as untitled special elements that summarize major points; titled special elements that present related topics of interest; key terms and concepts; and study questions. In addition, the chapters are well illustrated with tables and figures.

We hope that *Physical Activity and Health* will generate increased interest for the role that regular physical activity can play as a part of a lifelong, comprehensive preventive medicine plan.

Acknowledgments

The impetus for *Physical Activity and Health* began with a conversation between Dr. Rainer Martens, president of Human Kinetics Publishers, and one of us (CB) at the American College of Sports Medicine Annual Meeting held in St. Louis in 2002. After much discussion about a proposed publication and its content, we reached a consensus that the most pressing need was for a textbook for advanced undergraduate and graduate students that would cover a broad spectrum of topics in the area of physical activity and health. This is the goal that the three editors have tried to achieve with *Physical Activity and Health*.

We have been able to assemble a distinguished panel of collaborators to ensure that this textbook is a state-of-the-art publication. We are grateful to our colleagues from the United States, Canada, Europe, and Australia who accepted our invitation. We hope that they will be as pleased as we are with the quality of the publication.

We all are personally indebted to the pioneers in this field who continue to influence our research and that of others around the world. In this regard, we would like to recognize the enormous impact that Professors Ralph Paffenbarger Jr. and Jeremy Morris have had on us personally and on the field as a whole. As an expression of our gratitude, we would like to dedicate this book to them.

The three of us have enjoyed the generous and sustained support of the National Institutes of Health for our research over the years. One of us (CB) has also been the beneficiary of numerous grants from the Medical Research Council of Canada (as it was known before his relocation to the Pennington Biomedical Research Center in Louisiana) and other agencies from the government of Canada and the province of Quebec. We thank these agencies for their support over the years.

We also owe a great deal of gratitude to the colleagues, postdoctoral fellows, and students who have contributed so much to our research and productivity over the last few decades. It is impossible to recognize them individually here, but their names can be retrieved from our publications. They have greatly influenced our vision and thinking about issues of critical importance in preventing diseases and preserving health. For their contributions and patience with us, we are very grateful. Also, we thank our respective institutions, the Pennington Biomedical Research Center (and before, Laval University in Quebec City), the Cooper Institute, and Stanford University for providing environments conducive to exploring our academic interests.

The staff at Human Kinetics Publishers has been very supportive of this endeavor. They also have been very patient with us despite the fact that we missed some early deadlines. In particular, we would like to recognize and thank Mike Bahrke, acquisitions editor, and Elaine Mustain, developmental editor. Both Mike and Elaine nurtured this project as if it were their own and were very patient with the authors and editors. We would also like to thank Lee Alexander, assistant editor, and John Laskowski, knowledge management coordinator at Human Kinetics.

Finally, this book would not have been produced for 2006 and would not have been edited with extraordinary attention to countless details without the dedication and competent contribution of Mrs. Nina Laidlaw from the Pennington Biomedical Research Center. Nina made our responsibility as editors so much more enjoyable. Her incredible attention to detail in all aspects of the book, including interactions with the network of collaborators and the staff at Human Kinetics, is in the end responsible for much of its quality.

Even though Nina and the staff at Human Kinetics took great care in correcting errors and inconsistencies that may have existed in the manuscript, we take full responsibility for any omission and errors that may remain in the publication.

I

History and Current Status of the Study of Physical Activity and Health

Part I includes an overview of the evolution and emergence of physical activity and health as an area of scientific investigation. This is a young field; systematic research on physical activity and health has been under way since the middle of the 20th century. The field is maturing, the research database is extensive, and sedentary habits are now identified as a major public health problem in many countries of the world.

There are three chapters in part I that set the stage for what is to follow. These three chapters provide the organizational and conceptual framework for the book, a historical review of key developments, and the cur-

rent status of physical activity in the general population. Chapter 1, by the editors, provides our overall view of the purposes of the book. Chapter 2 includes an overview of key events in the development of physical activity and health as a scientific discipline in biomedical science and how this accumulated research helped make physical activity and health an important public health issue. Chapter 3 discusses physical fitness and activity with age among sex and ethnic groups. We hope that these introductory chapters will show you how the subdisciplines in exercise science and sports medicine are interrelated and are moving into an integrated field.

© PhotosForMe.com

Why Study Physical Activity and Health?

■ Claude Bouchard, PhD ■ Steven N. Blair, PED ■ William Haskell, PhD

CHAPTER OUTLINE

Human Evolution, History, and Physical Activity
- *Homo Sapiens* Is the Product of an Active Mode of Life
- Physical Activity: From the Advent of Agriculture to This Millennium

Burden of Chronic Diseases

Health and Its Determinants
- Health and Morbidity
- Genetic, Behavioral, and Environmental Determinants of Health

Aging and Health

Defining Physical Activity and Physical Fitness
- Defining Physical Activity
- Defining Physical Fitness

Physical Inactivity Versus Physical Activity
- Physical Inactivity As a Risk Factor
- Benefits of Regular Physical Activity

Summary

Review Materials

Because of the dramatic changes in the lives of people in industrialized countries over the last century, the necessity for most people to engage in challenging **physical activity** has disappeared. As physical activity has diminished, a host of physical ills related to inactivity have become manifest. Thus, intentional physical activity has become an important component of a healthy lifestyle. As you explore this issue, you will discover that *Homo sapiens* won the war against physical work of all kinds but has become afflicted by diseases brought about by a physically inactive lifestyle.

Your journey of exploration begins with this introductory chapter. In it you will learn about how health is defined in the 21st century compared with past periods; the concepts of health, quality of life, and longevity; the global burden of chronic diseases related to inactivity; some of the challenges posed by the aging of the population; the definitions of physical activity, fitness, and health paradigm; and why physical activity and fitness are poised to occupy a central place in preventive medicine and the public health agenda.

It is often said that the human body is designed for activity. Even though it is difficult to test this hypothesis in a formal experimental setting, at least three lines of evidence support this view.

- First, the human organism can adapt to a wide range of physical demands imposed by work and exercise. A young adult can easily increase his metabolic rate by tenfold when exercising and can sustain this rate of energy expenditure for a few minutes. It is not uncommon to find people who increase their energy output 100-fold above resting in maximal performance of very short durations. So, the human body architecture and physiology appear to be well organized to perform muscular work over a wide range of metabolic rates.

- Second, a low level of physical activity has been associated with a poor risk profile for common diseases, loss of functional capacity, and premature death.

- Third, the early humans could not have survived in life-threatening environments without having both adequate motor skills and the ability to perform demanding physical work.

Human Evolution, History, and Physical Activity

Evolution teaches us that those carrying genetic alleles favoring motor skills, strength, speed, stamina, and other physical attributes at relevant genes were more likely to have enjoyed greater reproductive fitness because of their greater probability of attracting mates and staying alive long enough to have children. The current chapter develops this line of reasoning. Other aspects of genetics are dealt with in greater detail throughout the book.

Homo Sapiens Is the Product of an Active Mode of Life

Although debate persists about the time and circumstances of the emergence of *Homo sapiens*, it is commonly recognized that the emergence of human beings was intimately related to progressive molecular changes in genes affecting brain functions of closely related nonhuman primates. The evolution of the brain meant not only greater brain capacity, progressive mastery of language, and more refined intelligence but also growing control over an expanded movement repertoire. From the research of paleontologists, anthropologists, anatomists, archeologists, and molecular biologists, the main events in the evolution of our species can be briefly outlined with an emphasis on those that have implications for physical activity.

The Human Body Is Designed for Activity

- The human organism can adapt to a wide range of metabolic demands imposed by work or exercise.
- A low level of physical activity is associated with risk for common diseases and premature death.
- Evolutionary history teaches us that early humans could not have survived without the ability to perform very demanding physical work.

Physical activity and **physical fitness** have been major factors in the evolutionary history of *Homo sapiens*. The most important events in the evolution of modern *Homo sapiens* occurred within the last 10 million years. The exact details of the molecular events that fueled this evolution are still a matter of debate. However, the end results of this complex journey and the molecular distances between *Homo sapiens* and closely related species are faithfully registered within the human genome. In brief, molecular alterations in the DNA of germ cells of our primate ancestors, combined with the effects of natural selection, led over millions of years to small creatures, clearly hominid in appearance, that are collectively referred to as *Australopithecus*. Several types of *Australopithecus* were uncovered on the African continent and dated as far back as 6 million years ago. In general, *Australopithecus* had a stature of about 1.5 m (5 ft) with a brain size of about 40% of that of modern *Homo sapiens*. Darwinian selection and undoubtedly random events progressively shaped humanlike forms of life until the modern *Homo sapiens* emerged about 100,000 years ago.

It is difficult to establish with precision the role played by motor ability and physical performance capacity in the evolutionary journey of our species. Comparative studies at the DNA and protein level between *Homo sapiens* and closely related nonhuman primates indicate that the genomic differences are generally on the order of 2% to 5%. However, these relatively small molecular differences carry major functional and behavioral implications, as is obvious to anyone who has observed today's primates in their natural habitat or in zoos around the world. Thus, human beings have optimized several traits that carried evolutionary advantages such as upright posture, bipedal locomotion, well-articulated thumbs for better hand prehension, vertical head position facilitating visual scanning, and refined language capacity, to name but a few. All of these characteristics had enormous selective advantages in the hostile environment prevailing during the evolutionary journey. They conferred an improved capacity to walk and run, grasp, carry, catch, and throw and to perform well in activities requiring quick responses, precision, speed, strength, and endurance.

Traits With Selective Advantages in the Hostile Evolutionary Environment

- Upright posture
- Bipedal locomotion
- Well-articulated thumbs for better hand prehension
- Vertical head position to facilitate visual scanning
- Refined language capacity

The emergence and spread of *Homo sapiens* involved conditions that required a high level of habitual physical activity, especially relative to today's standard. Furthermore, performance capacity and motor skill played a major role in survival. The best performers had a clear advantage in the quest for food and in defense against animal predators and times of conflict. Thus, the best male performers are likely to have contributed more genes to the next generations than the other males. Likewise, fitness and performance capacities played an important role in the success of females, who were called on to bear children as well as to help in gathering food, firewood, and other necessities while caring for their offspring.

Physical activity, then, has been a major force in the evolution of *Homo sapiens*. Hunting, gathering, escaping, and fighting were essential actions for the survival of our ancestors. They had to throw, lift, carry, climb, walk, run, and perform all kinds of basic motor skills throughout their lives. Thus, it is hard to imagine that physical activity and performance capacity were not important features during the evolutionary history of our species, conferring mating advantages to the carriers of the genetic alleles associated with these traits. Our ancestors would not have reached the age of reproduction if they had poor endurance, lacked speed and power, or had been clumsy. In other words, survival and reproductive success over tens of thousands of years required our ancestors to be physically active and good performers. Darwinian fitness was closely associated with physical fitness in the early ages of our species. Human reproductive capacity today, however, is less likely to depend on the level of physical activity and fitness than it did in the past (Malina 1991).

Physical Activity: From the Advent of Agriculture to This Millennium

With the advent of agriculture and animal domestication, when humans began to live in larger settlements, muscular work remained important. Strength, endurance, and skill must have been associated with economic success and survival in those times. Our ancestors learned to use various metals, the wheel was conceived, and tools of all kinds were developed to ease the burden of hunting, agriculture, and various domestic chores. Soon, some people began to have more leisure time. Indeed, archeological records have provided us with ample evidence of leisure activities in communities 5,000 to 8,000 years ago in several parts of the world. It is evident from museum artifacts that among these leisure activities, physical activities were quite popular. Images of foot races, throwing contests, wrestling, dances, and hunting are well represented in the archeological findings of this era.

By about three millennia ago, physical activities had become popular, in part because people believed that these activities influenced normal development and health. This was such a strong idea that elite performers were deified. In ancient Greece, the Olympic Games were started about 776 B.C. and were intimately associated with the civilization of the time.

During the next millennium, men and women remained interested in physical activities, as shown by the various types of games, tournaments, dances, and hunting expeditions that captured the attention of the nobles and the richest people. For most, however, physical activity meant long, hard days of physical labor that was necessary to subsist and satisfy a demanding master. Living conditions obviously had improved, but muscular work remained absolutely essential for the survival of the majority of people. Numerous wars also tested the fitness levels and performance capacities of the soldiers and of the people caught between rival factions.

The Renaissance period, with its taste for beauty and knowledge, changed the Western world. Physi-

cal activities remained quite popular and not only among the rich and noble. Games evolved but remained largely influenced by preoccupations with war and hunting. Large tournaments were regular occurrences on the aristocrats' agenda. Dances were also highly popular with the nobility. Peasants continued working hard, but they also enjoyed wrestling matches, horse racing, archery competitions, and dances. Jean-Jacques Rousseau introduced proposals to reform the education of children; these were quite compatible with the teachings of those who thought that physical activity should be part of the educational system. Widespread interest in sport soon appeared on the scene.

Throughout this journey, the struggle to free human beings from muscular work and physical exertion was a constant feature. It made impressive gains during the industrial revolution and even more in the last century through the technological progress achieved in industrialized countries. Lately, however, the alarm bell has begun to sound, proposing that the reduction in the amount of physical work may have gone too far (figure 1.1). The benefits of a physically active life style have been compared with those of an inactive mode of life, and although all the evidence is not in, it seems that human beings are better off when they maintain a physically active lifestyle. This is certainly one of the most striking paradoxes of the evolutionary and historical journey of *Homo sapiens*.

A Paradox From the Evolutionary Journey of *Homo Sapiens*

Now that the industrialized world has eliminated much of the need for hard physical labor to survive, the benefits of a physically active lifestyle have been compared with those of an inactive mode of life. Although not all the evidence is in, it seems that human beings are better off when they maintain a physically active lifestyle. This is certainly one of the most striking paradoxes emanating from the evolutionary and historical journey of *Homo sapiens*.

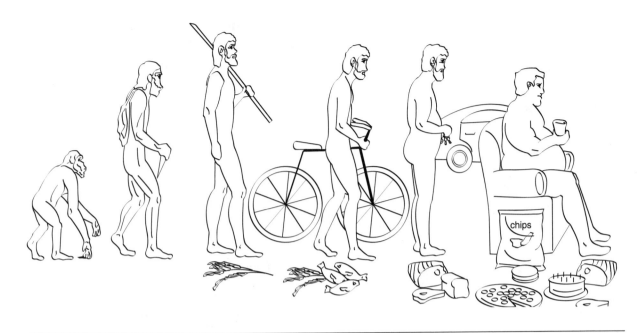

Figure 1.1 Is evolution leading *Homo sapiens* to *Homo sedens,* with a panoply of undesirable behavioral and biological features?

Courtesy of Nina Laidlaw.

Burden of Chronic Diseases

Chronic diseases are the most serious public health burden that the world faces today. Among these chronic diseases, cardiovascular diseases and cancer are the most important—and they are, as we shall see throughout this text, related to physical activity or lack of it. Moreover, the burden of chronic diseases is rapidly increasing around the world. It has been estimated that, at present, chronic diseases contribute approximately 60% of the 56 million reported annual deaths in the world (World Health Organization 2002). Almost half of the total deaths related to chronic disease are attributable to cardiovascular diseases. Obesity and diabetes are also on the rise. This trend is worrisome because they are both strong **risk factors** for vascular diseases and have started to appear earlier in life—even before puberty. It has been projected that by 2020, chronic diseases will account for almost three fourths of all deaths worldwide (World Health Organization 1998). The number of people with diabetes in the developing world will increase from 84 million in 1995 to 228 million in 2025 (Aboderin et al. 2001).

As for overweight and obesity, they are already epidemic in the developed nations and continue to increase in prevalence around the world, with more than one billion adults affected. The public health implications of these trends are staggering.

The situation for chronic diseases as leading causes of death in the United States is also quite striking. There are approximately 2.4 million deaths per year in the United States according to data from the Centers for Disease Control and Prevention, as summarized recently (Mokdad et al. 2004). Table 1.1 lists the 10 leading causes of deaths. Together, they account for more than 1.9 million of the deaths registered in 2000 in the United States. Among these leading causes, heart disease (29.6%), cancer (23.0%), and cerebrovascular disease (7.0%) are the dominant causes of premature death. They are the three diseases with the highest rates of occurrence per 100,000 persons. Physical activity and diet are thought to play an important role for at least four of these leading causes of death, that is, heart disease, malignant neoplasm, cerebrovascular disease, and diabetes mellitus. These four diseases were responsible for more than 1.5 million or more than 60% of the deaths in 2000.

Table 1.1   Leading Disease-Related Causes of Death in the United States in 2000

Causes of death	Number of deaths	% of all deaths	Rate/ 100,000
Heart disease	710,760	29.57	258.2
Malignant neoplasm	553,091	23.01	200.9
Cerebrovascular disease	167,661	6.98	60.9
Chronic lower respiratory disease	122,009	5.08	44.3
Unintentional injuries	97,900	4.07	35.6
Diabetes mellitus	69,301	2.88	25.2
Influenza and pneumonia	65,313	2.72	23.7
Alzheimer's disease	49,558	2.06	18.0
Nephritis, nephrotic syndrome, and nephrosis	37,251	1.55	13.5
Septicemia	31,224	1.30	11.3
Other	499,283	20.77	181.4
Total	2,403,351	100	873.1

The burden of mortality from cardiovascular causes is predicted to continue to increase in developed countries, as illustrated in figure 1.2. However, the projected increase in the number of deaths attributable to the same causes is much more dramatic in the developing world. For instance, there were nine million deaths per year from cardiovascular causes in the developing countries around 1990. In 2020, this number will reach approximately 19 million deaths per year. These numbers taken together with the already high prevalence rates of obesity and the predicted increases in cases of type 2 diabetes mellitus suggest that a devastating epidemic of common chronic diseases is currently in the making.

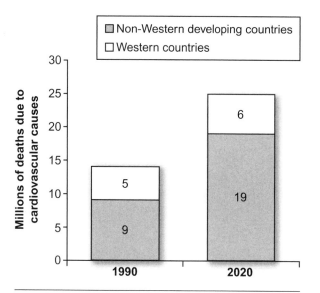

Figure 1.2 Projected death rates in developed and developing nations.

Burden of Common Chronic Diseases

- Currently, there are about 56 million deaths per year in the world.
- Chronic diseases account for 60% of these deaths.
- Cardiovascular disease accounts for half of the latter.
- Other causes of death from chronic diseases include those associated with cancer, diabetes, and obesity.

Health and Its Determinants

To take meaningful action on any social issue, one must first be clear on what that issue is, as well as the factors that affect it. Thus, before discussing the influence of physical activity and physical inactivity on health and morbidity, we will consider the definitions and components of health and morbidity, and the factors that affect them. In doing so, we will examine the concepts of active life expectancy, disability-free life expectancy, and wellness; and we will quickly survey the roles that genetic factors, behavioral traits, socioeconomic class, and quality and availability of medical care play in health and morbidity.

Health and Morbidity

Defining health remains a major challenge, despite the progress made in treating diseases and increasing the average life duration in Western societies. The World Health Organization described **health** as "a state of complete physical, mental, and social well-being and not merely the absence of disease or infirmity" (World Health Organization 1948). Health is "a human condition with physical, social and psychological dimensions, each characterized on a continuum with positive and negative poles. Positive health pertains to the capacity to enjoy life and to withstand challenges; it is not merely the absence of disease. Negative health pertains to morbidity and, in the extreme, with premature mortality" (Bouchard and Shephard 1994, p. 9).

Because health is complex and multifactorial and is not merely the absence of disease, traditional illness and mortality statistics do not provide a full assessment of health. A more comprehensive approach requires that the profile of the individual be established in terms of common health endpoints, risk factor profile, morbidities, temporary and chronic disabilities, physical and mental functional level, absenteeism, overall productivity, health-related fitness status, objective and perceived level of well-being, and use of all forms of medical services, including prescribed and nonprescribed drugs.

If the **health-related quality of life** is less than optimal, **life expectancy** should be adjusted to reflect a quality-adjusted value. Important concepts are those of **active life expectancy** and **disability-free life expectancy.** Active life expectancy is simply the age that a given person is expected to live free of conditions that may restrict his or her activities. On the other hand, disability-free life expectancy is the number of years of life remaining at a given age with no limitations attributable to physical or mental function impairments. Both concepts can be predicted based on age, gender, education, socioeco-

nomic circumstances, ethnic background, current health status, and other characteristics.

Morbidity can be defined as any departure from a state of physical or psychological well-being, short of death. Morbidity can be measured as

- the number of persons who are ill per unit of population per year,
- the incidence of specific conditions per unit of population per year, and
- the average duration of these conditions.

On the other hand, **wellness** is a holistic concept, describing a state of positive health in the individual and comprising physical, social, and psychological well-being.

Genetic, Behavioral, and Environmental Determinants of Health

Chronic diseases have complex etiologies. They are also heterogeneous in the sense that the paths leading to a disease manifestation vary from disease to disease and are also characterized by considerable individual differences. A large body of evidence indicates that genetic differences, behavior, and the physical and social environment all contribute in varying degrees to the burden of chronic diseases in any country.

Heart disease, stroke, cancer, type 2 diabetes, obesity, and other chronic conditions aggregate in families. The level of familial aggregation varies from condition to condition with a range from about 30% to 50% of the age- and gender-adjusted variance. This strongly suggests that genetic factors are involved. And indeed, a good number of molecular genetics studies have identified specific genes and mutations contributing to the burden of these common chronic diseases (see chapter 22). However, there is also strong evidence that behavioral factors contribute to the etiology of these diseases. Chronic diseases are largely preventable. Smoking, poor nutritional habits, excessive alcohol consumption, a sedentary lifestyle, substance abuse, and high-risk sexual behavior are among the behaviors typically associated with an increased risk of death or morbidity. The exact contribution of these behavioral traits to the global burden of disease is not easily quantified but appears to account for as much as one third of the variance.

Beyond genetic factors and behavioral traits, which together explain most of the predisposition to the major chronic diseases, the social and physical environment plays an important role. For example, people in low socioeconomic classes or with less education are more likely to be economically disadvantaged and are at a greater risk of being affected by chronic diseases and dying prematurely. Additionally, limitations in the health care delivery system, medical errors, and other situations out of personal control contribute to the fact that some affected individuals die prematurely.

Even though much more research is needed before an evidence-based breakdown of the causes of common chronic disease or of premature deaths can be established, the evidence suggests that genetic factors and behavioral traits contribute the most, perhaps as much as 80%, to the level of risk.

In this regard, the contributions of behavioral and selected environmental factors to the number of deaths in the year 2000 in the United States are summarized in table 1.2. It is quite clear from these data that smoking (18%), poor diet, and physical inactivity are the major behaviors associated with a greater risk of dying prematurely. (Estimates for poor diet and physical activity vary from 5% to 17% of all deaths.) These computations are admittedly soft, because they are based on prevalence data, relative risk, and population-attributable risk and thus require that several assumptions be made. Nonetheless, they emphasize that smoking, diet, and physical activity are important determinants of health.

The results from the INTERHEART study provide strong support for this notion. A total of 30,000 men and women from 52 countries were enrolled in this study (Yusuf et al. 2004). Half of these participants had experienced a myocardial infarction event. Irrespective of ethnic background and country of residence, the risk factor profile associated with a cardiac event was the same. This risk profile included smoking, sedentarism, and poor nutrition.

Poor nutritional habits play a key role in the etiology of several chronic diseases. A high-fat, high-sugar, energy-dense diet, with a substantial content of animal foods, appears to be the origin of many current health problems. However, diet is only one of the risk factors. Physical inactivity is also an increasingly important determinant of health. A sedentary lifestyle and poor nutritional habits, together with such other risk factors as tobacco use and stressful aspects of modern life, are potent enough to accelerate the development of chronic diseases, accentuate their severity, and contribute to the loss of function leading to frailty that often accompanies aging. More research is clearly needed on the mechanisms linking dietary habits and physical activity level to health. However, the available scientific evidence is already sufficiently strong to justify implementing preventive measures right now. The public health approach is likely to be the most cost-effective approach to coping with the chronic disease epidemic.

Public health is concerned not only about reducing smoking rates, improving the diet, and promoting physical activity. It is also concerned about the fact that as many as 85,000 deaths in the year 2000

Table 1.2 Leading Behavioral and Environmental Causes of Death in the United States in 2000

Behavior or external agent	Number of deaths	%
Smoking	435,000	18
Sedentarism and poor nutrition	112,000-400,000[a]	5-17
Alcohol consumption	85,000	4
Microbial agents	75,000	3
Toxic agents	55,000	2
Automobile accidents	43,000	2
Firearms	29,000	1
Risky sexual behavior	20,000	1
Illicit drug use	17,000	1
Total[b]	1,159,000	44

[a]Upper value from Mokdad et al. 2004; lower value from Flegal et al. 2005.

[b]Based on 2.4 million deaths in the United States in 2000.

Mokdad et al., 2004. *JAMA* 291:1238-1245. Copyright © 2004, American Medical Association. All rights reserved.

were alcohol-related; 43,000 deaths were caused by accidents or were related to the use of a motor vehicle (or refusal to buckle up in cars); and 29,000 deaths were caused by firearms in the United States alone. A global public health approach needs to take into account all of these factors plus all the other behaviors and agents identified in table 1.2.

Understanding what underlies the pervasiveness of such risk factors is the first step in reducing them and consequently diminishing their negative effects.

Aging and Health

The fact that the population is progressively getting older has an enormous impact on the importance of the physical activity and health paradigm (see chapter 18). The average life expectancy has substantially increased over the last 100 years. On average, women are expected to live about 80 years in some countries of the world, whereas men will live about 75 years. There are obviously considerable differences in these lifespan estimates even among developed countries.

Although Americans are living longer because of recent declines in heart disease and strokes, chronic diseases such as high blood pressure and diabetes are becoming more common among older adults. Aging can be defined as a progressive decline in the ability of an organism to resist stress, damage, and disease. It is characterized by an increase in the incidence of degenerative disorders. The United States population over 65 is projected to grow from 35 million in 2000 to 70 million in 2030. By 2030, one in five Americans will be age 65 or older. The largest increase in the aging population is occurring among those 85 years and older. By the year 2020, there will be close to 10 million Americans above the age of 85 (figure 1.3).

More than 20% of U.S. adults over the age of 65 live with at least partial disability, defined as some degree of difficulty in performing activities of daily living, according to the National Center for Health Statistics. Above the age of 85, 45% of persons need some assistance with one or more basic activities of daily living. This trend has enormous implications for health care systems. It also represents a challenge for all those who believe that regular physical activity is essential to preserve autonomy as we age. The statistics showing that the prevalence of common chronic diseases is on the rise worldwide, together

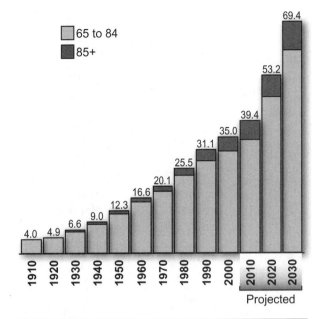

Figure 1.3 Changes since 1910 and the projection until 2030 of the number (in millions) of U.S. adults aged 65 to 84 years and those 85 and older.
Data from the U.S. Census Bureau.

with the fact that people live longer, suggest that societies will likely face a major escalation of health care costs and a public health crisis in the coming decades.

Aging is a very complex process and is only just beginning to be understood. Many factors are involved in the aging process including endogenous cellular processes, environmental insults, and interactions among environmental factors—including those associated with nutrients and the demands of physical work—with one's genome. No single factor can explain the variation in the way we look, feel, or behave as we get older. But it remains clear that regular physical activity is one of the most important lifestyle components for preventing the age-related decline in overall physical independence and well-being.

Defining Physical Activity and Physical Fitness

In this section, we consider the definitions of physical activity and fitness. Both are complex concepts that cover a number of components with potential applications to several fields of study and practice.

Defining Physical Activity

Physical activity comprises any bodily movement produced by the skeletal muscles that results in a substantial increase over resting energy expenditure. Under this broad concept, we need to consider leisure-time physical activity, exercise, sport, transportation, occupational work, and chores. The energy expenditure associated with physical activity is the only discretionary component of total daily energy expenditure. Energy expenditure of activity is typically only about 25% of daily energy expenditure in a sedentary person, whereas it may be as high as 50% in an endurance athlete on a training day or in persons performing heavy labor for many hours during the day.

Components of Total Daily Energy Expenditure

- Basal and resting metabolic rate account for about 65% of daily energy expenditure.
- Because of their high metabolic rates, cardiac muscle, liver, brain, kidney, pancreas, and other organs account for about 70% of the energy expended at rest.
- The thermic response to food (absorption, digestion, transport, and storage) accounts for about 10% of daily energy expenditure.
- Physical activity and movement of all types account for about 25% of the energy expended in a typical day by a sedentary person.

Leisure-Time Physical Activity

In most developed societies, after completion of work, traveling, domestic chores, and personal hygiene, the average person has 3 to 4 hours of "free," leisure, or discretionary time per day. However, there is wide interindividual variation, depending in part on such personal circumstances as the duration of paid work, the division of labor in the home, the need for self-sufficiency activities, daily travel time, and the number and age of dependents.

- *Leisure-time physical activity* is an activity undertaken in the individual's discretionary time

that increases the total daily energy expenditure. The element of personal choice is inherent to the definition. Activity is selected on the basis of personal needs and interests. When the motivation is to improve health or fitness, the pattern of activity undertaken will be consonant with this objective. But there are many other possible motivations (Dishman 1988) including aesthetic motivations (pursuit of a desired body type or an appreciation of the beauty of movement), ascetic issues (the setting of a personal physical challenge), thrill of fast movement and physical danger, chance and competition, social contacts, fun, mental arousal, relaxation, and even addiction to endogenous opioids (Bouchard and Shephard 1994).

- *Exercise* is a form of leisure-time physical activity that is usually performed repeatedly over an extended period of time (exercise training) with a specific external objective such as the improvement of fitness, physical performance, or health. When prescribed by a physician or exercise specialist, the regimen typically covers the recommended mode, intensity, frequency, and duration of such activity. For example, figure 1.4 shows the six intensity levels of endurance exercise, ranging from very light to maximal. Figure 1.4a displays the relationship between intensity of exercise and heart rate expressed as a percentage of the maximal attainable heart rate. Figure 1.4b illustrates the power output in **metabolic equivalents (METs)** across the six categories of exercise intensity for two individuals, one with a **maximal oxygen uptake ($\dot{V}O_2$max)** of 10 METS (i.e., equivalent to 10 times the resting energy expenditure) and the other with a $\dot{V}O_2$max of 5 METs, as in some elderly people.

- *Sport* is a form of physical activity that involves competition. In general, a sport is a competitive activity undertaken in the context of rules defined by an international regulatory agency. However, in some parts of the world, the term *sport* may also embrace exercise and recreation (as in the UNESCO "Sport for All" movement) (McIntosh 1980).

Work

Work is also an important component of daily activities. In the past, energy expenditures required by occupational work and the associated demands of transportation (on foot or on a bicycle) or work around the house accounted for a major fraction of the total daily metabolism in a large segment of the labor force.

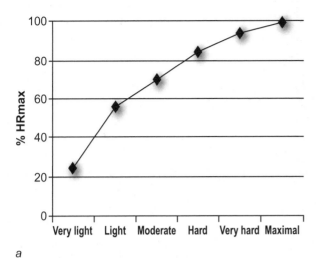

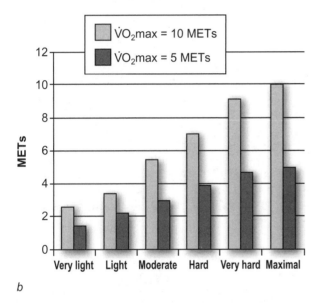

b

Figure 1.4 The definition of exercise intensity varies according to the level of fitness and can be expressed in terms of (*a*) % HRmax or (*b*) METs.

Table 1.3 Intensity of Occupational Work

Intensity	Energy expenditure in kcal·min⁻¹
Sedentary	<2.0
Light	2.0-3.5
Moderate	3.5-5.0
Heavy	5.0-7.5
Very heavy	>7.5

Data from J.R. Brown and G.P. Crowden, 1963, "Energy expenditure ranges and muscular work grades," *British Journal of Industrial Medicine* 20: 227.

the duration of individual activity bouts is usually prolonged, often there are other adverse circumstances (such as a high environmental temperature, awkward posture, or a heavy loading of small muscle groups), and normally the pace of working is set by such factors as a machine, a supervisor, or a union contract, rather than the individual (Bouchard and Shephard 1994).

• Household and other chores also need to be considered. Automation has progressively reduced the energy demands associated with the operation of a household in developed societies. Although some individuals may deliberately seek out heavy activities, most necessary domestic chores now fall into the "light" category of the industrial scale. Possible exceptions are the care of dependents (playing with young children and caring for elderly relatives) and vigorous gardening, which can on occasion involve quite heavy work.

Defining Physical Fitness

There is no universal agreed upon definition of fitness and of its components. The World Health Organization defined fitness as "the ability to perform muscular work satisfactorily" (World Health Organization 1968). Fitness implies that the individual has attained those characteristics that permit an acceptable performance of a given physical task in a specified physical, social, and psychological environment. Fitness is primarily determined by variables including the individual's pattern and level of habitual activity and heredity. Fitness is typically defined with a focus on two goals: performance or health.

• *Performance-related fitness* refers to those components of fitness that are necessary for optimal work or sport performance (Bouchard and Shephard 1994; Pate 1988). It is defined in terms

• Heavy occupational demand has had considerable epidemiological interest (Paffenbarger et al. 1990) in the past, because it was typically sustained for 30 to 40 hr/week over many years. This is still true in some developing societies, and even in the Western world there are still occupational categories with a high energy demand. However, today in the industrialized world, "heavy" employment is frequently accompanied by low levels of leisure-time physical activity. The standards defining a high or a very high intensity of occupational activity differ from those applicable to "exercise" (table 1.3). For instance, heavy work is defined as an energy expenditure of 5 to 7 kcal/min. This is because in industry, of the individual's ability in athletic competition. Performance-related fitness depends heavily on motor skills, cardiorespiratory power and capacity, muscular strength, speed, power or endurance, body size, body composition, motivation, and nutritional status. Performance-related fitness is not considered in detail in this book.

• *Health-related fitness,* in contrast, refers to those components of fitness that are affected favorably or unfavorably by habitual physical activity habits and that relate to health status. Health-related fitness has been defined as a state characterized by an ability to perform daily activities with vigor and by traits and capacities that are associated with a low risk for the development of chronic diseases and premature death (Pate 1988). Important components of health-related fitness include those listed in *Health-Related Fitness Components and Traits* (see below). These biological traits relate to health outcomes as assessed by the profile of risk factors and by morbidity and mortality statistics. They are grouped under five major components: morphological, muscular, motor, cardiorespiratory, and metabolic fitness. These components and embedded traits are addressed in various chapters of this book.

Definition of Fitness and Related Concepts

- Fitness is the ability to perform muscular work satisfactorily.
- Performance-related fitness refers to the components of fitness that are necessary for maximal sports performance.
- Health-related fitness refers to those components of fitness that benefit from a physically active lifestyle and relate to health.

Physical Inactivity Versus Physical Activity

The main emphasis of this book is on the deleterious effects of physical inactivity and the benefits of a physically active lifestyle. The central problem is briefly defined in the following paragraphs.

Physical Inactivity As a Risk Factor

Homo sapiens has attempted for millennia to reduce the amount of muscular work and physical activity required in daily life. The war on muscular work has been a remarkable success. Thus, the amount of

Health-Related Fitness Components and Traits

- Morphological component
  - Body mass for height
  - Body composition
  - Subcutaneous fat distribution
  - Abdominal visceral fat
  - Bone density
  - Flexibility
- Cardiorespiratory component
  - Submaximal exercise capacity
  - Maximal aerobic power
  - Heart functions
  - Lung functions
  - Blood pressure

- Muscular component
  - Power
  - Strength
  - Endurance
- Motor component
  - Agility
  - Balance
  - Coordination
  - Speed of movement
- Metabolic component
  - Glucose tolerance
  - Insulin sensitivity
  - Lipid and lipoprotein metabolism
  - Substrate oxidation characteristics

Adapted, by permission, from C. Bouchard, R.J. Shepherd, T. Stephens, 1994, *Physical activity, fitness and health: International proceedings and consensus statement* (Champaign, IL: Human Kinetics), 81.

energy expended by individuals to ensure sustained food supply, decent housing under a variety of climatic conditions, safe and rapid transportation, personal and collective security, and diversified and abundant leisure activities has decreased substantially. The decline in the amount of physical activity has been so dramatic that a variety of health problems, nurtured by a sedentary mode of life, increased considerably in the 20th century. These health problems were referred to as "hypokinetic diseases" more than 40 years ago (Krauss and Raab 1961).

Many agents reduced our overall amount of muscular work and increased our sedentary time. Motorized transportation is undoubtedly at the top of the list. Labor-saving devices and systems in the work environment also play a major role. Computers and a large number of electrically powered tools and gadgets have dramatically reduced the need to rely on muscular work. Television, video games, and domestic labor-saving devices have all contributed to the increase in sedentary time. Elevators, escalators, and other convenient modes of moving up and down in the urban environment have made the situation even more serious. Urban design generally favors the use of the automobile and has made it more challenging for people to be physically active.

In industrialized countries, most citizens who want to adopt a physically active lifestyle have to do so using some of their leisure time. However, in some of these countries, a substantial fraction of the people get plenty of physical activity simply because they choose to cycle or walk as their preferred mode of transportation. Table 1.4 lists travel modes in 12 countries of Western Europe and North America. Note that 40% or more of the adult populations are walking or bicycling as their mode of transportation in four of these countries: Austria, Denmark, Netherlands, and Sweden. In contrast, only about 10% of the adults in Canada and the United States do so. These observations suggest that promoting a physically active lifestyle represents a greater challenge in North America. These issues are discussed in subsequent chapters of this book.

Benefits of Regular Physical Activity

The relationship between physical activity and health is more complex than is apparent with a cursory glance. Figure 1.5 depicts the simplest path linking physical activity and health. The message from this diagram is quite straightforward: Physical activity is associated with health benefits. A low level of physical activity is likely to translate into unfavorable health outcomes, whereas the converse would be true for a high level of physical activity. However, the reality is more complex. For instance, consider the path diagram in figure 1.6. On average, and in most people, regular physical activity increases

Table 1.4 Travel Modes in Europe and North America

Country	Bicycle	Walking	Public transport	Auto	Other
Netherlands	30	18	5	45	2
Denmark	20	21	14	42	3
West Germany	12	22	16	49	1
Switzerland	10	29	20	38	3
Sweden	10	39	11	36	4
Austria	9	31	13	39	8
East Germany	8	29	14	48	1
England and Wales	8	12	14	62	4
France	5	30	12	47	6
Italy	5	28	16	42	9
Canada	1	10	14	74	1
United States	1	9	3	84	3

Column header spans: Percentage of trips by travel mode

Adapted with permission of the Eno Transportation Foundation, Washington DC, from "Bicycling Boom in Germany: A Revival Engineered by Public Policy," *Transportation Quarterly* 51: 31-36. Copyright 1997 Eno Transportation Foundation.

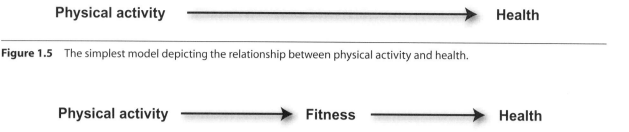

Figure 1.5   The simplest model depicting the relationship between physical activity and health.

Figure 1.6   A model in which the effects of physical activity on health are mediated by increases in fitness.

health-related fitness. This implies an increase in cardiorespiratory endurance and in insulin action in tissues such as skeletal muscle, an increase in high-density lipoprotein cholesterol, a decrease in blood pressure, and a decrease in whole-body adiposity, to name but a few. Such improvements in fitness are likely to have favorable effects on overall health.

However, a body of data shows that some health benefits are derived from being physically active even though there may be no or little associated gain in fitness as it is traditionally measured. Thus, two paths potentially contribute to the relationship between regular physical activity and health, one independent of changes in fitness and the second mediated by the gains in physical fitness (figure 1.7).

The reality is even more complex than suggested by the path diagram depicted in figure 1.7. For instance, the healthier are typically more active and are the fittest in a population. Thus the paths from activity or fitness to health are not necessarily causal paths. Furthermore, habitual physical activity can influence fitness, which in turn may modify the level of habitual physical activity (figure 1.8). For example, with increasing fitness, people tend to become more active, and the fittest become the most active. These potentially confounding relationships need to be taken into account to understand the relationships between regular physical activity and health.

Needless to say, the relationships among levels of physical activity, health-related fitness, and health are even more complex than suggested by the previous models. Figure 1.9 illustrates additional complexities in these relationships. This model shows that habitual physical activity can influence fitness, which in turn may modify the level of habitual physical activity. It not only shows that people tend to become more active with increasing fitness and that the fittest individuals tend to be the most active; it also specifies that fitness is related to health in a reciprocal manner. That is, fitness influences health, and health status also influences both habitual physical activity level and fitness level. Other factors are associated with individual differences in health status. Likewise, the level of fitness is not determined entirely by an individual's level of habitual physical activity. Other lifestyle behaviors, physical and social environmental conditions, personal attributes, and genetic characteristics also affect the major components of the basic model and determine their interrelationships.

Subsequent chapters discuss in greater detail how habitual physical activity and health-related fitness are related to various health outcomes. In this context, understanding the extent and causes of human variation is an important issue that is addressed in chapter 22. However, it is useful to appreciate early in this book that the causes of human variation are

Figure 1.7   In this model, physical activity has direct fitness benefits but also improves health.

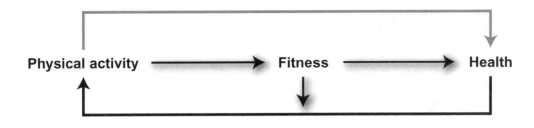

Figure 1.8 The model specifies not only that physical activity and fitness are positively associated with health but also that healthier individuals are more inclined to be physically active.

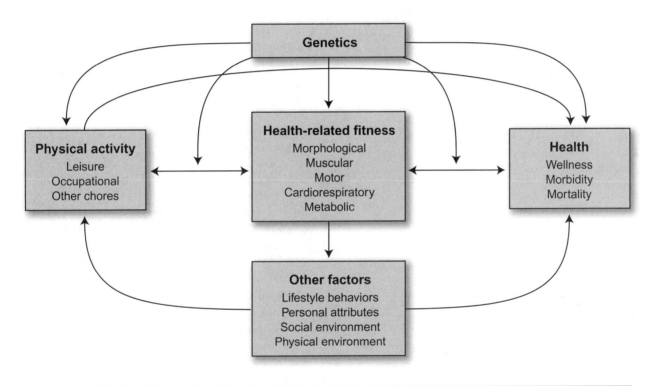

Figure 1.9 A more complete model defining the relationships among physical activity, health-related fitness, and health status. This model allows for contributions of inherited factors and of other lifestyle behaviors, personal attributes, and social and physical environmental factors.

legion. For instance, cardiorespiratory fitness, a key component of health-related fitness, results from the contributions of a number of effectors. This is illustrated in figure 1.10. Here we focus on the interindividual differences in cardiorespiratory fitness typically observed in a large population of sedentary adults. If the overall variance in cardiorespiratory fitness adjusted for body mass is set at 100%, a substantial fraction (up to 10%) will be accounted for by measurement errors and

other uncontrolled factors. Age, sex, and ethnic differences may account for as much as 25% of the variation seen among adults. The remaining 65% can be divided into two major components. About 15% of the individual differences in cardiorespiratory fitness can be explained by the slight variation in habitual physical activity even among sedentary people. Finally, cardiorespiratory fitness is determined in part by familial and genetic characteristics. The heritability level of this attribute is thought to

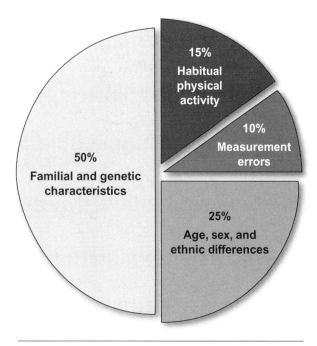

Figure 1.10 Partitioning of variance in cardiorespiratory fitness in a heterogeneous population of sedentary individuals.

reach almost one half of the overall variance. This simple partitioning of the population heterogeneity in a key component of health-related fitness has considerable implications for understanding the physical activity–fitness–health paradigm. These issues are considered in more detail in the chapter devoted to genetic differences (chapter 22).

There is no doubt that regular physical activity is accompanied by substantial benefits on health and quality of life indicators. Most chapters of this book are devoted to a review of the systems and attributes as well as risk factors and morbid conditions that are favorably influenced by a physically active lifestyle.

Summary

Almost half of all yearly deaths in the United States are attributable to common chronic diseases, which are related to lifestyle, social environment, and built environment characteristics. In this regard, as physical activity has diminished dramatically in our lives, a host of physical ills have become more prevalent. This chapter provided a flexible framework relating physical activity to health. Key concepts such as health, wellness, morbidity, and mortality were defined. Subsequently, the global burden of common chronic diseases was described, and the effects of biological, behavioral, and environmental determinants as well as aging were highlighted. To ensure that terms such as *physical activity* and *physical fitness* are interpreted in a consistent fashion in the remainder of this book, their complexities were considered. Physical inactivity as a risk factor was briefly discussed. A practical model linking physical activity, physical fitness, and health was introduced and serves as the basic conceptual framework for the organization of the text.

KEY CONCEPTS

active life expectancy: Age to which a person is expected to live free of disabling diseases or conditions.

disability-free life expectancy: Average number of years of life remaining to a person at a particular age without limitations in physical or mental functions.

exercise: Planned, structured, and repetitive bodily movement done to improve or maintain one or more components of physical fitness.

health: A state of complete physical, mental, and social well-being and not merely the absence of disease or infirmity.

health-related fitness: Components of fitness that are affected favorably or unfavorably by habitual physical activity and related to health status. This term has been defined as a state characterized by (a) an ability to perform daily activities with vigor and (b) demonstration of traits and capacities that are associated with a low risk of premature development of hypokinetic diseases and conditions.

health-related quality of life: The quality of one's personal and mental health and the ability to react to factors in the physical and social environments. Years lived with the full range of functional capacity.

leisure-time physical activity: Activity undertaken in the individual's discretionary time that substantially increases total daily energy expenditure. The element of personal choice is inherent to the definition.

life expectancy: Average number of years of life remaining to a person at a particular age.

maximal oxygen uptake ($\dot{V}O_2$max): Maximal capacity for oxygen consumption by the body during maximal exertion. It is also known as aerobic power and is considered a valid measure of cardiorespiratory fitness.

metabolic equivalent (MET): A unit used to estimate the metabolic cost (oxygen consumption) of physical activity. One MET equals the sitting metabolic rate of approximately 3.5 ml of oxygen per minute.

morbidity: Measures that pertain to disease states and conditions that have not achieved a mortal end point. Common measures of morbidity are disease- or condition-specific incidence rates, hospital admissions, bed-days, treatment costs, loss of physical function and independence, and lost days of work for specific causes.

performance-related fitness: Components of fitness that are necessary for optimal work or sport performance. This is defined in terms of the individual's ability in athletic competition, a performance test, or occupational work and depends heavily on motor skills, cardiorespiratory power and capacity, muscular strength, power or endurance, body size, body composition, motivation, and nutritional status.

physical activity: Bodily movement that is produced by the contraction of skeletal muscle and that substantially increases energy expenditure.

physical fitness: A set of attributes that people have or achieve that relates to the ability to perform physical work.

risk factor: An aspect of personal behavior or lifestyle, an environmental exposure, or an inborn or inherited characteristic that is known to be associated with health-related conditions. When present over an extended period of time, a risk factor can significantly either increase the probability of developing a common degenerative disease such as cardiovascular disease, type 2 diabetes mellitus, or osteoporosis or increase the probability of premature death.

wellness: A holistic concept describing a state of positive health in the individual and comprising physical, social, and psychological well-being.

STUDY QUESTIONS

1. Identify three lines of evidence in support of the view that the human body is designed for physical activity.

2. Name five of the traits that distinguish *Homo sapiens* from other primates and favored the emergence of humans despite a hostile environment.

3. What are the major causes of death around the world?

4. What is the prevalence of lack or partial lack of autonomy among Americans 85 years of age and older?

5. A fraction of total daily energy expenditure is accounted for by the energy costs of all forms of movement and activity. Describe how variable this fraction is among people who are sedentary and physically active.

6. Compare the classifications of exercise intensities and occupational work intensities and define the ways in which they differ.

7. Define morbidity and show its relationship to health-related fitness.

8. Compare the use of the automobile between countries of Western Europe and the United States and Canada.

9. Explain the path from physical activity to health versus that from physical activity to fitness to health.

10. Define the major source of variation in cardiorespiratory fitness, as assessed by $\dot{V}O_2$max adjusted for body mass, among a population of sedentary individuals.

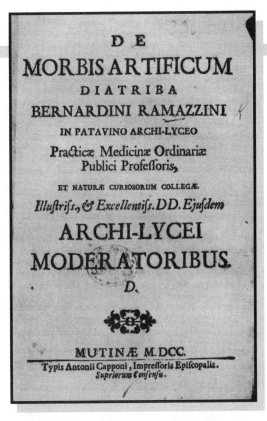

De
MORBIS ARTIFICUM
DIATRIBA
BERNARDINI RAMAZZINI
IN PATAVINO ARCHI-LYCEO
Practicæ Medicinæ Ordinariæ
Publici Professoris,
ET NATURÆ CURIOSORUM COLLEGÆ.
Illustriss., & Excellentiss. DD. Ejusdem
ARCHI-LYCEI
MODERATORIBUS.
D.

MUTINÆ M.DCC.
Typis Antonii Capponi, Impressoris Episcopalis.
Superiorum Consensu.

Historical Perspectives on Physical Activity, Fitness, and Health

Russell R. Pate, PhD

CHAPTER OUTLINE

Early Beliefs About Physical Activity and Health

Scientific Inquiry on Exercise and Health
- Exercise Physiology
- Epidemiology
- Clinical Science
- Behavioral Science
- Molecular Biology and Genetics

Evolution of Physical Activity Guidelines
- Groundwork
- Initial Attempts to Speak to the Public

- Exercise Prescription for the Public
- Importance of Moderate-Intensity Physical Activity
- Physical Inactivity—An Independent Risk Factor for Cardiovascular Disease
- Public Health Messages on Physical Activity
- U.S. Surgeon General on Physical Activity and Health
- Daily Physical Activity to Prevent Excessive Weight Gain

Summary

Review Materials

The body of knowledge in any scientific discipline evolves gradually over time, each new bit of information building on a base established by the results of earlier investigations. So it is with exercise science. Our current understanding of the relationship between physical activity and health is heavily influenced by the findings of studies completed rather recently—mostly in the last 20 years. But the key studies completed in recent decades used methods refined by earlier investigators and tested hypotheses that were suggested by the results of previous investigations. Without a doubt, today's exercise scientists stand on the shoulders of the pioneers who first applied the principles of scientific inquiry to the exercise–health relationship.

The goal of this chapter is to provide a historical context for our current beliefs about the effects of physical activity on health. One specific purpose of this chapter is to overview the broad developmental trends that are evident in the scientific study of physical activity and its effects on human health. Particular attention is given to the gradual expansion of exercise science into several distinct subdisciplines, each with its own unique set of scientific methods. A second purpose is to identify the seminal events that shaped our understanding of the effect of exercise on health. Finally, in recent decades numerous agencies and expert panels have applied the growing body of knowledge on physical activity and health by advancing guidelines on the types and amounts of physical activity needed to maintain or promote health. Therefore, a third and central purpose of this chapter is to track the evolution of physical activity guidelines that have been issued to promote public health.

Early Beliefs About Physical Activity and Health

Many ancient cultures, scientists, and physicians recognized the role of physical activity in promoting the health of mind and body. In China and India, concepts of health and prevention were developing as early as 3000 B.C. Ancient Chinese writings promoted harmony and prevention as keys to longevity. A medical document written in India between 1000 B.C. and 800 B.C., the *Ayur Veda*, recommended massage and exercise in the treatment of rheumatism. Over thousands of years both cultures developed philosophies, including Taoism and yoga, that emphasized the importance of a system of exercise on health.

Among Western cultures, the ancient Greeks dominated the study and understanding of the effects of physical activity on health, quality of life, and life span. As early as the 5th century, Greek physicians promoted the "laws of health"—breathe fresh air, eat good foods, drink the proper beverages, participate in exercise, and get adequate sleep (U.S. Department of Health and Human Services 1996). Herodicus (5th century B.C.) was the first of the Greek gymnasts (a type of medical practitioner) to prescribe therapeutic exercise. **Hippocrates** (5th century B.C.), the father of preventive medicine, wrote extensively about the benefits of exercise for a variety of ailments, including mental illnesses. Although he criticized Herodicus for prescribing exercise that was too strenuous, Hippocrates recommended walking and other forms of moderate-intensity exercise. Many Greek physicians who practiced medicine and taught in the medical schools during the 4th through 2nd centuries B.C., including Herophilus, Eristratus, Asclepiades, and Celsus, prescribed either moderate or vigorous exercise to maintain health and treat a variety of diseases (figure 2.1).

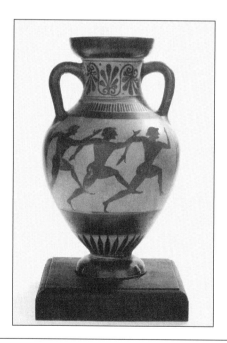

Figure 2.1 The value that the ancient Greeks placed on exercise is clearly illustrated by the fact that many of their surviving vases depict athletes. This amphora dates from the time of Hippocrates (5th century, BCE) and features three nude athletes running under the watchful eye of Athena, the patron goddess of Athens, who is pictured on the side of the vase that is not shown.

Courtesy of The Spurlock Museum, University of Illinois at Urbana-Champaign, accession number 1977.01.1685.

"Eating alone will not keep a man well; he must also take exercise. For food and exercise, while possessing opposite qualities, yet work together to produce health." Hippocrates, *Regimen*

Early Beliefs About Physical Activity and Health

- Scientists and physicians in China and India recognized a link between physical activity and health more than 5,000 years ago.
- Ancient Greek physicians, including Herodicus and Hippocrates, prescribed exercise to prevent and treat a variety of ailments as early as the 5th century B.C.
- During the 1500s, Italian physicians prescribed exercise for the healthy growth and development of children and for the treatment of elderly and ill people.
- In the early 1700s, Ramazzini identified the negative health effects of certain occupations. He noted that runners (messengers) avoided many of the health problems that affected cobblers, tailors, and other sedentary workers.

Claudius Galenus (Galen), born about A.D. 131, was a brilliant Greek physician who worked in Rome and whose teachings and views dominated European medicine for 1,000 years. He wrote and lectured extensively on anatomy and epidemiology and was the first scientist to systematically describe the human body and to recognize that contraction is the primary action of muscle. He believed that some form of exercise could be used to treat virtually every form of disease, and he classified exercises according to their purposes. In his medical practice and his writings, most notably *On Hygiene*, Galen promoted his belief that everyone—athletes, healthy adults and children, invalids, even babies—could benefit from exercise.

"The uses of exercise, I think, are twofold; one for the evacuation of the excrements, the other for the production of good condition of the firm parts of the body. For since vigorous motion is exercise, it must needs be that only these three things result from it in the exercising body—hardness of the organs from mutual attrition, increase of the intrinsic warmth, and accelerated movement of respiration. These are followed by all the other individual benefits which accrue to the body from exercise." Galen, *On Hygiene*

The influence of the ancient Greek physicians on European medicine faded during the Middle Ages but reemerged during the Renaissance, when some of the original Greek manuscripts were rediscovered. In the 15th century, Vergerius and Vittorino da Feltre of Italy became early advocates of regular exercise for children. da Feltre established a school in which children participated in exercises designed to meet their individual needs and in which all children participated in gymnastics and played many sports. In 1553 in Spain, Cristobal Mendez, a physician, published one of the earliest printed books on exercise, in which he prescribed

exercise for elderly people and those who were ill, noting that "the easiest way of all to preserve and restore health without diverse peculiarities and with greater profit than all other measures put together is to exercise well." Mercurialis, one of the most influential European physicians of the Renaissance, published the six-volume *Art of Gymnastics* in 1569. Mercurialis classified exercises as preventive or therapeutic and recommended that all sedentary people begin to exercise. His recommendations for exercises for people who were sick or infirm established the foundation for the future development of rehabilitation medicine.

Although scientific support for the belief that exercise could prevent or ameliorate disease did not emerge until the 20th century, investigations of that possibility began much earlier. Italian physician Bernardo Ramazzini wrote **Diseases of Workers,** the first published work on occupational diseases, in 1713. Ramazzini observed that runners who worked as messengers avoided the health problems suffered by cobblers and tailors and recommended that sedentary workers exercise on their holidays. In the 1840s, **W.A. Guy** compared the morbidity and mortality rates of men in active and sedentary occupations and noted the superior health of those in active occupations. He recommended that those in sedentary jobs perform physical exercise during their leisure time to improve their health (Paffenbarger et al. 2001).

> "Those who sit at their work and are therefore called 'chair workers,' such as cobblers and tailors, become bent and hump-backed and hold their heads down like people looking for something on the ground.... These workers, then, suffer from general ill health caused by their sedentary life ...they should be advised to take physical exercise, at any rate on holidays. Let them make the best use they can of some one day, and so to some extent counteract the harm done by many days of sedentary life." Bernardo Ramazzini, *Diseases of Workers*

Scientific Inquiry on Exercise and Health

Although great scholars had recognized the importance of physical activity to health as far back as the ancient cultures, it was not until the early 20th century that scientists began systematic study of the effects of exercise on the human body. The development of "exercise science" began with the work of physiologists who became interested in the body's functional responses to different types of exercise. Although the discipline of exercise physiology has contributed much to our knowledge of the relationship between physical activity and health, many other fields of scientific inquiry have come into play over the past century. In this section, we review the contributions made by several scientific disciplines to our rapidly expanding knowledge of the impact of exercise on human well-being.

Exercise Physiology

Exercise physiology is a scientific discipline dedicated to the study of the body's function during exercise and its physiologic adaptations to regular participation in exercise. In the early 20th century one of the earliest exercise physiologists, **R. Tait McKenzie** of the University of Pennsylvania, began studying the effects of regular physical activity on the body, using a system of regular medical examinations of athletes before and after sport participation. In 1909,

A.V. Hill of Cambridge University began studying the physiology of muscle contraction. He also conducted pioneering studies of the thermal changes associated with muscle function. August Krogh and Marie Jorgensen of Denmark studied carbon dioxide transport in the lungs, metabolism, and the role of insulin. Krogh also designed a bicycle ergometer with which he studied exercise intensity.

One of the first exercise laboratories, the **Harvard Fatigue Laboratory,** was founded in 1927. Although the lab conducted research on a number of topics, including nutrition, blood chemistry, and the stresses imposed on the body by altitude and climate, the lab is perhaps best known for its work in exercise physiology. Scientists at the lab made some of the first measurements of the body's capacity to consume oxygen. The director, D.B. Dill, studied thermoregulation during exercise and the effects of environment on exercise. Researchers who trained at the Harvard Fatigue Laboratory established many of the first exercise physiology programs at universities in the United States.

The landmark work of McKenzie, Hill, Krogh, Dill, and other founders of exercise physiology established certain basic principles and procedures of exercise physiology, but most of the early research of exercise physiologists focused on human performance, not on the health effects of exercise. But as the field developed during the middle to late decades of the 20th century, exercise physiology came to focus on the effects of acute and chronic exercise on factors that were thought to be associated with health. For example, Haskell identified the positive effects of exercise on plasma triglycerides, high-density lipoproteins, and other blood lipids (Haskell 1984). Other investigators identified the potential of exercise to lower blood pressure in both normotensive and hypertensive individuals (Tipton 1984).

Epidemiology

Epidemiologists study the rates at which diseases occur in populations and identify factors that are associated with the incidence of specific diseases. For example, in 1897, a British officer in the Indian Medical Service first demonstrated that mosquitoes transmit malaria to birds. Between 1898 and 1910, epidemiologists and other scientists uncovered the mechanism of transmission in humans and developed eradication and treatment procedures. One of the first large-scale tests of eradication procedures was implemented during construction of the Panama Canal. More recently, epidemiologists established cigarette smoking as a cause of lung cancer and coronary heart disease. The era of modern physical activity epidemiology really began in 1949, when Jeremy N. Morris and colleagues began to carefully examine the health effects of active occupations on workers' health (Paffenbarger et al. 2001). In the **London Bus Study** the Morris group found that conductors on London's double-decker buses, who climbed the buses' stairs many times each day and spent 90% of their shifts on their feet, had lower rates of coronary heart disease than the bus drivers, who were almost entirely sedentary (Morris et al. 1953). Morris also led a large necroscopy study of British workers that provided additional evidence of the health benefits of a physically active occupation (Morris and Crawford 1958). From these important beginnings, evidence of the link between physical activity and health continued to build throughout the 20th century.

Particularly notable is the work of **Dr. Ralph Paffenbarger,** who has conducted long-term epidemiological studies focusing on the relationship between physical activity and outcomes such as death attributable to cardiovascular disease (figure 2.2). His studies have included groups as diverse

AMERICAN
Journal of Epidemiology

Formerly AMERICAN JOURNAL OF HYGIENE

© 1978 by The Johns Hopkins University School of Hygiene and Public Health

VOL. 108	SEPTEMBER, 1978	NO. 3

Original Contributions

PHYSICAL ACTIVITY AS AN INDEX OF HEART ATTACK RISK IN COLLEGE ALUMNI[1]

RALPH S. PAFFENBARGER, JR., ALVIN L. WING, AND ROBERT T. HYDE

Paffenbarger, R. S., Jr. (Stanford University School of Medicine, Stanford, CA 94305), A. L. Wing and R. T. Hyde. Physical activity as an index of heart attack risk in college alumni. *Am J Epidemiol* 108:161–175, 1978.

Risk of first heart attack was found to be related inversely to energy expenditure reported by 16,936 Harvard male alumni, aged 35–74 years, of whom 572 experienced heart attacks in 117,680 person-years of followup. Stairs climbed, blocks walked, strenuous sports played, and a composite physical activity index all opposed risk. Men with index below 2000 kilocalories per week were at 64% higher risk than classmates with higher index. Adult exercise was independent of other influences on heart attack risk, and peak exertion as strenuous sports play enhanced the effect of total energy expenditure. Notably, alumni physical activity supplanted student athleticism assessed in college 16–50 years earlier. If it is postulated that varsity athlete status implies selective cardiovascular fitness, such selection alone is insufficient to explain lower heart attack risk in later adult years. Ex-varsity athletes retained lower risk only if they maintained a high physical activity index as alumni.

coronary disease; hypertension; kilocalories; obesity; physical fitness; smoking; sports

Figure 2.2 Abstract of the landmark 1978 study in which Paffenbarger and colleagues first quantified a relationship between health and physical activity. They found that the risk of first heart attack is inversely related to physical activity up to a level corresponding to approximately 2,000 kcal of energy expenditure in physical activity per week.

Reprinted, by permission, from R.S. Paffenbarger, Jr, A.L. Wing and R.T. Hyde, 1978, "Physical activity as an index of heart attack risk in college alumni," *American Journal of Epidemiology* 108(3): 161.

as former longshoremen, who as young men performed heavy physical labor on the docks in San Francisco, and alumni of Harvard College, most of whom were engaged in rather sedentary occupations but some of whom were very physically active in their leisure time. Also central to the development of physical activity epidemiology as a scientific discipline has been the research of Steven N. Blair. Blair and colleagues at the Cooper Institute for Aerobics Research have conducted a series of landmark studies examining the relationship between physical fitness, measured as performance on a treadmill exercise test, and a wide range of chronic disease outcomes. Considered collectively, these studies have demonstrated that achieving and maintaining at least a moderate level of physical fitness provide very important health benefits, including reduced risk of death attributable to cardiovascular disease and increased longevity. The work of Paffenbarger, Morris, Blair, and other epidemiologists has been extremely influential because it established the impact of physical activity on health at the population level. Also, as is discussed in more detail later in this chapter, the work of epidemiologists has been important in establishing the types and amounts of physical activity needed to provide health benefits.

Clinical Science

Clinical research is typically designed to observe the effects of an experimental treatment on patients who have been diagnosed with a disease or clinical condition. For example, before a new drug can be brought to market, it must be subjected to clinical trials in which the effects of the drug are compared with the effects of a placebo on persons who have a condition that the drug is designed to treat. Exercise was first used as a form of treatment for diseased patients in the 1950s, when pioneering cardiologist **Paul Dudley White** began prescribing exercise as part of a rehabilitation program for patients with coronary heart disease. At about the same time, Herman Hellerstein and other physicians and scientists recognized that bed rest, the common prescription following a heart attack, was detrimental to people with heart disease. Hellerstein promoted many ideas that helped form the emerging field of cardiac rehabilitation, including exercise and a multidisciplinary approach to recovery from cardiac events.

By the mid-1970s, clinical research supported the benefits of exercise for cardiac patients; the American Heart Association (AHA) noted in 1975 that "data from several sources suggest that survivors of myocardial infarction who participate in physical training programs have a mortality rate approximately one-third below that of patients who remain inactive" (AHA 1975, p. 22). These studies demonstrated important health benefits of physical activity and paved the way for clinical studies of exercise as a treatment for a wide range of chronic diseases, including type 2 diabetes, cancer, pulmonary disease, and many other maladies.

Behavioral Science

Physical activity is central to normal human function. Accordingly, psychologists and other behavioral scientists have long been interested in physical activity. However, much of the initial interest of behavioral scientists was focused on the learning of motor skills and enhancement of performance in sports, neither of which relates directly to the effects of physical activity on health. But in recent decades, with the growing interest of the scientific community in the physical activity–health relationship, health psychologists have begun to focus on topics that bear directly on physical activity as a behavior that influences the health of both individuals and entire populations.

Psychologists have studied the psychosocial factors that influence participation in physical activity, and as a result it is now known that numerous personal, social, and environmental factors combine to determine a person's habitual physical activity level. In addition, behavioral scientists have conducted studies of interventions to increase physical activity. These studies have been conducted with individuals, small groups, and communities. For example, Marcus and other investigators applied the concept of stages of change to physical activity to better understand the responses of individuals to exercise programs (Marcus et al. 1992; Marcus and Simkin 1993). King and others developed community approaches to physical activity promotion that included environmental, organizational, and policy strategies in addition to individual strategies (King 1991, 1994).

Molecular Biology and Genetics

Essentially all human characteristics are determined, at least in part, by the individual's genetic background. That is, much of what we are and can become is programmed in the genetic material, the DNA that we inherit from our parents. This principle has been developing as a scientific tenet since Gregor Mendel, an Augustinian monk born in Moravia, began exploring the science of genetics in the 19th century. In 1953, Watson and Crick published their landmark description of DNA, and since then molecular biology has exploded into countless lines of scientific inquiry. Exercise biochemistry is one of the fields that has been profoundly influenced by the rapidly developing technology for studying the genetic basis of human characteristics.

Claude Bouchard has been a pioneer in the application of genetics to the study of human responses and adaptations to exercise. In the 1980s, Bouchard and his colleagues began studying inheritance of exercise characteristics by observing identical and fraternal twins. These studies demonstrated that the responses to exercise were much more similar in identical twins than in fraternal twins. For example, in a study of 31 pairs of identical twins and 22 pairs of fraternal twins, Bouchard found significant genetic effects on the energy cost of submaximal exercise and the respiratory exchange ratio at low power outputs. Other studies found genetic influences on resting metabolic rate and relative rate of carbohydrate to lipid oxidation (Bouchard 1991). This line of investigation has continued, and the large-scale **HERITAGE Family Study** has shown that the health-related physiological adaptations to exercise training are heavily dependent on genetic background. For example, the HERITAGE Study found that changes in $\dot{V}O_2max$ in response to an exercise training program are highly variable and that the variability across families is much greater than within families (Bouchard et al. 1999). It seems

certain that molecular biology will play a central role in the future of exercise science. Many exercise scientists believe that elucidating the role of genetics in explaining the individual variability in exercise characteristics will be a key direction for research for the next several decades. For example, we now have evidence that a fit and active way of life prevents or delays the development of hypertension. A next phase of investigation on this topic involving genetics will be to determine if individuals with a certain genotype are more susceptible than individuals with other genotypes to develop hypertension as a result of a sedentary lifestyle.

Evolution of Physical Activity Guidelines

Scientific knowledge about physical activity and health is of little value if people cannot understand it and apply it to their lives. For the past 3 decades, there has been a gradual but steady development in the effort to present information on physical activity and health to the general public. This has come through public health messages known as physical activity guidelines.

Groundwork

Health scientists and practitioners have long believed that regular physical activity is essential to maintain good health. So it is not surprising that for a very long time, individual health professionals and health organizations have been making recommendations regarding the types and amounts of physical activity needed for health and fitness. As emphasized in the preceding section of this chapter, scientific support for the impact of physical activity on health has developed rapidly in recent decades. As the relevant knowledge base has grown, physical activity recommendations for the public have been modified to maintain consistency with the existing research evidence. In this section we track the evolution of physical activity guidelines as they have been presented to the public since the 1950s.

In 1957, Finnish researcher **Marti Karvonen** published the findings of a study that has become a classic in exercise science. Karvonen observed the effects of exercise training by treadmill running on endurance fitness in a small number of male medical students. He reported that training intensity corresponding to a heart rate of at least 60% of the

---

Scientific Inquiry on Exercise and Health

- The modern field of exercise science began to develop in the early 20th century.
- Physiologists were the first scientists to study systematically the effects of exercise on the human body.
- Other disciplines that contributed to understanding the relationship between physical activity and health included epidemiology, clinical science, behavioral science, and molecular biology and genetics.

Pate

heart rate range (maximum heart rate minus resting heart rate) was required to produce significant gains in cardiorespiratory fitness. Although Karvonen's study was very small and quite limited in its research design, his findings became the platform for exercise guidelines for the ensuing three decades. Karvonen's program was presented in terms of minima for frequency, duration, and intensity of training. Nearly a half century later, it seems remarkable that such a small and limited study could have had such a powerful influence on health practice.

> "Heart rate during training has to be more than 60% of the available range from rest to the maximum attainable by running...in order to produce a change in the WR [working heart rate]...A decrease of the WR is understood to indicate an increase of the maximum oxygen uptake." (Karvonen et al. 1957, p. 314)

Initial Attempts to Speak to the Public

During the 1960s, two American men, one a track coach and the other a physician, published books that brought practical physical activity guidelines to the masses. In 1963, Bill Bowerman, coach of the University of Oregon's track team, visited a coaching colleague in New Zealand, where he witnessed many middle-aged adults running for health and fitness. He was so impressed by what he had seen that, on his return to the United States, he wrote *Jogging*, a small paperback volume that has often been credited with launching a fitness revolution. Bowerman described a slow running program that emphasized gradual, progressive increases in distance and frequency of exercise. His basic recommendation was that almost everyone can benefit from "an exercise program of relaxed walking and running" and that jogging is something that almost everyone can do.

In 1968, only a year after Bowerman popularized jogging as a specific form of exercise, Dr. Kenneth Cooper, then an Air Force physician, published *Aerobics*, a book in which he laid out a simple point system for determining how much exercise should be accumulated on a weekly basis. His Aerobics Point System recommended that adults accumulate a minimum of 30 points per week. Cooper recommended that sedentary adults begin an exercise program by starting at a level compatible with

Table 2.1 Examples of Point Values in Cooper's Original Point System

Activity	Points
Run 1 mile in <8 min	5
Walk 3 miles in <43 min	6
Cycle 5 miles in <20 min	5
Swim 600 yards in <15 min	5

Data from K.H. Cooper, 1968, *Aerobics* (New York, NY: Bantam Books, Inc.).

their current fitness (perhaps earning as few as 10 points per week for the first few weeks, for those at the lowest levels of fitness), choose an activity they enjoy, and exercise with others when at all possible. Table 2.1 provides examples of point values assigned to exercises by Cooper. Although neither Bowerman nor Cooper was able to base these recommendations on extensive bodies of directly relevant scientific evidence, both were talented practitioners and gifted communicators who were able to draw on their extensive experience in educating the public about how much physical activity is needed for health and fitness.

Exercise Prescription for the Public

Concurrent with the popularization of the exercise-for-fitness movement and the so-called running boom of the late 1960s and 1970s, exercise scientists began to systematically explore the effects of various types, intensities, durations, and frequencies of endurance exercise on cardiorespiratory fitness. A leader of this extensive scientific effort was Dr. Michael Pollock. During the 1970s, Pollock and his colleagues undertook a series of experimental exercise training studies that, when considered collectively along with the work of other researchers, produced the knowledge needed to recommend exercise in a precise, detailed, and individualized manner. This method became the central dogma in efforts to communicate to the public the types and amounts of exercise needed to promote health and fitness. The American College of Sports Medicine (ACSM) first formally endorsed this detailed approach to recommending exercise in its exercise guidelines book in 1975 and in a position statement issued in 1978 (ACSM 1975). The key recommendations presented in the **ACSM position statement** are summarized in table 2.2.

During the same period that exercise scientists were systematically studying endurance exercise training and its impact on cardiorespiratory fitness

Table 2.2   1978 ACSM Position Statement: The Recommended Quantity and Quality of Exercise for Developing and Maintaining Fitness in Healthy Adults

Quality	Quantity
Frequency	3-5 days per week
Intensity	50-85% $\dot{V}O_2$max (60-90% maximum heart rate)
Duration	15-60 min

Data from American College of Sports Medicine, 1978, "Position statement—The recommended quantity and quality of exercise for developing and maintaining fitness in healthy adults," *Medicine and Science in Sports and Exercise* 10: vii-x.

Table 2.3   AHA's First Guidelines

Quality	Quantity
Frequency	3-4 times per week
Intensity	70-85% maximum heart rate
Duration	20-60 min

Data from American Heart Association, 1975, *Exercise testing and training of individuals with heart disease or at high risk for its development* (Dallas: American Heart Association).

Table 2.4   ACSM Guidelines for Exercise Prescription

Quality	1975	1995
Frequency	3 times per week	3-5 times per week
Intensity	60-90% $\dot{V}O_2$max	40-85% $\dot{V}O_2$max
Duration	20-30 min	20-30 min

Data from *Physical Activity and Health: A Report of the Surgeon General*. U.S. Department of Health and Human Services, 1996.

in healthy adults, cardiologists and clinical exercise physiologists were studying the effects of exercise training in patients with cardiovascular disease. This research demonstrated the critical and now well-accepted role that exercise can play in rehabilitation of patients with compromised cardiovascular function. But furthermore, this research and the clinical guidelines that it spawned established a medical approach to recommending exercise that came to be referred to as "exercise prescription." This technique drew on the research on normal healthy adults as well as research performed on heart patients. In 1975, the AHA published guidelines on exercise prescription for patients with cardiovascular disease. This document helped to establish a place for exercise in the practice of medicine and was influential in communicating to the public the significant health benefits that accrue to physically active persons, even those with already established cardiovascular disease. Table 2.3 summarizes the AHA's first guidelines for physical activity in people with heart disease or at risk for heart disease (AHA 1975).

Importance of Moderate-Intensity Physical Activity

ACSM's *Guidelines for Exercise Testing and Prescription* has undergone revision approximately every 5 years since its initial publication in 1975 (ACSM 1975). Note that the first version of this was called "Guidelines for Graded Exercise Testing," but subsequent revisions dropped the word *graded*. Each volume included a primary recommendation on prescription of exercise that reflected the current body of knowledge regarding the types of exercise needed to provide health and fitness benefits to initially sedentary adults. Between the first edition

published in 1975 and the sixth edition released in 2000, an interesting trend is evident. As shown by table 2.4, most elements in the exercise prescription guideline remained unchanged. The exception is the recommended range for exercise intensity, the lower end of which decreased from 60% $\dot{V}O_2$max to 40% $\dot{V}O_2$max. The earlier editions of this influential book indicated that rather vigorous exercise was needed to provide benefits, and this concept was widely communicated to the public during the 1970s and 1980s.

Recognition of the importance of moderate-intensity physical activity, as reflected by the changing exercise prescription guidelines of ACSM, evolved gradually during the 1980s and early 1990s as the result of a growing and changing body of research evidence. Two lines of research led to the conclusion that moderate-intensity physical activity (the equivalent of brisk walking) provided important benefits to health and fitness. First and most important, the science of physical activity epidemiology matured during the 1980s and produced a series of important investigations. These studies strongly suggested that regular performance of moderate-intensity physical activity provided important health benefits. Not only did these studies show that regularly active persons were less likely than sedentary persons to develop or die from cardiovascular disease, but in addition the studies demonstrated that much of the active population's physical activity came from walking and other forms of moderate-intensity physical activity. For example, results of the Third National Health and Nutrition Examination Survey (Crespo et al. 1996) showed that most of the physical activities preferred by American adults were moderate-intensity lifestyle activities, such as walking, gardening, and cycling (table 2.5).

As discussed previously, the exercise prescription method for recommending physical activity to the public was based primarily on the findings of a large number of experimental exercise training studies. The results of these studies had generally been interpreted as indicating that vigorous physical activity (6 METs or 60% or more individual functional capacity) was required to produce benefits. The assumption became that moderate-intensity physical activity did not provide those benefits. However, the epidemiologic studies published in the 1980s and early 1990s forced a reexamination of the experimental studies. A closer look revealed that, in studies that compared moderate- and vigorous-intensity physical activity, the moderate level produced increased fitness, although often not to the same extent as the vigorous level. Also, it was seen that moderate-intensity physical activity often provided comparable or even greater beneficial effects on health outcomes such as blood pressure and high-density lipoprotein (HDL) cholesterol. For example, Duncan et al. (1991) found that both

women who participated in a vigorous exercise program and those who participated in a moderate exercise program had significant improvements in their lipoprotein profiles (figure 2.3). Although women in the vigorous program had significantly greater gains in fitness, as measured by $\dot{V}O_2$max, increases in HDL were similar in the two groups.

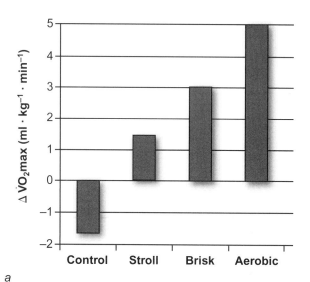

a

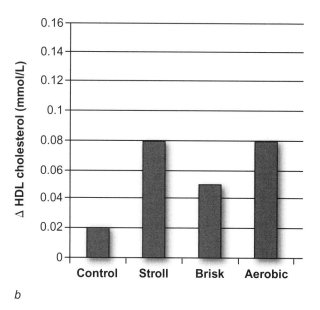

b

Figure 2.3 Effect of walking intensity on changes from baseline in (*a*) maximal oxygen uptake ($\Delta \dot{V}O_2$max) and (*b*) high-density lipoprotein cholesterol (Δ HDL cholesterol) after 24 weeks of exercise training.

Data from J.J. Duncan, N.F. Gordan and C.B. Scott, 1991,"Women walking for health and fitness: How much is enough?" *Journal of the American Medical Association* 266: 3295-3299.

Table 2.5 Physical Activities Preferred by American Adults and Preference Rankings

	Preference ranking	
Activity	**Men**	**Women**
Walking	2	1
Gardening and yard work	1	2
Calisthenics	3	3
Cycling	4	4
Jogging and running	5	8
Weightlifting	6	9
Swimming	6	6
Dancing	8	5
Aerobics and aerobic dance		6
Basketball	10	
Tennis		10
Golf	9	

Data from C.J. Crespo, S.J. Keteyian, G.W. Heath and C.T. Sempos, 1996, "Leisure-time physical activity among US adults. Results from the Third National Health and Nutrition Examination Survey," *Archives of Internal Medicine* 156: 93-98.

Physical Inactivity—An Independent Risk Factor for Cardiovascular Disease

The AHA is a large and influential organization, and many of its members are physicians and others who provide medical services related to cardiovascular disease. The official positions of the AHA are important because they influence medical practice, health care policy, and public health investments. Before the 1990s, the AHA, although supportive of physical activity through some of its programmatic initiatives, had not officially recognized physical inactivity as a risk factor for cardiovascular disease. That changed in 1992 with the publication of AHA's *Statement on Exercise: Benefits and Recommendations for Physical Activity Programs for All Americans* (Fletcher et al. 1992). This position statement declared that a sedentary lifestyle is a major and independent risk factor for premature development of atherosclerotic cardiovascular disease, the number one cause of death in the United States and other developed countries. This statement has been updated several times over the past decade to reflect new findings on the role of physical activity in the prevention and treatment of heart disease.

> "Regular aerobic physical activity increases exercise capacity and plays a role in both primary and secondary prevention of cardiovascular disease. . . . Inactivity is recognized in the AHA Statement on Exercise as a risk factor for coronary artery disease." (Fletcher et al. 1992)

The AHA position stand made several important recommendations. It stated that people of all ages could benefit from a regular exercise program and that activities such as walking, hiking, swimming, tennis, and basketball, when performed at 50% or more of a person's work capacity, were especially beneficial. The statement also noted that "even low-intensity activities performed daily can have some long-term health benefits and lower the risk of cardiovascular disease." However, the document did not provide a specific recommendation regarding the types and amounts of physical activity that should be performed to minimize risk of cardiovascular disease.

Public Health Messages on Physical Activity

The AHA's endorsement of physical inactivity as a risk factor for cardiovascular disease had many important effects. One of these was to heighten the awareness of public health leaders that there was a need to launch large-scale programs to promote physical activity in the American population. But in considering how such programs should be designed, these public health leaders noted a problem. Although many organizations had recommended that Americans become more physically active, there was a lack of clarity regarding the types and amounts of physical activity that should be recommended to promote health in the population. Furthermore, there was a concern that traditional exercise guidelines were not the best ones for broad public health purposes. It was concluded that a new "public health message" on physical activity should be developed.

In 1993, the U.S. Centers for Disease Control and Prevention (CDC) partnered with the ACSM in drafting a statement on physical activity that would be useful in communicating to the American public how much and which types of physical activity are needed to maintain good health. An expert panel was formed and charged with developing a clear, concise statement on physical activity for enhancement of public health. This panel included exercise scientists, epidemiologists, physicians, and health psychologists. In pursuing its task, the panel applied several criteria. First, the panel required that any recommendation be extensively supported by the scientific literature. Second, there was a need to craft a statement that would communicate clearly to the public. And third, there was a desire to recommend a model for physical activity that, while providing important health benefits, would be readily attainable by most people. Following is the recommendation that was issued by the panel.

> "Every U.S. adult should accumulate 30 minutes or more of moderate-intensity physical activity on most, preferably all, days of the week." (Pate et al. 1995, p. 402)

The release of this statement created considerable controversy, in part because several elements of the recommendation were novel and perceived by some as inconsistent with previous guidelines. Before issuance of the **CDC–ACSM recommendation on Physical Activity and Public Health,** there was a broad perception that exercise had to be continuous and rather vigorous to provide important health benefits. The new statement, by endorsing accumulation of moderate-intensity physical activity in bouts as short as 8 to 10 min, was intended to present the public with an approach to a physically active lifestyle that would include a weekly dose of physical activity that provides important health benefits and would be potentially attainable and attractive to the average person. A key principle underlying the recommendation was that the greatest public health gain would result from moving the large sedentary segment of the population into a regularly physically active pattern. Hence it was seen as important that the recommendation be viewed as attainable by persons who were currently sedentary.

U.S. Surgeon General on Physical Activity and Health

Although the CDC–ACSM recommendation on physical activity was seen as controversial in some quarters, its core elements were quickly endorsed by several highly credible organizations. The National Institutes of Health (NIH) convened a consensus conference on the health effects of physical activity, and the conclusions of that conference supported the position taken by the CDC and ACSM, that the public health would be greatly enhanced if most American adults performed at least 30 min of moderate-intensity physical activity daily (NIH Consensus Development Panel on Physical Activity and Cardiovascular Health 1996). The World Health Organization, the public health arm of the United Nations, also issued a statement supporting the health benefits of regularly performing 30 or more minutes of moderate-intensity physical activity (WHO–FIMS Committee on Physical Activity for Health 1995).

These important statements were followed by a landmark event, the release of *Physical Activity and Health: A Report of the Surgeon General* (U.S. Department of Health and Human Services 1996), for which Steven Blair served as senior scientific editor. This extensive document summarized the evidence supporting the health benefits of physical activity and drew several major conclusions.

The Surgeon General's report included several other key conclusions (see *Key Conclusions of the*

> "Significant health benefits can be obtained by including a moderate amount of physical activity (e.g., 30 minutes of brisk walking or raking leaves, 15 minutes of running, or 45 minutes of playing volleyball) on most, if not all, days of the week. Through a modest increase in daily activity, most Americans can improve their health and quality of life." (U.S. Department of Health and Human Services 1996, p. 4)

Surgeon General's Report). These recommendations, coming from a prestigious government agency, constituted the strongest support to date for the public health benefits of physical activity.

Daily Physical Activity to Prevent Excessive Weight Gain

The fundamental recommendation that adults should participate in 30 min of physical activity daily, as recommended in the CDC–ACSM guideline and the Surgeon General's report, was widely supported by public health authorities around the globe. Also,

Key Conclusions of the Surgeon General's Report

- People of all ages benefit from regular physical activity.
- Moderate physical activity (equivalent to 30 min of brisk walking on most days of the week) can provide significant health benefits.
- Greater amounts of physical activity can provide additional health benefits.
- Physical activity reduces the risk of premature mortality and of coronary heart disease, hypertension, colon cancer, and diabetes.
- More than 60% of American adults are not regularly active, and 25% are not active at all.
- Nearly half of American youth (ages 12-21) are not vigorously active on a regular basis, and physical activity declines significantly during adolescence.

during the decade following the initial release of that recommendation, its validity was further substantiated by a significant amount of new research. Nonetheless, some in the scientific community questioned whether 30 min of daily physical activity was enough to provide all the desired benefits of an active lifestyle. In particular, in 2003 a panel of the Institute of Medicine (IOM) concluded that 60 min of daily physical activity is needed to prevent excessive weight gain and obesity. That panel's conclusion was based on an examination of a much narrower body of scientific evidence than that considered by the Surgeon General's report or the committee that issued the CDC–ACSM recommendation. The IOM panel considered only the relationship between weight status and physical activity, whereas the other panels had taken a much more comprehensive view of the impact of physical activity on health.

A reexamination of these issues was undertaken by the U.S. Dietary Guidelines Advisory Committee in 2004, and that panel concluded that 30 min of daily physical activity provides important public health benefits by substantially reducing risk of chronic diseases such as coronary heart disease, type 2 diabetes mellitus, and osteoporosis (U.S. Department of Agriculture 2004). However, the panel also noted that many people may require more than 30 min of daily physical activity to prevent excessive weight gain and that more physical activity is needed to induce and maintain weight loss in formerly obese individuals than is needed to prevent excessive weight gain in the first place.

Physical Activity Recommendations of the 2005 U.S. Dietary Guidelines Advisory Committee

- Thirty minutes of at least moderate-intensity physical activity on most days provides important short- and long-term health benefits for adults.

- Up to 60 min per day of at least moderate-intensity physical activity may be needed to avoid unhealthy weight gain.

- Adults who have lost weight may need to participate in 60 to 90 min of moderate-intensity physical activity per day to avoid regaining weight.

Evolution of Physical Activity Guidelines

- Karvonen's investigation of the heart rate level needed to produce an increase in cardiorespiratory fitness set the stage for the development of physical activity guidelines.

- In 1978, the American College of Sports Medicine issued its first position statement on the quantity and quality of exercise required for developing and maintaining fitness.

- In 1992, the American Heart Association stated officially that a sedentary lifestyle is a major and independent risk factor for the development of cardiovascular disease.

- In 1995, the Centers for Disease Control and Prevention and the American College of Sports Medicine issued the first public health-oriented statement on physical activity and health. The statement said that every American adult should accumulate 30 or more minutes of physical activity on most, and preferably all, days of the week.

- **The Surgeon General's Report on Physical Activity and Health,** published in 1996, clearly established that regular, moderate physical activity provides health benefits to people of all ages and that additional activity can provide additional benefits.

Summary

The work of scientists, medical practitioners, and philosophers over several thousand years forms the foundation on which modern exercise science stands. Modern scientists from a number of disciplines, including exercise physiology, epidemiology, the clinical sciences, the behavioral sciences, and molecular biology and genetics, have contributed to our current understanding of the relationship between physical activity and health. Their work has led to breakthroughs in the measurement of physical activity and physical fitness, exercise testing and prescription for people of all ages and abilities, development of interventions that help people adopt and maintain active lifestyles, and a better understanding of the ways that family, community, and environment influence physical activity. Exercise scientists and their colleagues in related fields have also stepped outside their laboratories and clinical settings and worked to translate their findings into messages that the general public can understand. The result is physical activity guidelines, a series of public health messages designed to help people be active at levels that will improve their health and quality of life. Leading scientists and science organizations have participated in developing and publicizing these guidelines and in revising them to keep up with the pace of modern exercise science. The ideas and hypotheses of ancient Chinese and Greek scholars and European Renaissance physicians—that an active life promotes good health and prevents disease and disability—have been supported by more than a century of modern scientific studies. It seems certain that the scientific disciplines that study physical activity and exercise will continue to grow and change and to apply new techniques to understanding the effects of physical activity on the human body.

KEY PEOPLE AND EVENTS

ACSM Position Statement, 1978: The first statement by the American College of Sports Medicine on the quantity and quality of exercise required for developing and maintaining fitness in healthy adults. ACSM recommended that adults exercise 3 to 5 days per week for 15 to 60 min at 60% to 90% of $\dot{V}O_2$max.

CDC–ACSM Recommendation on Physical Activity and Public Health: Issued in 1995, a recommendation that differed from previous recommendations. It focused on moderate physical activity and on accumulation of activity throughout the day, on most if not all days of the week.

Diseases of Workers: The first published work to identify the health risks and benefits of certain occupations (Ramizzini 1713).

Claudius Galenus (Galen): Born circa A.D. 131; a brilliant Greek physician whose teachings dominated European medicine for 1,000 years. He believed that some form of exercise could be used to treat every form of disease and that all people, healthy and sick, could benefit from regular exercise.

W.A. Guy: In the 1840s, researcher who noted the superior health of men in active occupations. He recommended that men in sedentary occupations perform physical exercise in their leisure time to improve their health.

Harvard Fatigue Laboratory: Lab established in 1927 that conducted research on a variety of subjects, including exercise, nutrition, blood chemistry, and the effects of altitude and climate on the body. Pioneering exercise research included the first measures of the body's capacity to consume oxygen, thermoregulation during exercise, and the effects of environmental factors on exercise.

HERITAGE Family Study: Large-scale study of the role of genetics in cardiovascular and metabolic responses to aerobic training and regular exercise. The study was funded in 1992, with Claude Bouchard as the principal investigator, and now includes researchers from five field centers in the United States. One of the earliest findings of the HERITAGE Family Study was that changes in $\dot{V}O_2$max in response to exercise training are highly variable and that variability across families is much greater than within families.

Hippocrates: The "father" of preventive medicine, who wrote extensively in the 5th century B.C. about the benefits of exercise for a variety of illnesses.

Marti Karvonen: Finnish researcher who identified the intensity of exercise training required to produce gains in cardiorespiratory fitness.

London Bus Study: The first physical activity epidemiology study, conducted by Jeremy Morris and colleagues in the 1940s. The study showed that London's double-decker bus conductors, who climbed stairs and were on their feet throughout their shifts, had lower rates of coronary heart disease than bus drivers, who were almost entirely sedentary during their shifts.

R. Tait McKenzie: One of the first physiologists to study exercise. McKenzie investigated the effects of exercise on athletes before and after sports participation.

Ralph Paffenbarger: A pioneering physical activity epidemiologist. Paffenbarger conducted large-scale studies on the relationship of physical activity to health outcomes. Perhaps the best known of his studies, the Harvard Alumni Health Study, clearly established that physical activity reduces the rate and risk of developing coronary heart disease in men.

Surgeon General's Report on Physical Activity and Health: Landmark document, published in 1996, that summarized the science on physical activity and health to date and recommended that all adults accumulate 30 or more minutes of moderate physical activity on all or most days of the week.

Paul Dudley White: Cardiologist who began, in the 1950s, prescribing exercise as part of a treatment program for people with coronary heart disease.

STUDY QUESTIONS

1. Discuss the views of the ancient Greek physicians on the relationship between physical activity and health.

2. What contributions did the Greek physician Galen make to understanding the relationship between physical activity and health?

3. What information did observational studies of sedentary workers in the 18th and 19th centuries reveal?

4. Name five scientific disciplines that have contributed to the development of the field of exercise science.

5. Describe the work of three scientists who contributed to the development of exercise science in the 20th century.

6. What were the recommendations of the 1978 ACSM position statement on exercise for developing and maintaining fitness? What change in the ACSM position on exercise is reflected in recent versions of the position statement?

7. Why was the 1992 AHA *Statement on Exercise* an important milestone in establishing the physical activity–health connection?

8. Why was the 1995 CDC–ACSM recommendation controversial?

9. List four key conclusions of *Physical Activity and Health: A Report of the Surgeon General*.

10. What did the 2004 U.S. Dietary Guidelines Advisory Committee determine about the relationship of physical activity to weight loss and maintenance of a weight loss?

© Bill Crane

Physical Activity and Fitness With Age Among Sex and Ethnic Groups

▪ Peter T. Katzmarzyk, PhD

CHAPTER OUTLINE

Physical Activity
- Age
- Sex
- Ethnicity
- Recent Trends

Physical Fitness
- Age
- Sex
- Ethnicity

Summary
Review Materials

This chapter explores age, sex, and ethnic differences in **physical activity** and **physical fitness.** The chapter examines data collected from several countries, and, where available, studies using representative population samples are highlighted. The limitations of the existing databases and studies are also described, and important areas for future research are provided throughout the chapter. By the end of the chapter, you should be able to describe differences in physical activity and fitness between males and females and across different age and ethnic groups. Given the relationships among physical activity, fitness, and health described in part III of this book, we discuss specific population groups that are considered to be at a high risk of morbidity and mortality because of low levels of physical activity or fitness.

It is clear that individuals differ in physical activity levels and physical capabilities. For example, many people have difficulty climbing a flight of stairs without getting winded, whereas others have the ability to complete Ironman-distance triathlons, which involve swimming 4 km (2.4 miles), cycling 180 km (112 miles), and running 42 km (26 miles). There are many reasons for this incredible range of human variation in physical activity levels and physical abilities; social and behavioral influences as well as biological factors play a role. Although physical activity and fitness levels vary considerably from person to person, there are systematic differences according to sex, age, and ethnicity. The field of **epidemiology** is concerned with studying the distribution and determinants of disease, injury, and risk factors in society. As risk factors for chronic disease, physical inactivity and fitness are important variables in this regard. Thus, the purpose of this chapter is to present the **descriptive epidemiology** of physical activity and physical fitness levels across age, sex, and ethnic groups.

Although sex and age are self-explanatory concepts, race and ethnicity are more difficult to define. There are no widely accepted definitions of race or ethnicity; however, there are distinctions between the two. For example, the notion of race implies that biological traits can be used to categorize people into subgroups or races. On the other hand, ethnicity implies cultural similarities among individuals rather than strictly biological relatedness. The terms *race* and *ethnicity* are often used interchangeably in the scientific literature or combined into a single dimension such as *race/ethnicity* (Comstock et al. 2004). Throughout this chapter, the term *ethnicity*

is used consistently to refer to racial and ethnic differences in physical activity and physical fitness; however, the authors of the original studies that are described in the chapter may have used either one term or the other in their original reports.

The studies described in this chapter all used either cross-sectional or longitudinal research designs. A **cross-sectional research design** is one in which a sample of the population is examined at a given point in time and data from specific groups can be compared. This type of design is commonly used in national population surveys; however, it is difficult to infer changes in traits or behaviors within individuals; rather differences between individuals can be determined. A **longitudinal research design** is one in which participants are measured initially and then followed over time to directly measure changes. Longitudinal studies are generally more difficult and time consuming to conduct; however they allow researchers to measure changes within individuals and establish temporal sequences of events.

Physical Activity

As defined in chapter 1, physical activity is a behavior. Like most behaviors, it displays a wide range of variability between males and females, across age groups, and in different ethnic groups. This section presents the descriptive epidemiology of physical activity levels using the most recently available representative population data from several countries.

Age

This chapter provides a general overview of age-related differences in physical activity and fitness. A more detailed treatment of the topics of physical activity and fitness in childhood and in the aged is provided in part III, chapters 17 and 18.

Several international studies using a cross-sectional research design have documented clear age-related differences in physical activity levels. In general, physical activity levels are highest in childhood and decrease throughout adolescence and into adulthood. Figure 3.1 demonstrates the age-related differences in physical activity levels in Canada (figure 3.1*a*, Statistics Canada 2002) and the United States (figure 3.1*b*, Ham et al. 2004). Compared with younger adults, older adults have lower overall levels of **leisure-time physical activity** and

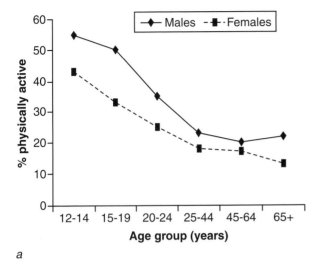

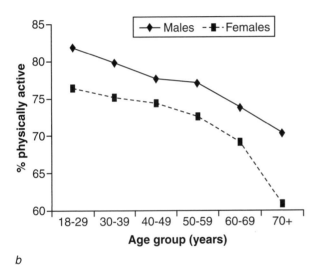

a

b

Figure 3.1 Age- and sex-related differences in physical activity levels across the life span in *(a)* Canada and the *(b)* United States. In Canada, physically active was defined as expending at least 3 kcal·kg⁻¹·day⁻¹, and in the United States it was defined as participating in at least some leisure-time physical activity in the last month.

Data from the 2000/2001 Canadian Community Health Survey and the 2002 U.S. Behavioral Risk Factor Surveillance System.

typically engage in less vigorous physical activities. Although the trends are similar, it is difficult to compare across the countries because the measures of physical activity are different. These age differences in physical activity levels were also evident in the Baltimore Longitudinal Study of Aging, in which age-related differences in leisure-time physical activity levels were observed, with a shift from high-intensity to moderate and low-intensity activities with advancing age (Talbot et al. 2000).

It is hypothesized that the observed age-related differences in physical activity levels in humans are the result of complex changes in psychological and social factors that occur with age; however, biological mechanisms may also play a role. Age-related declines in physical activity in laboratory and zoo animals, including invertebrates, rodents, dogs, and nonhuman primates, have been well documented. For example, the number of episodes of walking and jumping in three species of female monkeys observed in a zoological park showed a clear age-related trend, as demonstrated in figure 3.2 (Janicke et al. 1986). Biological mechanisms that have been proposed to explain the age-related declines in physical activity include the general physiological declines that occur with aging, declines in motor abilities, and neurobiological changes that may be involved in the motivation to be physically active.

> Physical activity levels decline across the life span, and evidence from nonhuman studies suggests that in addition to social and psychological factors, biological mechanisms may play a role.

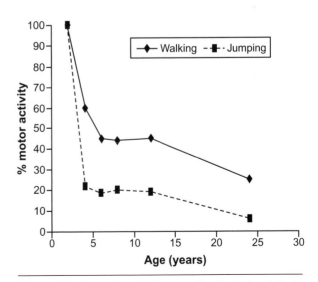

Figure 3.2 Age-related differences in physical activity in three species of female monkeys observed in a zoological park.

Adapted from Janicke, Coper and Janicke, 1986, courtesy of S. Karger AG, Basel.

In addition to trends in physical activity levels *across* age groups, an important consideration is **tracking,** or stability of physical activity with advancing age. Public health efforts aimed at increasing physical activity levels of the population assume that physical activity will become a habitual part of an individual's lifestyle. Thus, the degree to which individuals maintain their relative rank or position over time is of interest. In other words, how well does physical activity track from one time to the next? It is often assumed that physical activity levels during childhood track into adulthood, such that active children turn into active adults. However, the available evidence suggests that the relationship between childhood and adulthood physical activity levels is poor. On the other hand, some aspects of childhood physical activity, such as sport participation, may be more related to the persistence of physical activity levels into adulthood than general leisure-time physical activity levels. Future research should be directed at identifying those aspects and contexts of childhood and adolescent physical activity that best predict adult levels of participation. This information will be important to identify those children and youth who are at an elevated risk of becoming physically inactive adults who in turn will be at an elevated risk of chronic diseases later in life, and it would allow for targeted interventions that are aimed at changing behavior early through physical activity and health promotion.

> Physical activity levels do not track very well between childhood and adulthood. More research is required to determine those components of physical activity in childhood that best predict continued participation in physical activity in adulthood.

Sex

There are clear sex differences in physical activity levels that arise during childhood and persist throughout the life span. On average, males are more physically active than females, and they also tend to engage in more vigorous physical activities. Data from the 1997-98 Health Behavior in School-Aged Children survey, a cross-national survey of children and youth 11 to 15 years of age from 28 different countries, documented sex differences in physical activity levels that persisted across all

countries surveyed (World Health Organization 2000). In each country, more males than females reported regularly participating in physical activity on at least two occasions a week or for 2 hr or more per week. There were also lower physical activity levels across incremental age groups in adolescence, and the declines were more apparent in females than males.

In addition to demonstrating age differences in physical activity levels, figure 3.1 also shows the sex differences in leisure-time physical activity levels in Canada and the United States. At each age, males engage in higher levels of total leisure-time physical activity than females. These sex differences in physical activity levels were also evident in the Baltimore Longitudinal Study of Aging, and the higher levels of physical activity overall in males were partially explained by the fact that males engaged in higher levels of high-intensity physical activities and lower levels of low-intensity physical activities than females (Talbot et al. 2000).

Several hypotheses exist to explain the observed sex differences in physical activity levels. For example, a greater socialization toward sport participation in boys versus girls may explain some of the observed differences in level and intensity of physical activity. Other social factors such as peer pressure, family and childcare commitments, and access to opportunities for physical activity may also be involved.

> Across the life span, males are more physically active than females, and they are more likely to participate in more vigorous forms of physical activity.

Ethnicity

Ethnicity may play a role in explaining variation in physical activity levels. For example, cultural practices and religious beliefs may affect one's level of physical activity, and access to recreational resources may differ by ethnicity, attributable to potential differences in socioeconomic status across ethnic groups. There is limited information on ethnic differences in physical activity levels in most countries, and the validity of the information that is available may be tainted by cultural differences in reporting physical activities or the use of questionnaires that are not culturally specific. Figure 3.3 presents

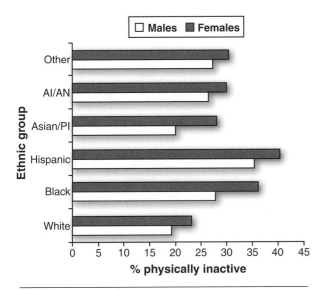

Figure 3.3 Prevalence of physical inactivity among ethnic groups in the United States. Data are from the 2002 Behavioral Risk Factor Surveillance System (BRFSS). Physical inactivity is defined as not engaging in any leisure-time physical activities in the past month. PI = Pacific Islander; AI = American Indian; AN = Alaska Native; other = other ethnicity or multiethnic.

Data from the 2002 Behavioral Risk Factor Surveillance System (BRFSS).

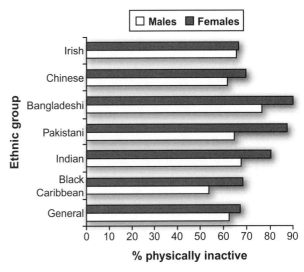

Figure 3.4 Prevalence of leisure-time physical inactivity among ethnic groups in England. Data are from the 1999 Health Survey for England. Physical inactivity is defined as no participation in sports or exercise of at least moderate intensity in the past 4 weeks.

Data from R. Teers, 2001, Physical activity. In *Health survey for England- The health of minority ethnic groups '99,* edited by B. Erens, P. Primatesta and G. Prior (London: The Stationary Office).

the prevalence of physical inactivity (reporting no leisure-time physical activity) across various ethnic groups in the United States (Ham et al. 2004). Black and Hispanic groups tend to have higher levels of physical inactivity than other groups, and within each ethnic group, females are more physically inactive than males. These ethnic differences in physical activity appear to begin in adolescence. Data from the 2001 U.S. Youth Risk Behavior Surveillance System indicate that black and Hispanic high school students are less likely than white students to achieve sufficient levels of moderate, vigorous, and total physical activity (Grunbaum et al. 2002).

Ethnic differences in physical activity levels have also been reported in other countries. For example, figure 3.4 presents the prevalence of physical inactivity in the major ethnic groups in England (Teers 2001). The ethnic differences in physical inactivity are not great, but there is a trend for Indian, Pakistani, and Bangladeshi groups to be more physically inactive than other groups in England. Caution must be used when comparing differences in physical activity levels among ethnic groups across countries, because from country to country, different methods are used, overall physical activity levels vary, and the cultural contexts and level of cultural assimilation differ.

Recent Trends

Up until the end of the 19th century, physical activity was an integral component of people's daily lives, because a high level of physical activity was required for many occupations and for the demands of daily living. However, the recent technological revolution has eliminated the requirement for humans to be physically active to make a living or to survive. On the other hand, technological advances have theoretically increased the availability of leisure time, and the potential for people to engage in physically active pursuits during their leisure time has increased. The recent trends in leisure-time physical activity in several countries for which temporal data are available are presented in figure 3.5. Although the survey questionnaires and sampling designs varied among the countries, there is generally a trend toward decreasing levels of physical inactivity in leisure time among adults within each of the countries. In other words, it appears that people are becoming more physically active. These reported trends for decreasing physical inactivity in the face of increasing levels of obesity in the same countries are somewhat of a paradox. However, the available data are for "leisure-time" physical activity, which generally excludes physical activity

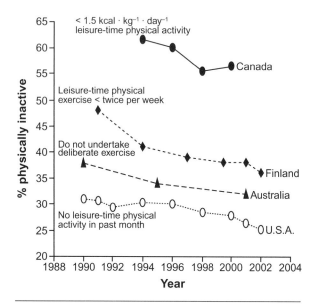

Figure 3.5 Recent trends in the prevalence of physical inactivity among adults in different countries.

Data sources include Canada, National Population Health Surveys and Canadian Community Health Survey (Statistics Canada 2002); United States, Behavioral Risk Factor Surveillance System (Ham et al. 2004); Australia, National Health Surveys (Australian Bureau of Statistics 2003); Finland, Finnish Health Behavior Surveys (Helakorpi et al. 2002).

undertaken at work or in other chores. Thus, it could well be that total daily energy expenditure levels have decreased while at the same time leisure-time physical activity levels have increased because technological advances have made it easier for us to perform our activities of daily living with a minimal expenditure of energy. These issues are explored further in chapter 11.

Another important factor that could partially explain the increases in reported physical activity levels is the increased public awareness of the health benefits of physical activity and what constitutes physical activity. As with any health behavior, a certain amount of social desirability is associated with reporting favorable levels of physical activity, and this effect may have increased in recent years. As well, there has been an increasing awareness that physical activities such as walking and gardening "count" as leisure-time physical activities, whereas in the past people may have only considered traditional activities such as running, cycling, or swimming as physical activities. Given that the most commonly reported leisure-time physical activities in the United States and Canada are indeed walking and gardening or yard work, this may have had an impact. Although these factors may have played a role in the observed increases in reported physical

activity levels, the extent to which the estimates have been affected is unknown.

> Studies conducted in the 1990s indicated little change in leisure-time physical activity levels in several countries.

Physical Fitness

As outlined in chapter 1, physical fitness has many components and can be conceptualized as either performance related or health related. Because there are so many facets of physical fitness, it is beyond the scope of this chapter to describe age, sex, and ethnic differences of all fitness components. Therefore, this section focuses on the descriptive epidemiology of **cardiorespiratory fitness** and **musculoskeletal fitness,** two of the primary components of physical fitness.

The measurement of physical fitness is more time consuming and expensive than the measurement of physical activity, which is typically done in national surveys using questionnaires. For example, in the Behavioral Risk Factor Surveillance System in the United States, the prevalence of no leisure-time physical activity is assessed by a negative response to the question, "During the past month, other than your regular job, did you participate in any physical activities such as running, calisthenics, golf, gardening, or walking for exercise?" (Ham et al. 2004). On the other hand, objective measures of physical fitness are rarely included in nationally representative population surveys because of the greater expense of having personal contact with the respondents and the greater amount of time required to take the measurements. In the 1981 Canada Fitness Survey, a variety of fitness tests were performed in the participants' homes, including a submaximal exercise step test on a portable set of stairs, hand-grip dynamometry, and sit-and-reach flexibility (Fitness Canada 1983). This section presents age, sex, and ethnic differences in physical fitness using population-level data when available; however, this information is supplemented with laboratory data when appropriate.

Age

Among adults, there is a general age-related physiological decline that affects many organs and organ

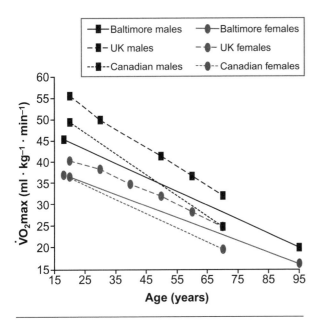

Figure 3.6 Age-related trends in V̇O₂max in males and females from the Baltimore Longitudinal Study of Aging (Talbot et al. 2000); England (Sports Council and Health Education Authority 1992); and Canada (Bonen and Shaw 1995).

systems. In turn, this has an effect on physical fitness levels. Both cardiorespiratory fitness and musculoskeletal fitness decrease with advancing age in males and females. Figure 3.6 presents **maximal oxygen uptake (V̇O₂max)** values across age among adult males and females from several cross-sectional studies. It is clear that older individuals have lower levels of cardiorespiratory fitness compared with younger individuals. The absolute levels of fitness should not be compared across the three different studies, because different methods were used to measure fitness in each study that are not directly comparable. What is of interest is that the age-related trends within studies are remarkably similar, and although males and females differ in absolute levels of fitness, the observed age differences are similar within a given population. The data from the Baltimore Longitudinal Study of Aging are arguably the best, because V̇O₂max was directly measured using a treadmill test; however, the sample was not nationally representative (Talbot et al. 2000). On the other hand, both the studies from Canada and England used representative population samples; however, V̇O₂max was predicted from submaximal exercise tests in both Canada and England (Fitness Canada 1983; Sports Council and Health Education Authority 1992). A further limitation of the Canadian data is that the equation used to predict V̇O₂max included age as a component; thus, some of

the observed declines with age could be an artifact of the prediction method. Nevertheless, all of the available data indicate significant age differences in aerobic fitness across the life span. On average, the age-related decrease across incremental age groups in V̇O₂max is approximately 1% per year in adults.

These results have important implications for elderly persons. In general, any given task, such as climbing a flight of stairs, requires a certain amount of energy expenditure, regardless of age. Thus, a given task for elderly people requires them to work at a higher percentage of their maximal aerobic capacity compared with younger people and will require more physiological effort. Even simple tasks can be very demanding for elderly persons.

The age-related differences in V̇O₂max presented in figure 3.6 are derived from studies using cross-sectional research designs. There is also evidence from studies using a longitudinal research design that aerobic capacity decreases with age. A series of studies were conducted on elite long-distance runners in the late 1960s and early 1970s to determine factors associated with success in this endurance event. Figure 3.7 presents the results of an interesting analysis of V̇O₂max in these same runners 22 years later, based on their current physical activity levels (Trappe et al. 1996). The runners were grouped according to whether they were currently sedentary (untrained), were training for physical fitness, or were maintaining the highest level of training and competition. Even in those runners

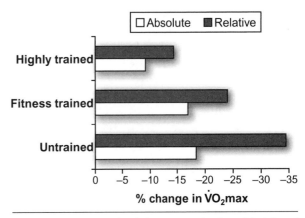

Figure 3.7 Longitudinal changes in absolute (L/min) and relative (ml·kg⁻¹·min⁻¹) V̇O₂max in elite long distance runners. Runners were currently untrained (UT), were training for physical fitness (FT), or were highly trained for competition (HT) after 22 years.

Data from S.W. Trappe, D.L. Costill, M.D. Vukovich, J. Jones and T. Melham, 1996, "Aging among elite distance runners: A 22-yr longitudinal study," *Journal of Applied Physiology* 80: 285-90.

who maintained high levels of training, there was a decline in $\dot{V}O_2$max with advancing age. However, the decreases in $\dot{V}O_2$max in those who were training for fitness or were completely untrained were progressively greater. These results highlight the importance of maintaining physical activity levels with advancing age to attenuate the decline in cardiorespiratory fitness levels.

> As part of the general physiological decline with advancing age, cardiorespiratory fitness levels decrease throughout adulthood. There is some evidence from longitudinal studies that keeping physically active may attenuate this observed decline in fitness.

Musculoskeletal fitness generally declines with age among adults, although the pattern of decline varies depending on the measurement or component of fitness studied. Figure 3.8 presents cross-sectional age-related differences in **muscular strength** (hand grip, figure 3.8a) and **trunk flexibility** (sit and reach, figure 3.8b) among Canadians (Fitness Canada 1983).

Dramatic increases in muscular strength occur during normal growth and maturation in childhood and adolescence and reach a peak among adults 20 to 29 years of age, after which strength decreases gradually through the rest of the life span. These cross-sectional observations regarding age-related declines in muscular strength have been confirmed in studies using a longitudinal research design. Figure 3.9 shows changes in grip strength in Japanese American males participating in the Honolulu Heart Program over 27 years (Rantanen et al. 1998). The results are presented as average changes per year over the entire follow-up period. Although there are reductions in grip strength in all age groups, there is a general trend for greater changes in males who were older rather than younger at the outset of the study. In other words, the rate of decline in muscular strength tends to accelerate with advancing age. These decreases in muscular strength are in large measure related to the loss of muscle mass associated with the aging process.

Flexibility is an important component of musculoskeletal fitness and tends to be joint specific. In other words, if you are quite flexible at the hip, there is no guarantee that you will be flexible at the

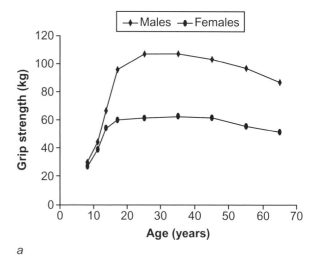

a

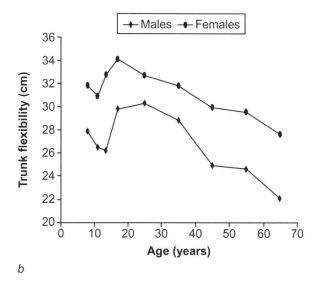

b

Figure 3.8 Differences in measures of fitness by age and sex. *(a)* Handgrip strength (combined left and right) and *(b)* trunk flexibility in Canadian males and females from 7 to 69 years of age.
Data from the 1981 Canada Fitness Survey.

knee, shoulder, or other particular joint. Enhanced flexibility is associated with maintaining the range of motion in a given joint and a reduced occurrence of back pain in the case of hip or trunk flexibility. Trunk flexibility, measured using a sit-and-reach test, is a composite measure of low back and hamstring flexibility, which tends to show age-related variation (figure 3.8b). Performance on the sit-and-reach test tends to decrease across age groups during adolescence. However, this decrease may be more associated with differential changes in the lengths of the arms and legs relative to the trunk as youth go through this period of rapid growth rather than

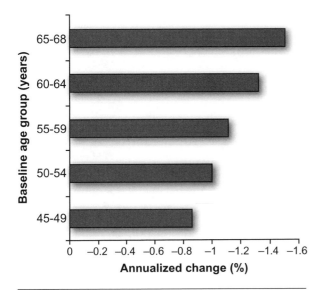

Figure 3.9 Longitudinal changes (per year) in maximal handgrip strength over 27 years in Japanese American males.

Data from T. Rantanen, K. Masaki, D. Foley, D. Izmirlian, L. White and J.M. Guralnik, 1998, "Grip strength changes over 27 yr in Japanese-American men," *Journal of Applied Physiology* 85: 2047-53.

actual reductions in the range of motion around the hip joint per se. This observed decline in sit-and-reach performance is followed by an increase into adulthood, after which it decreases throughout the rest of the life span.

Sex

Males and females differ in physical fitness levels, but the extent of the difference depends on the specific component of fitness. On average, males have higher levels of cardiorespiratory fitness at a given age. This is quite evident in figure 3.6, where it is shown that within a particular study, males have greater $\dot{V}O_2$max than females across the life span. The sex differences in cardiorespiratory fitness have generally been attributed to differences in lean body mass, blood hemoglobin levels, and levels of physical activity participation between males and females.

Males are generally stronger than females at any given age (figure 3.8*a*). This sex difference is mainly attributable to differences in the absolute amount of muscle mass between males and females. When muscle strength is corrected for muscle size, the difference between males and females becomes smaller and sometimes disappears completely. Although males may be stronger than females, females per-

form better on the sit-and-reach test and tend to be more flexible than males (figure 3.8*b*). The sex differences in flexibility are likely due to morphological differences in the architecture of the hip joint.

> Throughout the life span, males are generally stronger than females, whereas females are more flexible than males.

Ethnicity

Data on ethnic differences in physical fitness from population surveys are very limited. However, there is some evidence from relatively large laboratory-based studies that ethnicity may play a role in explaining variation in physical fitness levels. The HERITAGE Family Study was a large multiple-center trial conducted in the United States and Canada with the primary aim of determining genetic factors related to exercise training. The researchers recruited a large sample of sedentary (no regular exercise in previous 6 months) black and white adults 17 to 65 years of age from the population. Figure 3.10 presents the average $\dot{V}O_2$max values in black and white males and females when they entered the study (Skinner et al. 2001). In both

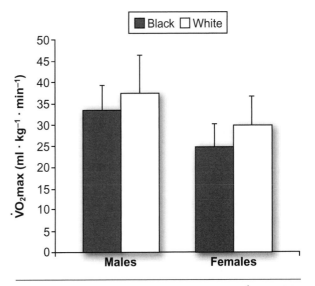

Figure 3.10 Maximal oxygen consumption ($\dot{V}O_2$max) in black and white participants in the HERITAGE Family Study. The error bars represent 1 standard deviation.

Data from J.S. Skinner, A. Jaskolski, A. Jaskolska, J. Krasnoff, J. Gagnon, A.S. Leon, D.C. Rao, J.H. Wilmore and C. Bouchard, 2001, "Age, sex, race, initial fitness, and response to training: The HERITAGE Family Study," *Journal of Applied Physiology* 90: 1770-1776.

sexes, white participants had significantly higher $\dot{V}O_2$max values than black participants.

Further evidence of ethnic differences in cardiorespiratory fitness comes from a large multiple-center cohort study in the United States that was designed to investigate the development of coronary heart disease risk factors in young adults. The CARDIA (Coronary Artery Risk Development in Young Adults) study enrolled black and white participants 18 to 30 years of age (Carnethon et al. 2003). Participants performed a maximal treadmill exercise test and then were divided into low (lower 20%), moderate (middle 40%), and high (upper 40%) fitness categories in males and females (figure 3.11). Clear trends toward a decreasing proportion of black participants across low, moderate, and high cardiorespiratory fitness groups were observed in both males and females. This suggests that the overall level of cardiorespiratory fitness was lower in young black adults compared with young white adults.

> The available evidence suggests that cardiorespiratory fitness levels are lower among black persons compared with white persons in North America. However, there is a lack of population-level data with which to further explore this issue.

Although the information described in this section is not based on representative population surveys, it provides evidence that there may be systematic differences in cardiorespiratory fitness between black and white adults in North America. The finding that black participants have lower levels of cardiorespiratory fitness compared with white participants is not surprising, given that black youths and adults are also the least physically active group, as described previously. More research is required to determine the underlying causes of ethnic differences in cardiorespiratory fitness and whether they are mainly related to differences in behavior or attributable to inherent biological differences. Furthermore, there is currently little information about differences in cardiorespiratory fitness among other ethnic groups, which limits our understanding of the range of variability in fitness in the general population.

Unfortunately there is very little information on ethnic differences in musculoskeletal fitness in the population. This is largely because the national

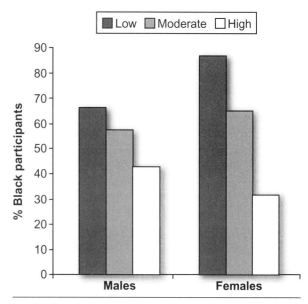

Figure 3.11 Distribution of black participants across cardiorespiratory fitness categories in the Coronary Artery Risk Development in Young Adults (CARDIA) Study.

Data from M.R. Carnethon, S.S. Gidding, R. Nehgme, S. Sidney, D.R. Jacobs and K. Liu, 2003, "Cardiorespiratory fitness in young adulthood and the development of cardiovascular disease risk factors," *Journal of the American Medical Association* 290: 3092-3100.

population surveys that have included representation from the various ethnic groups have not included measures of musculoskeletal fitness in their battery of tests. This is an important area of future research, because musculoskeletal fitness is related to the health and independent living among the elderly, and ethnic differences may have implications for identifying high-risk groups.

Summary

In this chapter you have explored the descriptive epidemiology of physical activity and fitness, with a particular emphasis on age, sex, and ethnic differences. The discussion has shown that there are systematic differences in physical activity and fitness across various demographic groups. Further research is required to further elucidate these differences and the impact they may have on health, particularly among minority populations. The mechanisms behind the observed differences remain largely unknown, and researchers should investigate the role of biological, social, and psychological factors in explaining age, sex, and ethnic variation in physical activity and fitness.

KEY CONCEPTS

cardiorespiratory fitness: Component of physical fitness that reflects the integrated function of the heart, lungs, vasculature, and skeletal muscles to deliver and use oxygen during dynamic physical activities. Also referred to as aerobic power or cardiorespiratory endurance, it is measured under laboratory conditions as maximal oxygen uptake ($\dot{V}O_2$max) based on ventilatory gas exchange and indirect calorimetry during maximal exercise testing. It can also be estimated from maximal exercise time on a treadmill or bike test, the final speed and grade of a maximal treadmill test, the maximal workload achieved during a bike test, or heart rate responses to multiple submaximal work stages on a bike or treadmill ergometer.

cross-sectional research design: Research design in which a cross-section of the population is examined at a given point in time and data from specific groups can be compared.

descriptive epidemiology: Assessment of variation in the prevalence of a trait or behavior by age, sex, ethnicity, and other geographic or demographic factors.

epidemiology: A scientific discipline that involves the study of the distribution and determinants of disease in human populations. In this chapter, epidemiologic studies are used as the basis to directly assess whether physically active individuals have lower rates of cancer than inactive persons (see also *case–control study, cohort study,* and *randomized clinical trial*).

leisure-time physical activity: For definition, see page 19.

longitudinal research design: Research design in which participants are measured initially and then followed over time to directly measure changes over time.

maximal oxygen uptake or power ($\dot{V}O_2$max): For definition, see page 19.

muscular strength: Ability of a muscle to produce force.

musculoskeletal fitness: Aspects of physical fitness that are related to muscular strength, muscular endurance, flexibility, and bone health.

physical activity: For definition, see page 19.

physical fitness: For definition, see page 19.

tracking: Stability, or the maintenance of relative rank or position in a group over time.

trunk flexibility: The suppleness and mobility of the hip joint, generally measured using a sit-and-reach test.

STUDY QUESTIONS

1. Differentiate between cross-sectional and longitudinal research designs.

2. Describe the age- and sex-related differences in physical activity observed in many population surveys.

3. Distinguish between age-related differences in physical activity and the tracking or stability of physical activity with age.

4. What evidence is there that the observed age-associated declines in physical activity may have a biological basis?

5. Describe the age-related changes that occur in cardiorespiratory fitness among adults. How do the changes in trained individuals compare with those in sedentary people? What are some potential limitations of the population data that are used to determine these trends?

6. What areas in the descriptive epidemiology of physical activity are still not well understood? Suggest some important areas for future research.

II

Effects of Physical Activity on the Human Organism

When a person engages in a sustained bout of physical activity, nearly all the systems or organs in the body come into play to support the contracting skeletal muscle. The cardiovascular and respiratory systems immediately respond to increase the availability of oxygen and substrate (glycogen and fat) for energy production in the muscle and to remove metabolic waste products. Free fatty acids are released from adipose tissue, and the liver produces more glucose in response to the increase in energy production. The nervous system is activated along with various hormones and enzymes to help coordinate and regulate all of these functions, and the force generated by muscle contractions and gravity puts strain on bone, ligaments, and tendons. When such bouts of exercise are repeated over weeks or months (exercise training), many of these systems begin to adapt to the activation by increasing their capacity or efficiency.

Chapter 4 reviews many of the major changes that occur during a single bout of exercise. The other four chapters in this section present key changes that typically take place in specific systems or organs as a result of regular physical activity or exercise training. The emphasis is on responses in generally healthy adults with some important exceptions noted. The value of examining these biological and structural responses to acute and chronic exercise is their critical role in the effect of regular activity on health. Part II of this book provides a foundation for issues that are studied by researchers around the globe.

Metabolic, Cardiovascular, and Respiratory Responses to Physical Activity

Edward T. Howley, PhD

Knowledge of the metabolic, cardiovascular, and respiratory responses to physical activity and exercise is important to understand the epidemiology of physical activity. This information provides a context in which to evaluate the relative or absolute intensity of the activity as well as the associated caloric expenditure—important elements in a conversation about "how much exercise is enough." This chapter provides an overview of these issues while discussing the cardiovascular and respiratory responses required to meet the metabolic demands associated with physical activity and exercise. Finally, the impact of endurance training on these responses is discussed. This chapter can only provide a brief overview of these topics; the reader is directed to one of the many exercise physiology texts for a more complete treatment.

Relationship of Energy to Physical Activity

Energy is required for movement. Adenosine triphosphate (ATP) is the energy source used by skeletal muscles to generate tension (force), which is, in turn, transmitted by connective tissue to the skeletal system to cause movement. The ATP must be replaced as quickly as it is used for muscles to continue to generate force. The greater the force or the rate at which force is developed (power), the greater is the need for ATP. Muscles have multiple systems for replacing ATP that allow athletes to run at very high speeds for short periods of time (e.g., 100 m dash) or at slower (but still fast) speeds for longer distances (e.g., marathon) or that allow farmers to work all day long in a field. Energy for muscle force generation is made available when ATP is split to adenosine diphosphate (ADP) and inorganic phosphate; energy from other sources is required to regenerate the ATP. Two broad classes of reactions are available to regenerate the ATP: **anaerobic** reactions, which can regenerate ATP rapidly and without oxygen, and **aerobic** reactions, which are slower to activate but can sustain ATP regeneration for long periods of time through oxygen utilization. In both cases, an increase in the concentration of ADP in the working muscles (from ATP breakdown) activates the enzymes that ultimately lead to the regeneration of ATP.

Anaerobic Energy Sources

There are one-enzyme reactions and one multiple-enzyme pathway that can replace ATP at a very fast rate. These energy systems allow muscle to continue to generate force during strenuous activities or help make the transition from rest to exercise as the slower aerobic system comes up to speed. In the most important of these one-enzyme reactions, catalyzed by creatine kinase, **phosphocreatine (PC)** reacts with ADP to form ATP. The PC concentration in muscle is limited and can provide only about 5 s of support during all-out activity. In the multiple-enzyme pathway called **glycolysis,** glucose is metabolized at a high rate and ATP is generated without oxygen. Lactic acid is produced in the process, leading to hydrogen ion accumulation in the muscle and blood. Glycolysis provides ATP at a high rate and is a substantial contributor to the ATP needed in all-out activities lasting less than about 2 min. In contrast, when glucose is metabolized aerobically, it represents a long-lasting source of ATP, producing about 18 times more ATP per glucose molecule than when metabolized anaerobically.

Aerobic Energy Production

Oxidative metabolism of carbohydrates (muscle glycogen and blood glucose) and fats (from both adipose tissue and intramuscular sources) provides the long-term source of ATP for physical activities and exercise that we typically associate with health-related outcomes; it also accounts for the majority of the ATP used in all-out performances lasting more than a few minutes (see figure 4.1). The complete oxidation of these fuels takes place in the mitochondria of the cell, which increase in number with endurance training. The larger number of mitochondria increases the capacity of the muscle to use fat as a fuel, because fat can only be metabolized via aerobic pathways.

When fats and carbohydrates are metabolized aerobically, oxygen (O_2) is consumed and carbon dioxide (CO_2) is produced. The ratio of the volume of CO_2 produced ($\dot{V}CO_2$) to the volume of oxygen ($\dot{V}O_2$) consumed by the cell is called the respiratory quotient (RQ). Because we measure $\dot{V}O_2$ and $\dot{V}CO_2$ at the mouth, this ratio is also called the **respiratory exchange ratio (R).** If fat is the only fuel used for energy during steady-state exercise, the R = 0.70, and 19.6 kJ (4.7 kcal) of energy is produced for each

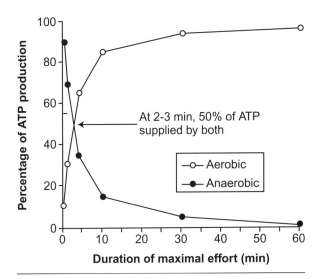

Figure 4.1 Percent of energy from aerobic and anaerobic sources for all-out activities of various durations.

Reprinted, by permission, from E.T. Howley and B.D. Franks, 2003, *Health fitness instructor's handbook,* 4th ed. (Champaign, IL: Human Kinetics), 478.

liter of oxygen consumed; when carbohydrates are the only fuel, the R = 1.0 and 20.9 kJ (5.0 kcal) is produced per liter of oxygen. Typically, a mixture of carbohydrates and fats is used to provide the ATP for most activities; however, as the intensity of exercise increases, carbohydrate becomes the pri-

Mechanisms for Energy Production During Exercise

ATP is the energy source used by muscles to generate tension or force. ATP must be replaced as quickly as it is used for exercise to continue. Muscles can produce ATP rapidly using anaerobic mechanisms (i.e., phosphocreatine and glycolysis), but their contribution lasts only 5 s to 2 min during all-out exercise, respectively. Aerobic metabolism of carbohydrates and fats in the mitochondria of the muscle is the primary source of ATP in all-out exercise lasting more than 2 to 3 min. Carbohydrates become a more important fuel as exercise intensity increases, and fat provides a greater percent of the energy as the duration of moderate exercise increases.

mary fuel, as noted by the increase in R (see figure 4.2*a*). In contrast, for moderate-intensity activities, the percent of energy derived from fat increases as the duration of the activity increases, as shown by the decrease in R (see figure 4.2*b*). The R can go over 1.0 when lactic acid that is generated in heavy exercise is buffered by plasma bicarbonate and CO_2 is produced; the elevated R is used as an indicator that a person has achieved maximal oxygen uptake ($\dot{V}O_2$max, see later). Oxygen must be delivered to

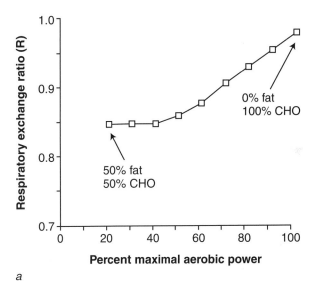

a

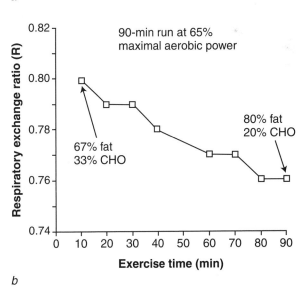

b

Figure 4.2 Effect of exercise *(a)* intensity and *(b)* duration on the pattern of carbohydrate and fat use.

Reprinted, by permission, from E.T. Howley and B.D. Franks, 2003, *Health fitness instructor's handbook,* 4th ed. (Champaign, IL: Human Kinetics), 474, 485.

and consumed by the mitochondria to produce the ATP from the metabolism of carbohydrates and fats. The roles that the cardiovascular and respiratory systems play in that regard are discussed in the next section.

Oxygen Consumption and Cardiovascular and Respiratory Responses to Exercise

Energy derived from oxygen consumption accounts for the vast majority of energy used in physical activity and exercise, compared with that from anaerobic sources. Consequently, knowledge of how $\dot{V}O_2$ changes during the transition from rest to exercise and then back to rest is helpful in understanding how the intensity and duration of exercise influence the caloric expenditure associated with exercise. Furthermore, although the cardiovascular and respiratory systems must respond appropriately to deliver the correct amount of oxygen to the muscles, the magnitude of these responses is influ-

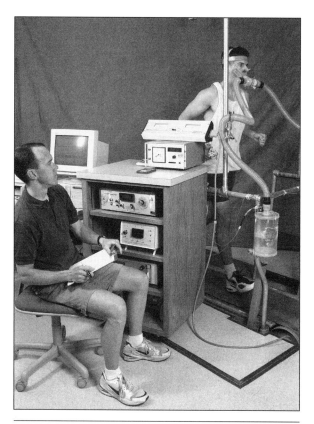

$\dot{V}O_2$max testing using a treadmill is an increasingly popular method of measuring oxygen uptake and other physiological variables during exercise.

enced by the individual's **maximal oxygen uptake ($\dot{V}O_2$max)**. The first part of this section describes briefly the $\dot{V}O_2$, **heart rate (HR),** and ventilation responses to a submaximal exercise task; the second part addresses in more detail the responses to a graded (incremental) exercise test taken to the point of volitional exhaustion. Within the latter context the effects of physical training are presented.

Rest to Exercise Transitions

An individual standing alongside a treadmill set at 200 m/min steps onto the belt and with that one step accelerates to 200 m/min; failure to do so would result in the individual drifting off the back of the treadmill. The rate at which ATP is used increases instantly from that needed for standing to that required to run at 200 m/min. However, figure 4.3 shows that $\dot{V}O_2$ did not increase instantaneously to the level required to generate the ATP for the task; instead there was a gradual increase over the first 2 to 3 min until a **steady-state** $\dot{V}O_2$ was achieved. The steady-state value is taken as the oxygen requirement for the task in which the oxidative generation of ATP meets the ATP requirement of the task. In fact, the steady-state oxygen uptake is used to calculate the caloric cost of the activity. If the steady-state $\dot{V}O_2$ is 2.0 L/min and we use 20.9 kJ (5.0 kcal) as the caloric equivalent of 1 L of oxygen, then 41.8 kJ (10 kcal) is produced per minute, and for 5 min of activity, 209 kJ (50 kcal) of energy is expended. Measurements

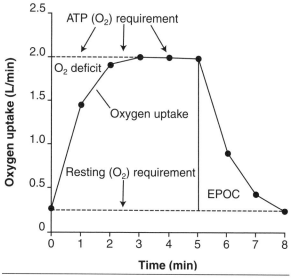

Figure 4.3 Oxygen uptake response to submaximal exercise.

Reprinted, by permission, from E.T. Howley and B.D. Franks, 2003, *Health fitness instructor's handbook,* 4th ed. (Champaign, IL: Human Kinetics), 486.

of oxygen consumption obtained during different physical activities are used to generate tables of the energy cost of those activities.

If one draws a line from the steady-state $\dot{V}O_2$ response back to the start of exercise, it is clear that the $\dot{V}O_2$ does not meet the ATP requirement during the first few minutes of exercise. An **oxygen deficit** is said to exist, and other energy sources have to provide the needed ATP—notably PC and glycolysis. The magnitude of the oxygen deficit increases with exercise intensity. When the individual steps off the treadmill at the end of 5 min of exercise, the $\dot{V}O_2$ does not return immediately to the resting level; there is a gradual return back to rest over several minutes. This has been called the oxygen debt, the recovery oxygen, or, most recently, the **excess postexercise oxygen consumption (EPOC)**. In the first 2 to 3 min of recovery, the $\dot{V}O_2$ falls rapidly as the PC and oxygen stores in muscle are quickly replenished. After that time, the $\dot{V}O_2$ falls more slowly to the resting baseline value; this portion of the EPOC is associated with the resynthesis of lactic acid to glucose and the elevated HR, ventilation, and hormone levels that have not yet returned to resting values (Gaesser and Brooks 1984).

Why does oxygen uptake increase slowly over the first 2 to 3 min of submaximal exercise when making the transition from rest? Figure 4.4 shows that HR and ventilation increase in a manner simi-

lar to $\dot{V}O_2$, which helps explain, in part, the lag in oxygen delivery to muscle at the onset of exercise. However, the muscle itself also contributes to this lag. It takes some time for the activity of the various mitochondrial enzyme systems involved in oxygen consumption to increase to the level required to meet the ATP demand of the tissue.

Transition From Rest to Exercise

In the transition from rest to submaximal exercise, ATP is delivered to the muscle contractile elements as fast as needed to maintain the intensity of the exercise. Because aerobic systems do not come on line instantly, anaerobic sources of ATP (phosphocreatine and glycolysis) provide the needed ATP at the onset of the exercise and an oxygen deficit is said to exist. By 2 to 3 min into a submaximal exercise task, the oxygen uptake ($\dot{V}O_2$) is adequate to meet the ATP demands and a steady state (balance between the supply of ATP by aerobic means and the demand for ATP by the muscle) exists. The lag in $\dot{V}O_2$ at the onset of exercise is related to a lag in cardiovascular and respiratory responses and to the time needed for mitochondria to increase their rate of ATP production. When exercise stops, the $\dot{V}O_2$ does not fall immediately back to the resting value. There is at first a very rapid decline, followed by a slower response back to the resting baseline value. The elevated oxygen uptake during recovery from exercise is called the excess postexercise oxygen consumption (EPOC).

Responses to Incremental Exercise Taken to Volitional Exhaustion

A graded exercise test (GXT) is used to evaluate a person's metabolic, cardiovascular, and respiratory responses to systematic increases in exercise intensity. Usually the test begins at a low intensity and progresses in equal-intensity increments. For example, during a treadmill test the belt speed might remain constant while the grade is increased 2% or 3% each 2 or 3 min. During a cycle ergometer test, the work rate might begin at 50 or 100 W and increase in increments of 25 or 50 W over the

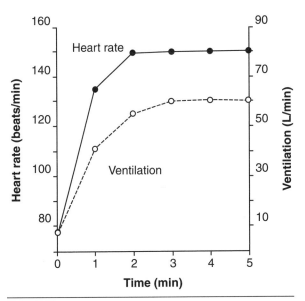

Figure 4.4 Heart rate and pulmonary ventilation responses to submaximal exercise.

Data from B. Ekblom, P.-O. Åstrand, B. Saltin, J. Stenberg and B. Wallström, 1968, "Effect of training on circulatory response to exercise," *Journal of Applied Physiology* 24: 518-528.

same time interval. If athletes are being tested, the initial stage is set at a higher intensity and the increment between stages is larger. During a GXT, the individual might be monitored for a wide variety of variables including electrocardiogram, HR, $\dot{V}O_2$, blood lactate concentration, and blood pressure. Measurements are typically made in the last 30 s of each stage of the test to approximate the steady-state response for that stage; data are then plotted against stages of the test. In this section, we begin with metabolic responses to such a test and then discuss the associated cardiovascular and respiratory responses.

Oxygen Uptake and Blood Lactate

Figure 4.5*a* shows the oxygen uptake response of an untrained individual to a GXT on a treadmill, in which the speed was set at 80 m/min and the grade increased 3% every 3 min until the participant experienced volitional exhaustion (i.e., participant terminated the test because of the perception of maximal effort). The $\dot{V}O_2$ values measured in the last 30 s of each stage were plotted against the percent grade. The $\dot{V}O_2$ increased with increasing grade, except at the end of the test, where the $\dot{V}O_2$ measured at 15% grade did not increase when the grade was increased to 18%. This "plateau" in $\dot{V}O_2$ with an increase in the exercise intensity has been taken, historically, as the criterion for having achieved $\dot{V}O_2$max. However, because only about 50% of individuals experience a plateau in $\dot{V}O_2$ during this type of GXT, investigators have used other measures as indicators that the participant was working maximally when the highest $\dot{V}O_2$ was obtained during a GXT. These include a high blood lactate concentration ($\geq$8 mM), a respiratory exchange ratio $\geq$1.10, and an HR response close to an age-predicted maximum value. The latter criterion is not as useful given the inherent variability in any age-predicted estimate of maximal HR (SD = ~10 beats/min) (Powers and Howley 2004). $\dot{V}O_2$max is used as a measure of cardiorespiratory fitness and is also called maximal aerobic power. Figure 4.5 also shows the changes in the $\dot{V}O_2$ response to the GXT following an endurance-training program. The $\dot{V}O_2$ at any submaximal stage was the same, but the $\dot{V}O_2$max increased.

There is considerable variability in $\dot{V}O_2$max values across the population. Cardiac patients may have $\dot{V}O_2$max values as low as 20 ml·kg^{-1}·min^{-1}, whereas elite endurance athletes may approach 80 ml·kg^{-1}·min^{-1}. Part of the differences found in the population are attributable to differences in the

quantity and quality of training, and part are attributable to genetic influences. In the HERITAGE Family Study, genetic factors explained about 50% of the variation that existed in $\dot{V}O_2$max in sedentary adults; however, the investigators believed that the value might be inflated because of nongenetic familial factors (Bouchard et al. 1998).

Figure 4.5*b* shows that the blood lactate concentration remains close to its resting value until 9% (pretraining measurement) or 12% (post-training measurement) grade, where it increases systematically. This occurs typically between about 50% and 80% of a person's $\dot{V}O_2$max. The work

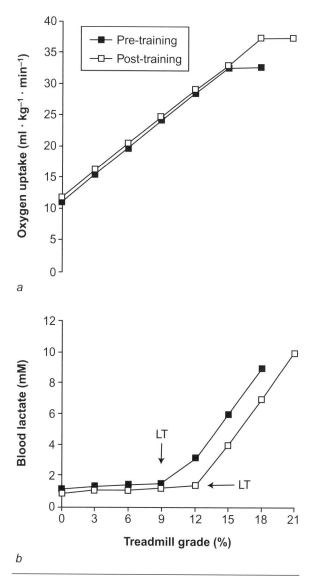

a

b

Figure 4.5 *(a)* Oxygen uptake and *(b)* lactate responses to a graded exercise test.

Reprinted, by permission, from E.T. Howley and B.D. Franks, 2003, *Health fitness instructor's handbook,* 4th ed. (Champaign, IL: Human Kinetics), 491-493.

rate (or oxygen consumption) at which the blood lactate concentration increases is called the **lactate threshold (LT).** The increase in the blood lactate concentration is attributable to its rate of appearance being greater than its rate of removal. The rate of appearance is linked to the recruitment of fast-twitch fibers (see chapter 7) as exercise intensity increases and to accelerated glycolysis that is driven by both intracellular factors (higher ADP levels stimulating key enzymes) and extracellular factors such as the rising concentration of epinephrine (Brooks 1985). The slower rate of removal is associated with a reduced blood flow to the liver (see later discussion on redistribution of blood flow). Regular endurance training delays the LT because of a reduced rate of lactate production, associated with an increase in mitochondrial number, and because of an enhanced capacity of muscle and other tissues to clear lactate (Brooks 1985). The LT has been used to predict endurance performance.

The $\dot{V}O_2$max can be expressed in absolute terms (L/min), relative to body weight (ml·kg^{-1}·min^{-1}), or as multiples of resting metabolic rate (**metabolic equivalents** or **METs**, where one MET is taken, by convention, to be 3.5 ml·kg^{-1}·min^{-1}). In the preceding example, the participant's pretraining $\dot{V}O_2$max was 31.5 ml·kg^{-1}·min^{-1} or 9 METs. The MET term is used as a simple way to express the more complicated term, ml·kg^{-1}·min^{-1} and is used extensively in cardiac rehabilitation and fitness programs,

---

Oxygen Uptake and Blood Lactate Responses During a Graded Exercise Test

Oxygen consumption ($\dot{V}O_2$) increases in a stepwise fashion with increases in exercise intensity during a graded exercise test (GXT), until the final stages where $\dot{V}O_2$ may not change with an increase in exercise intensity. The highest value achieved in such a test is taken as the individual's **maximal oxygen uptake ($\dot{V}O_2$max).** The blood lactate concentration remains close to resting values in the early stages of a GXT. The sudden increase in blood lactate concentration at about 50% to 80% of $\dot{V}O_2$max is called the lactate threshold (LT); the LT has been used to predict performance in endurance events.

---

physical activity interventions, and epidemiological investigations to express the intensity (e.g., moderate exercise is 3-6 METs) or volume (MET-hours) of physical activity or exercise (American College of Sports Medicine 2000). Next we look at the cardiovascular and respiratory responses that are linked to $\dot{V}O_2$max.

Heart Rate, Stroke Volume, Cardiac Output, and Oxygen Extraction

Oxygen consumption is the product of the **cardiac output,** expressed in liters of blood pumped from the heart per minute (L/min), and the volume of oxygen extracted from the arterial blood (CaO$_2$), in ml O$_2$/L blood. **Oxygen extraction** is the difference between the oxygen content of arterial blood and oxygen content of mixed venous blood (C$\bar{v}$O$_2$), the latter being measured in the right heart chamber. Cardiac output is the product of HR in beats/min and **stroke volume (SV)** in liters of blood pumped from the heart per beat, so

$$\dot{V}O_2 \text{ (L/min)} = HR \times SV \times (CaO_2 - C\bar{v}O_2).$$

The following sections describe how these variables respond during a GXT and indicate the effect of endurance training on each.

Figure 4.6 shows the changes in HR, SV, cardiac output, and oxygen extraction during a GXT. Cardiac output and HR increase in a linear fashion, but SV does not. In upright exercise, SV increases during the early stages of the test and levels off at about 40% $\dot{V}O_2$max. Consequently, further increases in cardiac output (beyond 40% $\dot{V}O_2$max) are attributable entirely to the increase in HR. The fact that HR tracks exercise intensity so well makes it a good indicator of exercise intensity. The increase in HR is attributable to both withdrawal of parasympathetic nerve activity and an increase in sympathetic nerve activity to the sinoatrial node (Rowell 1993).

Stroke volume is the difference between the volume of blood in the heart before contraction (**end-diastolic volume, EDV**) and the volume of blood in the heart after contraction (**end-systolic volume, ESV**). During exercise there is an increase in venous return to the heart that leads to an increase in EDV, causing a distension of the ventricles and an increase in the force of contraction. The increased venous return is linked to the alternate contraction and relaxation of skeletal muscles acting on the large veins in muscles (muscle pump); the lower intrathoracic pressure caused by the increased depth and rate of ventilation (respiratory pump); and the increase in abdominal pressure, compressing the large intra-abdominal veins (abdominal pump) (Rowell et al. 1996). The force of contraction of the ventricles is also increased by the increase in sympathetic nerve activity to the ventricle. End-systolic volume decreases with increasing intensity of exercise, and changes in both EDV and ESV contribute to the increased stroke volume during exercise. **Ejection fraction** (SV divided by the EDV) is a measure of ventricular function. It is about 0.65 (i.e., 65%) at rest and can increase to as high as 0.85 in peak exercise, attributable to an increase in cardiac contractility (Rowell et al. 1996).

Figure 4.6, *a-c*, shows that following an endurance training program, the cardiac output response to the submaximal stages of the GXT is very similar to the pre-training response but the manner in which it is realized is different: HR is reduced whereas SV is increased. The increased SV is caused by an increase in EDV. Maximal HR remains the same or is slightly lower as a result of training; however, when coupled with the increased SV, maximal cardiac output is increased and, with it, $\dot{V}O_2$max (Rowell 1993).

The volume of oxygen delivered into the systemic circulation per minute is a product of the cardiac output and CaO_2. The CaO_2 is determined primarily by the hemoglobin concentration and the pressure of oxygen (PO_2) in the arterial blood. The difference in the hemoglobin concentration between men and women (150 g/L vs. 130 g/L, respectively) accounts for some of the difference in the maximal oxygen uptake between genders. The lower PO_2 at altitude decreases the oxygen saturation of hemoglobin and, consequently, oxygen delivery to muscle, making $\dot{V}O_2$max lower at altitude compared with sea level. Figure 4.6*d* shows the changes in oxygen extraction ($CaO_2 - C\overline{v}O_2$ difference) from light to maximal work.

As exercise intensity increases, more oxygen is extracted by the muscles to support the oxidative generation of ATP, widening the difference between the CaO_2 and $C\overline{v}O_2$. The relative contributions of

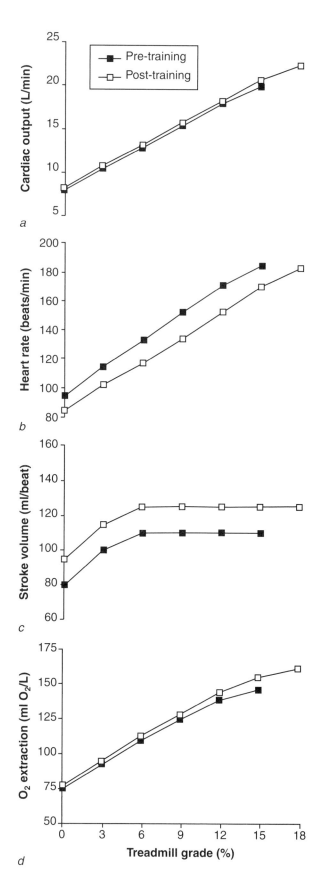

Figure 4.6 *(a)* Cardiac output, *(b)* heart rate, *(c)* stroke volume, and *(d)* oxygen extraction response to a graded exercise test.

cardiac output and oxygen extraction to oxygen consumption can be seen in the following calculations for rest and maximal work:

$$\dot{V}O_2 \text{ (L/min)} = \text{cardiac output} \times (CaO_2 - C\bar{v}O_2)$$

$$\text{Rest: } \dot{V}O_2 \text{ (L/min)} = 5 \text{ L/min} \times (200 \text{ ml } O_2/L \text{ blood} - 150 \text{ ml } O_2/L \text{ blood})$$

$$\dot{V}O_2 \text{ (L/min)} = 250 \text{ ml/min or } 0.25 \text{ L/min}$$

$$\text{Maximal work: } \dot{V}O_2 \text{ (L/min)} = 20 \text{ L/min} \times (200 \text{ ml } O_2/L \text{ blood} - 50 \text{ ml } O_2/L \text{ blood})$$

$$\dot{V}O_2 \text{ (L/min)} = 3,000 \text{ ml/min or } 3.00 \text{ L/min}$$

The 12-fold increase in oxygen consumption from rest to maximal exercise is due to a fourfold increase in cardiac output and a threefold increase in the oxygen extraction. Relative to $\dot{V}O_2$max, the importance of each of the three variables discussed in this section is described clearly in table 4.1. Three different groups are identified: athletes, normally active individuals, and individuals with mitral stenosis (narrow opening at the mitral valve between the left atrium and left ventricle that obstructs blood flow to the ventricle and reduces stroke volume). The $\dot{V}O_2$max values (in L/min) differ dramatically between groups, with those of the athletes being almost twice those of normally active individuals and those of normally active individuals being about 2.5 times those of mitral stenosis patients. What explains these large differences? The maximal HR values for all three groups are similar, as are the oxygen extraction values. The major factor accounting for the differences in $\dot{V}O_2$max is the cardiac output, attributable exclusively to the enormous differences in maximal SV between groups (Rowell 1993). The designation of $\dot{V}O_2$max as a measure of cardiorespiratory fitness is attributable to this link between $\dot{V}O_2$max and cardiac output.

The increase in cardiac output is not the only factor responsible for delivery of more oxygen to muscles with increasing exercise intensity. Another is the redistribution of the cardiac output. At rest, only about 20% (~1 L/min) of resting cardiac output (~5 L/min) is delivered to muscle; however, during maximal work, about 80% (~20 L/min) of maximal cardiac output (~25 L/min) is directed to muscles—a 20-fold increase (Rowell 1993). Blood flow is simultaneously increased to working muscles and decreased to the liver, kidneys, and gastrointestinal tract. This redistribution of blood flow not only is important in directing more blood to working muscles in heavy exercise but is crucial for maintaining blood pressure.

Blood Pressure

As exercise begins, local factors in the active muscles, such as increases in partial pressure of carbon dioxide (PCO_2), potassium ion (K^+), hydrogen ion (H^+), nitric oxide, and adenosine, cause a relaxation of arterioles serving those muscles; resistance is decreased and more blood flow is directed to these muscles. This autoregulation of blood flow provides a "supply and demand" arrangement that matches blood flow to the metabolic needs of the muscles. However, this increase in muscle blood flow has to be balanced by reductions in blood flow to other tissues, via sympathetic arteriolar constriction, to prevent blood pressure from falling. This also accounts for the redistribution of blood flow. **Mean arterial blood pressure (MABP)** is the driving force of the blood and is calculated as the sum of the **diastolic blood pressure (DBP)** and one third of the pulse pressure (difference between **systolic blood pressure [SBP]** and DBP). That is

$$MABP = DBP + 1/3(SBP - DBP).$$

Table 4.1 Physiologic Basis for Differences in $\dot{V}O_2$max in Three Groups

Participants	$\dot{V}O_2$max (ml·min^{-1})	=	Heart rate (beats·min^{-1})	×	Stroke volume (ml·beat^{-1})	×	(a-$\bar{v}$)O_2 difference (ml·L^{-1})
Athletes	6,250	=	190	×	205	×	160
Normally active	3,500	=	195	×	112	×	160
Mitral stenosis	1,400	=	190	×	43	×	170

"Table 5-1" from HUMAN CARDIOVASCULAR CONTROL by Loring B. Rowell, copyright © 1993 by Oxford University Press, Inc. Used by permission of Oxford University Press, Inc.

If MABP were to decrease, so would blood flow to the brain, with dire results. Consequently, MABP is "protected" by control systems that monitor blood pressure and invoke automatic responses to correct discrepancies. The MABP is dependent on both cardiac output and the resistance offered to that blood flow by the whole body (**total peripheral resistance, or TPR**):

$$MABP = cardiac\ output \times TPR.$$

For example, if blood pressure were to decrease at rest, baroreceptors (pressure or stretch receptors) in the carotid artery and arch of the aorta would sense the fall and signal the cardiovascular control center in the medulla of the brain stem. The control center directs an increased level of sympathetic nerve activity to the heart (to increase cardiac output) and to the arterioles (to increase resistance) to restore blood pressure (Rowell 1993). What happens to blood pressure during exercise?

Figure 4.7 shows the systolic and diastolic blood pressure responses to a GXT. The SBP increases throughout the test, whereas the DBP remains the same or decreases slightly. The MABP increases modestly throughout the GXT, from 90 mmHg at rest to 107 mmHg at maximal exercise, in this example. This small increase in MABP at maximal

exercise, in the presence of the large increase in cardiac output, is attributable to the large decrease in TPR (linked to the dilation of arterioles in working muscles). The simultaneous constriction of arterioles in other tissues and organs through the action of sympathetic nerves is sufficient to maintain adequate TPR to preserve the MABP. This appears to be in conflict with the regulation of blood pressure at rest (mentioned previously) where the cardiac output and arteriolar constriction are increased because of a decrease in blood pressure.

The cardiovascular control center in the medulla of the brain stem, however, receives input from a variety of sources during exercise. Clearly, the baroreceptors provide input about blood pressure. However, the combination of efferent input from higher brain centers (**central command**) and afferent input from receptors in the working muscles (**peripheral feedback**) drives the cardiovascular responses mentioned previously. The efferent activity is proportional to the motor unit recruitment and spills over to the medulla as the impulse traffic moves to the muscles. The afferent input comes from both mechanoreceptors and chemoreceptors to help fine-tune the cardiovascular response (Rowell 1993). As with other exercise responses, endurance training reduces the blood pressure responses to exercise compared with the pretraining responses at the same exercise intensity.

Respiratory Responses to Exercise

The respiratory system maintains the partial pressures of arterial blood gases (PO_2 and PCO_2) and assists in hydrogen ion (H^+) regulation. Venous blood, returning from the tissues, enters the right heart chambers and is pumped to the lungs. The pulmonary system is a low-resistance system, and the right ventricle does not have to generate as much force as the left ventricle to have the same stroke volume. As blood flows through the pulmonary capillaries, gas exchange occurs with the alveoli. Carbon dioxide diffuses rapidly down its partial pressure gradient from the blood to the lung and is easily equilibrated with alveolar gas within about 0.1 s; oxygen equilibration takes slightly longer. Obviously, as more oxygen is extracted from the arterial blood with increasing intensities of exercise, **pulmonary ventilation** must increase to bring more oxygen to the alveoli. Figure 4.8 shows the ventilation response to a GXT. Ventilation increases in an almost linear fashion from light through moderate exercise, at which point the rate

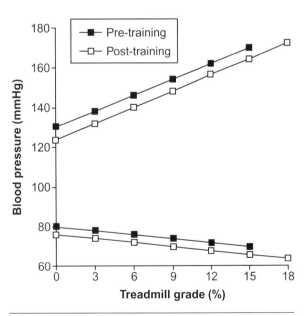

Figure 4.7 Changes in systolic and diastolic blood pressures during a graded exercise test.

Data from B. Ekblom, P-O. Astrand, B. Saltin, J. Stenberg and B. Wallstrom, 1968, "Effect of training on circulatory response to exercise," *Journal of Applied Physiology* 24: 518-528.

Cardiovascular Adjustments During a Graded Exercise Test

Heart rate and cardiac output increase linearly with exercise intensity during a GXT; stroke volume increases until about 40% $\dot{V}O_2$max in upright exercise and remains constant thereafter. Consequently, the increase in cardiac output after that intensity is attributable to heart rate alone. The oxygen extraction (difference between the arterial and mixed venous oxygen contents) increases with exercise intensity. The product of cardiac output and oxygen extraction determines $\dot{V}O_2$. Much of the variation in $\dot{V}O_2$max (L/min) among individuals is attributable to differences in maximal cardiac output, related primarily to differences in maximal stroke volume. $\dot{V}O_2$max is considered a good measure of cardiorespiratory fitness because of its relationship to maximal cardiac output. Systolic blood pressure increases with exercise intensity, whereas diastolic blood pressure remains the same or decreases. Mean arterial blood pressure increases modestly with exercise intensity. The simultaneous dilation of arterioles in working muscles (caused by local factors) and constriction of arterioles in the liver and kidneys (caused by sympathetic nerve activity) results in a redistribution of cardiac output to bring large volumes of blood to muscles during maximal work compared with resting conditions. The small increase in mean arterial blood pressure in the face of a four- to fivefold increase in cardiac output is attributable to the large decrease in total peripheral resistance associated with the dilation of arterioles in working muscles.

The respiratory control center is also located in the medulla of the brain stem and receives input from both central (within the central nervous system) and peripheral (carotid arterial and arch of aorta) chemoreceptors. Because the PO_2, PCO_2, and H^+ of arterial blood do not change during light to moderate exercise, what causes the ventilation to increase in a linear manner? The same input described previously for the cardiovascular responses appears to be important. Neural input from higher brain centers proportional to motor unit recruitment (central command) and peripheral feedback from receptors in working muscles to the respiratory control center appear to shape the ventilatory response. The sudden change in ventilation that signifies the ventilatory threshold is linked to the appearance of lactate (LT) and the need to buffer H^+ by respiratory compensation—exhaling the CO_2 produced in this reaction. Regular endurance training results in a reduction in the ventilatory response to the same submaximal work rate.

Pulmonary ventilation is the product of tidal volume or volume of air moved per breath (TV) and breathing frequency (breaths/min). Increases in both TV and frequency account for the increase in pulmonary ventilation; however, at very high rates of ventilation, TV levels off and the remaining increase in pulmonary ventilation is attributable entirely to an increase in breathing frequency (Dempsey et al. 1996).

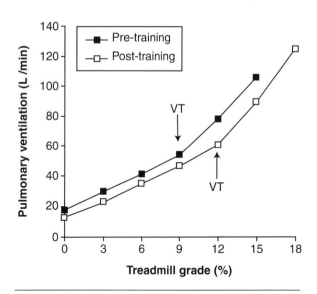

Figure 4.8 Changes in pulmonary ventilation during a graded exercise test.

Adapted from Ekblom et al. (1968).

of ventilation increases dramatically. This sudden break in the pattern of ventilation is called the **ventilatory threshold (VT)**; it is used as a noninvasive estimate of the LT. The connection between the LT and the VT is the buffering of the H^+ by the plasma bicarbonate (HCO_3^-); this buffering results in the generation of CO_2 that drives ventilation to a higher level.

How effective is pulmonary ventilation in maintaining the blood gases? Even at the maximal cardiac outputs seen in normally active individuals, the cross-sectional area of the pulmonary capillary bed is large enough to slow down blood flow and allow the PO_2 of the red blood cells to equilibrate with the PO_2 of the alveoli; the PO_2 of arterial blood is maintained within narrow limits. In contrast, about 40% to 50% of elite endurance athletes who have very large maximal cardiac outputs (see table 4.1) experience a true desaturation of hemoglobin during heavy exercise. At these very high cardiac outputs, the red blood cells move rapidly through the pulmonary capillaries, not allowing equilibration to occur with the alveolar PO_2. Consequently, the hemoglobin leaves the pulmonary capillaries before it is fully saturated with oxygen. This results in a lower arterial oxygen content and a reduction in oxygen delivery to muscle during heavy exercise in these individuals (Dempsey et al. 1996).

Respiratory Adjustments During a Graded Exercise Test

Pulmonary ventilation increases in a linear fashion with exercise intensity during light to moderate exercise; after that, it increases at a faster rate. This "break" from linear is called the ventilatory threshold (VT) and is used as a noninvasive estimate of the lactate threshold (LT). The LT and VT are linked to the buffering of the H^+ by the plasma bicarbonate (HCO_3^-), resulting in CO_2 generation that increases ventilation. The increase in ventilation is attributable to an increase in both tidal volume and breathing frequency, with the latter being more important at higher rates of ventilation. In normal persons and most endurance-trained athletes, pulmonary ventilation is sufficient to maintain the arterial PO_2 and arterial oxygen content. However, a proportion of endurance athletes with very high $\dot{V}O_2$max values experience a true desaturation of hemoglobin in near-maximal exercise. Part of the reason for this is related to the high rate at which their red blood cells move through the pulmonary capillaries—not allowing enough time for equilibration with the alveolar PO_2.

Effect of Training, Age, and Gender on Maximal Oxygen Uptake

The maximal oxygen uptake and associated cardiovascular responses are affected by training, age, and gender. This section summarizes important observations in regard to these variables.

Training

The average training-induced increase in $\dot{V}O_2$max in formerly sedentary individuals is about 15%; however, there is considerable variability in response. Data from the HERITAGE Family Study showed that although the average increase in $\dot{V}O_2$max attributable to an endurance training program was 16%, some participants did not change at all, and others had an increase of 100%. The variability in response appears to have a strong genetic link, with up to 47% of the gain in $\dot{V}O_2$max being tied to heredity (Bouchard et al. 1999). For more information on heredity and exercise, see chapter 22. The training-induced increase in $\dot{V}O_2$max in previously sedentary individuals is attributable to an increase in maximal cardiac output and maximal oxygen extraction, each contributing equally. The increase in maximal cardiac output is caused exclusively by an increase in maximal stroke volume, because maximal heart rate changes little with training (Rowell 1993).

Gender and Age

On average, $\dot{V}O_2$max is about 15% lower in adult women compared with adult men, attributable in part to a lower maximal cardiac output. The lower maximal cardiac output is linked to a lower maximal SV, secondary to a smaller heart size. Maximal oxygen extraction is also lower in women because of a lower hemoglobin concentration. In addition, when men and women do submaximal exercise at different intensities set at the same absolute $\dot{V}O_2$ in L/min (e.g., on a cycle ergometer), considerable differences exist in physiological adjustments to the exercise. At any given submaximal $\dot{V}O_2$, women have a higher heart rate response to compensate for the lower SV response; this results in a cardiac output response that is slightly higher than that of the males. The slightly higher cardiac output response compensates for the lower $CaO_2 - C\bar{v}O_2$ difference that is linked to the lower hemoglobin levels in women (Åstrand et al. 1964).

Effect of Training, Gender, and Age on Maximal Oxygen Uptake

In previously sedentary participants, the average training-induced increase in $\dot{V}O_2$max is about 15%, but with large variations in response among participants. Up to 47% of the variation in response can be traced to genetic factors. The increase in $\dot{V}O_2$max is attributable to increases in both maximal cardiac output and oxygen extraction. Women, on average, have a lower $\dot{V}O_2$max than men. In addition, during submaximal exercise set at the same absolute intensity ($\dot{V}O_2$ in L/min), women have a higher heart rate response to compensate for their smaller stroke volume. The cardiac output response is also slightly higher in women than in men, to compensate for their lower $CaO_2 - C\bar{v}O_2$ difference.

Maximal oxygen uptake decreases with age at the rate of approximately 1% per year in healthy, but sedentary, men and women. This decrease is attributable to both inactivity and weight gain as well as to any "aging" effect. Trained men experience a decrease of about 0.5% per year, whereas trained women decline at 1% per year. The age-related decrease in maximal heart rate is the major contributor to the decrease in maximal cardiac output, but maximal oxygen extraction also decreases with age. The latter is probably related more to the effect of inactivity on muscle mitochondria and capillary density than to any specific aging factor. The increase in $\dot{V}O_2$max in older adults in response to an endurance training program is similar to that observed for younger adults (Holloszy and Khort 1995). See chapter 18 for more on physical activity and aging.

Application to Exercise Training and Physical Activity Interventions

An exercise intervention usually specifies the intensity, duration (minutes per session), frequency (sessions per week), and mode of exercise (walk, cycle, swim). Intensity can be prescribed in absolute terms such as walking at a certain speed or MET level or cycling at a specific work rate or absolute $\dot{V}O_2$ in L/min. Exercise intensity can also be set in terms of an individual's maximal oxygen uptake (**percentage of maximal oxygen uptake, % $\dot{V}O_2$max**) or maximal HR (**percentage of maximal HR**). When this is done, the relative effort required is similar among individuals who may differ greatly in terms of $\dot{V}O_2$max. It should be no surprise that such an approach has been used extensively over the years to specify exercise intensity. The relative intensity can also be expressed as a percentage of the **heart rate reserve (HRR)** (difference between maximal HR and resting HR) or the **oxygen uptake reserve (VO₂R)** (difference between $\dot{V}O_2$max and resting $\dot{V}O_2$). The HRR has been used extensively as a means to express relative exercise intensity; %VO₂R is the newest expression of relative exercise intensity and is used interchangeably with %HRR (American College of Sports Medicine 2000).

Application of Physiological Data to Exercise Prescription

The intensity of exercise can be prescribed in numerous ways to reflect energy expenditure on an absolute (L/min) or relative ($ml \cdot kg^{-1} \cdot min^{-1}$) basis. However, in either case, where there are large differences in $\dot{V}O_2$max among individuals participating in exercise programs, the cardiovascular and respiratory responses, as well as the individual's perception of effort, will be markedly different at the same oxygen uptake. However, when the intensity of exercise is set as a percentage of the $\dot{V}O_2$max, percentage of maximal heart rate, or percentage of the heart rate reserve, variability in both physiological and perceptual responses is markedly reduced among individuals.

Summary

ATP is used by muscles to generate tension or force and is replaced rapidly by anaerobic reactions (phosphocreatine and glycolysis) and more slowly by aerobic reactions (metabolism of carbohydrates and fats in the mitochondria). Carbohydrates are more important in heavy exercise, and fats provide a greater percent of the energy during long-term moderate exercise. In the transition from rest to submaximal exercise, aerobic systems do not come on line instantly, so that anaerobic sources of ATP (phosphocreatine and glycolysis) provide some of the needed ATP. By 2 to 3 min the oxygen uptake ($\dot{V}O_2$) meets the ATP demands and a steady state exists. When exercise stops, the $\dot{V}O_2$ declines rapidly, followed by a slower transition back to the resting baseline value. The elevated oxygen uptake during recovery is called the excess postexercise oxygen consumption (EPOC).

Oxygen consumption ($\dot{V}O_2$) increases with exercise intensity during a graded exercise test (GXT). The highest $\dot{V}O_2$ value achieved in such a test is taken as the individual's maximal oxygen uptake ($\dot{V}O_2$max). The blood lactate concentration remains close to resting values in the early stages of a GXT but increases suddenly at about 50% to 80% $\dot{V}O_2$max; this is called the lactate threshold (LT). Heart rate and cardiac output increase linearly with exercise intensity, whereas stroke volume levels off at about 40% $\dot{V}O_2$max. Consequently, the increase in cardiac output after that intensity is attributable to heart rate alone. The oxygen extraction (difference between the arterial and mixed venous oxygen contents) increases with exercise intensity. Much of the variation in $\dot{V}O_2$max (L/min) among individuals is attributable to differences in maximal cardiac output, making $\dot{V}O_2$max a good measure of cardiorespiratory fitness. Systolic blood pressure increases with exercise intensity, whereas diastolic blood pressure remains the same or decreases. The simultaneous dilation of arterioles in working muscles (caused by local factors) and constriction of arterioles in the liver and kidneys (caused by sympathetic nerve activity) result in a redistribution of cardiac output to bring large volumes of blood to muscles during maximal work. The small increase in mean arterial blood pressure in the face of a four- to fivefold increase in cardiac output is attributable to the large decrease in total peripheral resistance associated with the dilation of arterioles in working muscles. Pulmonary ventilation increases in a linear fashion with exercise intensity during light to moderate exercise; after that, it increases at a faster rate. The "break" from linear is called the ventilatory threshold (VT) and is used as a noninvasive estimate of the lactate threshold (LT). The increase in ventilation is attributable to an increase in both tidal volume and breathing frequency. Pulmonary ventilation is sufficient to maintain the arterial PO_2 and arterial oxygen content in most individuals, with the exception of a subgroup of endurance athletes who experience a true desaturation of hemoglobin during near-maximal exercise. Part of the reason for this is related to the high rate at which their red blood cells move through the pulmonary capillaries—not allowing enough time for equilibration with the alveolar PO_2.

In previously sedentary participants, the training-induced increase in $\dot{V}O_2$max averages about 15% and is attributable to increases in both maximal cardiac output and oxygen extraction. Women, on average, have a lower $\dot{V}O_2$max than men. In addition, during submaximal exercise set at the same absolute intensity ($\dot{V}O_2$ in L/min), women have a higher heart rate response to compensate for their smaller stroke volume. The intensity of exercise can be prescribed in numerous ways to reflect energy expenditure on an absolute (L/min) or relative ($ml \cdot kg^{-1} \cdot min^{-1}$) basis. When the intensity of exercise is set as a percentage of the $\dot{V}O_2$max, percentage of maximal heart rate, or percentage of the heart rate reserve, variability in both physiological and perceptual responses is markedly reduced among individuals who differ in $\dot{V}O_2$max.

KEY CONCEPTS

aerobic: ATP-generating reactions that require oxygen and can sustain muscle activity longer; processes occur in the mitochondria of a cell.

anaerobic: ATP-generating reactions that do not require oxygen (e.g., phosphocreatine and glycolysis) and provide short-lived, high-intensity force quickly.

cardiac output: Volume (liters) of blood pumped from the heart per minute; product of heart rate and stroke volume.

central command: Neural activity in higher brain centers associated with the recruitment of motor units; this "spills over" to both the cardiovascular and respiratory control centers in the medulla of the brain stem, driving both cardiovascular and respiratory responses to meet the metabolic demands of the muscles.

diastolic blood pressure (DBP): Lowest pressure in a cardiac cycle when the ventricles are not contracting (diastole); lower number in blood pressure reading (e.g., 120/80 mmHg).

ejection fraction: Ratio of the volume of blood pumped from the heart per beat (stroke volume) to the volume of blood in the heart prior to contraction (end-diastolic volume).

end-diastolic volume (EDV): Volume of blood in the heart at the end of diastole, just before ventricular contraction; stretches the ventricle to increase the force of contraction and stroke volume.

end-systolic volume (ESV): Volume of blood in the heart after ventricular contraction.

excess postexercise oxygen consumption (EPOC): Also known as oxygen debt or recovery oxygen; oxygen uptake during recovery from exercise that remains above the pre-exercise resting baseline oxygen uptake.

glycolysis: Breakdown of glucose to lactate; occurs in the absence of oxygen (anaerobic) with lactate as the terminal product. The breakdown of glucose to pyruvate occurs via the Embden–Meyeroff pathway. The pyruvate can then be used to form lactate or acetyl coenzyme A, which can be used aerobically.

heart rate (HR): The number of times the heart contracts per minute; normally set by the "pacemaker" in the sinoatrial node.

heart rate reserve (HRR): The difference between maximal heart rate and resting heart rate; used as the denominator when expressing exercise intensity as a percentage of the HRR: (exercise HR – rest HR)/(maximal HR – rest HR) × 100%; used extensively as a means to describe exercise intensity.

lactate threshold (LT): Exercise intensity (expressed as actual speed, work rate, or oxygen uptake) at which the blood lactate concentration increases in a systematic manner above a baseline established at lower intensities of exercise.

maximal oxygen uptake ($\dot{V}O_2$max): The greatest rate at which oxygen is taken up during severe large muscle dynamic exercise; product of cardiac output and oxygen extraction.

mean arterial blood pressure (MABP): Driving force of the blood relative to blood flow; opposed by the total peripheral resistance; MABP = DBP + 1/3(SBP – DBP).

metabolic equivalent (MET): For definition, see page 19.

oxygen deficit: The difference between the steady-state oxygen uptake measured during a constant load exercise test and the actual oxygen uptake measured in the first minutes of the test; a measure of the amount of ATP that had to be provided from anaerobic sources (phosphocreatine and glycolysis).

oxygen extraction: Difference between the oxygen content of arterial blood and the oxygen content of the blood when it returns to the right heart chamber (mixed venous oxygen content).

oxygen uptake reserve (VO_2R): The difference between maximal oxygen uptake and resting oxygen uptake; used as the denominator when expressing exercise intensity as a percentage of the VO_2R: (exercise $\dot{V}O_2$ – rest $\dot{V}O_2$)/($\dot{V}O_2$max – rest $\dot{V}O_2$) × 100%.

percentage of maximal heart rate: Ratio of exercise heart rate to maximal heart rate, expressed as a percentage; used extensively as a means to describe exercise intensity.

percentage of maximal oxygen uptake (%$\dot{V}O_2$max): Ratio of exercise oxygen uptake to $\dot{V}O_2$max, expressed as a percentage; used extensively as a means to describe the relative intensity of exercise.

peripheral feedback: Afferent information from chemoreceptors and mechanoreceptors in working muscles provided to both the cardiovascular and respiratory control centers in the medulla of the brain stem to help shape the cardiovascular and respiratory responses needed to meet the metabolic demands of the tissues.

phosphocreatine (PC): Important anaerobic source of ATP; supports the ATP need during the transition from rest to exercise and can provide about 5 s worth of ATP during all-out activity.

pulmonary ventilation: The product of the tidal volume (liters per breath) and breathing frequency (breaths per minute); increases with exercise intensity to bring sufficient oxygen to the alveoli to saturate the hemoglobin and to clear the CO_2 generated by metabolic reactions.

respiratory exchange ratio (R): The ratio of CO_2 production to oxygen consumption ($CO_2/\dot{V}O_2$); ratio of 1.0 during steady-state work indicates 100% of the energy is from carbohydrate; ratio of 0.7 indicates 100% of energy from fat. In heavy exercise in which blood lactic acid accumulates, the R can exceed 1.0 because of buffering of the H^+ by plasma bicarbonate.

steady-state: A relatively unchanging response of a physiological measure (e.g., $\dot{V}O_2$, HR) during a submaximal exercise test, usually obtained several minutes into the test. The steady-state $\dot{V}O_2$ response is taken as the oxygen requirement for the task.

stroke volume (SV): Volume of blood pumped by the heart per beat; difference between the EDV and ESV.

systolic blood pressure (SBP): The highest pressure measured during a cardiac cycle when the ventricle is contracting (systole); top number in blood pressure reading (e.g., 120/80 mmHg).

total peripheral resistance (TPR): Whole-body resistance offered to blood flow by the arterioles; decreases dramatically during exercise to allow large blood flow to muscles with only small increases in the mean arterial blood pressure.

ventilatory threshold (VT): Sharp increase in pulmonary ventilation during a graded exercise test that is linked to the buffering of lactic acid by plasma bicarbonate; used as a noninvasive indicator of the lactate threshold.

STUDY QUESTIONS

1. Using figure 4.1, estimate the percent of energy coming from aerobic sources in an all-out 30 min activity. What do you think the value would be if the person did it at a submaximal effort rather than all-out?

2. In the transition from rest to submaximal exercise, why doesn't the oxygen uptake increase immediately to that level required for the activity?

3. What is the link between $\dot{V}O_2$max and cardiovascular function that makes $\dot{V}O_2$max a good measure of cardiorespiratory fitness?

4. What cardiovascular factor explains most of the differences in $\dot{V}O_2$max among individuals?

5. What is the ventilatory threshold, and how is it connected to the lactate threshold?

6. Why do women, on average, have a lower $\dot{V}O_2$max than men?

7. Given the large differences in $\dot{V}O_2$max among individuals, how can exercise intensity be set in a training program to cause most individuals to experience the same relative effort?

Acute Responses to Physical Activity and Exercise

Adrianne E. Hardman, MSc, PhD

CHAPTER OUTLINE

Lipids and Lipoproteins
- Postprandial Effect of Prior Exercise
- Influence of Intensity and Duration

Insulin–Glucose Dynamics

Blood Pressure

Hematological Changes

Immune Function Responses
- Cells of the Immune System
- Cytokines

Responses Related to Energy Balance
- Fat Oxidation
- Appetite Regulation

Augmentation of Acute Effects by Training

Summary

Review Materials

Research has repeatedly shown that exposure to regular, frequent bouts of physical activity stimulates physiological and metabolic changes that benefit health. It is helpful to classify these as either (a) **"chronic effects,"** i.e., adaptations to training acquired over weeks or months or (b) short-term, "acute" responses to each individual session of activity. Health-related adaptations to training have been dealt with in other chapters. This chapter describes selected acute responses that are clearly related to health outcomes, explaining their relevance. The extent to which these benefit health will depend on the type, frequency, and regularity of activity and the extent to which particular acute responses persist into the postactivity period. For people who adhere to current guidelines, acute responses should be stimulated on "most, preferably all, days of the week" (Pate et al. 1995 p 402).

For reasons often related to statistical power and logistical considerations, experimental models used to study acute health-related responses have relied mainly on planned, structured exercise, as opposed to physical activity performed during daily living. For this reason, the term *exercise* predominates in this chapter. Studies of short periods of detraining are included because they illustrate that the changes to health outcomes that are rapidly lost when regular training is interrupted are mainly attributable to **acute effects.** Intuitively, unstructured periods of physical activity may be expected to stimulate acute responses that are qualitatively similar to, but less conspicuous than, those arising from planned sessions of exercise.

Lipids and Lipoproteins

Lipoproteins are particles that transport triglycerides and cholesterol in the blood plasma. They have a hydrophobic lipid core and an outer surface layer that allows the particle to mix with the watery plasma. Four major groups of lipoproteins exist. These are based on their density, which in turn reflects their composition (see *Characteristics of the Main Classes of Lipoproteins* on this page). There are clear links between markers for disordered lipoprotein metabolism and heart disease: Plasma concentrations of both cholesterol (mainly low-density lipoprotein [LDL] cholesterol) and triglycerides are positively associated with the incidence of coronary heart disease, and there is a clear inverse relationship between high-density lipoprotein (HDL) cholesterol con-

centration and heart disease incidence. Changes to the concentration of lipoprotein lipids arising from a session of exercise are therefore of interest because of their implications for cardiovascular risk.

Characteristics of the Main Classes of Lipoproteins

There are four main classes of lipoproteins:

1. Chylomicrons, consisting mainly of dietary triglycerides, with a composition by weight of 90% triglycerides and 5% cholesterol.

2. Very low-density lipoproteins (VLDLs), which consist mainly of endogenous triglycerides from the liver. Triglycerides make up 65% of the weight of these particles and cholesterol 5%.

3. Low-density lipoproteins (LDLs), with cholesterol and cholesteryl ester as their main lipids. By weight, they comprise 45% cholesterol and 10% triglycerides.

4. High-density lipoproteins (HDLs), with main lipids cholesterol ester and phospholipids. Cholesterol makes up 18% of HDL and triglycerides 2%.

Note: Different subclasses of lipoproteins have been identified, and the distribution of plasma lipids between these is probably important for cardiovascular health outcomes. For example, small, dense LDLs that are particularly cholesterol-rich have been implicated in atherosclerosis.

Postprandial Effect of Prior Exercise

Changes to plasma concentrations of triglycerides, total cholesterol, or HDL cholesterol are observed immediately after an exercise session only when considerable amounts of energy have been expended. For example, men and women who completed the 1994 Hawaii Ironman Triathlon exhibited a nearly 40% decrease in triglycerides and decreases of around 10% in total and LDL cholesterol. Events such as a marathon can result in an immediate increase in HDL cholesterol of around 10%. On the other hand, a session of moderate-intensity exercise of relatively short duration does not lead to clear changes in lipoprotein variables when these are measured immediately after exercise. However, these measurements do not reveal the extent of the influence of exercise on lipoprotein metabolism.

Even a modest session of activity makes important inroads into the body's energy stores, leading to a prolonged period of metabolic "recovery." When blood samples are obtained hours (rather than minutes) after exercise, effects on lipoprotein metabolism are clear. Decreases in triglycerides and increases in HDL cholesterol have consistently been observed 24 hr after exercise, with some studies finding that these changes persist for 48 hr. The main factor influencing the extent of these changes is the amount of energy expended during the exercise session, irrespective of its intensity. Around 4.2 MJ (1,000 kcal or the equivalent of walking more than 10 miles) may need to be expended to elicit statistically significant changes.

These exercise-induced changes reflect the functional relationship that exists between plasma concentrations of triglycerides and HDL cholesterol. Hydrolysis of triglyceride-rich lipoproteins (chylomicrons and very low-density lipoproteins) by the enzyme lipoprotein lipase (LPL) is accompanied by the transfer of cholesterol and other surface materials from triglyceride-rich particles into HDL. Thus, rapid removal of triglyceride-rich lipoproteins is associated with an increase in cholesterol carried in HDL. Prior exercise enhances triglyceride clearance rates, probably by increasing the activity of LPL in muscle—the rate-limiting step in triglyceride clearance. Acute exercise has been reported to induce gene expression for LPL and thus its activity. There appears to be little cumulative effect of repeated daily sessions of exercise on LPL activity, so that this is essentially an acute effect of exercise that dissipates within 20 hr.

Changes to lipoprotein metabolism after an exercise session therefore derive mainly from enhanced clearance of triglyceride-rich lipoproteins, a phenomenon most clearly seen during the **postprandial** state when these particles are most numerous. A rich body of data illustrates that prior exercise markedly decreases the triglyceride response to a subsequent meal. The clinical relevance of this is that an exaggerated postprandial triglyceride response has been linked to the presence of coronary artery disease and to the atherogenic lipoprotein phenotype. Moreover, people spend the majority of their lives in the postprandial state.

Figure 5.1 shows the effect of prior exercise (90 min at 60% of $\dot{V}O_2$max) on serum triglyceride concentrations during the 6 hr following a standard, high-fat meal in normally active (figure 5.1*a*) and endurance-trained (figure 5.1*b*) middle-aged women (Tsetsonis et al. 1997). Prior exercise decreased the postprandial response, measured as the area under

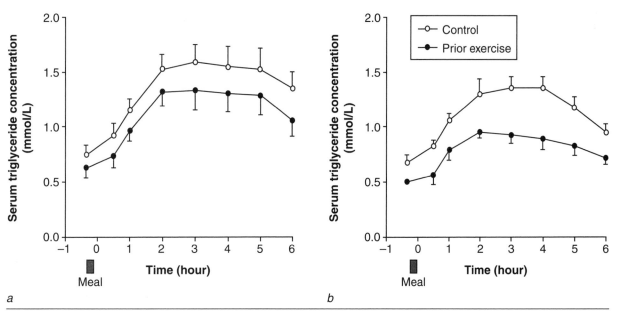

a   *b*

Figure 5.1   Effect of prior exercise on fasting and postprandial concentrations of serum triglycerides in *(a)* 13 untrained women and *(b)* 9 endurance-trained women. A high-fat mixed meal was consumed in the morning after a 12 hr fast. Control—participants refrained from exercise for 3 days; prior exercise—participants walked for 1.5 hr at 60% $\dot{V}O_2$max the previous afternoon. Values are mean and standard error.

the triglyceride concentration versus time curve, by 30% in trained women and by 16% in normally active women, compared with the control trial (no planned exercise for 3 days beforehand). Thus, a single session of exercise markedly decreases subsequent postprandial lipidemia. However, even in trained athletes, this effect is short-lived—and therefore an acute effect—as shown by a study of detraining. Endurance-trained athletes consumed a test meal on three occasions, 15 hr, 60 hr, and 6.5 days after their last training session. Compared with the 15 hr value, the postprandial triglyceride response was 35% higher after just 60 hr without exercise, with little further increase after nearly a week without training. Frequent exercise is therefore needed to maintain the benefits for cardiovascular risk that may be assumed to arise from its triglyceride-lowering effects.

Influence of Intensity and Duration

Prior exercise has such a clear effect on postprandial triglycerides that this model has allowed the influence of exercise intensity (figure 5.2) and pattern (figure 5.3) to be investigated. Researchers examined the effect of the intensity of prior exercise (controlling for energy expenditure), using a repeated-measures design (Tsetsonis and Hardman 1996). The same participants consumed a high-fat, mixed meal on three occasions: (1) control, no planned exercise for 3 days beforehand; (2) 15 hr after a 90 min treadmill walk at 60% $\dot{V}O_2$max; and (3) 15 hr after a walk that was twice as long (180 min) but at half the intensity (30% $\dot{V}O_2$max). In other words, the researchers asked, "Can intensity be 'traded' for duration?" As figure 5.2 clearly shows, postprandial lipidemia was decreased to the same degree (32%) after both exercise sessions, so that the answer to this question (for this particular health outcome) is "yes."

Different patterns of walking were compared in a study of day-long plasma triglyceride concentrations (figure 5.3). On three occasions middle-aged participants were followed throughout a day during which they consumed three ordinary meals (Murphy et al. 2000). Compared with a control trial where participants worked quietly, plasma triglyceride concentrations were decreased to the same degree (12%) by either one 30 min walk before breakfast or by three 10 min walks taken before breakfast,

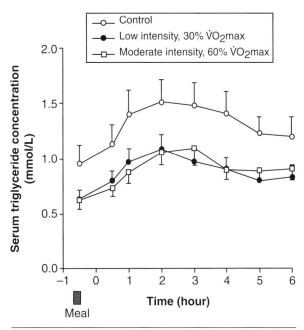

Figure 5.2 Influence of intensity of prior exercise on fasting and postprandial serum concentrations of triglycerides. A high-fat mixed meal was consumed in the morning after a 12 hr fast. Control—participants refrained from exercise for 3 days; prior low-intensity exercise—participants performed treadmill walking at 30% $\dot{V}O_2$max for 3 hr the previous afternoon; prior moderate-intensity—participants performed treadmill walking at 60% $\dot{V}O_2$max for 1.5 hr the previous afternoon. Values are mean and standard error for nine young adults.

Reprinted, by permission, from N.V. Tsetsonis and A.E. Hardman, 1996, "Reduction in postprandial lipidemia after walking: Influence of exercise intensity," *Medicine and Science in Sports and Exercise* 28: 1235-1242.

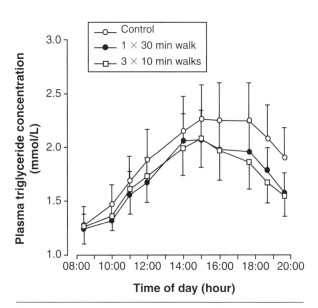

Figure 5.3 Influence of the pattern of brisk walking on postprandial plasma concentrations of triglycerides. Participants rested (control) or performed one 30 min walk before breakfast (long walk) or performed three 10 min walks before each meal. Values are mean and standard error for 10 middle-aged participants.

From M.H. Murphy, A.M. Nevill and A.E. Hardman, 2000, "Different patterns of brisk walking are equally effective in decreasing postprandial lipidemia," *International Journal of Obesity* 24: 1303-1309.

lunch, and the early evening meal. Thus, the acute decrease in postprandial triglycerides attributable to exercise appears to be determined by the associated energy expenditure rather than by its intensity or pattern. Because of the functional relationship between triglyceride clearance and HDL cholesterol concentration, it is reasonable to assume that this also holds true for the effects of frequent and regular exercise on the latter variable.

Insulin–Glucose Dynamics

Insulin resistance is the main pathology of type 2 diabetes, as explained in chapter 12. Skeletal muscle is the body's largest insulin-sensitive tissue and so it is not surprising that substrate deficits arising from a session of exercise influence whole-body insulin–glucose dynamics.

It has been known since the 1970s that endurance-trained individuals exhibit normal or improved **glucose tolerance** to a carbohydrate challenge, despite a markedly reduced insulin response. More recently, a raft of detraining studies have shown that the improved insulin action that underlies these characteristics is rapidly reversed with inactivity, suggesting that much of this benefit arises from acute rather than **chronic effects.** For example, King and colleagues investigated the effects of a 7-day interruption to training on glucose tolerance and insulin action (King et al. 1995). Their participants were middle-aged individuals with a habit of regular moderate exercise, and the measures used were derived from a simple oral glucose tolerance test so as to reflect normal homeostatic mechanisms that pertain in real life. During the 5 days before these tests, participants performed 45 min of exercise daily at about 70% $\dot{V}O_2$peak. Glucose tolerance was poor immediately after the last exercise session, possibly because elevated plasma concentrations of nonesterified fatty acids inhibit glucose uptake into muscle, but had improved markedly 24 hr later. This improved insulin action persisted for 3 days but not for 5 days, suggesting that the frequency of exercise needed to maintain the exercise-induced improvement in glucose tolerance is once every 3 days (figure 5.4).

As might be expected from these findings on detraining, a single exercise session enhances peripheral **insulin sensitivity.** For example, Young and colleagues gave untrained, normally active men a glucose tolerance test in each of three conditions: (1) control, (2) the morning after 40 min of exercise at 40% $\dot{V}O_2$max, and (3) the morning after a

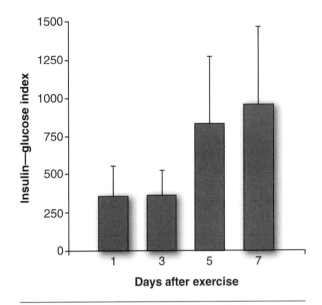

Figure 5.4   Time course of changes in insulin–glucose dynamics in moderately trained middle-aged people during a week without planned exercise. Values are mean and standard error for nine participants. Insulin–glucose index (a surrogate measure of insulin sensitivity) was calculated as the product of the areas under the glucose concentration and insulin concentration versus time curves above the baseline level during a 75 g glucose tolerance test.

Used, with permission, from D.S. King, R.J. Baldus, R.L. Sharp, L.D. Kesl, T.L. Feltmeyer and M.S. Riddle, 1995, "Time course for exercise-induced alterations in insulin action and glucose tolerance in middle-aged people," *Journal of Applied Physiology* 78: 17-22.

similar period of exercise at 80% $\dot{V}O_2$max (Young et al. 1989). The control condition was simply a day of inactivity, and the researchers ensured that diet was standardized during the days leading up to the tests. The plasma insulin response was 40% lower after a single bout of exercise than in the control trial (figure 5.5). Interestingly, the single bout of exercise essentially decreased the insulin response to the same level as that found in a group of participants who were engaged in regular, strenuous endurance training, illustrating the potency of the acute effect.

Studies using the "gold standard" **hyperinsulinemic, euglycemic clamp** technique (explained at the end of this chapter) have explained the basis of the effects of a single session of exercise on glucose and insulin concentrations (Wojtakzewski et al. 2003). These reflect an increase in whole-body **glucose disposal rate** for a given plasma insulin concentration (in other words, an improvement in the action of insulin) that can persist for as long as 2 days after a session of aerobic exercise. (Note two exceptions to this finding. First, immediately

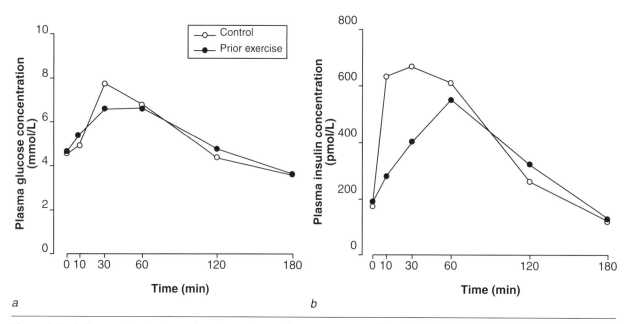

Figure 5.5 Influence of a single session of exercise on plasma responses of *(a)* glucose and *(b)* insulin to a glucose tolerance test. Values are means for seven untrained men. Control—at least 40 hour after any exercise; prior exercise—40 min on a cycle ergometer at 40% V̇O₂max the previous afternoon.

Used, with permission, from J.C. Young, J. Enslin and B. Kuca, 1989, "Exercise intensity and glucose tolerance in trained and untrained subjects," *Journal of Applied Physiology* 67: 39-43.

after exercise, insulin action on glucose disposal is impaired. Second, exercise involving predominantly eccentric contractions results in a prolonged decrease in insulin action, possibly because muscle damage is incurred.)

Seventy to ninety percent of the glucose ingested during an oral glucose tolerance test is cleared by skeletal muscle. Therefore, the increased rates of glucose disposal after exercise mainly reflect increased stimulation by insulin of nonoxidative glucose metabolism, primarily glycogen synthesis, in this tissue. Changes to two regulatory processes are involved: glucose transport across the muscle plasma membrane and the activity of the enzyme glycogen synthase. Glucose transport is increased because of an enhanced recruitment of GLUT4, the insulin-sensitive glucose transporter in skeletal muscle, to the muscle plasma membrane. This facilitates glycogen synthesis by increasing glucose availability. Further stimulus to glycogen synthesis arises from increased activation of glycogen synthase, a rate-limiting enzyme in this process. Both changes demonstrate the adaptive potential of muscle—these changes not only allow muscle to recover to a pre-exercise state but also can facilitate a degree of glycogen supercompensation, enabling the muscle to perform better during the next exer-

cise bout. Studies using the one-leg exercise model have shown that these are local responses in the previously exercised muscle.

Few studies have systematically examined the effects of the intensity, duration, and pattern of exercise in relation to acute effects on insulin–glucose dynamics. Limited evidence suggests that the degree of improvement in insulin sensitivity may be independent of the intensity of the exercise bout. For example, the effects of prior low-intensity (50% V̇O₂max) and high-intensity (75% V̇O₂max) exercise were compared using a repeated-measures design in women with type 2 diabetes (Braun et al. 1995). The duration of the sessions was adjusted so that energy expenditure was the same in both exercise conditions. Participants' plasma glucose (figure 5.6*a*) and insulin (figure 5.6*b*) responses to a mixed meal were determined as a test of changes to insulin–glucose dynamics in circumstances where the normal physiological interrelationship between glucose and insulin was preserved. Postprandially, plasma glucose profiles did not differ among the three conditions, but plasma insulin responses were significantly lower after both low- and high-intensity exercise, compared with the control (no exercise) condition. These changes reflected the fact that the rate of glucose disposal per unit of plasma

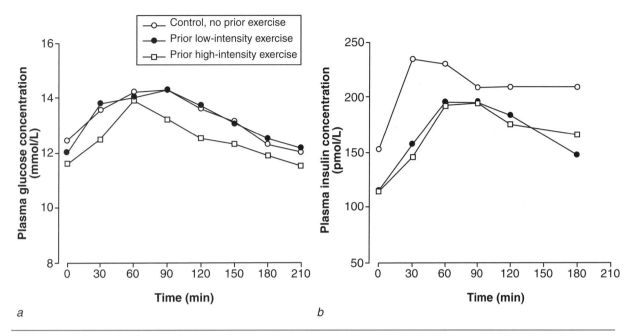

Figure 5.6 Influence of the intensity of prior exercise on plasma glucose *(a)* and insulin *(b)* responses to a mixed meal in eight women with type 2 diabetes. Control—no prior exercise; low intensity—treadmill walking at 50% $\dot{V}O_2$max on each of the two preceding days; high intensity—treadmill walking at 75% $\dot{V}O_2$max on each of the two preceding days. The duration of walking was adjusted so that the energy expenditure was the same for low- and high-intensity trials. Values are means for eight women.

Used, with permission, from B. Braun, M.B. Zimmerman and N. Kretchmer, 1995, "Effects of exercise intensity on insulin sensitivity in women with non-insulin-dependent diabetes mellitus," *Journal of Applied Physiology* 78: 300-306.

insulin (measured on a separate occasion during infusion at fixed rates of glucose and insulin) was enhanced by almost exactly the same degree after prior low- or high-intensity exercise, compared with the control condition.

The paucity of data describing the relative importance of intensity, duration, and pattern as determinants of exercise-induced benefits to insulin–glucose dynamics shows the need for applied studies of these topics. Complementary studies are needed to explore the nature and time course of the cellular mechanisms behind exercise-induced changes in insulin action.

In summary, the regulation of insulin-stimulated glucose disposal and glycogen synthesis in skeletal muscle is not altered in nature by prior exercise; this is simply more potently activated by insulin. These effects are restricted to exercised muscle and will therefore be maximized if exercise is undertaken with the body's large muscles. Exercise involving predominantly concentric, as opposed to eccentric, contraction is indicated. Available, but limited, evidence indicates that the benefits will be optimal after an exercise session that expends a large amount of energy.

Blood Pressure

Systolic blood pressure rises during dynamic exercise and returns to normal with the cessation of exercise. There is often a transient pressure "undershoot" caused by the pooling of blood in the dilated, previously exercised muscle beds. This can lead to light-headedness, particularly if the individual stands still. Moving about facilitates venous return through the muscle "pump." Baroreceptor reflexes counter this undershoot effect to reestablish homeostasis within 10 min. However, more recent studies of the prolonged postexercise period have shown that resting blood pressure may be decreased for several hours following a single session of exercise. This effect may therefore be an important determinant of blood pressure in people with a habit of frequent exercise.

Postexercise hypotension is more consistently observed in individuals with a degree of hypertension than in those with blood pressure in the normal range (MacDonald 2002). In those studies that have found a decline in blood pressure following a session of exercise in people without clinical hypertension, this has typically been 8/9 mmHg (systolic/diastolic). Greater decreases, in the range of 10 to 14/7 to 9 mmHg, have been observed in hypertensive individuals.

Studies in controlled laboratory settings have shown that the decreases in blood pressure after acute exercise persist for at least 3 hr. More recently, the observation period has been extended, using ambulatory blood pressure monitoring over 24 hr. For example, researchers obtained two 24 hr recordings, one immediately preceded by an exercise session (45 min of treadmill walking at about 70% $\dot{V}O_2$max) and another on a control day not preceded by exercise (Taylor-Tolbert et al. 2000). The participants were 11 obese, sedentary men aged around 60 with mild to moderate hypertension (140-179/90-109 mmHg). As figure 5.7 shows, systolic blood pressure was decreased by between 6 and 13 mmHg for the first 16 hr following exercise and by an average of 7.4 mmHg over the 24 hr period. Diastolic blood pressure was decreased by around 5 mmHg for 12 of the first 16 hr after exercise and for an average of 3.6 mmHg over the 24 hr period. Thus, a single session of dynamic

exercise leads to substantial and consistent decreases in average systolic and diastolic blood pressures that are sustained over many hours. The particular study just referred to used walking as the exercise modality, but postexercise hypotension lasting for several hours has also been reported after running and after leg exercise on a cycle ergometer. It also appears to persist when, after the session of exercise, participants engage in mild activity designed to simulate activities of daily living. This last finding strengthens the argument for the therapeutic utility of exercise in the management of the large numbers of individuals who have "high-normal" blood pressure.

What features of an exercise session determine the magnitude of the ensuing hypotension? Most but not all direct comparisons have concluded that postexercise hypotension is independent of exercise intensity. However, few of these studies have used 24 hr monitoring, and none have controlled for the energy expended at different intensities. Few data are available to provide evidence of the influence of exercise duration or the muscle mass engaged in exercise. Moderate-intensity exercise for a 10

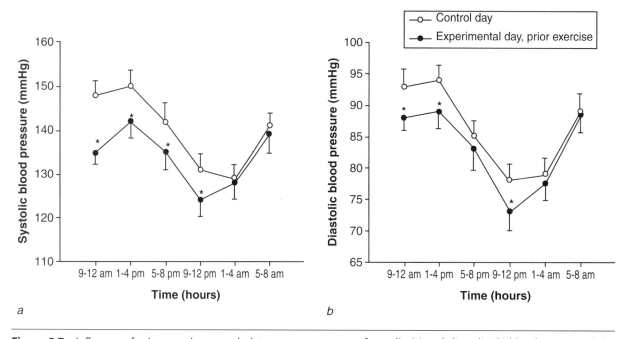

Figure 5.7 Influence of prior exercise on ambulatory measurements of systolic *(a)* and diastolic *(b)* blood pressures. Prior Exercise—Participants performed 45 min of treadmill walking at 70% $\dot{V}O_2$max immediately before measurements commenced; Control—no exercise.

*Difference between the two recordings statistically significant ($p < .05$). Values are mean and standard error for 11 sedentary men with mild to moderate hypertension.

Reprinted from *American Journal of Hypertension*, Vol. 13, Taylor-Tolbert et al., "Ambulatory blood pressure after acute exercise in older men with essential hypertension," pp. 44-51, Copyright 2000 with permission from the American Journal of Hypertension Ltd.

min period has, however, been found to decrease postexercise blood pressure.

The etiology of postexercise hypotension is poorly described, but it is probably not attributable to a single factor. Postexercise indexes of vascular resistance are decreased below pre-exercise values, but studies of potential mechanisms are contradictory, and no conclusion can be drawn from these. There are racial differences in participants' propensity to exhibit postexercise hypotension; it is not observed in black American women in whom hypertension is more prevalent and associated with greater comorbidities. Genetic differences in polymorphisms at critical loci involved in systems that affect blood pressure probably explain some of the variability in response. Future research into these topics may enable clinicians to refine exercise prescriptions for hypertensive persons or those with high-normal blood pressure. More information on aspects of the dose-response relationship is also needed.

Hematological Changes

The formation of an intravascular clot is a key event in the pathophysiology of acute cardiovascular complications such as heart attack and stroke. The likelihood that this will happen depends on the balance between several processes, such as platelet function, coagulation, and fibrinolysis. Thus, acute exercise-induced changes to these processes are of clinical importance. Evidence is available for platelet function, markers of coagulation and fibrinolytic potential (Womack et al. 2003).

> **Blood coagulation** is the process by which fibrin strands create a mesh that binds blood components together to form a fibrin clot. Two pathways may lead to the formation of such a clot—in each a cascade of reactions occurs by which one activated factor activates another. **Fibrinolysis** is the dissolution of fibrin (and therefore of a clot). A key stage in this process is the conversion of inactive plasminogen into active plasmin, an enzyme that digests fibrin and thus brings about dissolution of the clot.

Platelets play a key role in the immediate response to vascular injury. A platelet plug is formed when these cell fragments are activated and start to build up and adhere to the site of damage. Blood subsequently coagulates via a complex process in which fibrin strands create a mesh that binds blood components together to form a blood clot. This process may occur along either of two pathways, both of which involve a cascade of reactions in which each activated factor triggers the next one. The consensus is that strenuous exercise may provoke an increase in platelet activation or aggregation but that this is more likely in sedentary than in regularly active individuals. Platelet activation is enhanced by infusion of epinephrine or norepinephrine, so the interaction with training status may be explained by the higher catecholamine response to exercise in sedentary participants. There is some evidence that moderate exercise at 50% to 55% $\dot{V}O_2$max may inhibit platelet adhesion or aggregation, at least in sedentary participants.

Most markers of coagulation potential increase with acute exercise. For example, activated partial thromboplastin time (an overall indicator of the intrinsic pathway of blood coagulation) is shortened following exercise. This effect has been demonstrated for different exercise modes and in both men and women. Several other markers of coagulation potential known to be associated with cardiovascular disease are also increased after exercise. One of them, factor VIII (a component of the final common path of the coagulation cascade), can remain elevated for several hours after a strenuous session. Exercise-induced increases in this and other components of the cascade are largely dependent on exercise intensity. One marker for thrombin formation has been reported to increase following triathlons, high-intensity running, and cycling.

Fibrinogen, the substrate ultimately converted to a fibrin clot, plays a central role in the final phase of the blood coagulation cascade. It also promotes coagulation by enhancing platelet aggregation and is a key determinant of blood viscosity. Despite the inverse relationship found in cross-sectional population studies between fibrinogen and levels of physical activity or fitness, evidence concerning the effect of acute exercise on this protein is inconclusive. Several methodological problems may explain this. These include the timing of blood sampling (important because fibrinogen is an acute-phase protein) and variability attributable to genetic polymorphisms.

Fibrinolysis, the dissolution of fibrin, removes clots. Two components of this system can be measured: the enzyme that catalyzes the conversion of plasminogen to plasmin and the main circulating inhibitor (plasminogen activator inhibitor-1). These markers indicate that fibrinolysis increases acutely with exercise and that the magnitude of this response is dependent on the intensity and, to a lesser extent, on the duration of exercise. Values typically return to pre-exercise levels within 24 hr. As with the effect on platelet function, there are reports that these changes are provoked more readily in sedentary than in trained individuals.

Thus, in concert with the increase in blood coagulability, the activity of the fibrinolytic system that opposes this is also enhanced. However, because the majority of studies have examined only one aspect of thrombogenesis, the overall effect of a session of exercise on this multifactorial process is difficult to evaluate. Thrombogenesis itself has, however, been examined in sedentary men using an experimental model intended to measure the net effect of a session of moderate (50% $\dot{V}O_2$max) or hard (70% $\dot{V}O_2$max) exercise (Cadroy et al. 2002). Platelet thrombus formation and fibrin deposition were determined as arterial blood interacted with collagen (a molecule present in atherosclerotic plaques and primarily responsible for thrombus formation in vivo). Blood flow conditions mimicked those in moderately stenosed small arteries. Moderate exercise did not affect arterial thrombus formation. In contrast, platelet thrombus formation was increased by 20% after 30 min of hard exercise. These findings suggest that the net effect of a session of strenuous—but not moderate—exercise is probably to increase the risk for arterial thrombogenesis in sedentary men. This proposition is consistent with epidemiological findings that heavy exertion is a potent trigger to myocardial infarction in people unaccustomed to such exertion (discussed in chapter 19).

Immune Function Responses

Since the mid-1980s, researchers have systematically examined the effects of a session of exercise on immune function and related signaling molecules. Stimulated initially by anecdotal reports that athletes engaged in intense training regimens exhibited enhanced susceptibility to respiratory infections, research has focused on exercise of high intensity.

Cells of the Immune System

The body's immune system protects against infection by recognizing and actively attacking invading organisms. Many of its components exhibit change after prolonged, intense exercise (Klarlund Pedersen and Hoffman-Goetz 2000). The reader will recall that the immune system may conveniently be divided, according to function, into the innate immune system and the adaptive (or acquired) system. The former recognizes general characteristics of foreign organisms, whereas the latter can detect a particular organism among the thousands of possible candidates and recognize it again on subsequent exposure; in other words, the adaptive system is specific and has "memory."

Several populations of cells of the immune system respond to acute exercise: macrophages, neutrophils, and natural killer (NK) cells. Macrophages have phagocytic and **cytotoxic** capacities. Neutrophils, also phagocytic, are important in the nonspecific killing of bacteria. Natural killer cells seek out and destroy virus-infected cells nonspecifically. Macrophage antiviral function, neutrophil

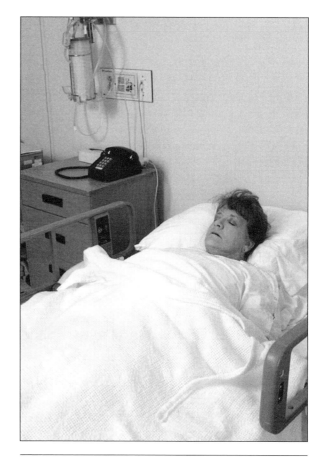

A striking illustration of the adverse effects of inactivity is that of prolonged bed rest, which results in loss not only of bone and muscle mass, but in negative changes in hemostasis as well.

function, and NK cell activity are all impaired for several hours after a session of high-intensity exercise, particularly if it is prolonged. Most functions of neutrophils decrease significantly after maximal exercise so that these cells are probably in an unresponsive state. Evidence on the effects of exercise of moderate intensity or duration is less consistent, but, in general, this shows that some functions of neutrophils—for instance phagocytosis—may be enhanced whereas others are unaffected.

Natural killer cells are very responsive to exercise, which induces their recruitment to the blood. After short-term (<60 min), high-intensity exercise, the concentration of these cells is typically 150% to 300% higher than pre-exercise values. However, this increase is transient and large numbers of these cells quickly exit the circulation so that, during the hours after intense exercise of long duration, their concentration declines to 25% to 40% below pre-exercise values (figure 5.8), as does their **cytolytic** activity (Malm et al. 2004). The maximal reduction is apparent 2 to 4 hr after exercise. Although evidence is sparse, neither fitness level nor sex appears to influence the magnitude of exercise-induced changes in NK cells.

In summary, prolonged heavy exertion may lead to transient but clinically important changes in immune function. During ["opportunity" of altered immunity] [between 3 and 72 hr, depending on t] [measured), opportunistic infections] foothold. On the other hand, a session o[ ]ate exercise may stimulate changes to the i[m] function that could be beneficial if such exerci[se] frequent and regular. There may be a summatio[n] effect from these acute positive changes, but little is known about effects of repeated bouts.

Cytokines

The local response to infection or tissue injury involves the production of cytokines (intercellular signaling molecules) that are released at the site of inflammation. These cytokines facilitate an influx of several populations of cells of the immune system, which participate in clearing the antigen and healing the tissue. The local inflammatory response is accompanied by a systemic response known as the acute phase response, mediated largely by the so-called proinflammatory cytokines, specifically tumor necrosis factor-α, interleukin-1β, and interleukin-6 (IL-6).

Exercise induces the release of a cascade of anti-inflammatory cytokines, particularly IL-6 (Febbraio and Klarlund Pedersen 2002). After a marathon race, for example, the plasma concentration of IL-6 has been reported to increase 50-fold. Data from the Copenhagen Marathon between 1995 and 1997 show a correlation between the intensity of exercise and the increase in IL-6.

It was commonly thought that the IL-6 response to exercise represented an immune response to local muscle damage. It is now known, however, that this cytokine is produced locally in skeletal muscle in response to contraction per se. Biopsy studies in a rat one-leg exercise model have shown that IL-6 gene expression is increased in the exercised skeletal muscle but not in rested muscle. The time scale of the increase in the release of IL-6 during exercise, among other features of the evidence, suggests that this cytokine may be a metabolic regulator signaling low muscle glycogen. This research suggests that, as well as intensity, both the muscle mass involved in exercise and the type of contraction (eccentric or concentric) probably influence the level of response.

The health implications of exercise-induced changes to cytokines are not clear. Speculatively, IL-6 production and subsequent release from skeletal muscle may play a role in regulating glucose homeostasis in insulin-sensitive tissue and thus

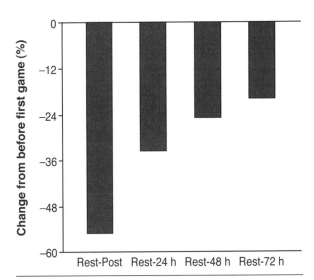

Figure 5.8 Decrease in numbers of natural killer (NK) cells from before to after two soccer games separated by 20 hr. Data are average values for 4 different subpopulations of NK cells.

*Difference from before to after the soccer games statistically significant ($p < .01$). Values are means for 10 elite Swedish soccer players aged between 16 and 19.

Reprinted, by permission, from C. Malm, O. Ekblom and B. Ekblom, 2004, "Immune system alteration in response to two consecutive soccer games," *Acta Physiologica Scandinavica* 180: 143-155.

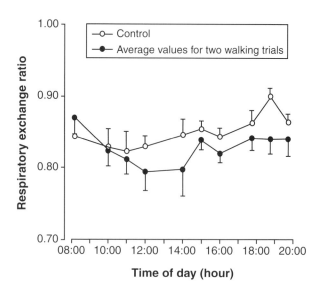

... nced insulin sensi-

... in this area include the ... ercise-induced changes to ... he incidence or progression ... nistic links between exercise ... dexes of immune function, and ... okine response between exercise ... s of temporary immunosuppression ... septic shock).

Responses Related to Energy Balance

An individual's overall level of physical activity exerts a major influence on his or her energy balance and hence, over time, weight regulation. As well as the energy expended during activity, acute effects apparent during the hours after each individual session of activity are important in this regard. Findings on two effects—fat oxidation and appetite—are described next; potential effects on leptin and ghrelin are topics of current research interest but are not sufficiently well developed to be included here.

Fat Oxidation

Carbohydrate and protein stores are closely regulated by adjusting oxidation to intake. It follows that fluctuations in fat balance mainly determine day-to-day fluctuations in energy balance and that fat balance determines energy balance. The evidence summarized next shows that the repeated effects of frequent, regular sessions of activity probably help to maintain neutral or negative fat balance.

Fat oxidation is enhanced for some hours after an exercise session, compared with a control condition without prior exercise. This is the case for moderate amounts and intensities of exercise as well as for exhaustive exercise. For example, figure 5.9 shows the influence of just 30 min of brisk walking on the respiratory exchange ratio throughout a "normal" day. (This ratio reflects whole-body substrate oxidation, low values indicating predominantly oxidation of fat.) The study protocol was described in the preceding section on lipids and lipoproteins (Murphy et al. 2000). During a daylong observation period, participants consumed breakfast, lunch, and an early evening meal. In the control condition, they

Figure 5.9 Influence of 30 min of brisk walking on the respiratory exchange ratio throughout an ordinary day. Values are mean and standard error for 10 middle-aged participants. For details of protocol, see legend to figure 5.3. Average respiratory exchange ratio values for the two walking trials (one 30 min walk or three 10 min walks, treated as a single condition) significantly lower than control.

From M.H. Murphy, A.M. Nevill and A.E. Hardman, 2000, "Different patterns of brisk walking are equally effective in decreasing postprandial lipidemia," *International Journal of Obesity* 24: 1303-1309.

did minimal activity; during the exercise conditions they walked briskly for 30 min—on one occasion in three 10 min sessions and on the other occasion in one 30 min session. In figure 5.9, the two exercise conditions are treated as one because the effect on the respiratory exchange ratio was independent of the pattern of walking. Researchers estimated that, compared with the control trial, an additional 5 g of fat was oxidized over the 11 hr observation period, decreasing fat storage by 4% to 5%.

Thus, an exercise session shifts the pattern of substrate utilization at rest toward fat oxidation. This effect has been found up to 24 hr after the exercise session and is apparent even if the increase in energy expenditure associated with exercise is balanced by an increase in dietary intake. Short-term studies also suggest that physical activity opposes the positive fat balance that occurs during a shift to an iso-energetic high-fat diet under sedentary conditions.

Appetite Regulation

Two questions are relevant: What is the effect of a session of exercise on energy intake? Does prior exercise alter the proportional contribution to energy metabolism from different substrates? There is no clear answer to the second question. By contrast, findings addressing the first question are rather consistent; a session of exercise usually stimulates an increase in energy intake but this increase only partially compensates for the additional energy expenditure of exercise, leading to an energy deficit.

Methodologically sound studies of this topic over periods of a few days are available. (These are included here because imprecision reduces the usefulness of studies of the effect of a single exercise session.) For example, Stubbs and colleagues observed the effects in women on energy balance and sensations of hunger and appetite of two 40 min sessions of cycle ergometer exercise on each of 7 days (Stubbs et al. 2002). Energy intake was higher on the exercise days than on the preceding control days, but the increment in intake was equivalent to only one third of the additional expenditure, leading to an energy deficit. This finding was consistent with the small increase in hunger reported by the participants.

The effect on energy intake of the opposite intervention, that is, decreasing physical activity within the sedentary range, has also been studied. In one carefully designed study, where participants were studied twice in 7-day protocols in a whole-body indirect calorimeter, energy expenditure was reduced from an average of 12.8 MJ/day (3,050 kcal/day) to 9.7 MJ/day (2,310 kcal/day) (Stubbs et al. 2004). (These levels were selected as the high and low ends of the "sedentary" range of activity for healthy, free-living adults in Western societies.) The difference was achieved through varying cycle ergometer exercise. When activity was restricted, there was no compensatory decrease in energy intake so that participants stored more energy than while on the more active regimen. In this study, the greater positive energy balance in the restricted activity condition was largely accounted for by greater fat storage.

This is a difficult field, and available literature is not extensive. Moreover, for methodological reasons, research is largely restricted to cycle ergometry and inevitably confounded by effects on the behavior of human volunteers. Finally, lean individuals—more commonly studied—may behave differently than those who are overweight or obese.

Augmentation of Acute Effects by Training

In a now-classic article, published in 1994, Haskell coined the term the "last bout effect," proposing that some of the health benefits of activity—for example, lowering of blood pressure, improved plasma lipoprotein profiles—might be attributable more to acute biological changes following each bout of activity than to a true training response (Haskell 1994). Does this mean that performing repeated bouts of exercise over weeks or months, progressing by increasing frequency, duration, or intensity—in other words, obtaining a training response—has no synergistic effects on such benefits?

First, if a specific effect lasts for more than 24 hr, this effect may be enhanced when exercise is undertaken daily, as occasional exercisers progress toward a habit of regular exercise. This is shown schematically in figure 5.10a. An example would be that, in men with hypertriglyceridemia, fasting plasma triglyceride concentration is progressively decreased over a 5-day period when exercise is performed daily. However, the degree of this augmentation of an acute effect must inevitably decrease over time as values for the variable in question approach the end of the physiological range.

As originally suggested by Haskell, the most important way in which training augments acute effects is by enabling more intense, longer, or more frequent exercise to be performed. This is depicted schematically in figure 5.10b. The total energy expenditure of an exercise session is an important determinant of the level of the ensuing beneficial effects on several (speculatively many or most) health-related outcomes, as described previously. Thus, training enhances acute effects by enabling a greater overall level of exercise energy expenditure.

The data portrayed in figure 5.1 (on page 69) illustrate the augmentation by training of an acute effect for one health-related outcome—postprandial lipidemia. The decrease in lipidemia attributable to a prior 90 min session of treadmill walking was twice as great in trained as in untrained women. Moreover, prior exercise greatly enhanced the difference in lipidemic response between these two groups. A likely explanation is that, although all the

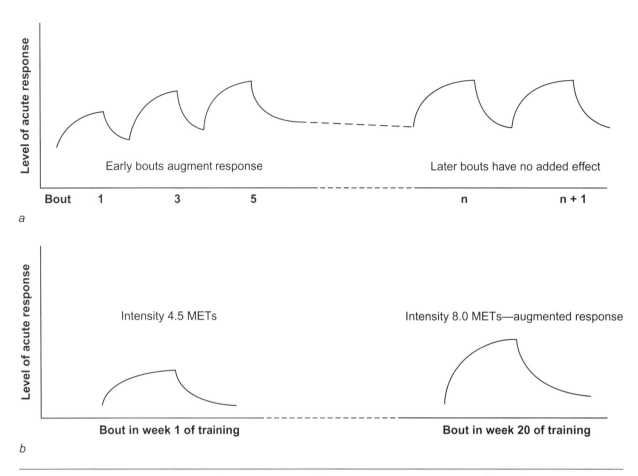

Figure 5.10 Schematic representation of potential ways in which training may enhance acute effects of a single exercise session. Augmented acute response, repeated bouts of exercise produce increasingly greater acute effects *(a)*. Acute response is augmented by training because exercise bouts are at a higher intensity—and thus higher energy expenditure *(b)*.

Reprinted, by permission, from W.L. Haskell, 1994, "Health consequences of physical activity: Understanding and challenges regarding dose-response," *Medicine and Science in Sports and Exercise* 26: 649-660.

women walked at the same relative intensity (60% $\dot{V}O_2$max), the trained women expended 50% more energy because of their higher $\dot{V}O_2$max values.

Summary

The acute effects of individual bouts of activity are varied and have the potential to make an important contribution to health. Effects on metabolism and cardiovascular functions in particular are relevant to the modern epidemics of coronary heart disease, type 2 diabetes, and obesity. Much remains to be learned about the extent and persistence of acute effects and their interaction with the intensity and duration of activity. Nevertheless, current evidence is sufficient to demonstrate that a regimen of frequent, regular bouts of activity at a moderate intensity is likely to improve the health of presently sedentary people.

KEY TERMS

acute effects: Short-term responses to an individual session of exercise or activity linked to the body's recovery from that session.

blood coagulation: The process by which fibrin strands create a mesh that binds blood components together to form a fibrin clot.

chronic effects: Adaptations to training (exercise progressing in intensity or frequency) acquired over weeks or months and that persist for days or weeks when such a regimen is interrupted.

cytolytic: Destruction or degeneration of cells.

cytotoxic: Toxic to cells.

fibrinolysis: The dissolution of fibrin (and therefore of a clot).

glucose disposal rate or metabolic clearance rate for glucose: Rate at which glucose is removed from the circulation.

glucose tolerance: Extent of the increase in blood glucose concentration when glucose is ingested.

hyperinsulinemic, euglycemic clamp: Technique used to determine insulin sensitivity. Insulin is infused intravenously to produce the same plasma insulin concentration in all individuals. Simultaneously, glucose is infused to achieve the same blood glucose concentration in all individuals. The greater the rate of glucose infusion required to achieve this state, the more insulin sensitive the individual.

insulin sensitivity: Responsiveness to the stimulation of glucose uptake by insulin.

lipoproteins: Particles with a highly hydrophobic lipid core and a relatively hydrophilic outer surface.

postprandial: During the hours after a meal.

STUDY QUESTIONS

1. Why has recent research focused on the effect of exercise on plasma triglycerides measured in the postprandial—as opposed to the fasted—state? Summarize the evidence showing that the energy expenditure of an exercise session mainly determines the extent of the subsequent decrease in postprandial triglycerides.

2. What is meant by the term *glucose tolerance*? Explain why prior exercise improves glucose tolerance.

3. Explain what is meant by the term *postexercise hypotension*. During what period of postexercise would you expect an individual to demonstrate this?

4. Why are the acute effects of exercise on hemostatic factors often described as complex? Under what circumstances is an exercise session likely to cause an overall increase in blood coagulability?

5. Describe the role of natural killer (NK) cells in immune function. What are the main effects of an exercise session on this population of cells?

6. By what means may an exercise session influence energy balance?

7. Explain how training and the acute effects of individual sessions of exercise act synergistically to provide optimal benefits for health.

Hormonal Response to Regular Physical Activity

■ Peter H. Farrell, PhD

CHAPTER OUTLINE

Defining Hormones
- Hormone Activity
- Hormone Functions Related to Physical Activity
- Some Control Mechanisms Related to Physical Activity

Importance of Hormonal Regulation
- Regulation of Hormones at Very Low Concentrations
- Regulation of the Rhythmic Release of Hormones

Regular Physical Activity and Hormonal Adaptations
- Epinephrine and Norepinephrine
- Glucagon
- Hormonal Adaptations to Chronic Physical Activity and Long-Term Food Intake
- Cross-Adaptations

Summary

Review Materials

The endocrine system regulates the physiological and psychological functioning of the human organism. Hormones share this job with the autonomic nervous system, and both hormonal and neural controls are subject to the basic structural and functional constraints that abound in the body. In response to acute physical activity, the concentrations of most hormones in the plasma increase, whereas some do not change and others decrease. This response to physical activity changes as a person progresses from being habitually sedentary to becoming physically active, and those adaptations allow the organism to function at a higher capacity. One limitation to summarizing the literature on health aspects of physical activity is that the predominant objective of most previous investigations in this area was to describe adaptations to chronic physical activity without referring to the potential positive health benefits of such adaptations. This chapter provides speculation and comments that should stimulate more research in this area because the adaptations thus far reported can be interpreted as only potentially positive for human health.

Defining Hormones

Defining what is and what is not a hormone used to be rather simple. Hormones were considered to be endogenous biochemicals that were secreted from ductless glands and then traveled through the interstitial fluid to the blood and then to all tissues. Hormones affected only some tissues because those tissues possessed a receptor specific for that molecule. Hormones were classified as peptides (made up of many amino acids linked by peptide bonds, also classified as water soluble), amines (structures derived from tyrosine, classified as lipid soluble), or steroids (basic biochemical structure similar to cholesterol, classified as lipid soluble). We now find the definition of hormones to be much more complicated. No longer does a ductless gland need to be considered because we know that fat tissue secretes leptin, which circulates and modifies satiety. Fat also secretes adiponectin (which partially regulates both lipid and carbohydrate metabolism and is found in the brain), tumor necrosis factor, interleukin-6, and resistin. Ghrelin is secreted principally by the stomach, where it stimulates acid secretion and gastric motility, but it also travels to the brain to stimulate appetite

and growth hormone (GH) secretion under basal conditions. Individual cells such as macrophages secrete cytokines, which travel throughout the body to affect many physiological functions such as immune responses. The picture of hormones traveling to other tissues to affect function is further complicated by the fact that many biochemicals that are considered classical hormones are made in one cell and affect either the secreting cell itself (**autocrine** function) or cells in close proximity (**paracrine** function).

A good example of this complexity is insulin-like growth factor 1 (IGF-1), which has classic endocrine function. It is secreted from the liver in response to GH, travels bound to specific binding proteins, and then attaches to IGF-1 receptors in muscle and other tissues to stimulate growth. However, IGF-1 is also made in muscle cells and affects the same cell or cells close by. A paracrine function of IGF-1 may consist of stimulating adjacent satellite cells to enter into the cell cycle and then to proliferate as new myoblasts and then myofibers. From an autocrine standpoint, IGF-1 stimulates transcription of muscle proteins that make up that cell's structure. Proof that these functions, autocrine versus hormonal, can be completely separate is that marked increases and decreases in muscle IGF-1 after exercise can occur with no change in circulating IGF-1. Thus, it is clear that our understanding—and thus our definition—of hormones changes constantly.

Hormone Activity

Another view that must be modified is that circulating concentrations of a hormone accurately reflect "functional activity" of that hormone. Rather, the concentration of a hormone reflects the net result of secretion, **volume of distribution** (the volume of the body into which the hormone is distributed), quantity of hormone attached to receptors or binding proteins, hemoconcentration, rate of blood flow through the endocrine gland, and hormone catabolism. Some have argued that elevations of hormones consequent to physical activity merely reflect decreases in blood volume attributable to large amounts of fluid leaving the extracellular compartment. Yet hemoconcentration cannot account for all of the increase in plasma concentrations observed during physical activity, the usual estimate being that hemoconcentration accounts for no more than 10% of the elevation in a hormone concentration during exercise.

In addition to these characteristics, which are external to the target cell, we now know that the status of intracellular pathways can modify the actions of a certain amount of hormone that has attached to a receptor. Intracellular events (biochemical signaling pathways) can make the attachment of a hormone to its receptor either greatly or mildly stimulatory, inhibitory, or ineffective. Furthermore, there is a wide range of endocrine responses to acute exercise, with some hormones increasing (catecholamines), some decreasing (insulin), and some not changing, such as luteinizing hormone (LH). Finally, there are marked differences in the temporal pattern of exercise responses. As one example, plasma insulin concentrations begin to decline almost immediately after the start of even mild exercise, yet glucagon, which is made by and secreted from the same organ as insulin and is catabolized by the same organ (liver), normally does not begin to increase until after 30 min of sustained mild to moderate exercise.

Another complicating issue is the intimate relationship (codependence) between the classic endocrine system and the nervous system. It is more appropriate to think in terms of neuroendocrine systems because hormones modify neural activity and nerves alter hormone secretion. It is clear that exercise stimulates many hormones, and the final effect of that deluge of hormones resides in the way they function in the aggregate, not individually. Despite the complexity described here, much is known about how regular physical activity alters hormonal status, and this chapter summarizes selected parts of that large body of information. This brief summary relies primarily on data obtained from human studies; however, some experimental manipulations performed in animal models are included.

Hormone Functions Related to Physical Activity

Certain general physiological functions are influenced by single bouts of endurance and probably resistance exercise, and many of these functions are at least partially modulated by hormones. They include the following:

- Mobilization of fuels for use by contracting muscles
- Uptake of fuels
- Regulation and distribution of blood flow
- Regulation of electrolyte stability
- Regulation of hydrogen ion status
- Some aspects of thermoregulation (likely related to changes in blood flow)
- Cardiorespiratory function (Viru 1985a, 1985b; Warren and Constantini 2000)

Changes in rate of growth attributable to chronic physical activity also have an endocrine basis (e.g., resistance exercise leads to muscle hypertrophy). GH, insulin, IGFs, testosterone, estrogens, and other hormones have a role in this adaptation. Additionally, in response to chronic endurance exercise, thyroid hormones are important for muscle fiber protein phenotype transformation.

Each hormone typically responds to a specific major stimulus and also to secondary stimuli. The intimate relationship between insulin and glucose is a good example. As glucose concentrations increase in the arterial blood perfusing the pancreas, the beta-cells increase secretion of insulin. As glucose concentrations decrease, insulin secretion subsides. During acute exercise, however, other factors override this relationship; most notably, exercise causes an increase in α-adrenergic stimulation of the pancreatic beta-cells, which decreases insulin

Because fat produces various hormones, one's body composition will significantly affect the physiological effects of exercise in a myriad of ways from lipid metabolism to satiety.

secretion even when glucose concentrations remain constant or increase during brief supramaximal exercise. Additionally, glucose ingestion or infusion during exercise does not elevate insulin. Thus, the suppression of insulin secretion during physical activity is difficult to override because the intimate relationship between glucose and insulin is modified during exercise by the sympathetic nervous system. This interruption allows plasma insulin to decline during exercise, which has the following positive effects on fuel mobilization: (a) enhanced hepatic glucose production, (b) enhanced rates of lipolysis (mobilization of stored fat for use during exercise), and (c) enhanced rates of gluconeogenesis. Although it seems counterintuitive that insulin should decrease when more glucose needs to be transported into muscle cells, this seeming aberration is easily explained by the fact that muscle contractions per se augment glucose uptake in the contracting muscle. The need for insulin for glucose uptake during exercise diminishes greatly in active muscle. Thus, regulators that control the hormone–stimulus relationship at rest may not be the principal regulators during exercise.

Another aspect of the endocrine response to physical activity is that, in addition to physical fitness, body composition can influence the hormonal response to physical activity. A classic example of this was Hansen's demonstration that, in lean participants, GH increased during and after exercise whereas this response was absent in obese participants (Hansen 1973). Gustafson and colleagues (1990) expanded on this observation by showing that the epinephrine, norepinephrine, and glucagon responses were markedly blunted in massively obese women during submaximal exercise.

Some Control Mechanisms Related to Physical Activity

Significant evidence suggests that the endocrine response to physical activity is regulated by feedforward, feedback, and central command mechanisms, which makes the endocrine system similar to the cardiovascular and respiratory systems (Galbo 1983). Proof of this is that increases in hepatic glucose production and in the hormones needed to accomplish those increases are observed before any decline in circulating glucose during physical activity in rats. Specifically, plasma insulin declines within the first minute of exercise, which leads to a disinhibition of hepatic glucose production. Thus,

the endocrine system responds before metabolic cues, such as reduced glucose, can occur. This suggests that feedback, for example from active muscle, is important to the endocrine response. Afferent nerve feedback from muscle fibers also contributes to the regulation of GH, adrenocorticotropin (ACTH), and β-endorphin (BeP) responses to contractions. Moreover, the aldosterone response to acute exercise using two legs is about half the magnitude found when one-leg exercise is performed at the same absolute oxygen consumption. Thus, feedback from the contracting muscle at least partially regulates the endocrine response of some hormones to activity.

> A significant gap exists in our understanding of how hormonal adaptations to regular physical activity contribute (or not) to improved health.

The response of several hormones to physical activity is partially regulated by what is referred to as "central command." Proof of this resides in studies that have used neuromuscular blockade, which weakens the muscle and consequently requires greater effort (central drive) to produce a given amount of work. Under these conditions, secretion of catecholamine and anterior pituitary hormones (GH, BeP, and ACTH) is greater than during control conditions even though the same absolute work is performed. Thus, some hormones respond to the perceived effort of the task rather than to the actual work performed. Evidence that traditional feedback mechanisms are also important is provided by the glucagon response in that glucagon increases primarily when glucose begins to decline, usually late in exercise.

Therefore, a variety of regulatory mechanisms must be considered as we evaluate endocrine adaptations to regular physical activity. Because the motor cortex partially controls some endocrine responses to physical activity, it is reasonable that adaptations should occur in that part of the brain. Unfortunately, very little is known about adaptations to regular exercise that occur in the brain. With the preceding background, several general concepts about how the endocrine system adapts to regular physical activity can be discussed along with some comments about possible health implications of those changes.

Importance of Hormonal Regulation

The human body has developed complex and redundant mechanisms that keep circulating concentrations of all hormones within a very narrow range. For example, during and after exhausting supramaximal treadmill exercise, circulating concentrations of β-endorphin increase only from 3 to 12 fmol/ml and ACTH increases from 30 to 40 fmol/ml in untrained participants. The need for this regulation over a narrow range is evident when we consider that only moderately elevated thyroid hormone concentrations can significantly elevate metabolic rate. Thus, there is a critical need for precise regulation of the balance between secretion and catabolism of all hormones so that the net result (circulating concentration) remains in the physiological range.

Regulation of Hormones at Very Low Concentrations

Other hormones besides insulin could illustrate this concept; however, insulin is used here because this hormone is involved in the Metabolic Syndrome, which is proving to be a major health concern. Another chapter in this book (chapter 12) discusses physical activity in diabetic populations; therefore, the following section covers some of the same hormones, insulin and glucagon, but concentrates on responses in nondiabetic populations.

A major health-related adaptation of insulin to regular physical activity is a decline in its plasma concentration, which seems to be precisely matched by an increase in insulin sensitivity. Insulin sensitivity is the term for a given amount of insulin resulting in a greater amount of glucose being transported into an insulin-sensitive cell—primarily fat, muscle, and liver. Both acute and regular physical activity increase insulin sensitivity for glucose uptake. Epidemiological studies show that risk for several diseases increases as basal insulinemia increases. Relative insulinemia is also of concern. Relative insulinemia is the amount of insulin compared with its effectiveness. For example, if an increase of insulin from 30 to 90 pM elevates glucose uptake from 2 to 5 mg·kg body weight^{-1}·min^{-1} in one person but, in another person, the insulin must increase from 30 to 120 pmol for the same glucose uptake, then the second person has a relative hyperinsulinemia, which is characteristic of insulin resistance. Even modest reductions (and elevations) in relative insulinemia can have a significant impact on many physiological phenomena such as fat mobilization. A decline in insulinemia of only 10 to 20 pM markedly disinhibits fat metabolism, which elevates fat oxidation (Bonadonna et al. 1990).

Numerous studies have shown that insulin concentrations in the plasma are lower either after regular exercise training (longitudinal designs) or when groups that differ in habitual activity levels are compared (cross-sectional designs). Lower basal insulin concentrations are attributable to reduced secretion rather than enhanced insulin clearance. The lower insulin attributable to chronic exercise, coupled with the higher insulin sensitivity, probably allows better glucose control.

The preceding discussion concentrates on the effects of chronic physical activity on basal insulinemia. However, studies investigating the insulin response to hyperglycemia (nonbasal conditions) also report a reduced insulin secretion in the trained state (Farrell 1992). This reduced insulin response is found at low, moderate, or high levels of glucose stimulation. Exercise-training-induced reductions in basal insulinemia and glucose-stimulated insulin secretion are demonstrable in the whole body (humans, rats, dogs), isolated pancreas (dogs and rats), isolated pancreatic Islet of Langerhans (rats), and even single isolated pancreatic beta-cells (rats) (Farrell 1992). However, the fact that this adaptation (reduced insulin secretion in response to glucose) is measurable at lower levels of inquiry (cells vs. whole body) does not mean that the entire adaptation can be explained at the cellular or subcellular level. Associated adaptations such as altered total or directional blood flow through the pancreas must also be considered as well as elevated sensitivity to the inhibitory effects of somatostatin. Somatostatin is a small peptide hormone made in the pancreas (and brain) that has inhibitory effects on insulin as well as many other hormones. Therefore, as the concentration of somatostatin increases in either the blood or pancreas, insulin secretion will decline.

Figure 6.1 provides a compilation of several studies by Mikines and colleagues that demonstrate several basic healthful adaptations of insulin in response to both chronic and acute physical activity.

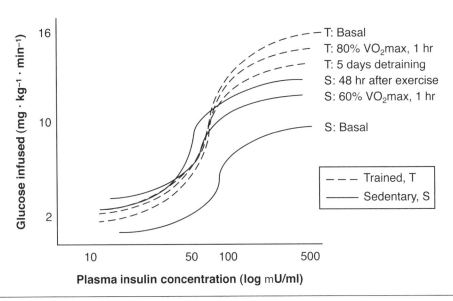

Figure 6.1 Positive adaptations of insulin to acute and chronic exercise.

Data from Mikines et al. (1988a, 1988b, 1989).

Questions on Figure 6.1*

- **Fact 1.** In the basal state, a trained person has elevated insulin sensitivity (insulin-stimulated glucose uptake at midrange insulin concentrations) and responsiveness (glucose uptake at maximal insulin concentrations).

  Question 1: Which two curves (lines) prove this point?

- **Fact 2.** A single bout of moderate-intensity exercise increases both insulin sensitivity and responsiveness in sedentary people.

  Question 2: Which two curves (lines) prove this point?

- **Fact 3.** The increase observed in question 2 lasts for at least 48 hr in sedentary people.

  Question 3: Which two curves (lines) prove this point?

- **Fact 4.** A single bout of exercise in trained people does not alter insulin sensitivity or responsiveness.

  Question 4: Which two curves (lines) prove this point?

- **Fact 5.** A period of 5 days of total inactivity (bed rest) does not change insulin sensitivity but decreases insulin responsiveness.

  Question 5: Which two curves (lines) prove this point?

*Answers to questions on page 90.

Very little information is available on the temporal sequence of adaptations to regular physical activity in endocrine secretion versus target tissue adaptations. Increases in insulin sensitivity occur with the first bout of exercise. However, an identical acute bout does not change glucose-stimulated insulin secretion (Mikines et al. 1987). Thus, anatomically distinct organs (muscle and the pancreas) adapt to chronic physical activity in a manner that ultimately results in precisely controlled glucose uptake requiring less insulin, but the individual tissues adapt with different temporal patterns. We know very little about the way this coordinated adaptation between organs is accomplished. It is unlikely that glucose per se is the common message, because resting concentrations of glucose are identical in trained and untrained people, as is the rate of hepatic glucose production.

This same pronounced sensitivity to changes in insulinemia around basal and elevated levels is not evident in some other functions in which insulin has a role. Insulin has a permissive role in allowing rates of protein synthesis to increase after acute resistance exercise. However, insulinemia must be reduced to less than 20% of basal concentrations before rates of protein synthesis do not respond (increase) appropriately after exercise. A general working hypothesis could center on the fact that adaptations to regular physical activity elevate or reduce basal hormone levels, affecting only certain actions of these hormones.

In contrast to our understanding of changes in glucoregulation by insulin with physical activity, the effects of changes in insulin sensitivity attributable to chronic physical activity on functions such as protein synthesis, protein catabolism, lipolysis, and vasodilation are much less clear. Moreover, the effect of chronic physical activity on plasma concentrations of hormones other than insulin is even less clear. Basal plasma epinephrine concentrations may be slightly higher in trained athletes; however, this is not a consistent finding. When higher concentrations are found either at rest or during intense exercise (very high workloads) in trained individuals, sufficient data exist to indicate that such elevations are attributable to an increase in epinephrine secretion rather than a decrease in epinephrine clearance. Resting concentrations based on single blood samples of ACTH, BeP, norepinephrine, and glucagon seem to be similar between active and inactive people.

Regulation of the Rhythmic Release of Hormones

Many, if not most, hormones are secreted from glands in a rhythmic manner. Those oscillations can last from minutes to months. The consequence of such secretion is that plasma concentrations of most hormones rise and fall with predictability, and we can assume that such patterns enhance survivability and function of the organism (Borer 2003). Hormone pulses have many complex characteristics, and only two—pulse frequency (time between pulses) and pulse amplitude (peak amount of hormone secreted during a pulse)—are discussed here. The time between pulse intervals can range from minutes (**circadian,** e.g., insulin) to hours (**ultradian,** e.g., GH, cortisol, LH) to days (**infradian,** e.g., reproductive hormones, such as follicle-stimulating hormone [FSH], LH, and estrogens).

> Many (or most) hormones are released from endocrine glands in a pulsatile manner, and regular physical activity causes adaptations in those pulse profiles that are compatible with better function.

For reasons that are not completely known, a pulsatile release (and consequent presentation to target tissues) of hormones augments target tissue action. Therefore, we would predict that regular exercise should augment the pulsatile release of hormones. In some cases (GH in young women), the literature supports an augmented pulsatility, but in the case of insulin and melatonin, the opposite is found.

Knowledge of the pattern of a pulse profile is required to correctly interpret exercise data. Reliance on a single blood sample is at best hazardous because any exercise-related response could be masked or accentuated because of a naturally changing baseline. An example from many years ago is that researchers found that if exercise begins when a normal elevation in cortisol occurs, an exercise-induced increase in circulating cortisol concentrations does not occur; but if the same exercise starts when cortisol is at a nadir (lowest level), then circulating cortisol increases significantly. Thus, this effect of the baseline concentrations and patterns was often neglected in interpreting research results and probably accounts for a large amount

Answers to Questions on Figure 6.1

Answer 1: Compare T: basal versus S: basal

Answer 2: Compare S: basal versus S: 60% $\dot{V}O_2$ max, 1 hr

Answer 3: Compare S: basal versus S: 48 hr after

Answer 4: Compare T: basal versus T: 80% $\dot{V}O_2$ max, 1 hr

Answer 5: Compare T: basal versus T: 5 days detraining

of the confusion in the physical activity–hormone literature.

Insulin

Insulin is secreted from the pancreas in a pulsatile manner and will be used as an example of how regular physical activity can change the pattern of hormone release in a nonstimulated state. Reproductive hormones and growth hormone are also secreted in a pulsatile manner and are also reviewed here.

In physically trained people, a reduction in the pulse amplitude of insulin has been reported and this makes sense as follows: Providing a hormone in a pulsatile manner increases the effectiveness of that hormone, but because the insulin-sensitive tissues such as muscles become more sensitive to insulin, there is less of a need to present insulin to those tissues in a pulsatile manner. Thus, a reduced pulse amplitude of insulin is needed in well-trained people.

Under nonstressful conditions, hormones are present in the plasma in very low concentrations, and small elevations or reductions in those basal concentrations may affect health significantly. Regular physical activity has the potential to cause such small deviations.

In concert with reduced plasma concentrations of insulin assessed by single point sampling, minute-by-minute sampling for 90 min shows that the pulse amplitude is markedly lower in exercise-trained men and women whereas pulse frequency is not changed (figure 6.2). This is an interesting observation because it is generally accepted that hormones are more effective (greater target tissue effects) when delivered in a pulsatile manner. Data in figure 6.2 (Engdahl et al. 1995) were interpreted to suggest that insulin action at target tissues was elevated to the point that delivering insulin in a pulsatile manner was not as necessary to maintain a certain rate of glucose uptake in trained individuals. The logical but unproven conclusion is that because of the greater tissue sensitivity, there is less need for insulin to be secreted (and delivered) in a pulsatile manner.

From an evolutionary standpoint, the pattern for the trained group may, in fact, be the desirable pattern, because there may have been a gradual elevation in burst amplitude over the thousands of years in which we have become less active and less insulin-sensitive. The coordinated changes in insulin pulsatility and sensitivity with regular physical activity are quite distinct from those observed in people with type 2 diabetes mellitus. In both people with type 2 diabetes mellitus and glucose-intolerant first-degree relatives of those people, the insulin pulse profile is characterized as either absent or erratic, with poorly formed pulse amplitudes and wide disruptions in pulse frequency.

Although meager, the information available on **hormone pulsatility** and chronic exercise suggests that pulse amplitude (especially for insulin, GH, and LH, described subsequently) but not pulse frequency is adaptable to chronic physical activity. Interestingly, this is similar to changes found in some diseases in that pulse amplitude changes but pulse frequency may be normal.

Growth Hormone and IGF

Another hormone that is secreted with an ultradian periodicity is GH. Secreted from the anterior pituitary, this hormone plays a critical role in somatic growth. GH has many isoforms, but the 22 kd isoform is probably the most studied from an exercise perspective, and it increases in plasma in an exercise intensity-dependent manner, especially during very heavy physical activity. GH circulates bound to binding proteins, and the role of these binding proteins in the exercise response is not clear. After puberty, an exercise-induced increase in GH availability may have its greatest impact on maintaining skeletal muscle mass in the elderly.

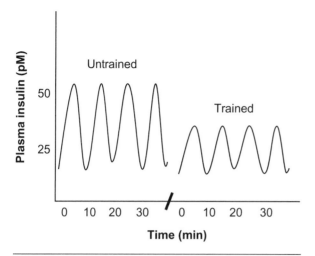

Figure 6.2   Effect of regular physical activity on insulin pulsatility.

Data from J.H. Engdahl, J.D. Veldhuis and P.A. Farrell, 1995, "Altered pulsatile insulin secretion association with endurance training," *Journal of Applied Physiology* 79(6): 1977-1985.

The complexity of GH regulation becomes quickly evident when we realize that two of the most potent physiological stimuli are sleep (no movement) and exercise (movement). High plasma concentrations occur 2 to 3 hr after sleep begins, but even higher concentrations are recorded after intense acute aerobic and resistance exercise. During nonstimulatory periods, GH output from the pituitary depends on the balance between the stimulatory effects of the hypothalamic stimulatory peptide GH-releasing hormone and the inhibitory peptide somatostatin. The stimulus for elevated GH during acute exercise is not known. However, cholinergic stimulation, nitric oxide, hydrogen ions, muscle afferent activation (which directly stimulates the anterior pituitary), and muscle vibration have been implicated in various studies (e.g., Gosselink et al. 2004). Cholinergic, dopaminergic, and serotoninergic blockade each inhibit the GH response to exercise. Exercise-induced increases in GH are probably not attributable to catecholamines (β-or α-adrenergic), opioids (at least those antagonized by naloxone), or changes in plasma ghrelin. Plasma ghrelin does not change during exercise when elevations in GH are evident. This finding (no effect of ghrelin on GH), specific to the exercise stimulus, conflicts with studies in which ghrelin infusion stimulated increases in plasma GH, ACTH, and prolactin.

Chronic physical activity elevates GH pulse amplitude but not frequency in premenopausal women when the training stimulus is above the lactate threshold. Elevations of GH pulse amplitude, as well as the total amount of the hormone secreted over a 24 hr period following a single bout of exercise, are found in young women but not young or old men. Data on middle-aged men do not seem to exist. For several hours after heavy resistance exercise, GH pulse amplitude and GH concentrations are significantly lower in young men (22 years). Eventually, however, GH concentrations seem to be higher than under nonexercise conditions. It is obvious that much more work must be done to document whether and how acute exercise alters the pulse profile of GH. Elevations in GH amplitude may have implications for maintaining muscle mass in elderly people because reductions in GH concentrations are observed in both obesity and aging.

GH stimulates production and release of IGF from the liver. IGF-1 and mechano growth factor (MGF) increase in plasma and in the skeletal muscle after resistance exercise, whereas combining exogenous GH with resistance exercise leads to even larger increases in messenger RNA (mRNA) for MGF. Hepatic-derived IGF-1 is stimulated by GH, however, elevations in tissue concentrations of IGF-1 in response to contractions can be independent of circulating GH and plasma IGF-1. An increase in muscle mass attributable to resistance exercise probably involves both GH-stimulated increases in IGF-1 and elevations in intramuscular MGF caused by increased gene expression (mRNA eventually translated into MGF protein). Some lines of evidence suggest differential splicing of the IGF-1 gene in response to chronic muscle contractions. However, this general finding may be more applicable to rodents than humans, because some reports suggest no exercise-training-induced increase in human muscle IGF-1. Acute resistance exercise of moderate to heavy intensity increases serum IGF-1, whereas concentrations of IGF binding protein-3 decrease and IGF binding protein-1 does not change.

Mounting an appropriate anabolic response to resistance exercise probably involves many hormones, possibly in a redundant fashion. For instance, moderately diabetic rats can increase rates of protein synthesis after resistance exercise, and the reduced insulinemia seems to be compensated for by increased IGF-1 concentrations in the exercised muscle in the immediate postexercise period (up to 24 hr). This does not occur in nondiabetic rats until

2 days after exercise. Thus, a relationship seems to exist between IGFs in the muscle and insulin availability during periods of anabolism.

Female Reproduction

Reproductive hormones are also secreted episodically, and chronic excessive exercise that causes chronic energy deficits also suppresses the pulse profile of several reproductive hormones. It must be noted, however, that normal women can engage in long-term (>1 year) moderate- to high-intensity endurance exercise without affecting reproductive function. Some exercise is probably beneficial to reproductive hormone status, but excessive exercise (very high levels of physical activity that result in an energy deficit) is not compatible with producing offspring (Loucks 2001). Therefore, it is not surprising that a small percent of female athletes (approximately ≤6%) who engage in frequent, high-volume, and high-intensity physical activity will temporarily lose reproductive function (cease menstruation) either fully or partially. Although the cause of this decreased function is not completely clear, overlapping lines of evidence support the working hypothesis that it is not the stress of exercise per se that causes the dysfunction but rather energy deficits caused by the combination of increased energy expenditure from activity as well as undernutrition.

Exercise requiring significant energy expenditure reduces LH pulse amplitude in women. Such reproductive dysfunction must involve areas in the hypothalamus that determine LH secretion. LH is under complex control, but the major stimulatory regulator is gonadotropin-releasing hormone (GnRH), which is made in and secreted from the arcuate nucleus of the hypothalamus. A one-to-one relationship exists between GnRH and LH release, and both are disrupted by energy deficits associated with high levels of physical activity. Consequences of reduced LH pulsatility can include luteal phase defects such as shortening of the luteal phase and lengthening of the follicular phase or insufficient secretion of progesterone by the corpus luteum.

Another anterior pituitary hormone, FSH, is also adversely affected (reduced circulating concentrations) consequent to energy deficits caused by physical activity. FSH attaches to receptors on ovarian follicles and stimulates their maturation. These follicles in turn make and secrete estrogens that are also lower in women who perform extensive exercise. Disruption of the hypothalamic anterior pituitary gonadal axis caused by chronic high levels of exercise (or the consequent energy deficit) may occur at any level of the axis. However, gonadotropic receptor sensitivity to GnRH is not altered by chronic exercise. Reproductive dysfunction in women may also be attributable to excessive secretion of corticotropin-releasing factor, ACTH, cortisol, androgens, or BeP. Each has been linked to reduced LH secretion (Borer 2003). At the same time, these women have higher concentrations of GH and androstenedione. Another hormone, leptin, may also be involved in reproductive dysfunction in female athletes because it is also suppressed by low energy status.

Chronic and excessive physical activity early in life can adversely affect sexual maturation. Young girls who perform strenuous high-volume exercise can have delayed menarche of about a year. Characteristics of this delay include a prepubertal pattern of low-frequency LH pulses, increased FSH/LH secretory ratios, and a lack of the enhanced estradiol secretion that is necessary to progress through puberty (Borer 2003).

The ultimate effects on health of disrupted reproductive function in women are not known. If regular exercise chronically lowers circulating estradiol, this may provide protection against heart disease and breast cancer. On the other hand, lack of a normal reproductive cycle will cause bone loss and osteoporosis. Again it must be emphasized that menstrual disturbances are neither a common nor an expected consequence of regular nonexcessive physical activity levels.

Male Reproduction

Acute exercise results in elevations of free and total testosterone and androstenedione in males; however, heavy chronic training decreases testosterone. A reasonable working hypothesis is that male reproductive hormones respond negatively to energy deficits in a manner similar to that found in females. Supporting this hypothesis, chronic, prolonged, high-intensity physical activity decreases testosterone and spermatogenesis. Whether chronic moderate-intensity physical activity improves reproductive function in either males or females has not been documented.

Regular Physical Activity and Hormonal Adaptations

The suggested relationship between IGFs and insulin availability illustrates the significant concept of redundancy of hormone action. This seems to be especially important in the area of fuel mobilization. Regular physical activity alters hormone availability, and this may alter the pattern of fuel utilization during exercise.

The hormones most involved in the regulation of fuel utilization and mobilization during exercise are epinephrine, glucagon, insulin, and, to a lesser extent, GH, norepinephrine, ACTH, and cortisol. Epinephrine, cortisol, and glucagon are referred to as counterregulatory hormones in that they balance the effects of insulin and promote glucose availability. Because of the importance of catecholamines, this section concentrates on certain catecholamines (epinephrine and norepinephrine) because they alter exercise metabolism and adapt to regular physical activity. Because of the difficulty in differentiating between adrenal-derived catecholamines and those originating from sympathetic nerves, most texts refer to the sympathoadrenal system.

Epinephrine and Norepinephrine

The catecholamines most studied during exercise are epinephrine, norepinephrine, and dopamine. Epinephrine is made in and secreted from the adrenal medulla; norepinephrine is also made in the adrenal medulla, but the major elevations in plasma found with exercise are attributable to spillover from activated sympathetic nerve terminals. A major neurotransmitter in the central nervous system, dopamine also appears in the peripheral circulation. However, the function and source of peripheral dopamine remain unclear. Plasma epinephrine and norepinephrine concentrations increase exponentially with exercise intensity (figure 6.3), and elevations are most closely associated with the intensity of work relative to maximal aerobic capacity. The cause of this increase has been shown to originate both from peripheral factors (local factors in the working muscle) and from central command in the brain. Major physiological functions regulated by catecholamines during exercise are fuel mobilization (epinephrine), partitioning of blood flow (norepinephrine), increased cardiac contractility (norepinephrine), and elevation of blood pressure (norepinephrine). Circulating catecholamines (as opposed to norepinephrine derived from cardiac sympathetic nerves) are probably not critical to the heart rate, systolic blood pressure, and respiratory responses to exercise because these physiological responses are identical when biadrenalectomized participants exercise to exhaustion. Circulating catecholamines are rapidly degraded, and their concentrations decline quickly after exercise.

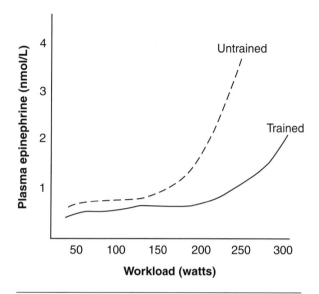

Figure 6.3 Plasma epinephrine response to exercise specific to absolute workloads.

Catecholamines account for a major proportion of the exercise-induced increases in adipose tissue lipolysis, but differential increases in blood flow can cause one adipose tissue bed to increase lipolysis to a greater extent than another bed during exercise. For example, sympathetic nerves are not important for exercise-induced increases in lipolysis from umbilical and clavicular beds. Although epinephrine stimulates both hepatic glycogenolysis and adipose tissue lipolysis, the adaptations to training may differ depending on the tissue. Regular physical activity reduces epinephrine-stimulated hormone-sensitive lipase activation in muscle but elevates it in adipose tissue. Both epinephrine and adrenocorticotropic hormone stimulate greater lipolysis in isolated adipocytes as an adaptation to regular physical activity. Catecholamines also stimulate lipolysis indirectly by helping to suppress circulating insulin and stimulating the release of GH, cortisol, and glucagon, which have lipolytic effects.

Catecholamine response to exercise is a well-established example of the fact that the exercise response of many hormones depends on the relative effort of the exercise and not the amount of work actually being performed, as shown in figures 6.3, 6.4, 6.5, and 6.6 (Kjaer et al. 1986). Most glands that secrete hormones do not sense how much total work is being done but rather how much of the organism's capacity is being used to accomplish that work. This concept has been derived mostly from studies that required large muscle contractions (running, bicycle exercise), and the applicability of this concept to

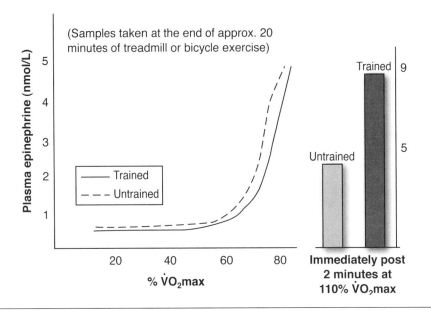

Figure 6.4 Plasma epinephrine changes during exercise expressed as relative workload.

Adapted, by permission, from M. Kjaer and H. Galbo, 1988, "Effect of physical training on the capacity to secrete epinephrine," *Journal of Applied Physiology* 64(1): 11-16.

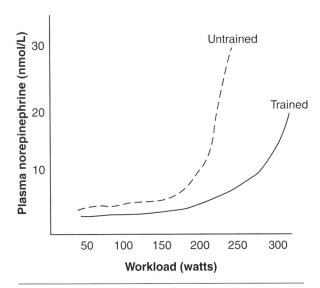

Figure 6.5 Plasma norepinephrine response to exercise, specific to absolute workloads.

movements using only a small muscle mass requires documentation.

Adequate fuel mobilization is critical to physical activity, and hormones have an important role in this physiological function. It is also clear that redundant endocrine systems affect fuel mobilization during exercise.

Whether adaptations of the endocrine system result in "better" fuel mobilization and utilization during exercise is complicated. Fat mobilization and utilization are appropriate at low work intensities, but carbohydrate mobilization is appropriate at high intensities. Epinephrine has been shown to be a primary stimulator of adipose tissue lipolysis but not muscle glycogenolysis during exercise (Kjaer and Galbo 1988), and fuel mobilization during exercise depends on these functions to a significant degree. Endurance-trained individuals have an increased lipolytic response to infused epinephrine; however, this adaptation is short lived (lost in only a few days) once training ceases. This elevated lipolytic response is observed at both the whole-body level as well as when human fat cells are stimulated in vitro. Some of this adaptation must reside in an enhanced sensitivity of adrenergic receptors (both α and β), as shown in the late 1970s and the 1980s (Krotkiewski et al. 1983). Such an adaptation is consistent with elevated rates of fat oxidation during submaximal exercise in trained individuals. Thus, under several conditions, hormone-stimulated lipolysis from several depots is enhanced by regular physical activity.

Well-trained endurance athletes can secrete greater amounts of epinephrine at high work intensities compared with untrained people and when work is expressed relative to maximal aerobic capacity. This ability has advantages, such as the

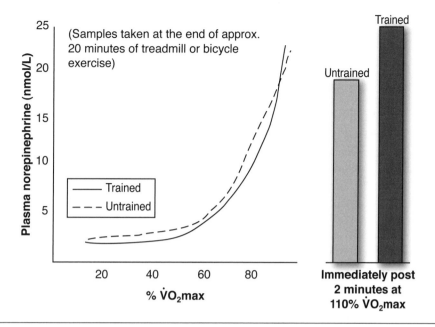

Figure 6.6   Plasma norepinephrine changes shown in relative workload.

Adapted, by permission, from M. Kjaer and H. Galbo, 1988, "Effect of physical training on the capacity to secrete epinephrine," *Journal of Applied Physiology* 64(1): 11-16.

mobilization of glucose, as suggested by a linear relationship between hepatic glucose production and plasma epinephrine during exercise when plasma epinephrine increases from 1 to 6 nmol/L. Further support for the necessity of β-adrenergic stimulation during hard exercise is that β-adrenergic blockade with propranolol reduces work output at high but not low work intensities. High glucose production would be advantageous during times of high athletic competition. This adaptation of the adrenal medulla to training may also be true for selected anterior pituitary hormones because BeP, GH, and ACTH all increase in plasma to a greater extent in trained athletes during very hard exercise.

Figure 6.5 shows that plasma norepinephrine also increases with work intensity in an exponential manner, and figure 6.6 demonstrates that well-trained athletes have higher concentrations during very high intensity exercise. The metabolic implications of these changes are not clear.

> Positive hormonal adaptations to regular physical activity occur shortly after the commencement of training and probably disappear as quickly.

The rate at which hormonal adaptations to regular physical activity disappear during detraining is not clear. The exercise training–induced adaptation of the pancreas is lost quickly (within days), but the adrenal medulla maintains the ability to secrete a high level of epinephrine even after many weeks of detraining. This is a markedly understudied area of exercise endocrinology.

Glucagon

A primary function of glucagon is to oppose the glucose-lowering actions of insulin. Glucagon is made in and secreted from the alpha-cells in the Islets of Langerhans of the pancreas. Twenty-four-hour averaged glucagon concentrations are not different between physically active men and inactive men; however, some studies report that resting concentrations of glucagon are lower after chronic physical activity. Although there is a good correlation between plasma concentrations of epinephrine and glucagon during exercise, β-adrenergic blockade does not affect the glucagon response to exercise. The glucagon response to submaximal exercise is highly dependent on reductions in plasma glucose but is dependent on neither parasympathetic nor α-adrenergic blockade. As with epinephrine, the glucagon response to supramaximal exercise is markedly higher in trained than untrained people, yet the glucagon response to insulin-induced hypoglycemia is lower in trained than untrained people.

Hormonal Adaptations to Chronic Physical Activity and Long-Term Food Intake

Related to fuel utilization during exercise is the issue of fuel storage replenishment after exercise. Leptin is a 146-amino-acid peptide that circulates in both bound and free forms. Early studies clearly demonstrated that leptin availability was closely tied to energy balance; that is, negative energy balance reduced circulating leptin concentrations whereas energy surplus elevated concentrations. Simultaneously, studies showed that obese individuals had higher concentrations of leptin, whereas athletes had reduced levels. Studies progressed from these correlational designs to those using manipulations of energy balance, which advanced and supported the working hypothesis that circulating leptin concentrations do not change with exercise unless a marked energy deficit occurs. Thus, it may be the exercise-related energy deficit rather than the activity per se that reduces circulating leptin concentrations. Such conclusions need to be refined.

Leptin's effects on body weight are mediated through effects on hypothalamic centers that control feeding behavior, hunger, body temperature, and energy expenditure. Daily injections of recombinant mouse or human leptin into ob/ob mice (i.e., obese mutants unable to synthesize leptin) led to a dramatic reduction in food intake within a few days and to roughly a 50% reduction in body weight within a month. Weight loss resulting from administration of leptin results from decreased hunger and food consumption and increased energy expenditure. Unfortunately, leptin injection into obese humans has not resulted in consistent weight loss.

Because leptin is acutely sensitive to negative energy balance (fasting or caloric restriction), studies are needed that can distinguish the effects of exercise per se from any attendant change in energy balance or, perhaps more specifically, energy availability. Thus, leptin may act as a sensor of the difference between energy intake and expenditure. The mechanisms by which leptin exerts its effects on metabolism are largely unknown and are likely quite complex.

Related to long-term food intake are the questions of whether basal metabolic rate (BMR) changes with chronic physical activity and, germane to this chapter, whether changes in hormones attributable to physical activity can account for such a change. Unfortunately there is no consensus on whether BMR is different between active and inactive people. Numerous articles both support and refute such changes. Because of the potent effects of thyroid hormones on basal metabolism, physical activity–induced changes in thyroid hormones should also be considered when evaluating adaptations that may alter long-range energy balance. Hormones that control BMR may also adapt to regular physical activity. Thyroid hormones are secreted from the thyroid gland (located just below the larynx) in response to thyroid-stimulating hormone (TSH), which is secreted by the anterior pituitary gland. TSH itself is released because of thyrotropin-releasing hormone (TRH) made in and secreted from the hypothalamus. Triidothyronine (T3) is the active thyroid hormone, and it binds to receptors in the nucleus of specific cells. A key to understanding a role for T3 and physical activity is that thyroid hormones change physiology in general but muscle function in particular by changing gene expression (transcriptional regulation). Most prominent is that T3 increases the synthesis and activity of $Na^+–K^+$–adenosine triphosphatase (ATPase) pumps. If chronic physical activity causes increases in T3 that translate to increased $Na^+–K^+$–ATPase pumps, then basal metabolism should increase. Increases in BMR have been reported, but other reports on 24 hr energy expenditure suggest no difference in trained female athletes compared with sedentary women.

Acute endurance exercise results in increases (in the basal state) in TSH and unbound tetraiodothyronine (T4), whereas the average concentrations for **reverse T3 (rT3)** and T4 are elevated after chronic physical activity. Unbound T3 seems to remain unchanged after chronic physical activity. However, when excessive physical activity is performed, both T3 and T4 are reduced, as in the case of amenorrheic women athletes or athletes performing very high intensity training at the peak of a season. This may be a manifestation of chronic energy deficits because starvation also reduces these thyroid hormones. In terms of a time course for these changes, early training results in initial declines in T3 and T4, which should cause a greater level of TSH, but then as moderate training proceeds, there is an increase in T4 and rT3 but no change in T3. Some reports suggest that the TSH response to TRH is blunted in active people. Uncoupling proteins such as UCP-3 are important for stimulating energy metabolism, and T3 stimulates UCP-3 mRNA expression. Thyroid hormones are not the only hormonal system that could account for changes in BMR if in fact increases or decreases actually occur with chronic physical activity. Obviously, the endocrine control of individual components of 24 hr energy expenditure must be assessed.

Cross-Adaptations

Generally, the first exposure to a stress results in the greatest response, and subsequently the magnitude of the response declines. Thus, regular exercise lowers the organism's response to subsequent exercise sessions, and this may have positive health benefits.

An underlying assumption concerning health benefits of physical activity is that regular exercise is not performed to an excessive degree but rather on a regular basis and in moderate to high but not excessive volumes. Examples of deleterious effects of excessive training are that the menstrual cycle or testicular function can become dysfunctional, and basal catecholamines and cortisol can become chronically elevated. However, when training is not excessive, high levels of chronic physical activity cause adaptations that can be viewed as positive. Teleologically, it is reasonable that the endocrine system in a well-trained person functions at a "better" level than that of a person who is completely sedentary. There are no reports of deleterious hormonal adaptations to regular moderate-intensity physical activity. From an evolutionary perspective, this makes sense because the current human genome developed during periods that required high levels of physical activity.

Because hormones increase in plasma according to the percentage of the maximal aerobic capacity, a trained person can perform a certain absolute amount of physical work with less of a hormonal response yet still meet the metabolic, cardiovascular, and psychological demands of that work. Therefore, as regular physical activity increases, one's functional capacity to complete normal tasks and challenges should engender less of a stress response. Many animal studies support this concept, but much more work must be done to confirm this phenomenon in humans.

This concept probably crosses several adaptations because adaptations to one stress usually allow reduced responses to other stresses. As one example, the catecholamine response to mental tasks is lower in trained participants compared with sedentary people. Although eating is not typically thought of as a stress, many hormones increase or decrease after a meal, and it is clear that the insulin response, for example, to a meal is markedly reduced in trained persons. Considering the frequency of food intake in today's society, this consistently lowered insulin response could have major health benefits. Note, however, that athletes ingest more carbohydrates than nonathletes, and it has been shown that when 24 hr calorie intake is considered, integrated 24 hr insulin concentrations are not different between trained and sedentary people.

The anterior pituitary adrenal axis is activated during and after all known stresses. Hormones of both the adrenal cortex (cortisol, aldosterone, and adrenal androgens) and the adrenal medulla (epinephrine and norepinephrine) are activated by stress according to the deviation from homeostasis caused by the stress. The cortisol response is one of the few hormones (along with the catecholamines) in which the increase in plasma concentrations has been proven to be caused by an increase in secretion rather than a decrease in clearance. For cortisol, as exercise intensity increases, the rate of clearance also increases. With epinephrine, as the exercise intensity increases, the rate of clearance decreases at high work intensities, but during very light work the rate of clearance increases or does not change. These distinctions are important when we are trying to assess how much the organism is perturbed and how much an endocrine gland increases output to control that perturbation.

Summary

Because the endocrine system is a major regulator of body function, it is reasonable that some of the health adaptations known to occur with regular physical activity must be caused by changes in hormone action. Although this is teleologically sound, surprisingly few studies have reported an endocrine adaptation and then proven the health-related benefits of that adaptation. This chapter relies heavily on the review of adaptations of insulin to regular physical activity because of insulin's central role in the metabolic syndrome and the current epidemics of obesity, childhood type 2 diabetes mellitus, and other established chronic diseases that have insulin resistance as a common phenotype. Chronic physical activity reduces basal and glucose-stimulated insulin secretion, and this persistent relative hypoinsulinemia may prove to be the most important beneficial endocrine adaptation in terms of our public health.

KEY CONCEPTS

autocrine: Hormone that is made in a specific cell, is secreted by that cell, and changes the function of that same cell.

circadian: Hormone rhythms that occur with an approximately 24 hr cycle.

hormone pulsatility: Regular increase and decrease in the secretion of hormones by an endocrine gland (rather than in a constant rate). Intervals can range from minutes to months.

infradian: Hormone rhythms that are longer than a 24 hr cycle.

paracrine: Hormone that is made by a specific cell, is secreted by that cell, and affects the function of an adjacent cell.

reverse T3 (rT3): Another form of 3,5,3'-triiodo-thyronine (T3) with a structure of 3,3',5'-triio-dothyronine.

ultradian: Hormone rhythms that occur with a period of less than 1 day.

volume of distribution: Volume of the body (usually expressed as liters) into which a hormone is distributed once secreted from the gland.

STUDY QUESTIONS

1. Provide an example of a hormone that has classical hormonal actions but also acts in an autocrine and a paracrine manner.

2. Give one example of feedback, feedforward, and central command regulation of the hormone response to physical activity.

3. Name three changes in insulin action or secretion attributable to chronic physical activity that should have positive health outcomes.

4. Name one hormone with a pulse profile that is altered by chronic physical activity and describe the possible health implications of that alteration.

5. Describe three hormonal adaptations to regular physical activity that result in "better" fuel mobilization during times of stress.

6. Are the hormonal adaptations to regular exercise confined to this particular stress?

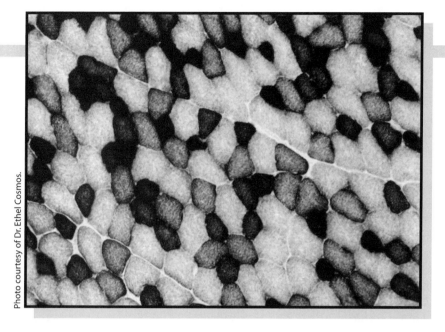

Photo courtesy of Dr. Ethel Cosmos.

Skeletal Muscle Adaptation to Regular Physical Activity

■ Howard J. Green, PhD

CHAPTER OUTLINE

A fundamental property of skeletal muscle is its ability to adapt; skeletal muscle is one of the most malleable tissues of the human body, capable of extensive remodeling if systematically challenged (Booth et al. 2002). The remodeling that occurs is to a large degree dependent on the type of contractile activity performed. Skeletal muscle can also rapidly change in response to inactivity and disuse, leading to pronounced deterioration in mechanical capabilities. This can be particularly problematic during aging, when an inevitable decline of muscle mass and function occurs.

Our evolutionary history and the central role of skeletal muscle in survival suggest that our natural state of **acclimatization** is one of regular daily activity. Such activity not only promotes optimal muscle function and task performance but also provides health benefits, both directly and indirectly. Research has shown that chronic exercise decreases the risk for both morbidity and all-cause mortality.

In this chapter, we examine the underlying structure and composition of muscle, the factors that allow humans to perform highly specialized tasks, the effects of regular activity on the structure and composition of muscle, and the functional consequences of the **adaptations** that occur. Finally, we examine the role of muscular adaptations in improving health and well-being.

Skeletal Muscle and Human Life

Skeletal muscles are the largest tissue mass of the human body, constituting in excess of 600 individual muscles. The most dramatic example of the crucial role of skeletal muscles in our survival comes from the "fight-or-flight" response, originally conceptualized by physiologist Dr. Walter Cannon. This term describes a predictable strategy to perceived danger, namely engaging in combative action or fleeing from the stressful agent. Although these responses are not commonly used today, given the technological and social advances that have occurred, evolutionary theorists have proposed that the number, mass, and composition of our muscles are the selective and adaptational consequences of preceding generations in which drawing and hauling were labors of survival and hostile encounters were common.

We each have inherited a large mass of tissue whose fundamental function is the ability to contract and generate force. Muscles allow us to perform an array of motor activities that range from the most delicate of tasks to those requiring large amounts of force and velocity. To allow us to function effectively and efficiently, an elaborate hierarchy of muscular organization has evolved. Muscles have subsets, each with a common synergistic action and each able to contribute depending on the demands of the task. Muscle groups exist for posture, locomotion, vision, mastication, breathing, and a range of other functions.

Another level of organization allows selected muscles to perform specific mechanical functions with great effectiveness. To a large degree, this capability can be attributed to the design and composition of the individual fibers that constitute the muscle. The individual muscle fibers can be distinguished by differences in the type and amount of proteins and substrates. The performance of skilled motor activities not only is crucial to independent living but also enables us to interact with our physical environment.

Muscles also perform functions that either directly or indirectly contribute to our health and well-being. Shivering thermogenesis, for example, in which contracting muscles produce heat, can directly help us maintain body temperature in cold environments where heat loss is increased. Even at rest, the heat liberated by muscle is significant in body temperature **homeostasis.** Muscles offer protection during falls and collisions, buffering the impact that could fracture bones and damage internal organs.

The control of body weight and adiposity can to a significant degree be managed by exercise-induced increases in caloric expenditure, which is mediated by the contracting muscles. Increasing evidence suggests that regular exercise promotes bone health and reduces hypertension and lipid abnormalities. Perhaps the best example of the direct role of muscles in disease prevention and management is type 2 diabetes. Type 2 diabetes, a disease in which blood glucose regulation and insulin resistance are abnormal, can to a large extent be controlled by regular physical activity. This is because blood glucose disposal is primarily regulated by skeletal muscle. Trained muscle can also assist in managing chronic heart failure and chronic obstructive lung disease by preventing excessive accumulation of the metabolic by-products of muscle contraction, which can negatively affect cardiovascular and ventilatory function and consequently the **strain** associated with these diseases.

As a characteristic of our species, we each inherit a large mass of muscle tissue. Muscles are specialized to contract, enabling us to interact with each other and with our environment. To function optimally, muscles must contract regularly. Regular muscle activity appears to provide enormous benefits in disease prevention and management.

Finally, an appealing but yet to be confirmed hypothesis is that selected amino acids released by the working muscle can act as precursors for different neurotransmitters, the formation of which can dampen undesirable mood states such as depression.

Muscle Cell—Composition, Structure, and Function

The **mechanical properties** of a single muscle are, to a large degree, the product of its constituent muscle fibers. Muscle fibers are regulated not independently but rather in subgroups, with each subgroup activated by a single motor nerve. Each subgroup with its nerve constitutes a motor unit. The number of motor units in a muscle depends on its function. Muscles that are specialized for large power output, as in locomotion, have large numbers of motor units and a large number of fibers within each motor unit. In contrast, muscles that are specialized for very precise tasks, such as eye movement, display a small number of motor units with relatively few fibers in each motor unit. By modulating the impulse frequency and the motor unit pools recruited and the timing of these events, the nervous system is able to coordinate graduated motor responses.

Excitation and Contraction Processes

The muscle cell or muscle fiber is the fundamental force-generating entity, capable of translating instructions from a nerve into a mechanical event. To do this, it has a variety of special structures that allow specific processes to occur (figure 7.1). The initiating event involves the conversion of the transmitter acetylcholine, released from the motor nerve into the neuromuscular junction, into an action potential. The action potential is, in turn, conducted onto the surface membrane of the fiber and ultimately, by invaginations of the surface membrane, into the interior of the fiber (figure 7.2 on page 103).

Muscle fibers are covered by a membranous network (the sarcolemma) that penetrates into the interior of the fiber at regular intervals as transverse tubules (T-tubules). The T-tubules form a network that extends throughout the muscle fibers, encircling the individual myofibrils. Once inside the fiber via the T-tubules, the action potential, if it is to increase the free calcium ion (or $[Ca^{++}]_f$) enough to cause a tetanic contraction (approximately 100-fold), must be able to regulate the release of stored Ca^{++} from the sarcoplasmic reticulum (SR) (Berchtold et al. 2000). The SR is a system of tubes that extend longitudinally throughout the muscle fiber, connecting the lines of T-tubules and lying along the longitudinal surfaces of the myofibrils. Calcium release from the SR is controlled by a protein called the calcium ion (Ca^{++})-release channel (CRC) or ryanodine receptor (RyR). According to current evidence, the action potential in the T-tubule is sensed by a dihydropyridine receptor (DHPR). The resulting changes in the DHPR appear to result in opening the CRCs by both mechanical and chemical coupling. The opening of the CRC leads to a rapid increase in $[Ca^{2+}]_f$ levels in the sarcoplasm (the gelatinous material between the myofibrils). Because the SR envelops the individual contracting myofibrils and allows close proximity between the T-tubules and CRCs, a near-synchronous opening of the CRCs occurs in a fiber in response to an action potential, allowing rapid increases in $[Ca^{2+}]_f$ and, consequently, rapid activation of the contractile apparatus. A common term for this first category of signaling processes which include the sequence of events beginning with the action potential and culminating with an increase in $[Ca^{++}]_f$ levels is **excitation–contraction (E–C) coupling**.

The second category of signaling processes that mediate contraction involves the sensing of elevated $[Ca^{++}]_f$ levels by a subunit of the regulatory protein, troponin, and movement of a second regulatory protein, tropomyosin. This removes inhibition between actin and myosin by the thin filament, allowing these proteins to move from the dissociated or weakly bound state into a strong-binding, force-generating state. Movement is generated when the number of actin and myosin in the force-generating state is sufficient to overcome an external load, allowing the individual myosin molecules to cycle to successive actin sites, resulting in shortening of the muscle fiber. The myosin molecule possesses a unique structure that allows it to interact with actin

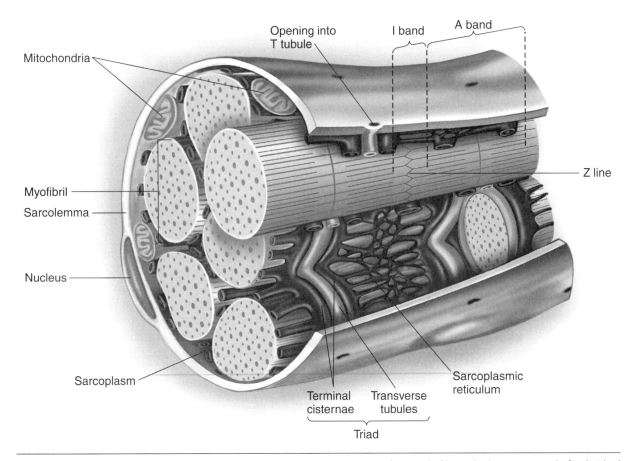

Figure 7.1 Mammalian muscle fiber. The diagram depicts the constituents of a muscle fiber, which is composed of individual myofibrils. The sarcoplasmic reticulum (SR) has been described as a waterjacket that surrounds each myofibril. The SR is composed of two regions, the terminal cisternae and the longitudinal reticulum. The terminal cisternae, which cross the myofibrils, lie on both sides of the T-tubules. A triad is composed of two terminal cisternae and one T-tubule. The longitudinal reticulum of the SR runs parallel to the myofibrils, inserting into the terminal cisternae. Most of the CRCs are located in the terminal cisternae, whereas calcium (Ca)–adenosine triphosphatase (ATPase), the enzyme responsible for pumping Ca^{++} back into the SR, is mainly located in the longitudinal reticulum.

and to convert chemical energy into work. Each myosin molecule contains two heavy chains (HC) and four light chains (LC), all of which regulate the force–velocity characteristics of the fiber (Schiaffino and Reggiani 1996).

Relaxation from the force-generating state in the muscle fiber begins with termination of the motor nerve impulses and reductions in $[Ca^{++}]_f$ levels. The reduction in $[Ca^{++}]_f$ levels occurs as a consequence of both inhibition of Ca^{++} release from the CRCs and uptake of Ca^{++} back into the SR, where it is stored. As a result of the reduction in $[Ca^{++}]_f$ concentration, Ca^{++} is released from the troponin subunit, allowing the thin filament to resume its inhibiting position, thereby allowing actin and myosin to dissociate or assume a weak-binding, non-force-generating

state. As with the rapid increase in $[Ca^{++}]_f$ during activation, relaxation also occurs rapidly, given the extensive SR network and the density of the Ca^{++}–adenosine triphosphatase (ATPase), the protein responsible for pumping Ca^{++} into the SR.

Processes Involved in Energy Production and Utilization

The processes involved in contraction and relaxation in the muscle cell are dependent on energy. The free energy is provided by the hydrolysis of adenosine triphosphate (ATP), which is generated from the combustion of foods such as fat and carbohydrate. The majority of the ATP used in working muscle is used for three specific functions, namely for trans-

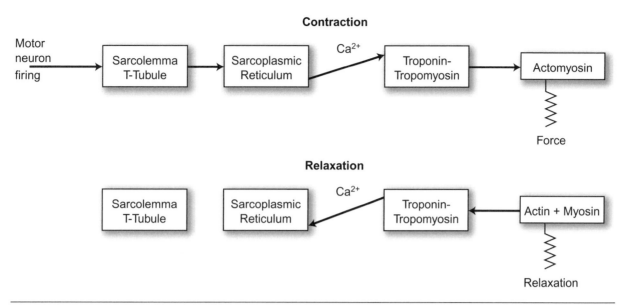

Figure 7.2 Excitation and contraction processes in muscle. The top panel illustrates the sequence of events involved in contraction beginning with the motoneuron impulses to the sarcolemma of the muscle fiber. The action potential is conducted to the interior of the fiber via the T-tubule, which is then transmitted to the calcium release channel of the sarcoplasmic reticulum. With activation, the channels open, resulting in an increase in $[Ca^{++}]_f$. The increase in $[Ca^{++}]_f$ results in Ca^{++} binding to a subunit of troponin, ultimately leading to movement of tropomyosin and removal of a preexisting inhibition between actin and myosin and contraction. Relaxation occurs when the motoneuron impulses stop, and the Ca^{++} is pumped back into the sarcoplasmic reticulum by Ca^{++}–ATPase, thereby lowering $[Ca^{++}]_f$ and reestablishing the inhibition between actin and myosin.

Reprinted, by permission, from H.J. Green, 2004, "Mechanism and management of fatigue in health and disease," *Can J. Physiol Pharmacol* 29(3): 264-273.

porting sodium ion (Na^+) and potassium ion (K^+) across the sarcolemma and T-tubule, for sequestering Ca^{++} into the SR, and for actomyosin cycling. The transport of Na^+ and K^+ is necessary to maintain membrane potential and excitability, allowing for the regeneration of action potentials (Clausen 2003). An action potential, which consists of both depolarization and repolarization of the membrane, begins with depolarization, which occurs when Na^+ channels open and allow Na^+ to flow from the outside of the cell, where the concentration is greatest, to the inside of the cell. Repolarization occurs when K^+ channels open, allowing K^+ to flow from inside of the cell down its concentration gradient to the outside of the cell. For action potentials to be generated at high frequencies, namely 40 to 50 impulses per second, which are necessary to raise $[Ca^{++}]_f$ to levels required to realize the full force potential of the cell, the transmembrane gradients for Na^+ and K^+ must be rapidly reestablished. This is accomplished via a specialized protein called Na^+–K^+–ATPase, which can both generate free energy from the hydrolysis of ATP and use the free energy to induce conformational changes in the protein and, in the process, transport Na^+ out

of the cell and K^+ into the cell. Na^+–K^+–ATPase is a membrane protein composed of α and β subunits that can regulate the rate of Na^+ and K^+ transport by changing the catalytic rate.

The transport of Ca^{2+} into the SR is also mediated by a specialized protein called Ca^{++}–ATPase. Ca^{++}–ATPase is similar to Na^+–K^+–ATPase in that the energy liberated by the hydrolysis of ATP causes a conformational change in the protein allowing it to pump $[Ca^{++}]_f$ back into the SR, where it can be stored in preparation for another contraction cycle. The regulation of $[Ca^{++}]_f$ cycling is dependent on both Ca^{++} release and Ca^{++} uptake.

In addition to Ca^{++}–ATPase and Na^+–K^+–ATPase, there is a third ATPase called actomyosin ATPase. The actomyosin ATPase uses the majority of energy in the working muscle cell. This ATPase forms part of the myosin HC component of the molecule. When activated, it provides the energy needed for actomyosin to cycle through a series of force-generating steps before dissociating to begin the next cycle. There are other ATPases in the cell serving other functions; however, the ATP requirements of the other ATPases, although important, are not pronounced.

Regeneration of ATP

To satisfy the needs of the ATPases, a large increase in ATP regeneration must occur. ATP is generated by an elaborate network of enzymes that form different metabolic pathways and segments (Hochachka 1994). These pathways are specialized to satisfy the energy requirements of the cell under special conditions or circumstances (figure 7.3). For example, if the requirements for ATP are relatively modest and if oxygen and other substrates are not limiting, the mitochondria, which contain the assembly of enzymes that form the citric acid cycle and the electron transport chain, can use either fat or carbohydrate (CHO) to generate ATP and, in the process, produce the by-products water and carbon dioxide. Energy generated by mitochondrial respiration is termed **oxidative phosphorylation.** Alternatively, if the force and power generated by the muscle are large and, consequently, the demand for energy is high, ATP can be rapidly generated by another metabolic pathway, namely **glycolysis.** In this pathway, ATP regeneration is restricted to the use of CHO. Because this pathway does not use oxygen, it can also become the primary short-term source of energy for the working muscle in oxygen-deprived environments such as at altitude. Although the recruitment of this pathway to meet the energy demands of the cell has several appealing features, the tradeoff is the final by-product that is produced, namely lactic acid. The accumulation of lactic acid, which is a strong acid, can cause several undesirable effects unless properly buffered and disposed of.

Substrates, either fats or CHO, are essential fuels for the metabolic pathways to respond to the energy needs of the cell. These substrates can be either stored in the cell as triglycerides or glycogen or delivered to the cell by the circulatory systems in the form of fatty acids and glucose. The blood fatty acids are liberated from the adipose tissue, whereas blood glucose is released from the liver.

The transport of substrates and metabolites in and out of the working muscle cell must be carefully regulated to protect the intracellular integrity of working muscle. In the case of substrates, a lack of availability can blunt ATP production depending on the metabolic pathway recruited. This is especially the case with CHO, which is in limited store in the liver and muscle. A depletion of this substrate from the liver, for example, can result in hypoglycemia and central nervous system dysfunction. It is also common for glycogen reserves to be essentially depleted in the working muscle. Depletion of muscle glycogen has

often been associated with an inability to sustain a desired force or power (fatigue).

For the muscle cell to perform the many functions necessary to support muscle contraction, it must be able to precisely regulate what enters and leaves the cell. A barrier to the free diffusion of various constituents exists in the form of the surrounding membrane, which by virtue of its phospholipid composition is essentially impermeable to most molecular species. The transmembrane movement of various constituents is facilitated by numerous proteins embedded in the phospholipid bilayer that form channels, transporters, and pumps. This allows for the highly specific regulation of numerous molecules. Such is the case with glucose and fatty acids, both of which have specially designed carrier proteins that facilitate their transport across the membrane (MacLean et al. 2000; Turcotte 2000). These proteins are expressed as families of fatty acid transporter proteins such as the fatty acid binding protein (FABP) and glucose transporters (GLUT). We have also seen that special channels exist for Na^+ and K^+ diffusion and a wide range of other cations and anions. Special proteins also exist for the transport of major metabolic by-products of glycolysis, namely lactate and hydrogen ions (H^+). Movement of lactate and H^+ across the plasma membrane is facilitated by MCTs (Juel and Halestrap 1999).

For muscle fibers to be able to contract and perform work, a balance must exist between those processes generating ATP and those processes using ATP. Moreover, this balance must be precisely regulated because, at rest, ATP exists at very low concentrations, sufficient for only a few contractions. Research has repeatedly shown that even with the most demanding tasks, ATP levels are extremely well protected, with reductions of greater than 25% rarely reported (figure 7.4 on page 106). The protection of ATP during periods of increased demand depends on a close integration between the catalytic activity of the ATPases, the enzymes that use ATP, and the flux rate in the metabolic pathways that produce ATP (Connett et al. 1990). According to current theory, the catalytic rate at which each of the ATPases perform increased work involving cation transport and actomyosin cycling increases the rate at which ATP is hydrolyzed. Stored levels of creatine phosphate (PCr) provide short-term protection of ATP, because the near-equilibrium nature of the enzyme creatine phosphokinase makes the hydrolysis of PCr very sensitive to decreases in ATP. Production of ATP can also be supplemented by the adenylate kinase reaction, also a near-equilibrium

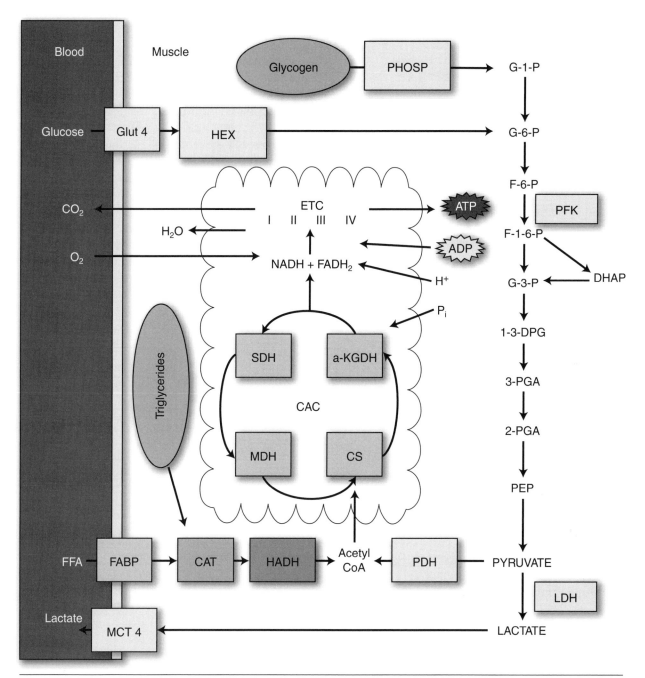

Figure 7.3 The different metabolic pathways with representative or rate-limiting enzymes that have been used to assess the potential of the pathway. Glycogenolysis, phosphorylase (PHOSP); glycolysis, phosphofructokinase (PFK); pyruvate oxidation, pyruvate dehydrogenase (PDH); glucose phosphorylation, hexokinase (HEX); oxidative potential, citrate synthase (CS), malate dehydrogenase (MDH); succinate dehydrogenase (SDH), election transport chain (ETC) fatty acid oxidation, β-hydroxacyl-CoA dehydrogenase (β-HADH); pyruvate reduction, lactate dehydrogenase (LDH). Also depicted are the transporters, namely the glucose transporter (GLUT4), the fatty acid transporter (FABP), and the lactate transporter (monocarboxylate transporter [MCT] 4). Carnitine acetyl transferase (CAT) is the enzyme involved in the transfer of fatty acids across the mitochondrial membrane.
From H.G. Green and T.A. Duhamel.

enzyme. However, rapid activation of glycolysis and oxidative phosphorylation is essential if ATP is to be protected beyond the first several contractions. The by-products of ATP hydrolysis, namely adenosine diphosphate (ADP) and inorganic phosphate (P_i),

in particular, are believed to be major stimuli in the recruitment of oxidative phosphorylation and rate-limiting enzymes of glycogenolysis and glycolysis, such as phosphorylase and phosphofructokinase, respectively. If the metabolic pathways are unable to respond with adequate production of ATP to meet demands, the activity of one or more of the ATPases is dampened, which in turn reduces ATP utilization. The dampening of the ATPases is believed to result from excessive accumulation of ADP and Pi as a consequence of reductions in ATP concentration (Allen et al. 2002; Sahlin et al. 1998). In addition, H$^+$, which accumulates as a metabolic by-product, and other by-products such as **reactive oxygen species (ROS)** and temperature are also potential inhibitory agents. Biochemists use the term **phosphorylation potential** to describe the concentration of high-energy compounds in the muscle. The status of the phosphate energy system, given the importance of its by-products in regulating other metabolic pathways, is

viewed as the principal integrator in coupling ATP demand to ATP supply.

An imbalance between ATP utilization and production limits the muscle mechanical response (Fitts 1994). Although fatigue, defined as an inability to produce a desired mechanical response, may be undesirable in the short term, it is protective in the long term, preventing the serious consequences associated with ATP depletion in the muscle cell. The need to protect ATP levels also offers some insight into the desired outcomes of training. A key question is, How can the phosphorylation potential of the cell, which is determined primarily by the levels of ATP and PCr, be protected while the muscle cell performs a greater amount of work following training?

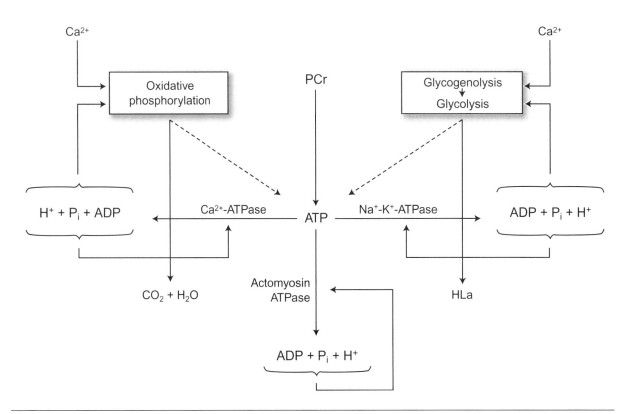

Figure 7.4 This schema depicts adenosine triphosphate (ATP) as the dependent variable whose concentration must remain protected. The hydrolysis of ATP by the different adenosine triphosphatases (ATPases) results in the production of metabolic by-products. The phosphorylation potential is sensed by the metabolic pathways (glycolysis, oxidative phosphorylation), which can stimulate increases in pathway flux rate and consequently increases in ATP production. Excessive accumulation of the by-products of ATP hydrolyses can inhibit the ATPases, resulting in reduced ATP utilization. Reductions in ATPase activity would be expected to result in fatigue. The terminal products for oxidative phosphorylation are carbon dioxide (CO_2) and H_2O, whereas lactic acid (HLa) is the terminal product of glycolysis.

Reprinted, by permission, from H.J. Green, 2000, Muscular factors in endurance. In *Encyclopedia of sports medicine: Endurance in sports,* edited by R. Shephard and P.O. Astrand (Oxford, UK: Blackwell Science Ltd.), 179.

For muscle cells to contract, the neural signal must be conducted into the interior of the fiber, ultimately enabling the contractile proteins, actin and myosin, to associate in a force-generating state. The viability of the individual processes involved in excitation and contraction depends on the activation of three enzymes, namely Na^+–K^+–ATPase, Ca^{++}–ATPase, and actomyosin ATPase. The activation of these enzymes results in rapid increases in ATP utilization. The products of ATP utilization increase the flux in the metabolic pathways, oxidative phosphorylation and glycolysis, which increase ATP production and defend ATP homeostasis. An inability of the metabolic pathways to produce sufficient ATP, as during heavy exercise, can result in progressive accumulation of metabolic by-products and can inhibit the ATPases. As a consequence, excessive reductions in ATP are prevented at the expense of fatigue.

Muscle Fiber Types and Subtypes

The vast array of motor skills that we are capable of performing, ranging from rapid velocities and large power outputs to sustained effort performed over many hours and even days, are possible because of differences in the structure and composition of the muscle cell. To achieve specialization in mechanical function, different fiber types and subtypes have evolved. Fiber type and subtypes display a coordinated set of properties, each supporting a particular function. For example, fibers adept at generating large velocities and power outputs must first be able to rapidly convert the neural command into increases in $[Ca^{++}]_f$ and then to quickly translate the Ca^{++} signal in rapid rates of cross-bridge attachment and detachment. Also crucial, given the rapid velocities, is the requirement for quick relaxation rates and consequently the ability to rapidly lower $[Ca^{++}]_f$ levels. Muscle fibers specialized for high velocities and high power outputs also have

unique energy needs. Metabolic pathways must be specialized for rapid activation and large flux rates to meet the demands for ATP generated by this type of work.

Fiber Types

Specialization at the level of distinct processes in the cell is primarily accomplished by the amount and type of protein and protein isoforms that exist and the regulatory factors that direct the response of these proteins (figure 7.5). A given protein can exist in several different molecular forms called an **isoform** or, in the case of enzymes, an isozyme. Isoforms generally display only minor differences in composition. However, these differences in composition allow varying degrees of specialization with regard to the mechanical events or functions in specific intracellular environments. To exploit the unique properties of the proteins and protein isoforms, differences in regulatory stimuli are needed. Regulatory stimuli can be localized in nature, generated specifically by events within the contracting cell or secondary to messages directed at the cell from external stimuli. Many of these external stimuli arise from changes in the level of blood hormones and neurotransmitters. These hormones act, in large part, on specialized receptors embedded in the surface membrane of the cell, triggering a cascade of responses that ultimately allow increased or decreased function of the targeted protein. **Signal transduction** processes are vital in activating the metabolic pathways and E–C processes during exercise.

The classification of muscle fibers into distinct types and subtypes is not without controversy given the many potential properties that can be used for classification. Indeed, many schema are currently in use, producing a confusing array of terminology. The most popular classification schema is based on the contractile protein, myosin, and specifically the HC component of the myosin that contains the ATPase (Reggiani et al. 2000). Each myosin molecule contains 2 HC and 4 LC. Myosin heavy chains contain a number of distinct isoforms (approximately nine in all) that are expressed in a species-specific and tissue-specific manner. In the case of human skeletal muscle, two major isoforms are recognized, namely HCI and HCII, which result in type I and type II fibers, respectively (table 7.1). In humans, HCII also exists

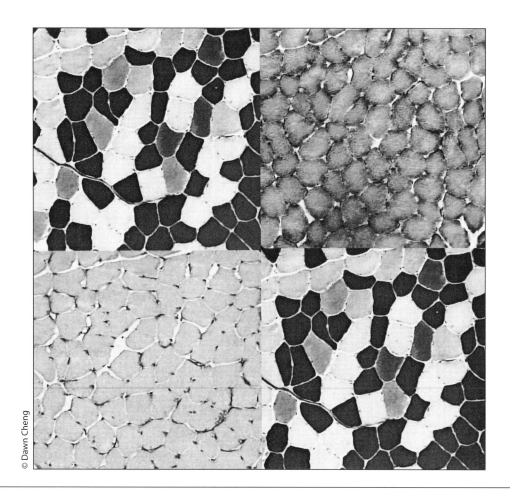

© Dawn Cheng

Figure 7.5 Cross-section of fibers obtained from a tissue sample extracted from the vastus lateralis. The upper left panel depicts the fiber types (type I, type IIA, and type IIX) as revealed by the different staining intensities. Type I fibers are the darkest staining, whereas type IIA and type IIx are light and intermediate, respectively. The lower left panel is a stain for the capillaries surrounding each fiber; the upper right panel features the oxidative potential as measured by succinate dehydrogenase activity. The lower right panel is the same as the upper left panel and is provided for comparison with the capillarization.

Table 7.1   Light and Heavy Chain Composition in Histochemically Identified Fiber Types in Humans

Fiber type	Light chains	Heavy chains
IIA	$(LC1f)_2 (LC2f)_2$	$(HCIIa)_2$
	$(LC1f) (LC3f) (LCS2f)_2$	$(HCIIa)_2$
	$LC(3f)_2 (LC2f)_2$	$(HCIIa)_2$
IIX	$(LC1F)_2 (LC2f)_2$	$(HCIIx)_2$
	$(LC1f) (LC3f) (LC2f)_2$	$(HCIIx)_2$
	$(LC3f)_2 (LC2f)_2$	$(HCIIx)_2$
I	$(LC1s)_2 (LC2s)_2$	$(HCI)_2$

Each myosin contains two heavy chains (HC) and four light chains (LC). In a given myosin, two of the LCs are essential (LC1, LC3) and two are regulatory or phosphorylatable (LC2). The HC isoforms include HCIIa, HCIIx, and HCI. Both fast (f) and slow (s) isoforms exist for the myosin LCs.

as two isoforms, namely HCIIa and HCIIx. These fibers are called type IIA and type IIX, respectively. It is also possible for various myosin HC isoforms to exist in a single fiber, which provides for additional subtypes such as type IIC, type IC, and type IIAX. Rodents contain a third HC isoform, namely HCIIb. Previously in humans, the HCIIx was erroneously labeled HCIIb. A myosin molecule also contains 2 classes of LCs, namely the essential LCs (LC1, LC3) and the regulatory LCs (LC2), each of which have slow and fast isoforms.

The use of the myosin HC for fiber type and subtype classification provides insight into the functional nature of the fiber. The myosin HC isoform, in conjunction with the LC isoform, which acts as a secondary modulator, is the primary determinant of the velocity and power that can be generated (figure 7.6). It is for this reason that

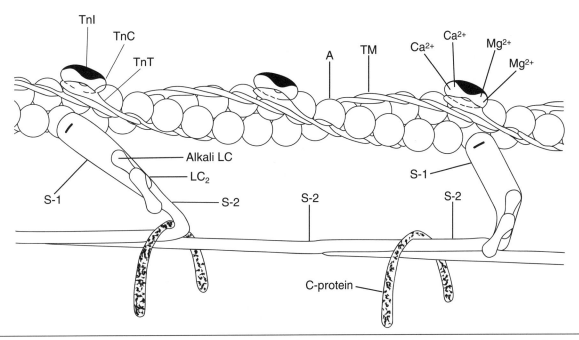

Figure 7.6 The thin filament consisting of the individual globular actin, the long-stranded tropomyosin (TM) and three troponin subunits (TnI, TnT, TnC), and the myosin molecule. The myosin contains the heavy chains, which consist of the S-1 and S-2 portions, and the different light chains (alkali or essential LC and LC2 or phosphorylatable LC).

Reprinted from R.L. Moss, G.M. Diffee and M.L. Greaser, 1995, "Contractile properties of skeletal muscle fibers in relation to myofibrillar protein isoforms," *Rev Physiol Biochem Pharmacol* 126: 1-63, with kind permission of Springer Science and Business Media.

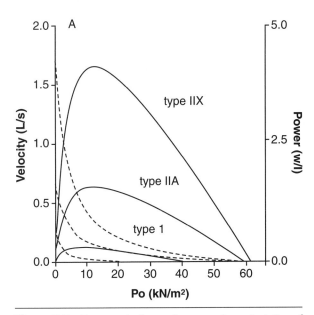

Figure 7.7 Force–velocity and power characteristics of selected fiber types. Velocity characteristics are represented by the dashed lines, whereas power (the product of force × velocity) is represented by the solid lines. Type 1, type IIA, and type IIx represent the different fiber types.

Reprinted from the *Journal of Electromyograph Kinesiology,* Vol. 9, R. Bottinelli, M.A. Pellegrino, M. Canepari, R. Rossi and C. Reggiani, "Specific contributions of various muscle fibre types to human muscle performance: An in vitro study," pp. 87-95, Copyright 1999, with permission from Elsevier.

the type I and type II fibers have also been labeled slow-twitch (ST) and fast-twitch (FT), respectively (figure 7.7).

Using the myosin HC isoform for classification highlights a coherent set of properties in the different fiber types. For example, type IIA and type IIX fibers also display a well-developed SR, with high densities of both the CRCs and Ca^{++}–ATPase proteins. This allows for both rapid increases and decreases in $[Ca^{++}]_{fi}$, which are necessary for quick contraction and relaxation cycles (figure 7.8). In contrast, type I muscles, which display the HC isoform, are incapable of rapid cross-bridge cycling and are consequently unable to generate large velocities and large power outputs. Expectedly, the SR in these fibers is not well developed. Differences in SR between fiber types also extend to a complex of regulatory proteins and isoform types (table 7.2 on page 111). In skeletal muscle, the Ca^{++}–ATPase exists as two sarcoplasmic–endoplasmic reticulum calcium ATPase (SERCA) isoforms. SERCA 1 is expressed in type II fibers, whereas SERCA 2 is expressed in type I fibers.

The potential of the various metabolic pathways is also generally consistent with the HC composition

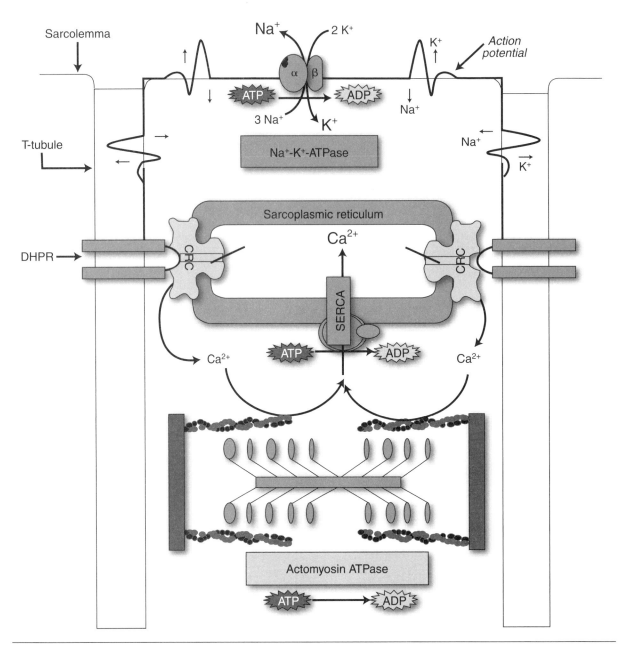

Figure 7.8 Essential elements of the Na⁺–K⁺–ATPase, Ca⁺⁺-release channel (CRC), and Ca⁺⁺–ATPase (sarcoplasmic–endoplasmic reticulum calcium ATPase, or SERCA) of the sarcoplasmic reticulum. The dihydropyridine receptor (DHPR), which is located in the T-tubules, is involved in communication between the T-tubules and CRC, as shown.
From H.G. Green and T.A. Duhamel

of the fiber (table 7.3). Fast-contracting fibers have high activities of enzymes that allow rapid rates of high-energy phosphate transfer, glycogenolysis and glycolysis. In contrast, slow-contracting type I fibers, which have relatively slow rates of ATP utilization as dictated by Ca⁺⁺–ATPase and actomyosin ATPase, do not have well-developed pathways of this nature. The **oxidative potential** of type I fibers is well developed (as might be expected, given their relatively low energy demands), providing for a high dependence on oxidative phosphorylation. Such is also the case with the type IIA fibers, which display both a high oxidative and a high glycolytic potential.

The specialization of the metabolic pathways to meet specific energy demands also extends to different isozymes. Both creatine phosphokinase and lactate dehydrogenase, for example, have multiple isoforms, allowing them to respond to unique challenges. Muscle fibers with a high oxidative potential are also liberally supplied with capillaries,

Table 7.2   Characteristics of Sarcoplasmic Reticulum in Human Fiber Types and Subtypes

	Fiber type		
	I	IIA	IIX
Ca^{++}–ATPase			
Protein	MOD	MOD–HI	HI
Isoform	SERCA 2	SERCA 1	SERCA 1
Maximal activity	MOD	MOD–HI	HI
Ca^{++} uptake	MOD	MOD–HI	HI
Ryanodine receptor (RyR)			
Isoform	RyR I	RyR I	RyR I
Ca^{++} release rate	MOD	MOD–HI	HI

Specific characteristics are based on those actually measured in human single fibers. Where no experimental data were available on humans, the characteristics were projected based on rodent muscle. Ca^{++} = calcium ion; ATPase = adenosine triphosphatase; MOD = moderate; HI = high; SERCA = sarcoplasmic–endoplasmic reticulum calcium ATPase. Ryanodine receptor (RyR) is also another term used for the calcium release channel.

Table 7.3   Potential of Metabolic Pathways and Segments in Human Fiber Types and Subtypes

	I	IIA	IIX
High-energy phosphate (CPK)	MOD	MOD–HI	HI
Glycogenolytic (PHOSPH)	MOD	MOD–HI	HI
Glycolytic (PFK)	MOD	MOD–HI	HI
Oxidative (CS, SDH)	HI	MOD	MOD–LO
β-oxidation (HADH)	HI	MOD	MOD–LO

Relative pathway potential is based on maximal activities of rate-limiting or representative enzymes. MOD = moderate; HI = high; LO = low; CPK = creatine phosphokinase; PHOSPH = phosphorylase; PFK = phosphofructokinase; CS = citrate synthase; SDH = succinic dehydrogenase; HADH = hydroxyacylCA dehydrogenase.

Table 7.4   Fiber Type Distribution, Area, and Capillarization in Human Male Adult Muscle

	Fiber type			
	I	IIA	IIX	IIAX
Distribution (%)	50	33	12	5
Area (μm^2)	4705	5305	4082	4175
Caps (n)	5.37	5.13	3.65	4.19
Caps/area (μm$^2 \cdot 10^{-3}$)	1.17	0.99	0.90	1.0

Fiber type characteristics are based on tissue samples extracted from the vastus lateralis of untrained male adults. Area = average area of each fiber type; Caps = average number of capillaries around each fiber type; Caps/area = average area of the fiber supplied by a capillary.

allowing delivery of essential oxygen to support mitochondrial respiration (table 7.4). Associated with the unique metabolic pathway structure in the different fiber types are the transporter proteins (table 7.5). GLUT4, MCT1, and FABP are highest in fibers with high oxidative potential, both type I and type IIA, whereas the lowest concentration is observed in type IIX. In contrast, MCT4 is higher in type IIX and type IIA compared with type I.

Somewhat at odds with the pattern of properties cited in defined fiber types is the Na$^+$–K$^+$–ATPase, the cation pump important in protecting membrane excitability (table 7.6). The density and isoform composition of the α and β subunit of this protein appear to be scaled to the oxidative potential of the cell and not to the speed of contraction of the cell. This is curious because type II fibers, both type IIA and type IIX, need a high frequency of impulses to

Table 7.5 Membrane Transporters and Human Muscle Fiber Types and Subtypes

	Fiber type		
	I	**IIA**	**IIX**
Glucose			
GLUT1	LO	LO	LO
GLUT4	HI	HI	LO
Lactate			
MCT1	HI	MOD	LO
MCT4	LO	HI	HI
Fatty acids			
FABP	HI	MOD	LO

Relative concentrations for humans projected based on rodent muscle fiber types. HI = high; MOD = moderate, LO = low; GLUT = insulin-sensitive glucose transporter; MCT = monocarboxylate transporter; FABP = fatty acid binding transporter.

generate a maximal tetanic contraction. In a rat, the isoforms α_1 and α_2 and β_1 are typically high in type I fibers, whereas the α_2 and β_2 predominate in type IIX fibers. Type IIA fibers contain a mixture of all isoforms.

Fiber Size

The final property that determines the ability of the muscle cell to perform work is the area of the fiber. To a large degree, the fiber area reflects the number of myofibrils and consequently the number of actin and myosin molecules available for force generation. Although the fiber types and subtypes show large differences in area in rodents, such is not the case in humans. Type IIX fibers, which are more specialized for large power outputs, are comparable in area to type I and type IIA fibers in humans. In general, type

IIX fibers, at least in the vastus lateralis, represent only 10% to 15% of the total fiber distribution in the untrained muscle.

Muscle fibers exist in various type and subtypes. A common property used to identify fiber types is the myosin ATPase, which is located in the HC component of the molecule. Different types of fibers contain different HC isoforms. The HC isoforms are the primary determinant of the velocity that the fiber can generate. Accordingly, type 1 fibers are slow contracting, whereas type II fibers are fast contracting. The contractile speed of the fiber also correlates with a variety of other properties. For example, type II fibers contain a well-developed sarcoplasmic reticulum, which allows for rapid regulation of $[Ca^{++}]_f$ and rapid force development and relaxation. In this type of fiber, the potential for high-energy phosphate transfer and glycolysis is high, allowing for large and rapid production of ATP. In contrast, type 1 or slow-contracting fibers have poorly developed sarcoplasmic reticulum and low high-energy phosphate transfer and glycolytic potentials. Oxidative potential is always high in type I fibers and may be high in type II fiber subtypes such as type IIA compared with type IIX.

Regular exercise changes the expression of the various proteins and protein isoforms and their regulatory mechanisms. We next examine the cellular adaptations that occur with different types of training stimuli and the functional consequences

Table 7.6 Characteristics of Na⁺–K⁺–ATPase in Human Fiber Types and Subtypes

	Fiber type		
	I	**IIA**	**IIX**
Protein	HI	HI	MOD
Isoform	$\alpha_1\beta_1 / \alpha_2\beta_1$	$\alpha_1\beta_1 / \alpha_1\beta_2$ $\alpha_2\beta_2 / \alpha_2\beta_2$	$\alpha_2\beta_2$
Maximal activity	HI	HI	MOD

Specific characteristics are based on those actually measured in human single fibers. Where no experimental data were available on humans, the characteristics were projected based on rodent muscle. Isoform distribution represents the dominant α and β subunit distribution. NA⁺ = sodium ion; K⁺ = potassium ion; ATP = adenosine triphosphatase; HI = high; MOD = moderate.

of the adaptations, with an emphasis on human muscle.

Muscle Adaptation and Functional Consequences

A distinguishing feature of skeletal muscle is its ability to adapt to altered patterns of contractile activity (Pette 2002). The extensive remodeling of the muscle cell that can occur extends to all levels of organization, including the morphologic, ultrastructural, biochemical, and molecular, which can all affect a wide range of processes. The specific changes that occur in cellular proteins, lipids, and carbohydrates are to a large extent determined by the challenges induced by the type of contractile activity, as modified by the environment in which the contractile activity is performed and the characteristics of the participant.

Studies investigating the muscular effects of regular exercise have mostly been performed over a relatively short time, normally not involving more than 10 to 12 weeks. The majority of these **acclimation** studies used three fundamentally different training models, each designed to stress different processes in the cell. The three models used are high-resistance exercise, consisting of brief periods of contraction involving high resistance or high power output; high-intensity, intermittent exercise; and prolonged low-intensity exercise. The majority of studies that have been published emphasized the changes that occur in the metabolic pathways and segments involved in ATP synthesis. It is only recently that studies examining the cellular adaptation to chronic activity have focused on the proteins and processes involved in excitation and contraction

processes. Further understanding of the adaptations that occur at this level is essential if we are to fully comprehend the mechanisms regulating ATP homeostasis in the trained state.

Prolonged Endurance Training (PET) Adaptations

Prolonged submaximal exercise training promotes a variety of cellular adaptations in support of oxidative phosphorylation (Holloszy and Booth 1976; Hood 2001). These adaptations are not surprising, given the dependency on aerobic-based ATP regeneration that occurs with this type of exercise (table 7.7). Among the most conspicuous changes are increases in the maximal activities of the enzymes of the citric acid cycle and the complexes of the respiratory chain, all of which occur secondary to increases in enzyme protein content. These adaptations increase oxidative potential or the maximal potential for oxidative phosphorylation. As a result, a greater rate of ATP synthesis is possible, secondary to increased oxygen utilization.

Substrate Utilization

The chronic adaptations to repetitive submaximal exercise also extend to the substrate used by the mitochondria for oxidative phosphorylation, namely fatty acids and glucose (table 7.8). Increases in facilitated diffusion of these substrates across the sarcolemma occur via increases in specific isoforms of their respective transport proteins, namely FABP and GLUT. Aerobic-based training elevates both the FABP and GLUT4, which allows fatty acids and glucose to be more readily transported from the blood to the inside of the muscle cell (Holloszy 2003; Turcotte 2000). In the case of fatty acids,

Table 7.7 Adaptations in Metabolic Pathway and Segment Potentials With Different Types of Training

	PET	HRT	HIIT
High-energy phosphate transfer	NC	NC–SI	LI
Glycogenolysis	NC	NC–SI	LI
Glycolysis	NC	NC–SI	LI
Oxidative	LI	NC	MI–LI
β-oxidation	LI	NC	NC
Glucose Phosphorylation	LI	NC	MI–LI

PET = prolonged endurance training; HRT = high-resistance training; HIIT = high-intensity intermittent training; NC = no change; SI = small increase; MI = moderate increase; LI = large increase.

Table 7.8 Adaptations in Membrane Transporters to Different Types of Training

	PET	HRT	HIIT
Glucose			
GLUT1	SI	NC	MI–LI
GLUT4	LI	NC–SI	LI
Lactate			
MCT1	SI	NC–SI	LI
MCT4	LI	NC–SI	LI
Fatty acids			
FABP	LI	NC	NC–SI

PET = prolonged endurance training; HRT = high-resistance training; HIIT = high-intensity intermittent training; GLUT = glucose transporter; MCT = monocarboxylate transporter; FABP = fatty acid binding transporter; NC = no change; SI = small increase; MI = moderate increase; LI = large increase.

increases in the maximal activities of the enzymes involved in transport, activation, and **β-oxidation** allow the fatty acids to be more readily available as a substrate for mitochondrial respiration.

For glucose to be metabolized in the cell, it must be phosphorylated. This is accomplished by hexokinase, whose maximal activity increases soon after the onset of training. The formation of glucose-6-phosphate allows this substrate to enter the glycolytic pathway, resulting in the formation of pyruvate. Pyruvate dehydrogenase occupies a central position in allowing pyruvate to be more readily available to the citric acid cycle. An additional hallmark of endurance training is an increase of the muscle substrates, both triglyceride and glycogen, which are derived from the fatty acids and glucose precursors, respectively. Additional muscular enzymatic adaptations allow ketone bodies such as β-hydroxybutyrate and acetoacetate, formed by the incomplete oxidation of fat by the liver, to be converted to forms that can also be used as mitochondrial substrates. Finally, adaptations in specific enzymes known as transaminases allow selected amino acids such as alanine and glutamine to be converted into intermediates ready for entry into the citric acid cycle.

Enzymes and Contractile Proteins

Additional adaptations that might be predicted with extensive aerobic training involve down-regulation of the maximal activities of the enzymes involved in high-energy phosphate transfer and in the glycogenolytic–glycolytic potential. Down-regulation of the potentials of these metabolic pathways in conjunction with the up-regulation that occurs in the oxidative potential would create

a metabolic organization strongly poised for oxidative phosphorylation similar to that observed for type I fibers. However, no training study, either with humans or animals, has been able to elicit a significant down-regulation of pathway potential for high-energy phosphate transfer and glycogenolysis–glycolysis. Such adaptations appear possible, however. Chronic low-frequency electrical stimulation to low oxidative type II–based muscles, which parallels submaximal exercise, when sustained for 12 to 24 hr a day over several weeks can induce a metabolic reorganization similar to that observed in a type I or ST fiber (Pette 2002). This finding suggests that voluntary exercise training programs have the potential to induce a response similar to that observed for chronic low-frequency electrical stimulation. However, the volume of voluntary training needed is probably beyond what can be accomplished.

Multiple cellular adaptations are also observed in the proteins involved in excitation and contraction processes in response to regular sustained submaximal exercise. At the level of the contractile protein, a myosin shift is observed in the isoform type from HCIIx to HCIIa, which results in an increase in the histochemically determined type IIA fibers at the expense of type IIX (Baldwin and Haddad 2002) (table 7.9). This is a relatively minor transformation, however, because muscles in untrained humans typically contain less than 15% of type IIx fibers. There is little consensus that training, regardless of type and particularly in humans, can significantly transform the myosin HC to slow from the fast isoform types, namely HCI from HCII, and consequently the percent distribution of type I or ST fibers. Endurance training, at least in rats, is also accompanied by changes in the myosin LCs and, in particular, a reduction in LC1f, LC2f, and LC3f, which occurs in association with an increase in LC1s and LC2s. In rats, these changes are essentially confined to muscles that contain an abundance of type II or fast-twitch fibers. In humans, it has been difficult to isolate the fiber type population affected because the locomotor muscles, such as the vastus lateralis, which have been the most frequently examined, contain a mixture of fiber types and because measurements are usually performed on mixed tissue samples extracted by biopsy and not at the single-fiber level. Low-intensity training can also induce a small hypertrophy in both the type I and type II populations as measured by the cross-sectional area of the fibers (table 7.10). The increase in cross-sectional area appears to be attributable to the increase in the myofibrillar content, the proteins that dominate the intracellular composition of the fiber.

Na$^+$–K$^+$–ATPase

Sustained repetitive activity places strain on the sarcolemma and T-tubule systems given the need to generate action potentials to conduct the neural signal to the interior of the fiber. Protection of membrane excitability, which is necessary for repeated transmission of the action potential, is dependent on Na$^+$–K$^+$–ATPase, the enzyme responsible for restoring transmembrane gradients for Na$^+$ and K$^+$. Regular exercise is a potent stimulus for the up-regulation of both the α and β subunits of Na$^+$–K$^+$–ATPase (Clausen 2003) (table 7.11). The increase in these subunits is also accompanied by increases in the maximal catalytic activity (V_{max}) of the enzyme, an adaptation that should increase the maximal transport rate of the cations. Current evidence, albeit limited, indicates that the α_1, α_2, and β_1 isoforms are up-regulated with short-term training. However, the changes that occur in specific

Table 7.9   Adaptations in Myosin Composition to Specific Types of Training

	PET	HRT	HIIT
Fiber type			
Type I	NC	NC	NC
Type IIA	SI	SI	SI
Type IIX	LD	LD	LD
Myosin HC			
HCI	NC	NC	NC
HCIIa	SI	SI	SI
HCIIx	LD	LD	LD

PET = prolonged endurance training, HRT = high-resistance training; HIIT = high-intensity intermittent training; HC = heavy chain; NC = no change; SI = small increase; LD = large decrease.

Data from K.M. Baldwin and F. Haddad, 2002, "Skeletal muscle plasticity: Cellular and molecular responses to altered physical activity paradigms," *Am J Physical Medicine and Rehabilitation* 81(Suppl): S40-51.

Table 7.10   Adaptations in Fiber Area and Capillarization to Different Types of Training

	PET	HRT	HIIT
Area			
Type I	SI	MI	MI
Type IIA	SI	MI	MI
Type IIX	SI	MI	MI
Capillaries, area			
Type I	LI	NC	SI–MI
Type IIA	LI	NC	SI–MI

PET = prolonged endurance training; HRT = high-resistance training; HIIT = high-intensity intermittent training; NC = no change; SI = small increase; MI = moderate increase.

Table 7.11   Adaptations in Na⁺–K⁺–ATPase Characteristics to Specific Types of Training

	Fiber types		
	PET	**HRT**	**HIIT**
Maximal activity	LI	NC–SI	LI
Content	LI	NC-SI	LI
Isoforms			
α_1	LI	NC–SI	LI
α_2	LI	NC–SI	LI
β_1	LI	NC–SI	LI
β_3	SI	NC	SI

Na⁺ = sodium ion; K⁺ = potassium ion; ATPase = adenosine triphosphatase; PET = prolonged endurance training; HRT = high-resistance training; HIIT = high-intensity intermittent training; NC = no change; SI = small increase; LI = large increase.

Data from T. Clausen, 2003, "Na+-K+ pump regulation and skeletal muscle contractility," *Physiological Reviews* 83: 1269-1324.

isoforms with extended submaximal training remain unclear. Moreover, it is also uncertain which α and β isoform combinations are responsible for the increases in V_{max} that are observed. It is possible, depending on the isoform population adapted with this type of training, that the sensitivity to K⁺ or Na⁺ could also be affected. Evidence indicates that training improves regulation of Na⁺ and K⁺ transport across the cell membrane and protects membrane excitability during sustained, low-intensity contractions. Although not yet studied, adaptations might also occur in the many channels and transporters involved in the regulation of Na⁺, K⁺, and Cl⁻, the principle channels involved in depolarization and repolarization of the membrane.

The precise regulation of intracellular $[Ca^{++}]_f$ is a prerequisite for purposeful mechanical responses in the cell as well as a variety of other functions (Berchtold et al. 2000). If mechanical performance is to remain protected in the face of repetitive activation, the SR must be able to preserve Ca⁺⁺ release and Ca⁺⁺ uptake functions of the SR to maintain $[Ca^{2+}]_f$ levels and, consequently, the behavior of the myofibrillar proteins. Aerobic-based training decreases the rates of both Ca⁺⁺ release and Ca⁺⁺ uptake (Green 2000) (table 7.12). The decrease in both of these functions appears to be attributable to decreases in the density of CRC, the density of the Ca⁺⁺–ATPase enzymes, and specifically the SERCA 1 isoform. In the case of the Ca⁺⁺–ATPase, the lower protein content appears responsible for the decrease in the enzyme V_{max} that appears to occur in the absence of changes in Ca⁺⁺ sensitivity. Interestingly, the changes that result in Ca⁺⁺ uptake and Ca⁺⁺ release with training are consistent with a decrease in Ca⁺⁺ cycling with each contraction,

while possibly protecting the $[Ca^{++}]_f$ transient level. The reduction in the Ca⁺⁺ cycling potential may be beneficial given the low intensity of aerobic-based training programs. There is also evidence that the Ca⁺⁺ uptake rate is better defended during exercise following training, given the structural damage that occurs to the enzyme in the untrained state.

High-Resistance Training Adaptations

Other training models designed to stress different elements of cellular function can also markedly alter the structure, composition, and function of muscle. Particularly intriguing are the adaptations that result from brief maximal or near-maximal contractions performed for relatively few repetitions per session. This type of resistance training, often classified as strength or power training, leads to a pronounced increase in muscle mass, reflected by increases in the cross-sectional area of both type I and type II fibers. The increase in protein synthesis mediated by high-resistance training (HRT) is believed to be primarily restricted to the regulatory and contractile proteins, increasing the number and size of myofibrils. As with prolonged submaximal exercise training in humans, HRT results in fiber type transformation that is restricted to the fast-twitch subtypes, where an increase in the percentage of type IIA fibers occurs in conjunction with a decrease in the percentage of type IIX fibers. These changes reflect a transformation of HCIIX toward HCIIA. Frequently, increases in Type IC and Type IIC fibers can also be identified, which indicate the presence of more than one myosin HC type in a fiber.

High-resistance training has minimal effects on the organization of the energy metabolic pathways

Table 7.12   Adaptations in Sarcoplasmic Reticulum to Specific Types of Training

	Training type		
	PET	**HRT**	**HIIT**
Ca^{++}–ATPase activity	MD	NC–SI	MI
Ca^{++} uptake	MD	NC–SI	MI
Isoforms			
SERCA 1	MD	NC	MI
SERCA 2	MI	NC	MD
Ca^{++} release	MD	NC–SI	MI
RyR content	MD	NC–SI	MI
Isoform			
RyR 1	MD	NC–SI	MI

PET = prolonged endurance training; HRT = high-resistance training; HIIT = high-intensity intermittent training; Ca^{++} = calcium ion; ATPase = adenosine triphosphatase; SERCA = sarcoplasmic–endoplasmic reticulum calcium; RyR = ryanodine receptor; NC = no change; SI = small increase; MI = moderate increase; MD = moderate decrease.

Data from H.J. Green, 2000, Muscular factors in endurance. In *Encyclopedia of sports medicine. Endurance in sports,* edited by R. Shephard and P.O. Astrand (Oxford, UK: Blackwell Science Ltd.), 156-163.

and segments in the cell. Most studies report no changes in maximal activities of representative enzymes involved in high-energy phosphate transfer, glycogenolysis, glycolysis, or oxidative phosphorylation. Although there is some suggestion of decreases in capillaries per muscle fiber area with HRT given the increase in fiber cross-sectional area that occurs with resistance training, current evidence indicates that, at least with moderate resistance training, increases in capillarization can offset the approximate 15% to 20% increase in cross-sectional area of the fibers that would be expected. Resistance training protocols vary widely and consequently evoke different neural and muscular challenges; not all protocols would produce the same adaptations. Few reports have addressed the effect of HRT on the membrane transporters and the excitation–contraction coupling processes. The current belief is that HRT has little impact on these properties.

High-Intensity Intermittent Training Adaptations

Another training strategy that emphasizes the extremes of physiologic function and consequently the unique capacity of the muscle to remodel is high-intensity intermittent training (HIIT). This type of exercise includes elements of high-resistance exercise but the frequency of repetitions is increased in conjunction with the number of work bouts. In contrast to HRT, where the brief duration of each

contraction and the small number of repetitions primarily depend on stored ATP and PCr, HIIT demands a major involvement of high-energy phosphate transfer and glycogenolytic–glycolytic systems to rapidly supply large quantities of ATP to meet the energy needs of the different cellular ATPases. Given the relatively high force levels generated and the sustained nature of each work bout, the capacity of the excitation and contraction processes for signal transmission and for force development and relaxation is also challenged. Unique to HIIT is the need to maintain viability of the excitation and contraction processes given the limited reserve remaining and the accumulation of a range of metabolic by-products, which could inhibit one or more of the processes.

HIIT can result in a considerable reconfiguration of the energy metabolic machinery. Unlike aerobic-based training, HIIT can increase the maximal activities of rate-limiting enzymes involved in high-energy phosphate transfer, glycogenolysis, and glycolysis. Accompanying the increase in these anaerobic pathways is an increase both in the maximal activity of hexokinase, the enzyme used to phosphorylate glucose, and in the glucose transporter GLUT4, the protein used to facilitate the transport of glucose across the sarcolemma. Depending on the specifics of the training protocol, this type of training may also increase oxidative potential and the ability to increase the rate of ATP production by mitochondrial respiration. Adaptations also occur at the level of glycogen synthase, the enzyme used to increase endogenous glycogen deposition in the cell. High-intensity training can markedly elevate glycogen deposition.

As expected, training of this nature leads to significant increases in the cross-sectional area of all fiber types. As with previously discussed training strategies, fiber type transformation appears to be primarily restricted to the transformation from type IIX to type IIA. Training of this nature generally also increases capillary number surrounding each fiber, leading to increases in capillary to fiber area ratio and increases in diffusion potential.

Increases in Na^+–K^+–ATPase protein expression are also observed with high-intensity training, which probably increases the Vmax of the enzyme. The increases extend to both the α and β subunits, with the specific isoforms affected as yet unclear. The increase in Vmax also appears to support the requirement for increased Na^+ and K^+ transport to protect membrane excitability, given the high frequency of action potential required to perform this type of exercise. Studies have shown that the loss of K^+ from the working muscle to the blood is reduced following this type of training.

The sarcoplasmic reticulum also displays adaptations consistent with the strain imposed on Ca^{++} cycling behavior. An up-regulation in both the CRC and the Ca^{++}–ATPase protein density has been reported in response to sprint-type training. The increase in the CRC has been correlated with a more rapid and larger release of Ca^{++} during activation. Increases in Ca^{++}–ATPase protein, both SERCA 1 and SERCA 2, have also been reported. However, unexpectedly, increases in maximal Ca^{++}–ATPase activity and Ca^{++} uptake were not observed. The failure to find increases in Ca^{++}-sequestering properties raises the possibility that the enzyme was structurally damaged or that regulatory signals controlling enzyme activation were deficient.

Additional adaptations relate to the management of the metabolic by-products that are generated by HIIT. One such by-product of major significance is lactic acid, which as a strong acid, dissociates into H^+ and lactate. Pronounced increases in the lactate transporters MCT1 and MCT4 have been observed with high-intensity training, which would be expected to facilitate the transport of lactate and H^+ both out of the cell (MCT4) and into it (MCT1). This may be an adaptation where the isoform affected is specific to a given fiber type. Several studies have reported increases in the buffer capacity of muscle after training, which could assist in the management of H^+ and, consequently, cellular acidity.

Functional Consequences of Training Adaptations

During prolonged submaximal exercise following training, compared with similar exercise before training, the phosphorylation potential of the cell is better protected as observed by a more preserved concentration of ATP and PCr, two high-energy phosphate compounds (figure 7.9). In effect, the

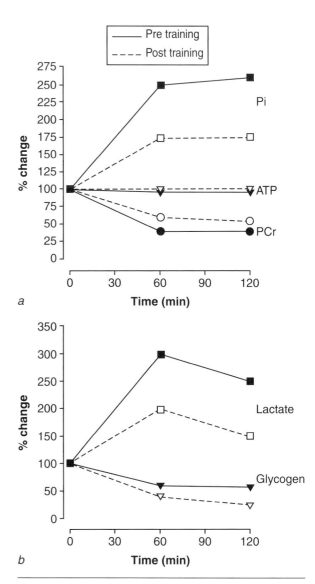

Figure 7.9 Percent changes in the high-energy phosphates (adenosine triphosphate [ATP], creatine phosphate [PCr]), and the related metabolites inorganic phosphate Pi *(a)*, lactate, and glycogen *(b)* during submaximal exercise before and after training. Percent changes are based on measurements performed prior to exercise (0 min) and at 60 min or 120 min of exercise. Although not included in the graph, reduction with training also occur in free adenosine diphosphate or free adenosine monophosphate.

energy state or phosphorylation potential of the contracting cell has been increased. As expected, there is less accumulation of the metabolic byproducts P_i, ADP, and lactic acid. Time-dependent increases in the oxidation of fatty acids and decreases in CHO oxidation are also well-characterized effects of this type of training (Brooks 1998). The decrease in CHO oxidation occurs as a result of decreases in the utilization of both blood glucose and endogenous glycogen stored in the muscle cell. It is also generally believed that decreases in glycolysis and lactate production accompany this type of training and, in conjunction with increased removal of lactate from the cell, result in a lower concentration.

The reduced utilization of CHO, both blood glucose and endogenous glycogen, has important functional consequences. Decreases in blood glucose utilization protect blood glucose levels during exercise, a substrate that is in limited supply and that is a vital fuel for central nervous system function. The improved preservation of muscle glycogen levels during prolonged exercise frequently has been linked to improved endurance by mechanisms that are as yet unclear. Developing evidence suggests that CHO may be an essential substrate for the function of both Na^+–K^+–ATPase and Ca^{++}–ATPase, the two enzymes involved in cation regulation and excitation–contraction (E–C) coupling. A failure in either of these cation pumps could impair impulse transduction in the cell and cause an inability to develop and sustain force production. Reduction in the use of CHO during exercise and, in particular, blood glucose also dampens the strategies invoked to defend glucose levels from approaching hypoglycemic levels. Acute increases in liver glycogenolysis and gluconeogenesis in conjunction with the promotion of increased fatty acid utilization by the working muscle are under elaborate hormonal regulation involving, in part, insulin, glucagon, and the catecholamines epinephrine and norepinephrine. The higher CHO levels observed during exercise after training improve homeostasis and decrease the need for hormonal support. This places less strain on the endocrine systems.

The adaptation that occurs in the E–C coupling processes with prolonged endurance training may also confer an important functional benefit. Unlike the enzymes of metabolic pathways involved in ATP synthesis, whose maximal activity appears to be preserved during prolonged exercise, the maximal catalytic activities of both Na^+–K^+–ATPase and Ca^{++}–ATPase are depressed during such exercise. The depression would be expected to reduce the rate of cation transport, particularly at high exercise intensities. In the case of Na^+–K^+–ATPase, training results in an overcompensation, which up-regulates the α and β subunit protein content, and the maximal catalytic activity could allow more precise and sensitive regulation of the enzyme according to the demands of the exercise. This adaptation not only would improve protection of Na^+ and K^+ transmembrane gradients and membrane excitability during exercise but would minimize the loss of K^+ from the muscle.

In the case of the SR, the **stress** imposed by prolonged submaximal exercise training results in a down-regulation of Ca^{++}–ATPase activity and Ca^{++} uptake. Because this type of training also decreases CRC content and Ca^{++}-release kinetics, Ca^{++} cycling rate is reduced, a seemingly appropriate adaptation given the relatively low force levels generated by this type of training. In the absence of other regulatory changes, the decrease in cycling rate could decrease the energy costs of SR Ca^{++} cycling and, in particular, Ca^{++} uptake. There is some evidence that this type of training protects the CRC and the Ca^{++}–ATPase proteins against the structural damage induced by repetitive exercise. The structural damage is believed to result from oxidation and nitrosylation processes, generated by reactive oxygen species (ROS). It is generally believed that although training cannot reduce the ROS produced from oxidative phosphorylation and purine nucleotide metabolism, which are the primary sources of production during exercise, training can more effectively scavenge ROS via adaptation of the antioxidant enzymes.

A potentially important factor in the adaptations that occur in skeletal muscle during aerobic-based training programs is the amount of muscle mass activated. In principle, the same adaptations should occur with small versus large muscle group protocols, providing the mechanical history of the muscle of concern is comparable. However, large muscle group activity places a greater strain on the respiratory and cardiovascular systems, potentially limiting blood flow and oxygen delivery to the working muscles. Regular aerobic exercise typically increases $\dot{V}O_2$max, also referred to as **peak aerobic power,** 10% to 20%. At low to moderate exercise intensities, no differences are generally observed in $\dot{V}O_2$ during steady state early in the exercise. This would imply that the net mechanical efficiency has not changed. At higher exercise intensities, however, $\dot{V}O_2$ can increase as a result of increases in $\dot{V}O_2$max, resulting in a higher level of oxidative phosphorylation, thereby reducing the contribution of glycolysis to ATP homeostasis.

With increases in $\dot{V}O_2max$, increases in $\dot{V}O_2$ kinetics also occur during the non-steady-state adjustment to exercise. Indeed, recent research has demonstrated that $\dot{V}O_2$ kinetics during the non–steady state may also occur early in training before increases in $\dot{V}O_2max$ and muscle oxidative potential. The consequence of these adaptations is that the muscle metabolic behavior observed during submaximal exercise resembles that of the trained state, with improved concentrations of ATP and PCr and less accumulation of metabolic by-products such as P_i, ADP, and lactate. Because at steady state the $\dot{V}O_2$ remains unchanged and because the phosphorylation state is believed to be of major importance in the recruitment of mitochondrial respiration, the adaptation has been interpreted as an increase in mitochondrial sensitivity. By mechanisms that are as yet unknown, training also results in less upward drift in $\dot{V}O_2$ as the exercise is prolonged.

Benefits of High-Resistance Exercise Programs

High-resistance training programs offer the obvious benefit of increasing the force and power that can be generated by muscle. The increases in force and power result primarily from increases in the cross-sectional area of the fiber, because the transformations in the myosin isoforms with training are relatively modest. The increases in force and power permit the performance of a wider complexity of tasks, many of which may be essential to daily living and quality of life. The increases in maximal force and power that occur after high-resistance training may also have additional benefits for the performance of prolonged submaximal exercise. Evidence exists that a traininglike effect may occur in the metabolic behavior of the working muscle as indicated by the improved energy homeostasis and reduced by-product accumulation. This could occur as a result of the increase in cross-sectional area of the muscle, allowing the same absolute force to be generated after training compared with before training, by reducing the firing frequency to the muscle or reducing the number of motor units that need to be activated.

The adaptations that occur with high-intensity intermittent training are consistent with those needed for burst-type tasks characterized by high mechanical power outputs and high ATP turnover rates. The increase in the cross-sectional area of the muscle fibers in conjunction with the increased capability of sarcolemma and T-tubule Na^+–K^+ exchange and SR Ca^{++} cycling provides for increased rates of signal transmission to the myofibrillar proteins and increases in the number and rate of actomyosin cycling and force generation. The energy needed for this type of adaptation is provided by the increase in potential in the high-energy phosphate transfer reactions and glycogenolysis–glycolysis. Increases in cellular glycogen and in lactate transport and buffering mechanisms provide the fuel and tolerance to by-product accumulation necessary to sustain this type of activity for longer periods following training. High-intensity training protocols can also be designed to benefit cellular aerobic metabolism, given the increase in mitochondrial potential and capillarization that can occur.

Skeletal muscles are capable of extensive adaptations in response to contractile activity if appropriately challenged. The adaptations to different proteins and processes that occur within the cell are highly dependent on the type of exercise and the demands imposed on excitation–contraction and metabolic pathways involved in energy production. Three different types of training programs have been selected to illustrate the different types of muscle adaptations that can result, namely prolonged endurance training (PET), high-resistance training (HRT), and high-intensity intermittent training (HIIT). The best characterized type of training is PET. This type of training, which primarily depends on energy supplied by aerobic-based metabolism, increases oxidative and β-oxidative potential, muscle fiber capillarization, and the proteins involved in glucose, fatty acid, and lactate transport. This type of training is also accompanied by an up-regulation of Na^+–K^+–ATPase and a down-regulation of the sarcoplasmic reticulum Ca^{++} cycling potential. Collectively, these adaptations increase aerobic capacity while improving phosphorylation potential, causing a greater dependency on fat oxidation and providing better management of metabolic by-products.

Health Benefits

Yet to be addressed is how the muscular adaptations resulting from different training programs can contribute to health and well-being throughout the life span. As emphasized, the adaptations that occur can allow a person to reclaim a more active lifestyle, permitting her or him to independently perform a greater diversity of the tasks of daily living. The adaptations can also have beneficial effects both at work and at leisure by increasing productivity and by allowing a person greater opportunity to interact with the environment. In effect, he or she can not only enjoy the benefits of greater task diversity but also experience less fatigue in the process. The reduction in strain that occurs in the different cellular processes as a result of training adaptations modulates the activation of neural and hormonal systems necessary to meet the various challenges of activity. For example, evidence demonstrates that feedback signals from the working muscles are reduced, which, in turn, result in less activation of hormonal, respiratory, and cardiovascular systems. The reduction in feedback is thought to occur as a result of the reduction in the accumulation of select metabolic by-products.

Some adaptations also have direct application to disease prevention. The depletion of muscle glycogen stores with exercise provides a sink for glucose disposal following meals. The increase in glucose transporter expression and glucose phosphorylation potential in adapted muscle facilitates the transfer and utilization of glucose, lessening insulin resistance and reducing the risk of type II diabetes.

Muscle contractile activity can also have a variety of indirect effects. Repetitive muscle contraction, particularly with large muscle groups, recruits a variety of support systems to satisfy the requirements of the working muscles. The systematic recruitment of these systems, in turn, improves their behavior, reducing strain. Muscle activity also provides the vehicle for better bone health and improved immune function, all of which may delay or prevent a number of diseases.

The question for all who want to include exercise in their daily routine is the form that it should take. The general practice has been to engage in formal training, designed to optimize benefits in the minimal amount of time. An alternative and perhaps more beneficial strategy is to add exercise to our lives by performing diverse physical tasks every day.

Assuming these tasks are of adequate volume and intensity, such an approach could promote health without formalized training.

Aging Muscle—The Role of Training

Aging is accompanied by profound changes in skeletal muscle and performance. Aging results in pronounced reductions in muscle force, velocity, and power and substantial reductions in muscle mass and the intrinsic properties of the muscle fibers. For most people, the advancing years are also associated with large reductions in physical activity. These interrelationships suggest that the performance decrements observed with age can be mechanistically linked to alterations in muscle and that inactivity and disuse may be part of the etiology. Alternatively, the reduction in regular activity with age might occur as a consequence of the muscular changes. Given the significance of these issues, they are a dominant interest in gerontology. Several impressive reviews have been written in this area (Carmelli et al. 2002; Doherty 2003; Margreth et al. 1999; Vandenvoort 2002). These reviews indicate that much research remains to be done.

Changes in Muscle With Aging

It is widely accepted that with aging the cross-sectional area of muscle cell decreases and the type II or fast-twitch fibers experience the greatest loss. There is also evidence that in the advancing years, the proportion of type I fibers increases and the proportion of type II fibers decreases. The shift in fiber type proportions may reflect not fiber type transformation but rather a loss of neurons and type II motor units. Although these changes would undoubtedly compromise the mechanical function of muscle, it is not known whether this can explain the full age-associated decline observed. Intrinsic cellular changes are also known to occur with aging; researchers typically refer to these as changes in muscle quality.

The E–C coupling processes are one area where intrinsic changes in muscle are observed in animals. In muscle from a variety of species, age-dependent declines are observed in Na^+–K^+–ATPase pump content and in maximum capacity for cation transport across the sarcolemma. Similar declines with age, at

Intelligently planned and consistently practiced resistance training can significantly slow the drastic loss of strength and coordination that is associated with aging in sedentary societies.

There is also evidence that the coupling between the T-tubule and CRC may be altered in older muscles, given the reductions in DHPR and the reductions in the ratio of DHPR to CRC that have been observed. The loss of DHPR could reduce the direct coupling between DHPR and CRC and the Ca^{++}-induced regulation of adjacent CRC. There is also a suggestion that one or more of the regulatory proteins in T-tubule–CRC coupling are altered with age.

There is a consensus that the intrinsic changes in the proteins involved in E–C occur as a result of oxidation–nitrosylation reactions, which in turn result from ROS. Many of the proteins involved in muscle E–C processes are very prone to free radical–mediated damage. Free radical–mediated damage is also believed to be responsible for the structural damage that occurs to the cation pumps and the CRC with acute exercise.

Age-related changes in muscles also occur in the metabolic pathways involved in ATP production (Carmelli et al. 2002; Short and Nair 2001). The process most affected appears to be the aerobic process, given the reduction in maximal activity of a number of mitochondrial enzymes in both the citric acid cycle and the electron transport chain. The maximal activities of many of the glycolytic enzymes also appear to be depressed by aging. Interestingly, even with the reduction in the muscle fiber cross-sectional area that is typically observed in older individuals, the capillary to fiber area ratio is also reduced. These changes suggest that coordinated reductions in blood flow may accompany the reduction in mitochondrial function.

Finding the Causes of Muscle Changes With Age

Because many of the muscle changes observed with age are similar to those observed with inactivity or disease, it has been speculated that activity level and not age is primarily responsible for the change. To some extent this may be true. Older individuals respond to aerobic-based training programs with increases in oxidative potential and in capillary to fiber area ratios, adaptations that are consistent with improved aerobic function. Older individuals also display increases in $\dot{V}O_2max$ after training, suggesting linked adaptations in both central and peripheral processes.

least in pump content, have not been documented in human muscle. However, decreases in maximal enzyme activity and in maximal cation transport have been shown to occur with age in humans. These findings suggest that intrinsic structural alterations in the Na^+–K^+–ATPase enzyme itself or in one or more of the complex regulatory factors are modified with age. The expected effect of these changes would be to impair membrane excitability, which could result in loss of Na^+ and K^+ homeostasis and early fatigue.

Sarcoplasmic reticulum Ca^{++} cycling also appears to be affected by age, because both Ca^{++} uptake and Ca^{++} release are decreased in older individuals. As with Na^+–K^+–ATPase, these decreases appear to be intrinsic because they are not accompanied by loss of either Ca^{++}–ATPase or CRC protein content.

Not expectedly, high-resistance training also increases fiber cross-sectional area and the force and power that can be generated by the trained muscle. There is conflicting evidence regarding the effect of regular exercise, regardless of protocol, on the proteins and processes involved in excitation and contraction. Few studies have been performed, and most that have been performed involved nonhuman species. If, as suggested, aging-associated changes are attributable to intrinsic changes in the cell mediated by oxidation–nitrosylation, training adaptations may revolve around management of ROS with regard to pathways involved in both production and scavenging. Proper exercise prescription may be essential to achieve desired benefits, because inactivity apparently is associated with increased ROS concentration, as are excessively demanding protocols of regular exercise.

There is little question that regular exercise can benefit the aging individual as assessed by improvements in both mechanical and aerobic capacities. Such improvements not only would permit a greater number of daily tasks to be performed but also would effectively reduce the strain associated with many of the activities of daily living. The essential experimental question remains, What mechanisms underlie the improved physical capabilities? It is encouraging that recent evidence indicates that the beneficial effects of regular exercise and proper nutrition on health and well-being can extend well into the advancing years.

Summary

Skeletal muscle occupies a central role in both the prevention and the management of a variety of disease states. Muscles must be primarily viewed as a machine, specialized for translating chemical energy into physical movement. As a machine, muscles have a number of mass and fiber-type properties that allow us to perform a vast array of motor tasks. However, this unique capability comes at a price. Muscles must be regularly challenged to contract or they lose their ability to function optimally. Inactive or sedentary muscles also affect a variety of other tissues and systems, compromising their functional potential and predisposing them to disease. Muscles are capable of extensive adaptations at all levels of organization. The nature of adaptations depends in large part on the type of contractile activity performed and the challenges imposed on the excitation and contraction process, which uses energy and metabolic pathways that supply the energy. This chapter describes the composition, structure, and functional properties of these different components of the muscle cell and the manner in which they adapt to different training stimuli. Insight is provided into how these different adaptations affect the performance of different motor tasks and perhaps, more important, how they defend against disease. An important feature of muscles is that they never lose their ability to adapt, even in advancing age.

KEY CONCEPTS

acclimation: Adaptation that occurs in response to a single factor, for example, the changes occurring in response to a short-term formalized training program.

acclimatization: Adaptations that occur to the natural environment.

adaptation: Process of modification or adjustment to a changing environmental element to minimize the strain needed to maintain homeostasis.

β-oxidation: Successive degradation of fatty acids that are attached to coenzyme A to acetyl coenzyme A, which can then be used by the mitochondria for oxidative phosphorylation.

excitation–contraction (E–C) coupling: Sequence of events starting with the spread of the action potential in the sarcolemma and culminating with the increase in free cytosolic Ca^{++} levels.

glycolysis: For definition, see page 65.

homeostasis: Constancy of the internal environment, achieved by maintaining many properties such as blood glucose, pH, and arterial oxygen tension at fixed levels or set points.

isoforms: Separate proteins expressed in molecular forms that generally display only minor variations in composition. Multiple isoforms provide functional versatility in mechanical function and in different environments. Isozymes are a class of isoforms that catalyze the same biochemical reaction.

mechanical properties: Variety of mechanical properties of muscle that have been classed according to whether the activity is isometric or dynamic. The isometric potential is measured by maximal tetanic force, whereas the dynamic properties are based on the force–velocity properties. This dynamic allows for characterization of the power and work capabilities of the muscle.

oxidative phosphorylation: Phosphorylation of adenosine diphosphate to adenosine triphosphate using the energy provided by transport of electrons to molecular oxygen.

oxidative potential: Maximal potential rate at which the mitochondria can generate adenosine triphosphate using oxygen.

peak aerobic power: Measure of the maximal rate at which adenosine triphosphate can be generated by oxidative phosphorylation; also known as maximal oxygen uptake ($\dot{V}O_2$max). $\dot{V}O_2$max is usually measured using tasks involving large muscle groups where intensity is progressively increased until fatigue.

phosphorylation potential: Content of high-energy bonds in the muscle; depends primarily on the level of the high-energy phosphagens, adenosine triphosphate and creatine phosphate.

reactive oxygen species (ROS): By-products of oxidative metabolism as well as other reactions that can induce cellular dysfunction. Examples of ROS include superoxide radical anion (O_2^-), hydrogen peroxide (H_2O_2), and hydroxyl radical ($\cdot OH$). ROS production can be increased by both exercise and aging.

signal transduction: Involves the pathways beginning with the binding of a chemical to a receptor and terminating with a cellular response, usually by altering the phosphorylation status of a target protein.

strain: Effort that has to be expended to resist stress forces.

stress: State of threatened balance to equilibrium or homeostasis.

STUDY QUESTIONS

1. Describe the different levels of organization that allow muscles to function effectively and efficiently.

2. Name the different types of myosin heavy chain isozymes in the different muscle fibers and outline how they regulate the force–velocity and power characteristic of the fiber.

3. What is meant by homeostasis, and how does exercise have the potential to disturb homeostasis?

4. Indicate the direct and indirect potential benefits of aerobic, trained muscle to health and well-being.

5. It has been stated that given the lifestyle of our ancestors, our natural state is one of physical activity, not inactivity. If such is the case, how would the composition of muscle be different in our ancestors compared with individuals today?

6. Outline the processes and proteins in muscle fiber that are involved in translating a neural command into a mechanical event.

7. What is meant by phosphorylation potential? How can it be protected during heavy intermittent exercise?

8. Describe the adaptations that occur in the excitation and contraction proteins and

processes in response to aerobic-based training and discuss the implications of these adaptations to energy supply in submaximal exercise.

9. Aerobic-based exercise, high-resistance exercise, and high-intensity intermittent exercise impose different degrees of strain on the excitation and contraction processes and the metabolic systems. Contrast the different forms of exercise in terms of the strain imposed on these systems.

10. Aging causes deteriorations in both muscle mass and muscle quality. Discuss the potential of aerobic-based training in offsetting these age-associated effects.

Response of Brain, Liver, Kidney, and Other Organs and Tissues to Regular Physical Activity

Roy J. Shephard, MB, BS, MD (London), PhD, DPE

CHAPTER OUTLINE

A great deal has been written about the health implications of regular physical activity for the heart, lungs, and muscles, but the impact of exercise on such important body systems as the brain, liver, kidneys, gut, and immune system has received much less attention. This chapter briefly notes the acute effects of vigorous exercise on these organs and tissues and examines chronic responses in more detail, where possible noting applications in preventive and therapeutic medicine. There is little evidence of gender specificity in response in these tissues, and most of the conclusions apply equally to the health of both men and women. Although normal, moderate exercise has minor and generally positive effects on function in most of these regions of the body; adverse consequences can arise from bouts of very heavy, prolonged exercise and excessive training.

Acute Effects of Physical Activity

In the short term, intensive physical activity decreases blood flow to the skin and viscera in an attempt to maintain blood pressure (Wade and Bishop 1962), particularly if the individual is simultaneously exposed to hot environmental conditions (figure 8.1). The total splanchnic blood flow decreases from a normal resting figure of 1.4 to 1.5 L/min to as little as 0.4 L/min (Rowell 1971). Intensive exercise can also increase the total energy requirements of the body 10- to 20-fold, with metabolic consequences for the viscera, and many physical pursuits subject the abdominal viscera to considerable mechanical forces.

Brain

Repetitive moderate exercise such as walking or jogging can have a sedating effect and thus has value for those who feel under pressure at work. In contrast, vigorous exercise—particularly if accompanied by

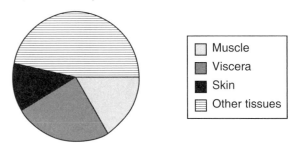
Rest, cardiac output 6 L/min

Legend:
- Muscle
- Viscera
- Skin
- Other tissues

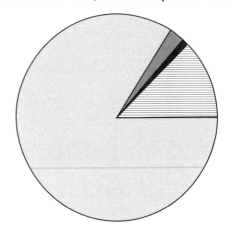
Maximum exercise, cardiac output 21.4 L/min

Figure 8.1 Approximate distribution of cardiac output under resting conditions and during maximal exercise.

loud music, as in some aerobics classes—has an arousing effect, which may be beneficial for those whose work is repetitive and boring. Vigorous exercise may also increase the secretion of **endorphins,** producing other favorable changes in mood state. If exercise is taken early in the day, the resulting relaxation helps to counter difficulties in sleeping, but if taken too late at night, sleeping difficulties may be increased, as shown in table 8.1.

There is only a limited potential for vasodilation in the cerebral blood vessels, so that the main

Table 8.1   Acute Effects of Exercise on the Brain

Type of exercise	Environment	Effect	Application
Moderate rhythmic exercise (e.g., jogging)	Peaceful, not too late at night	Relaxation, sedation	Stressful job, difficulty in sleeping
Vigorous exercise	Group activity or loud music	Arousal, postpones sleeping	Boring job, night time, highway driving

Everyone, from young students to seniors, can experience improvement in brain function through regular exercise that is not extreme.

determinant of cerebral blood flow is the systemic blood pressure. Moderate exercise may increase the overall blood flow to the brain by as much as 25%, with much of this increase being directed to the areas associated with motor activity.

However, if exercise progresses to the point where there is difficulty in maintaining blood pressure, cerebral blood flow is reduced. The first sign of cerebral **hypoxia** may be a loss of judgment—for example, the marathon runner who turns in the wrong direction on entering a stadium. The athlete has difficulty in maintaining a good posture, and if exercise continues, there is soon a complete loss of consciousness. In a hot environment, there may also be evidence of **hyperthermia,** irritability, hallucinations, and epilepsy-like contractions or **hemiplegia** preceding a potentially fatal loss of consciousness.

The brain can metabolize only glucose, so that if a bout of endurance activity is continued until **hypoglycemia** develops, the falling blood sugar can impair performance, with a deterioration of teamwork in games such as soccer and difficulty in making complex decisions in complex solo events, such as dinghy sailing. An associated decrease in plasma concentration of branch-chained amino acids and a rising concentration of **tryptophan**

Signs of Failing Cerebral Circulation in a Hot Environment

The signs of failing cerebral circulation are listed in likely order of presentation.

1. Loss of judgment and skills
2. Deterioration in teamwork
3. Deterioration in posture
4. Rise of core temperature
5. Irritability
6. Hallucinations
7. Epilepsy-like contractions
8. Hemiplegia and other forms of paralysis
9. Progressive loss of consciousness
10. Brain death
11. Death

increase the relative transfer of tryptophan across the blood–brain barrier, with a resultant increase in the cerebral synthesis of **serotonin.** In theory, the increased serotonin levels could augment the sensation of fatigue, although in practice this may be offset by a parallel increase in the production of endorphins.

Liver

Blood flow in both the hepatic arteries and the hepatic vein is reduced by vigorous exercise, and these circulatory changes have been advanced to explain a decrease in the clearance of substances such as **sorbitol** during exercise. There is a major widening of the hepatic arteriovenous oxygen difference, with the oxygen content of blood in the hepatic veins dropping as low as 5 ml/L; this allows the total cellular metabolism of the liver to remain unchanged or even to increase during vigorous exercise.

The hepatic tissues play an important role in removing lactate from the circulation during anaerobic activity, although ultimately the low rate of visceral blood flow may limit lactate metabolism. Exercise also has a selective effect on the gluconeogenic activity of hepatocytes, perivenous cells being up-regulated to a greater extent than **periportal cells.** Glucose production is enhanced as much as 14-fold, with depletion of local reserves of glycogen and **gluconeogenesis** from both lactate and amino acids (Rowell 1971). Finally, there is an increased expression of **insulin-like growth factor (IGF) binding protein-1** messenger RNA that persists for at least 12 hr postexercise, with obvious implications for systemic levels of IGF and thus muscle hypertrophy.

When rats perform a bout of prolonged exercise, data show a progressive reduction of liver **glutathione** to levels as low as 20% of the values found in a sedentary control group of the same species. This may reflect the large increase in oxidative metabolism during exercise, but it probably has implications for the total sulfhydryl content of the liver and thus its ability to buffer reactive species and other toxic ions.

If exhausting exercise is combined with excessive heat stress, acute hepatic failure may develop some 24 hr after the exercise bout. The affected individual becomes **jaundiced,** and hepatic function tests show typical markers of damage (elevated blood levels of the hepatic enzymes aspartate aminotransferase, alanine aminotransferase, γ-glutamyl transferase,

Impact of Vigorous Acute Exercise on Hepatic Metabolism

Total cellular metabolism unchanged or increased

Lactate clearance diminished

Clearance of hepatic function markers diminished

Glycogen stores depleted

Gluconeogenesis up-regulated (perivenous hepatocytes > periportal cells)

Expression of insulin-like growth factor binding protein-1 messenger RNA increased

and alkaline phosphatase). At this stage, biopsies or postmortem specimens show evidence of edema around the hepatic sinusoids and centrilobular **necrosis.**

Kidneys

The drastic reduction of renal blood flow during vigorous endurance exercise has adverse effects on many aspects of renal function. Intensive exercise has a marked antidiuretic effect, because increased concentrations of circulating **antidiuretic hormone** attempt to sustain plasma volume. An increase of **aldosterone** secretion helps reabsorption of sodium ions from filtered tubular fluid, thus compensating for the loss of sodium ions in sweat. Despite large increases in plasma lactate concentration, the renal mechanisms involved in transcellular transfer of lactate quickly become saturated during heavy exercise, and urinary excretion of lactic acid remains a minor component of anaerobic activity (Poortmans and Vanderstraeten 1994).

Glomerular filtration, measured as the capacity of the kidney glomeruli to clear **creatinine** from the bloodstream, is reduced by 30% or more during vigorous exercise, and there is often a substantial **proteinuria** (figure 8.2) that persists for about 4 hr postexercise (up to 50-fold increases over leakages of protein at rest) (Poortmans and Vanderstraeten 1994). The protein loss has been attributed to an

Impact of Vigorous Acute Exercise on Renal Function

- Antidiuresis attributable to increased secretion of antidiuretic hormone
- Increased reabsorption of sodium ions attributable to increased secretion of aldosterone
- Saturation of lactate transfer mechanisms that limit lactate excretion
- Reduced **glomerular filtration** attributable to reduced glomerular flow
- **Proteinuria** attributable to impaired glomerular electrostatic barrier, and decreased tubular reabsorption of protein

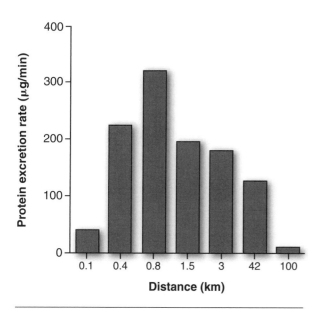

Figure 8.2 Relationship between running distance and protein excretion postexercise.

Reproduced with permission of Adis Publications, Auckland, New Zealand.

α-adrenergic mediated constriction of the efferent arterioles of the renal glomeruli, and it can be attenuated some 40% by administration of clonidine, which blocks the vasoconstrictor output from the brain stem, acting as an α_2-adrenergic antagonist. Heavy exercise decreases the glomerular electrostatic barrier, facilitating transfer of macromolecules into the renal tubules, and there may also be a decrease in the tubular reabsorption of protein.

The **prostaglandin** secretion associated with vigorous physical activity helps to counter renal vasoconstriction during exercise in the heat, and administration of a prostaglandin inhibitor such as ibuprofen exacerbates the exercise-induced depression of renal function. If physical activity is pushed to exhaustion, **hemoglobinuria** and **myoglobinuria** can develop in association with a potentially fatal total cessation of renal function (heat **oliguria, anuria,** or acute tubular necrosis). **Endotoxemia** and a generalized septic reaction may contribute to the failure of renal function in this situation, as discussed in the next section (Marshall 1998).

Stomach and Intestines

Evidence concerning exercise and esophageal mechanics is conflicting. One investigator found that treadmill running induced a modest increase in pressure at the lower esophageal sphincter, but abstracts from other laboratories have described decreases in esophageal contractions and sphincter pressures as the intensity of exercise was increased. Despite anecdotal complaints of heartburn, there is as yet no laboratory evidence that even high-intensity exercise seems to have no acute effect on the likelihood of esophageal reflux (van Nieuwenhoven et al. 1999).

The acute feelings of nausea that many runners experience are probably related to an excessive ingestion of "replacement" fluids or a slowing of gastric emptying over the course of a long-distance race. Immediate gastric emptying (particularly of solids) and small intestinal motility are depressed by intensive exercise (Moses 1994). Fluid intake should not exceed 600 ml/hr. Reports of diarrhea and abdominal cramping during endurance runs support the view that prolonged exercise subsequently speeds the overall rate of passage of food through the intestines (Moses 1994). This tendency is exacerbated following irradiation of the intestines (e.g., as a side effect of the treatment of prostate cancer). However, it is less clear that the diarrhea caused by a bout of intensive exercise is the same phenomenon as the faster transit time postulated in the regular moderate exerciser.

Prolonged and exhausting exercise can cause ischemic damage to the intestines, with gastrointestinal bleeding and a leakage of endotoxins from the gut into the bloodstream. One case report described exercise-induced bleeding from esophageal varices following administration of large doses of anabolic steroids, although it is unclear why anabolic agents should have such an effect except through an association with high-intensity resistance exercise. Minor gastrointestinal bleeding seems relatively common among long-distance runners. The reported incidence of stools that test positive for **occult blood** is quite variable (from as low as 1% to as high as 85%, depending in part on the timing of sampling and the intensity of effort that has been undertaken) (Moses 1994). Bleeding seems to peak 24 to 48 hr after the exercise bout and (probably because of intensity differences) is greater following competition than after a practice run. The stomach is the usual site of blood loss, although the colon is sometimes responsible.

Although bleeding was once blamed on a mechanical shaking of the viscera, the dominant reason for the bleeding seems to be ischemia of the gut wall, whether induced by the redistribution of visceral blood flow (Wade and Bishop 1962) or a subsequent septic reaction (Marshall 1998). In some athletes, ingestion of excessive quantities of **nonsteroidal anti-inflammatory drugs** may also be a contributing factor. The amount of blood loss is usually small, although it may sometimes contribute to "athlete's anemia." There have been suggestions that bleeding can be prevented by prophylactic administration of antihistamine medications; in any event, the blood loss generally resolves if the athlete takes 2 to 3 days of rest (table 8.2).

Any circulating endotoxins come from the cell walls of **Gram-negative bacteria** in the gut, giving rise to a portal vein endotoxemia, and if hepatic function is also compromised this quickly progresses to a systemic endotoxemia. The endotoxins stimulate release of the endogenous **pyrogens** tumor necrosis factor and interleukin-1; this sets in motion a chain of immunological disturbances that can progress to a form of endotoxic shock (Marshall 1998), widely disseminated intravascular coagulation, and the various clinical manifestations of exercise-induced heat stroke.

Circulation and Immune Function

The immediate effect of an acute bout of exercise is a decrease in blood volume, caused by a combination of sweating and exudation of fluid into the active muscles; this develops progressively over the first 30 min of exercise. The magnitude of change depends on the environmental temperature but is often of the order of 10%. It is thus important that apparent changes in plasma contents—both cell counts and concentrations of hormones and **cytokines**—be adjusted for this phenomenon.

It is difficult to evaluate responses of the human immune system to exercise, because most reports are based on changes in the numbers and functional activity of blood constituents; unfortunately, the bloodstream contains only 1% to 2% of the body's leukocytes, and apparent changes in functional status can arise quite quickly through exchanges between the bloodstream and sequestrated cells (Shephard 1997b). Moderate physical activity may enhance some aspects of resting immune function. On the other hand, a very strenuous bout of exercise leads to a transient (typically 2-72 hr) decrease in the circulating **natural killer (NK) cell** count and activity, a depression of salivary and nasal secretory immunoglobulin (Ig) A and IgM for up to 24 hr (Gleeson 2000), and often a neutrophilia but suppression of neutrophil function. It has been argued that such changes provide a brief "open window" (Nieman 2000) when the susceptibility to acute viral and bacterial infections is increased (Shephard 1997b) (figure 8. 3). Several studies have shown that athletic competition is associated with acute increases in the incidence of such infections, although it remains a challenge to tease out the effect of possible immunocompromise from other factors such as exposure to novel pathogens through travel, disturbances of sleep, mental stress, nutritional disturbances, and depletion of key nutrients, and it is questionable whether the criteria have been met for establishing a causal relationship.

Table 8.2 Characteristics of Gastrointestinal Bleeding in Endurance Athletes

Features	Site	Likely cause
Occult blood in stools peaking 24-48 hr postactivity	Esophageal varices	Mechanical trauma or NSAIDs
Blood loss small, resolves in 2-3 days	Gastric mucosa	NSAIDs
Possible prevention by antihistamines	Intestinal ischemia	Bleeding and leakage of endotoxins into bloodstream

NSAIDs = nonsteroidal anti-inflammatory drugs.

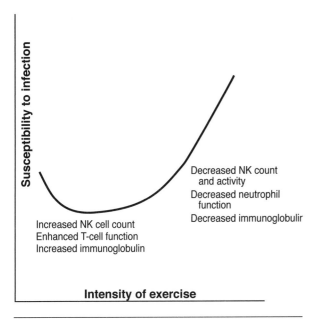

Figure 8.3 Influence of the intensity of acute or chronic exercise on susceptibility to infection. For the reasons indicated, risk is reduced by moderate physical activity but is increased by excessive activity.

Physical Activity and Immune Function

Following are effects of moderate to strenuous physical activity on immune function as seen 2 to 72 hr after a bout of such exercise.

- Decreased NK cell count
- Decreased NK cell activity (activity *per cell* unchanged)
- Decrease of secretory IgA
- Decrease of secretory IgM
- Increased neutrophil count
- Decreased neutrophil function

Chronic Effects of Physical Activity

Regular moderate exercise has a number of positive effects on the body systems considered in this chapter. However, adverse reactions may be anticipated if the intensity of effort is sufficient to cause

repeated bouts of severe tissue hypoxia, and there is growing evidence of pathological consequences from excessive endurance activity. It is thus important to curb the enthusiasm of young type A people for participation in extreme events for which they may be ill-prepared.

Brain

The repetition of movement patterns leads to a progressive learning of motor skills. As a movement becomes "automated," appropriate sequences of muscle activation become stored in the cerebellum, with a reduction in the energy cost of performing the movements concerned. There have also been reports that regular moderate exercise has a favorable effect on the development of academic skills in the growing child—an important argument in the ongoing debate over the allocation of curricular time to physical education (table 8.3). However, it remains unclear whether any academic benefits from increased exercise arise directly from the effects of movement on cortical organization, whether exercise-induced increases of **arousal** make students more attentive in class, or whether gains are simply incidental to a change of pace for teachers and pupils (Shephard 1997a).

Alzheimer's disease is a growing concern, given the rapid aging of the population in many developed societies. Regular exercise has been held to delay the

Table 8.3 Effects of an Experimental Increase of Physical Education by 5 Hr Per Week on Academic Performance of Primary School Students

Grade	Experimental classes (%)	Control classes (%)
1	78	80
2	81	76
3	80.5	77
4	80	78
5	81	78
6	79	76

Values are mean percentages for experimental and control classes, based on alphanumeric grades as reported by teachers. The control students received a standard 40 min gym class per week and followed normal physical activities in the community. Despite a small initial disadvantage, the experimental students attained a significantly higher score in grades 2 through 6.

Data from Shephard et al., 1984, Required physical activity and academic grades: A controlled study. In *Children and sport,* edited by J. Ilmarinen and I. Valimaki (Berlin: Springer Verlag), 58-63.

onset of deterioration in various aspects of cerebral function. But again, it remains unclear whether any benefit is a direct consequence of the higher blood pressures induced by physical activity; an effect of exercise on signaling molecules that regulate neurotrophin, levels of fibroblast growth factor, and neuroplasticity; or merely an incidental response to the greater social and environmental stimuli enjoyed by an active person. Animal studies have suggested that regular exercise augments concentrations of a brain-derived neurotrophic factor, increasing neuronal survival. Likewise, cross-sectional comparisons in humans show that executive functions involving the frontal and hippocampal lobes of the brain are selectively conserved in individuals with high levels of physical fitness (Churchill et al. 2002). Another potential mechanism of benefit may be an exercise- and training-induced reduction in the concentration of damaging reactive oxygen species.

Certain forms of moderate exercise have value in countering "stress," elevating mood, and countering both anxiety (Landers and Petruzello 1994) and depression (North et al. 1990). The secretion of endorphins contributes to this process, although it can also create an "exercise addiction" in some individuals. Excessive physical activity has adverse effects on the psyche (an increase of anxiety and a loss of vigor) that provide the most effective index of overtraining (Shephard 1997b).

One significant negative consequence of vigorous exercise is an increased risk of injury from concussion, whether from falls sustained during individual or contact sports, heading of the ball in soccer, or blows received in boxing. The accumulation of such injuries can eventually lead to deterioration in cognition.

Liver

Regular moderate exercise has beneficial effects on several aspects of hepatic function, and there have been suggestions that the greater metabolic activity of chronic endurance pursuits can cause a hypertrophy of hepatic tissues. The stimulation of hepatic metabolism can offset the age-related decline in glucagon signaling capacity and responsiveness, apparently by normalizing the ratio of inhibitory to stimulatory G protein and increasing adenyl cyclase activity. The cytochrome *c* oxidase activity of the liver is also enhanced by endurance activity, and in some studies increasing serum levels suggest an increased production of IGF. Some cross-sectional studies also suggest that resistance training protects

against the age-related decline in IGF. In contrast, recent studies of older men show that regular exercise, particularly if combined with a low-fat diet, increases serum concentrations of IGF-1 binding protein and decreases serum IGF-1; these changes in the IGF axis could contribute to a decrease in the risk of prostatic cancer.

A small-scale cross-sectional comparison between eight endurance runners and relatively sedentary medical students found no significant intergroup differences in such indexes of metabolic function as aminopyrine metabolism, galactose elimination, and indocyanine green clearance. On the other hand, a 3-month period of moderate training, sufficient to increase maximal oxygen intake by an average of 6%, increased the metabolism of antipyrine and aminopyrine by 12% to 13%, individual gains in maximal oxygen intake showing a correlation of .6 to .7 with the gains in hepatic metabolism. More vigorous exercise has less positive results on cellular metabolism. A study in rats noted that the increased production of nitric oxide associated with vigorous exercise augmented the activity of iron regulatory protein 1, thus down-regulating activity in one of the key enzymes of the Krebs cycle, cytosolic aconitase.

Regular exercise decreases the fat content of the liver, and moderate aerobic programs have thus been applied as a component of therapy in patients with chronic hepatic disease and fatty liver, apparently without adverse effect. In rats, at least, regular exercise decreased halothane-induced hepatotoxicity and attenuated the ethanol-induced decline in hepatic function. One large epidemiological study found an association between physical activity and a low incidence of gallbladder disease, after adjusting for other risk factors, but a 16 yr follow-up of Harvard alumni suggested that much of any association might be mediated indirectly through the influence of regular exercise on obesity levels.

Repeated bouts of hepatic ischemic hypoxia could predispose a person to an accumulation of the central lobular necrosis seen in patients with a restricted cardiac output (Rowell 1971), and in support of this view, a rigorous military training course was shown to induce both leukocytosis and the release of several hepatic enzymes. Liver disorders are frequently suspected in very active individuals because of abnormal serum concentrations of bilirubin and such enzymes as aspartate aminotransferase, alanine aminotransferase, and alkaline phosphatase. Laboratory findings of this type could indeed reflect chronic hepatitis, but in the case of the enzyme markers, it is difficult to be certain that muscle damage is not the primary cause. There are currently no reports linking an accumulation of hepatic enzymes or bilirubin with hepatic effects of intensive training. The abuse of steroids is a confounding factor in some classes of athletes; there have been occasional case reports that the abuse of large quantities of androgens not only increases hepatic enzyme levels but can also induce such conditions as peliosis hepatis (a benign form of purpura that affects the liver, with the development of blood-filled cysts), nodular hyperplasia (a rapid and disorganized but benign nodular growth of liver tissue), and (rarely) hepatic carcinoma.

Kidneys

During the early stages of training, when plasma volume is expanding, there is an increased reabsorption of sodium ions in the proximal tubules of the kidneys. The expansion of plasma volume, together with a worsening of hypoxemia, increases symptoms if a person with mountain sickness chooses to exercise during his or her first few days at simulated or real high altitude.

Even moderate physical activity has often been restricted in patients with chronic renal disease, but this seems unnecessary and may be counterproductive. In hypertensive rats with chronic renal failure, regular exercise appears to attenuate proteinuria and protect against progression of renal sclerosis. Likewise, a program of resistance exercise can be helpful in human chronic renal disease, helping to reverse the muscle wasting that may arise from a low-protein diet and possibly reducing the depression of mood state that is commonly associated with chronic renal disease.

Stomach and Intestines

A speeding of passage of food or changes in segmentation of the large intestine, perhaps mediated by prostaglandins, could reduce exposure of the colon wall to toxins, thus accounting for the reduced risk of cancers of the descending colon seen in those who perform regular physical activity through either their occupation or leisure pursuits (Shephard and Futcher 1997). Other possible factors that could influence the risk of colonic cancer include exercise-induced reductions in gastrointestinal blood flow and neuroimmunoendocrinological alterations.

Debate continues concerning the mechanical effects of exercise on the gastrointestinal tract;

there may be both acute and chronic effects on gastrointestinal motility. One uncertainty is whether running has a similar effect to seated exercise. Several reports have noted little effect: The mouth-to-cecum transit time was unchanged by running 9.6 km (6 miles), and walking a distance of 4.5 km (2.8 miles) in 1 hr did not change the total transit time in previously sedentary laboratory workers. A similar total colonic transit time was shown for laboratory technicians and soccer players who were training as much as 15 hr per week. However, other studies have suggested a substantial effect of habitual exercise. Six weeks of endurance training reduced the total carmine transit time to a range of 35 to 24 hr, whereas the transit for a control group remained at a more sluggish 45 hr. When the physical activity of healthy and recreationally active elderly subjects was restricted for 2 weeks, the colonic transit time was on average almost doubled. Likewise, 1 week of cycling or jogging at 50% of maximal oxygen intake almost doubled the speed of intestinal transit in healthy young adults. Thirteen weeks of resistance training sufficient to induce a 40% increase in the peak force developed by key muscle groups more than doubled the speed of total transit, this change being attributed almost exclusively to faster movement through the large intestine. Exercise-related changes in the type and quantity of food ingested may be important issues. This idea is supported by the fact that no changes in bowel transit time were seen when training experiments were carried out in a metabolic laboratory, where subjects adhered to a controlled diet and sophisticated radioactive markers of fecal movement were used.

The intensity of the training regimen adopted in these controlled diet experiments was only moderate, but a moderate intensity was also typical of a number of the epidemiological studies that found physical activity to protect against cancer. A change in local or general movement of the gut contents remains the most likely explanation why the risk of carcinoma is reduced by as much as 50% in a physically active person (figure 8.4), but there remains a need to conduct further careful observations of the acute and chronic responses of both motility and cancer risk to appropriately graded intensities of effort. Further study is also needed to examine the impact of regular physical activity on the risks of diverticulosis and inflammatory bowel disease, although one recent report found an increase of neutrophil activation and an exacerbation of zinc deficiency when patients in remission from Crohn's disease undertook a moderate exercise program.

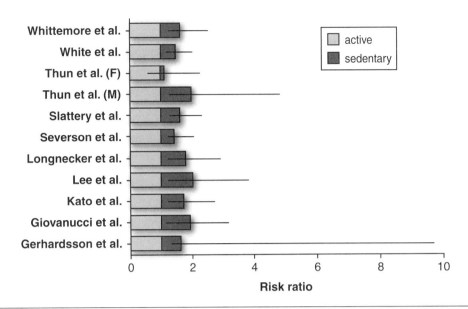

Figure 8.4 Eleven surveys, summarized by Shephard and Futcher (1997), presented here as risk ratios, with 95% confidence limits for the sedentary samples shown as black lines.

Circulation and Immune Function

Despite the occult hemorrhage frequently observed during acute bouts of endurance exercise, a large-scale 3-year prospective study of elderly persons found that regular walking, gardening, or vigorous physical activity was associated with a substantially reduced risk of severe gastrointestinal hemorrhage, even after statistical adjustment for associated risk factors.

Perhaps in part because of the depression of plasma volumes that occurs during acute exercise, a period of endurance training leads to an expansion of plasma volume. This effect is greater in the upright than in the supine posture. It is accompanied by increases in serum albumin and sodium ion concentrations and a decreased excretion of urine. One consequence of the plasma expansion is a reduction in hemoglobin concentration (the so-called athlete's anemia), although oxygen transport is well maintained. A true anemia is uncommon, although it can develop if food intake is inadequate or there is frequent exercise-induced bleeding from the alimentary tract.

The impact of prolonged endurance training on the immune response has received only limited attention. Moderate training appears to augment the number and activity of NK cells, possibly reversing age-related changes in both NK and **T-cell** function. There may also be increased salivary concentrations of **immunoglobulins,** whether expressed as absolute values or relative to grams of salivary protein. The main evidence of benefit is a decrease in the symptoms associated with upper respiratory infections. A 12-month study of 547 healthy adults estimated a 20% reduction in risk of upper respiratory infection among individuals who were moderately (>4 MET-hours per day) or vigorously (>12 MET-hours per day) active. However, in a second experiment participants were inoculated with rhinovirus, and the severity and duration of illness did not differ between those individuals who were undertaking moderate exercise (40 min on alternate days at 70% of heart rate reserve) and sedentary controls.

Extreme training such as the peak phases of preparation for international competition can temporarily reduce NK cell activity, neutrophil function, and serum, salivary, and membrane concentrations of IgA and IgM, particularly if the supply of nutrients and vitamins is inadequate. Such changes are in turn associated with an augmented susceptibility to acute respiratory infections (Mackinnon 2000). A deterioration of immune function is one of the markers of overtraining, although findings are less consistent than the associated deterioration in performance times and mood state (Shephard 1997b). Large doses of vitamin C may help to sustain neutrophil function during periods of intensive training by countering an increased release of reactive oxygen species.

Issues Requiring Further Research

Many of the responses to physical activity depend on the intensity of the effort that is undertaken, and there is thus an urgent need for more information on dose–response relationships. We also need to confirm suggestions that physical activity can enhance cerebral function, both in children and in old age; to assess potential adverse effects of prolonged bouts of heavy exercise on the viscera and immune system; and to assess what appear to be large beneficial effects of regular physical activity on an individual's biological age.

- *Dose–response issues.* As discussed in part IV of this book, there is surprisingly little information concerning either the minimum amount of physical activity needed to enhance health or the threshold beyond which health deteriorates rather than improves. The optimal dosage probably varies with the organ system or tissue under considerations, and dose–response information is particularly limited for the organs and tissues discussed in this chapter.

- *Cerebral development.* Teachers and parents often argue that schools should not allocate greater curricular time to physical education and sports programs because of the ever-growing volume of academic material to learn. There is thus an urgent need to confirm and extend the findings of the Trois Rivières study, which suggest that academic performance is enhanced rather than worsened by an hour of classroom physical education per day. Such research should explore whether the response is incidental, or is an integral consequence of motor activity; it should define the type of physical activity giving the greatest benefit and examine whether adult intellectual achievement is enhanced. We also need to undertake further long-term evaluations of the type begun in Trois Rivières, to determine whether childhood physical education programs have a favorable impact on attitudes of adults and their willingness to engage in preventive exercise programs.

- *Cumulative impact of repeated bouts of heavy exercise on hepatic and renal function.* Given suggestions that overly strenuous exercise can cause local tissue necrosis, there is need to extend comparisons of hepatic and renal function between those who have engaged in repeated ultraendurance competition and moderate exercisers, exploring whether excessive exercise has cumulative adverse effects on the viscera.

- *Resistance to viral infections.* Upper respiratory infections remain a major source of morbidity in otherwise healthy individuals. More research is thus needed to verify or disprove tantalizing suggestions that moderate physical activity can enhance immune barriers to such infections but that excessive training can exacerbate the risk. Such research should be designed to exclude associated modulations of viral exposure, stress, and nutritional status plus physical changes in the respiratory membranes associated with the oral inhalation of increased volumes of cold, dry, or polluted air.

- *Countering the effects of aging.* Deterioration of mental function is becoming an ever more important problem for older adults, as longevity is extended by the countering of chronic disease. The apparently beneficial effect of regular physical activity on the age-related deterioration in cognitive function thus merits careful documentation, making due allowance for associated influences such as increased social contacts and greater environmental stimulation among active members of the elderly community.

Summary

This chapter examines the responses of often neglected body systems—the brain, liver, kidneys, stomach, intestines, circulation, and immune system—to both acute bouts of exercise and programs of physical training. As in other parts of the body, moderate exercise and moderate-intensity training result in adaptations that have positive consequences for both immediate physical performance and longer term health. However, an intense and prolonged bout of exercise, particularly if it is performed under extreme environmental conditions, can have negative effects on the individual's health, occasionally progressing to a fatal pathology. Regular, moderate physical training confers a number of useful health benefits, apparently including a decreased risk of Alzheimer's disease and modifications of the IGF axis that decrease the likelihood of some cancers. On the other hand, repeated excessive training carries potential hazards that include exercise addiction, ischemic visceral damage, and immunosuppression. Moderation is thus an important tenet of exercise prescription.

KEY CONCEPTS

aldosterone: A hormone produced in the adrenal cortex that stimulates the reabsorption of sodium ions in the kidneys, thus countering sodium loss in the sweat and conserving or increasing blood volume.

Alzheimer's disease: A form of senile dementia associated with the accumulation of characteristic peptides and protein deposits within the brain. It has been argued that regular exercise retards the degeneration of cognitive function in older adults.

antidiuretic hormone: A pituitary hormone that reduces urine flow by stimulating water reabsorption in the kidneys. This hormone becomes active during prolonged physical activity, when large amounts of fluid are being lost in the sweat.

anuria: A rapidly fatal complete cessation of urine production, as may occur in severe cases of exercise and heat stress.

arousal: An increase of neural activity in certain regions of the brain, particularly the reticular formation of the brain stem. Arousal is increased by a bout of vigorous physical activity, making the individual feel more wide awake.

creatinine: An end product of creatine metabolism, normally cleared from the body in the urine. The rate of clearance of a test dose is used to assess changes in the efficiency of kidney function.

cytokines: Hormonelike compounds that act mainly on adjacent cells, modifying the proliferation and differentiation of cells, regulating immune responses, and initiating inflammation. Vigorous exercise greatly increases the production of various cytokines by various types of white blood cells.

endorphins: Hormones with a structural resemblance to morphine that the body produces in response to sustained and vigorous physical activity. The secretion of endorphins enhances a person's mood and may explain why some people become addicted to exercise.

endotoxemia: The penetration of the gut wall by toxins from bacteria present in the intestines. The penetration is commonly attributable to a large decrease in the blood flow to the intestines during severe exercise, and it can cause a dangerous generalized septic reaction.

glomerular filtration: The rate at which the renal glomeruli can clear substances from the bloodstream, commonly measured by the clearance of a test dose of creatinine. Glomerular filtration is often compromised by prolonged endurance exercise.

gluconeogenesis: The production of glucose from noncarbohydrate precursors such as amino acids. This process is important during prolonged endurance activity, as the carbohydrate reserves of the skeletal muscles are depleted and the blood glucose level falls.

glutathione: A sulfur-containing tripeptide that is a coenzyme but also acts as an antioxidant, protecting key enzymes of the body against degradation by reactive species of oxygen. Liver levels of glutathione drop during prolonged exercise, attributable to a combination of a high rate of metabolism and an increased production of reactive species.

Gram-negative bacteria: A type of bacteria found in the intestines that does not retain the basic dye in its wall during Gram staining. Such bacteria produce the toxins that can give rise to endotoxemia during exercise.

hemiplegia: Unilateral paralysis of the body muscles.

hemoglobinuria: Appearance of hemoglobin in urine, sometimes associated with exposure of the kidneys to combinations of heat and exercise. The pigments and red cells may form casts that can block urinary flow, leading to kidney failure.

hyperthermia: Originally a core body temperature in excess of a fixed and clinically dangerous threshold but now a term commonly applied to any substantial elevation of body temperature above normal resting values. A core temperature greater than 39°C has the potential to damage the brain, kidneys, and other visceral organs.

hypoglycemia: A pathologically low blood glucose concentration that may develop because of very prolonged endurance activity and an associated depletion of body carbohydrate reserves. Because the brain relies uniquely on glucose for its metabolism, hypoglycemia can cause a deterioration of judgment and eventually brain damage.

hypoxia: A lower than normal partial pressure of oxygen in a given tissue, usually leading to a depression of function in that tissue (e.g., a cerebral hypoxia caused by a decrease in cerebral blood flow can affect a person's judgment and eventually lead to loss of consciousness and death).

immunoglobulins: Proteins secreted by lymphocytes that have specific antibody activity. A depression of IgA secretion by mucous membranes is thought to contribute to the vulnerability of overtrained individuals to upper respiratory tract infections.

insulin-like growth factor (IGF): Homologues of insulin that are less important than insulin in the regulation of blood glucose but are more potent in the stimulation of growth and muscle hypertrophy.

IGF binding proteins: Proteins that bind IGF in the serum and other body fluids, thus regulating the action of IGF.

jaundice: A yellowing of the skin and the whites of the eyes, reflecting excessive concentrations of bilirubin in the blood. It may result from obstruction of the bile duct, excessive hemolysis of red cells, or impaired function of the liver.

myoglobinuria: The excretion of the red pigment of skeletal muscle (myoglobin) in the urine, associated with the muscle injury resulting from severe exercise.

natural killer (NK) cell: A type of white cell that provides the first line of defense against viruses and tumor cells, acting without any priming from other components of the immune system. Suppression of NK cell activity following a heavy bout of exercise such as a marathon run has been postulated as leaving an "open window" when upper respiratory tract infections are more likely to can develop.

necrosis: Cell breakdown and death, seen in the liver, for example, if the local blood flow becomes inadequate.

nonsteroidal anti-inflammatory drugs: Drugs such as aspirin and ibuprofen. These medications are commonly taken without prescription to relieve discomfort from minor injuries but can provoke gastrointestinal bleeding, particularly during exercise.

occult blood: Presence of pigments in the stools indicative of gastrointestinal bleeding. The pigments (breakdown products of hemoglobin) may not be visible to the naked eye but can be detected by appropriate chemical tests.

oliguria: A greatly reduced flow of urine, as may occur with dangerous combinations of exercise and heat stress.

periportal cells: Liver cells grouped around the portal veins.

prostaglandins: Group of biologically very active metabolites of arachidonic acid and related compounds secreted during exercise and injury, whose proinflammatory action is countered by nonsteroidal anti-inflammatory drugs such as ibuprofen.

proteinuria: Escape of plasma proteins into the urine, a phenomenon commonly seen with prolonged and vigorous exercise.

pyrogens: Substances that give rise to fever. Vigorous exercise leads to the production of several pyrogens such as tumor necrosis factor and interleukin-1 within the body.

serotonin: An amino acid produced in the brain from tryptophan, where it is associated with sensations of fatigue. Serotonin is also a strong vasoconstrictor substance and is released during tissue injury.

sorbitol: Sugar produced from glucose. The rate of clearance of a test dose of sorbitol from the bloodstream is used to assess changes in liver function.

T-cells: Class of lymphocyte responsible for cell-mediated immunity. The T-lymphocytes include cytolytic cells that break down cells identified for elimination and helper cells that facilitate this process. The number of T-cells diminishes with aging, but regular exercise may help to slow this trend.

tryptophan: An amino acid that is transported across the blood–brain barrier and then converted to serotonin, influencing the individual's mood state.

STUDY QUESTIONS

1. Can a bout of very heavy physical activity have adverse short-term consequences for organs such as the brain, liver, kidneys, and intestines? If so, what are the dominant mechanisms, what environmental circumstances are likely to exacerbate these effects, and what precautions can be suggested?

2. Discuss the long-term benefits of regular exercise for the brain in terms of cognition, movement patterns, and mood state at various points in the life span.

3. Would you agree that in the case of the liver, the main benefit of regular physical activity is to enhance metabolic function? If so, what are some of the more important reactions that are enhanced?

4. Why is it important to recommend regular physical activity to patients with chronic renal disease?

5. When regulating the preparation of athletes for top-level competition, would you look for immunological evidence of overtraining? If so, what?

6. Some educators argue that with the ever-increasing volume of academic material that must be taught, allocation of extracurricular time to physical education is impracticable. How would you counter this assertion?

7. Many exercise scientists focus mainly on the cardiorespiratory effects of physical activity. Do you think more attention should be directed to the viscera, and if so, why?

PART

Physical Activity, Fitness, and Health

You are now moving into the central topic of this book—the relationships among physical activity, fitness, and health. In part I, you learned about the evolution of the concepts concerning the relationships among physical activity, physical fitness, and health. You have also reviewed some of the known differences and similarities between women and men and among ethnic groups and the changes that take place with age. In part II, you learned about the acute and chronic effects of physical activity on the heart, lungs, blood vessels, skeletal muscles, endocrine and other hormones, kidneys, brain, and other organs and tissues.

This section provides information that underscores sedentary living habits as a major public health problem. The growing recognition of the threat to health, function, and well-being caused by inactivity largely fueled the development of exercise science in the latter half of the 20th century. Eleven chapters are devoted to this key area. Chapter 9 deals with the contribution of a sedentary lifestyle and low

levels of fitness to the risk of dying prematurely. Chapter 10 examines the effects of physical activity and fitness on cardiac, vascular, and pulmonary morbidities. Chapter 11 focuses on obesity, and chapter 12 deals with diabetes mellitus. These chapters are particularly important given the current dramatic increases in the prevalence of obesity and type 2 diabetes mellitus in industrialized countries. Chapter 13 reviews the evidence for a role of physical activity and fitness on the risk of developing a number of cancers. Chapter 14 addresses bone and joint disorders, and chapter 15 reports on muscular fitness and disorders. Mental health problems are discussed in chapter 16. Chapter 17 covers fitness and activity in children, whereas chapter 18 focuses on physical activity and aging. Chapter 19 discuses the risks of adverse musculoskeletal and cardiac events associated with regular physical activity. You will need to have a good grasp of this corpus of knowledge to take full advantage of the new material presented in subsequent parts of the book.

Physical Activity, Fitness, and Mortality Rates

■ Steven N. Blair, PED ■ Michael J. LaMonte, PhD

CHAPTER OUTLINE

Physical Activity, Fitness, and Mortality
- Physical Activity and Mortality
- Fitness and Mortality
- Comparison of Activity and Fitness As Exposures

- Activity or Fitness and Mortality in Different Subgroups of the Population

Biological Mechanisms

Summary

Review Materials

The research reviewed here is drawn primarily from the epidemiological literature, although some of the information on biological mechanisms is from clinical trials, laboratory investigations, and pathological studies. As reviewed in chapter 1, human evolution has been dependent on a physically active lifestyle. Thus, existence in a modern world where physical activity has largely been engineered out of daily living is an aberration from our evolutionary constitution. Logically, then, a sedentary way of life should be unhealthy for our species, and indeed that is what is shown by the data reviewed here.

Sedentary habits and low cardiorespiratory fitness are among the strongest predictors of premature mortality and thus, for many industrialized countries, are a major threat to public health in the first part of the 21st century. We review information on activity or fitness and mortality in several groups according to their demographic and health status. Although there is a brief discussion of potential biological mechanisms that may help explain the lower death **rates** observed among active and fit individuals compared with their sedentary and unfit counterparts, other chapters in this book cover these issues in more detail. An important distinction is that **physical activity** refers to a behavior, specifically body movement, and **physical fitness** is the **outcome** of that behavior. Although the term *physical fitness* represents a broad spectrum of physiological attributes such as aerobic power or "cardiorespiratory fitness," muscular strength and endurance, body composition, musculoskeletal flexibility, speed, and balance, this chapter focuses only on **cardiorespiratory fitness** (aerobic power) and **muscular fitness** (muscular strength and endurance).

> Physical activity is a behavior, specifically body movement that results from skeletal muscle contraction. Physical fitness is a set of physiological attributes that result from participation in physical activity and, to some degree, from genetic influences.

The primary objective of this chapter is to review the evidence relating physical activity or cardiorespiratory fitness to mortality. We summarize data from various epidemiological studies on these topics and draw heavily from our own research, the Aerobics Center Longitudinal Study (ACLS), which is a **prospective study** of physical activity, fitness, body habitus, clinical factors, and health that has been conducted since 1970 at The Cooper Institute in Dallas, Texas.

Physical Activity, Fitness, and Mortality

Research on physical activity and mortality did not begin in earnest until the second half of the 20th century. Professor Jeremy Morris of London generally is recognized as the scientist who conducted the first major studies in which physical inactivity was systematically investigated as a cause of premature mortality, specifically attributable to coronary heart disease. Professor Morris evaluated different levels of occupational physical activity as the exposure in several studies, comparing individuals with higher occupational activity levels with individuals who had sedentary jobs. For example, double-decker bus conductors were compared with bus drivers, and postal carriers were compared with postal clerks and telephone operators. Those in sedentary jobs had coronary heart disease death rates approximately twice as high as their peers who were more active, and this difference remained after adjustment for factors that might confound the relationship. A more thorough review of Professor Morris's contributions to physical activity **epidemiology** is given in a paper presented at his 90th birthday symposium (Paffenbarger et al. 2001).

Physical Activity and Mortality

Influenced by Professor Morris's findings, other investigators began to evaluate the role of physical inactivity and mortality. A key figure in the study of activity and mortality, and one who has made substantial contributions to our understanding of this issue, is Professor Ralph Paffenbarger, Jr., of Stanford University (Paffenbarger 1988). He initially investigated the role of occupational activity in coronary heart disease mortality in a group of San Francisco area longshoremen. At the time of his studies, the longshoremen who actually unloaded the ships performed hard physical labor such as lifting, pushing, carrying, and shoveling heavy cargo. Other members of the longshoremen's union, such as clerks and supervisors, had much more sedentary jobs. Just as Morris observed in

London, the sedentary clerks and supervisors had much higher death rates during follow-up than did the more strenuously active men who unloaded the ships. Collectively, the seminal studies of Professors Morris and Paffenbarger provided the first prospective, systematic examination of physical inactivity and adverse health effects and established the value of epidemiological research in exercise science.

Seminal population-based studies on the relationships of occupational and leisure-time physical activity habits with mortality outcomes, conducted by Professor Jeremy Morris among London busmen and postal workers and by Professor Ralph Paffenbarger, Jr., among San Francisco dockworkers and Harvard alumni, were the first large-scale epidemiological studies to show that physical activity is an important and independent predictor of mortality from cardiovascular disease and all-cause mortality. Together, Morris and Paffenbarger introduced the application of epidemiological research to the exercise sciences.

Physical Activity During Leisure Time

The nature of occupational work changed throughout the 20th century, and the pace of these changes accelerated during and after the post–World War II economic expansion within industrialized nations. Fewer people performed hard physical labor at work, and more workers spent the majority of their day sitting or standing and doing light work. Because these widespread shifts in occupational **energy expenditure** are unlikely to be reversed, it seemed logical to expand physical activity epidemiology studies to include leisure-time activity **exposures.**

Professor Morris examined physical activity during leisure time among British civil servants. In his first study he recruited 17,944 male office workers who were 40 to 65 years of age and initially free of known health problems. Morris followed these men over the next 8.5 years and recorded the **incidence** of deaths attributable to coronary heart disease. To estimate the cohort's physical activity patterns, each man completed a detailed physical activity diary on a weekday and a weekend day. The principal hypothesis of this study was that higher levels of total leisure-time physical activity would be related to lower rates of coronary heart disease

© Eyewire/Photodisc/Getty Images

Research suggests that because modern transportation and labor saving devices have so drastically reduced the amount of physical activity needed to conduct the daily activities of life, it is increasingly important for all people to include significant amounts of physical activity in their leisure time. One of the best strategies for doing this is to engage in active leisure pursuits as a family.

mortality. However, Morris found that total physical activity was not related to coronary heart disease death rates, but vigorous physical activity participation, defined as requiring an energy expenditure of 31.5 kJ (7.5 kcal) per minute, was associated with a substantially lower mortality rate than was seen in men who did not participate in vigorous exercise. This definition of vigorous exercise is based on absolute intensity and is probably in the range of 4 to 6 METs (multiples of resting energy expenditure), as exercise intensity is typically expressed in contemporary studies. This absolute intensity is now more likely to be labeled as moderate intensity or moderately vigorous, as often stated by Professor Paffenbarger. Another point worth noting is that an absolute intensity of 4 to 6 METs is classified as moderate, but in terms of relative intensity, it might be considered light for a young fit person and strenuous for an older unfit person. The primary results for the British Civil Servant Study are shown in figure 9.1.

To test the new hypothesis that vigorous exercise is required to protect against premature coronary heart disease, Professor Morris conducted a second investigation among 9,374 male British civil servants who were 45 to 64 years old and healthy at entry into the study. As in the first study, men who participated in vigorous exercise had much lower rates of fatal

coronary heart disease than men who did not. In fact, the data show a **dose–response** gradient across categories of vigorous exercise (figure 9.2). Men partaking of frequent vigorous exercise, defined as two or more times per week, had less than one half the coronary heart disease death rate compared with men who engaged in vigorous activity only one time per week and about one third the death rate of men reporting no vigorous exercise.

Professor Paffenbarger and his colleagues also made numerous important contributions to the study of leisure-time physical activity and mortality. In the early 1960s, he initiated a prospective investigation of lifestyle characteristics and health in men who entered Harvard College during the period 1916 to 1950. These men completed an extensive baseline health and lifestyle questionnaire in either 1962 or 1966 and have been followed for morbidity and mortality to the present. Many articles have been published from this study and collectively provide a substantial database supporting the hypothesis that regular physical activity protects against mortality from coronary heart disease and from all causes. Examples of the inverse association for several measures of physical activity with mortality are shown in table 9.1. The pattern of association for lower all-cause mortality incidence rates with higher levels of activity was

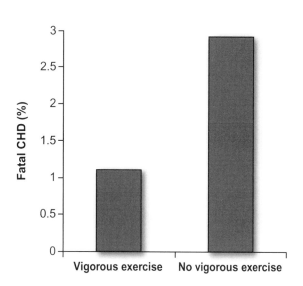

Figure 9.1 Fatal coronary heart disease (CHD) in 17,944 men in the British Civil Service for those reporting vigorous exercise, requiring at least 31.5 kJ/min (24 deaths), and those who did not report vigorous exercise (411 deaths).

Data from J.N. Morris, R. Pollard, M.G. Everitt and S.P.W. Chave, 1980, "Vigorous exercise in leisure-time: Protection against coronary heart disease," *Lancet* 11: 1207-1210.

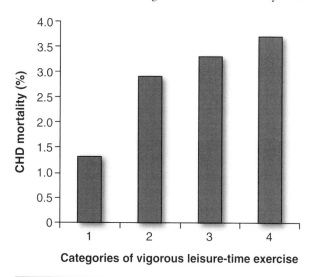

Figure 9.2 Coronary heart disease (CHD) deaths (n = 289) in male British civil servants by categories of vigorous exercise (VE) requiring at least 31.5 kJ/min. Category 1 = VE at least twice per week; category 2 = VE at least once but less than twice per week; category 3 = VE one to three times in 4 weeks; and category 4 = no VE.

Data from J.N. Morris, D.G. Clayton, M.G. Everitt, A.M. Semmence and E.H. Burgess, 1990, "Exercise in leisure time: Coronary attack and death rates," *British Heart Journal* 63: 325-334.

Table 9.1   Leisure-Time Physical Activity and Risk of Death in 17,815 Harvard Alumni Followed From 1977 to 1992

Physical activity	No. of deaths	Relative risk of death (95% confidence interval)
Physical activity index (kcal/week)		
<1,000	1,908	1.0
1,000-1,999	1,101	0.81 (0.75-0.87)
>1,999	1,299	0.75 (0.70-0.81)
Walking (km/week)		
<5	1,284	1.0
5-14	1,698	0.90 (0.83-0.96)
>14	1,249	0.87 (0.80-0.94)
Stair climbing (floors/week)		
<20	2,025	1.0
20-54	1,764	0.89 (0.83-0.95)
>54	506	0.86 (0.78-0.95)
Sports play (weekly)		
None	1,903	1.0
Light only (<4.5 METs)	628	0.84 (0.76-0.92)
Moderately vigorous ($\geq$4.5 METs)	1,777	0.73 (0.69-0.79)

Relative risks are adjusted for age, smoking habit, blood pressure status, body mass index, alcohol intake, premature parental death, and health status.

Data from R.S. Paffenbarger Jr. and I-M. Lee, 1998, "A natural history of athleticism, health and longevity," *Journal of Sports Sciences* 16: S31-45.

observed for several physical activity indexes, and these associations persisted after statistical adjustment for several potentially confounding variables. An impressive finding from these data is the observation that protection against premature mortality is seen among men at all ages from 45 to 90 years. The Harvard Alumni Health Study was one of the first large, prospective, epidemiological studies to demonstrate that health benefits can be accrued from an active way of living even among individuals of advanced age.

Physical Activity and Longevity

Epidemiological studies, such as those conducted by Morris and Paffenbarger, have consistently shown a graded inverse pattern of association between physical activity and mortality. These results led to the logical question whether active persons live longer than their sedentary peers. However, this hypothesis was not adequately evaluated until Professor Paffenbarger published his seminal article on the topic in 1986 (Paffenbarger et al. 1986). From 1962 to 1978, he followed 16,936 Harvard alumni aged

Data from the Harvard Alumni Health Study showed that men who expended $\geq$2,000 kcal/week in leisure-time physical activity lived an average of 2.15 years longer than men who obtained <500 kcal of weekly activity. Harvard men who exercised received approximately two additional hours of life for each hour they were active.

35 to 74 years and healthy at baseline. There were 1,413 deaths in the cohort during the follow-up period. Likelihood of surviving to age 80 years was calculated for various categories of physical activity. The results showed that men who expended $\geq$2,000 kcal/week in leisure-time physical activity lived an average of 2.15 years longer than men who obtained <500 kcal of weekly activity. It has often been stated that a person may live longer by being regularly physically active but that the additional years of life are taken up by exercise. However, Professor

Paffenbarger's data do not support that notion, because Harvard men who exercised received approximately two additional hours of life for each hour they were active.

Change in Physical Activity and Mortality

The studies reviewed thus far had a simple prospective epidemiological design where a single baseline measure of physical activity was related to mortality outcomes in follow-up. A perplexing issue for such studies is whether study participants changed their health habits during follow-up. That is, individuals smoking at baseline can stop at some point during the observation period, and others could start smoking. The same is true for other lifestyle habits such as physical activity. Those active at the start of the study might stop early or later in the follow-up period and thus might be misclassified on their activity patterns for much of the observational period, which introduces a biased or imprecise measure of association between the exposure and outcome variables of interest. Misclassification bias can reduce the likelihood of showing a true significant association between the exposure and health outcome if such an association does in fact exist.

The misclassification phenomenon seen in epidemiological research is similar to concerns over random measurement errors in experimental studies. Measurement error reduces the reliability of laboratory data and thereby increases the likelihood of missing a true association, which is referred to by measurement experts as a type II or β error—the failure to identify a significant association. These theoretical assumptions notwithstanding, it became apparent that a more complete evaluation of the causal hypothesis for physical inactivity and mortality could be achieved by evaluating the association of changes in physical activity and subsequent mortality.

Once again, Professor Paffenbarger led the way in this research paradigm. He examined mortality rates in 10,268 Harvard alumni who had completed physical activity questionnaires in either 1962 or 1966 and again in 1977 (figure 9.3, Paffenbarger et al. 1993). Study participants were then followed for mortality from 1977 through 1985, during which time 476 deaths were recorded. This study design allowed for calculation of changes in activity between 1962 or 1966 and 1977 and an evaluation of the relationship of changes in activity to mortal-

ity rates during follow-up. In this study, moderately vigorous sports were defined as those requiring an energy expenditure of at least 4.5 METs, and in this population included activities such as swimming, tennis, squash, racquetball, handball, and jogging. Study participants who did not engage in moderately vigorous sports (requiring ≥4.5 METs) at baseline but who started moderately vigorous sports by 1977 had a 23% lower risk of death than the men who remained sedentary throughout follow-up. Furthermore, men who were active at baseline but became sedentary had a 15% increase in mortality risk. Although not as rigorous as experimental data from a randomized controlled trial, the epidemiological findings reported among Harvard alumni on the association between changes in free-living physical activity habits and future risk of mortality provided important evidence for a causal association between physical activity and death from all causes.

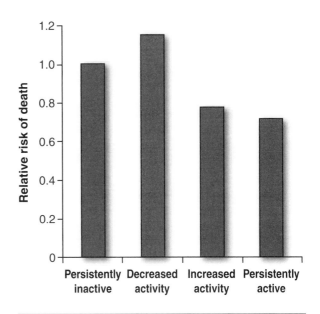

Figure 9.3 Changes in moderately vigorous sports activity and risk of death in Harvard alumni. Persistently inactive = no moderately vigorous sports at either assessment point; decreased activity = reported sports at the first assessment but not in 1977; increased activity = no sports at the first assessment but reported sports in 1977; persistently active = reported sports play at both assessments.

Data from R.S. Paffenbarger Jr., R.T. Hyde, A.L. Wing, I.-M. Lee, D.L. Jung and J.B. Kampert, 1993, "The association of changes in physical-activity level and other lifestyle characteristics with mortality among men," *New England Journal of Medicine* 328: 538-545.

Findings among Harvard men demonstrated the importance of changes in physical activity habits on future risk of mortality. Men who did not engage in moderate- to vigorous-intensity sports (requiring $\geq$4.5 METs) at baseline but who took up such activities during follow-up had a 23% lower risk of death than the men who remained sedentary throughout follow-up. Men who were active at baseline but became sedentary had a 15% increase in mortality risk.

Summary

The classic studies briefly presented here are a small fraction of the evidence on physical activity and mortality that has accumulated over the past 50 years. A more thorough review of this research is available (Kohl et al. 1996) and demonstrates the consistency of the findings in women and men, in various demographic and racial groups, and across the age range. Investigators have presented strong associations, temporal sequencing, consistency of results, plausible biological mechanisms, and graded dose–response, and meeting these criteria leaves little doubt that the association between inactivity and mortality is causal.

Simple prospective epidemiological analyses of a baseline physical activity exposure and future mortality may underestimate the true strength of association between these changes, because changes in physical activity habits during follow-up may result in misclassification on exposure, which may produce a biased or imprecise test of the hypothesis.

Fitness and Mortality

As mentioned earlier, physical fitness includes a wide variety of physiological responses, including speed, balance, and coordination, but for this chapter the discussion of fitness is limited to cardiorespiratory fitness and muscular fitness. Cardiorespiratory fitness as used here refers to maximal physical working capacity or aerobic power, which is an integrated assessment of the function of the heart, lungs, vascular system, and skeletal muscles. Muscular fitness as used here refers to strength and endurance of the skeletal muscles.

Cardiorespiratory Fitness

One of the most well-documented findings in exercise science is that physical activity, especially moderate- to vigorous-intensity aerobic activities, improves maximal cardiorespiratory fitness. Thus, it is reasonable to assume that cardiorespiratory fitness is a good indicator of recent habitual physical activity habits. Our research group has compared the relationship between detailed self-reports of daily physical activity recorded in a computer-based exercise log over a 3-month period with results from a maximal exercise test on a treadmill. Approximately 70% of the variation in treadmill test performance is accounted for by the physical activity data. This is consistent with data from other sources on the genetic contribution to maximal aerobic power, which is estimated to be in the range of 25% to 40% of the variation in an individual's aerobic power. Thus, just as a person's low-density lipoprotein (LDL) cholesterol level is primarily determined by habitual diet, especially the intake of saturated fat, as well as by genetic influences, cardiorespiratory fitness is largely a function of habitual physical activity patterns and to a lesser degree genetic influences.

Several investigators have evaluated cardiorespiratory fitness as a predictor of future mortality.

Cardiorespiratory fitness is a measure of the efficiency of the heart, lungs, vasculature, and skeletal muscles to deliver and use oxygen during heavy dynamic physical activity. The primary determinant of cardiorespiratory fitness is participation in moderate- to vigorous-intensity physical activities; however, genetics is also influential to some degree. Because cardiorespiratory fitness measures are more objective and reproducible than self-reported recall of past physical activity, these measures provide an accurate assessment of sedentary and irregularly active lifestyles for use as exposures in epidemiological studies of mortality outcomes.

As indicated earlier, we draw primarily on findings from the ACLS, but our findings are consistent with those from other populations. The ACLS is a prospective observational study of patients examined at the Cooper Clinic in Dallas, Texas, since 1970. The Cooper Clinic is a preventive medicine facility that conducts 8,000 to 9,000 periodic health examinations per year. The examination includes a maximal exercise test on a treadmill, which we use to estimate maximal aerobic power (maximal METs where 1 MET = 3.5 ml O_2 uptake·kg^{-1}·min^{-1}) from exercise time to exhaustion. Participants are women and men from middle to upper socioeconomic strata, with greater than 90% non-Hispanic whites, approximately 80% college graduates, and most employed or formerly employed in professional or executive positions. A diagram of the ACLS is shown in figure 9.4.

Maximal treadmill test times in the ACLS are used to create age- and sex-specific fitness categories. The least fit 20% of women and men in each age group are classified as unfit or low fit, the next 40% are classified as moderately fit, and the most fit 40% are classified as high fit. These arbitrary classifications arose from our initial reports on cardiorespiratory fitness and mortality, where we started with the study participants grouped into quintiles of fitness. Of course, aerobic power is similar to other risk exposures such as lipid levels or blood pressure values, which are continuous

variables that are often grouped into categories for research or clinical purposes.

The association of maximal cardiorespiratory fitness with all-cause mortality in the ACLS shows a steep, inverse, curvilinear pattern (figure 9.5). This curve is nearly a mirror image of a similar curve for LDL cholesterol and coronary heart disease mortality, and in both cases creating categories out of continuous data is somewhat problematical. Nonetheless, such cut points are often used for clinical applications and for public health recommendations. We do not claim that the maximal MET cut points developed in the ACLS are necessarily the best or the only possible cut points; however, in the case of cardiorespiratory fitness, and unlike physical activity exposures, there is no consensus public health recommendation to guide our analysis of fitness as a predictor of health outcomes. Until such recommendations are developed, we believe that maintaining consistency in our variable definitions is the most prudent methodological approach. The specific maximal MET cut points for women and men in different age groups are shown in table 9.2.

We have conducted numerous studies on cardiorespiratory fitness and all-cause mortality. The general finding is that age-adjusted death rates

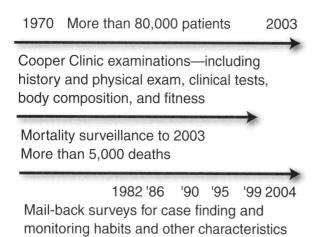

Figure 9.4 An overview of the Aerobics Center Longitudinal Study from 1970 to 2003. The top arrow indicates that more than 80,000 patients have been examined since the Cooper Clinic opened in 1970. The middle arrow indicates mortality surveillance through 1996, and the bottom arrow indicates that five mail-back surveys have been conducted in the cohort between 1982 and 1999.

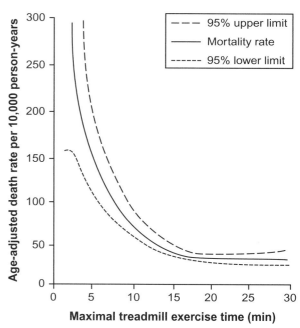

Figure 9.5 Age-adjusted death rates (601 deaths from any cause) in 25,341 men in the Aerobics Center Longitudinal Study by maximal treadmill exercise time in minutes. Death rates (solid line) and 95% confidence limits (dashed lines) fitted to individual fitness levels by curvilinear regression techniques.

Table 9.2   Maximal MET Cut Points for Low, Moderate, and High Cardiorespiratory Fitness in the Aerobics Center Longitudinal Study, 1970 to 2002

Fitness group	Age groups (years)			
	20-39	40-49	50-59	60+
Women				
Low	≤8.6	≤8.1	≤7.2	≤6.3
Moderate	8.7-10.8	8.2-9.9	7.3-8.9	6.4-8.1
High	>10.8	>9.9	>8.9	>8.1
Men				
Low	≤10.8	≤9.9	≤8.9	≤8.1
Moderate	10.9-13.3	10.0-12.4	9.0-11.3	8.2-10.3
High	>13.3	>12.4	>11.3	>10.3

Table values are maximal METs estimated from a treadmill test to volitional exhaustion.

Data from The Cooper Institute.

> Data from the Aerobics Center Longitudinal Study indicate that moderately fit men and women have approximately half the risk (e.g., 50% lower) of all-cause mortality compared with their low fit counterparts. Mortality risk is another 10% to 15% lower when highly fit study participants are compared with their moderately fit peers.

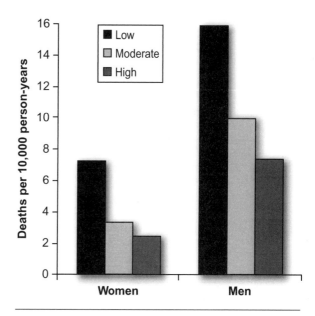

Figure 9.6   Cardiovascular disease (CVD) mortality death rates (21 CVD deaths in women and 226 CVD deaths in men) per 10,000 person-years of observation for low, moderate, and high cardiorespiratory fitness in 7,080 women and 25,341 men participating in the Aerobics Center Longitudinal Study. Rates are adjusted for age and examination year.

Data from S.N. Blair, J.B. Kampert, H.W. Kohl, C.E. Barlow, C.A. Macera, R.S. Paffenbarger Jr. and L.W. Gibbons, 1996, "Influences of cardiorespiratory fitness and other precursors on cardiovascular disease and all-cause mortality in men and women," *Journal of the American Medical Association* 276: 205-210.

during follow-up are about 50% lower in moderately fit women and men compared with their low fit counterparts. We typically also see incrementally lower death rates of 10% to 15% when comparing high fit with moderately fit study participants. An example of the relationship between cardiorespiratory fitness and cardiovascular disease mortality is shown in figure 9.6.

The pattern of association between objectively measured cardiorespiratory fitness and mortality from cardiovascular and all causes is generally consistent with extant data on associations between physical activity exposures and mortality outcomes. The association for fitness and activity with mortality persists after adjustment for potential **confounding factors** such as age, smoking, and conventional risk predictors. The magnitude of association with mortality outcomes tends to be higher for fitness rather than activity exposures (Kohl 2001; Lee and Skerrett 2001) and is likely attributable to the greater objectivity and lower likelihood of misclassification on exposure with laboratory-based quantification of fitness compared with self-reported recall of past physical activity habits. Available epidemiological evidence indicates that physical inactivity and low fitness are important independent predictors of premature mortality for men and women throughout the adult age range.

Muscular Fitness

Most studies on fitness and mortality have used cardiorespiratory fitness as the exposure; however, a few studies used some aspect of muscular fitness as the predictor. Most of these latter studies used grip strength as the exposure and were conducted in elderly populations. The findings generally show an inverse association between grip strength and mortality. One issue with these studies is the concern that grip strength in elderly populations may be an overall marker for frailty and health status and that strength per se is not a causal factor.

We recently evaluated a muscular strength and endurance index as a predictor of all-cause mortality in 9,105 women and men 20 to 82 years of age in the ACLS (FitzGerald et al. 2004). These individuals completed a three-item muscular fitness assessment in addition to the standard clinical examination and maximal treadmill exercise test. The three items were a 1-repetition maximum (1RM) bench press, a 1RM leg press, and the maximum number of bent-leg sit-ups performed in 1 min. We calculated sex-specific muscle fitness indexes by dividing the distribution for each test into thirds. We then assigned scores of 0 to 2 to each third (e.g., 0 = lowest third, 1 = middle third, and 2 = highest third) and then summed the individual scores to create an overall muscle fitness index with a range of 0 to 6. There were 194 deaths in 106,046 **person-years** of follow-up. Age- and sex-adjusted relative risks of all-cause mortality for low, moderate, and high muscular fitness were 1.0, 0.56 (95% confidence interval 0.40-0.80), and 0.65 (0.42-0.99), respectively. Additional adjustment for cardiorespiratory fitness and other clinical and behavioral risk predictors attenuated but did not eliminate the association between muscular fitness and mortality. The limited data on the association between muscular fitness and mortality that have been reported by our study and others suggest that muscular fitness may confer health benefits through mechanisms that are independent of cardiorespiratory fitness. Whether the protective effect is attributable to maximal muscular strength per se or to regular participation in resistance exercises is not known.

> Limited data on the association between muscular fitness and mortality suggest that muscular fitness may confer health benefits through mechanisms that are independent of cardiorespiratory fitness.

Comparison of Activity and Fitness As Exposures

The previous sections showed that both physical activity and cardiorespiratory fitness have an inverse association with all-cause mortality. The question often arises as to which of these exposures is the more important predictor. In some ways, this question is unwarranted or illogical because physical activity is the principal determinant of cardiorespiratory fitness and thus is the fundamental exposure. However, it is perhaps useful to give the comparison of activity and fitness more thought. Data from the ACLS for both cardiorespiratory fitness and physical activity as predictors for all-cause mortality are shown in table 9.3 for both women and men. The inverse gradient for mortality risk is steeper for cardiorespiratory fitness than it is for physical activity and in fact is not significant for physical activity for women. A similar pattern of results for fitness and activity is seen in combined analysis of men and women in the Canada Fitness Survey data (Arraiz et al. 1992). The observation that fitness is stronger than activity as a predictor of mortality in these studies is most likely attributable to the fact that fitness can be measured more objectively than activity. This leads to greater error variance for the activity measures, increases the likelihood of misclassification on the activity exposure, and, as indicated in an earlier section, tends to bias the results toward the null hypothesis. Another possibility is that other factors might be associated with both fitness and health, such as exposure to tobacco smoke or environmental pollutants.

Activity or Fitness and Mortality in Different Subgroups of the Population

Physical activity and cardiorespiratory fitness are inversely associated with mortality outcomes, as reviewed previously. One way to evaluate whether the relationships observed in observational studies are causal or attributable to chance associations is to check for consistency in the relationship within different subgroups of the population. This section includes a few examples of studies in which the exposures of activity or fitness were related to mortality in women and men across the age range and in various groups based on health status.

Gender

Most of the early studies on physical activity and mortality were carried out only in men and tended

Table 9.3   Relative Risk for All-Cause Mortality by Physical Activity and Cardiorespiratory Fitness Categories in the Aerobics Center Longitudinal Study, 1970 to 1989

Exposure	Women	Men
	Relative risk (95% CI)	Relative risk (95% CI)
Cardiorespiratory fitness categories[a]		
1	1.0[b]	1.0[b]
2	0.53 (0.30-0.95)	0.55 (0.44-0.70)
3	0.56 (0.31-1.01)	0.61 (0.48-0.78)
4	0.22 (0.10-0.49)	0.52 (0.41-0.66)
5	0.37 (0.19-0.72)	0.49 (0.37-0.64)
Physical activity categories[c]		
1	1.0[d]	1.0[e]
2	0.68 (0.39-1.17)	0.71 (0.58-0.87)
3	0.39 (0.09-1.65)	0.83 (0.59-1.16)
4	1.14 (0.27-4.80)[f]	0.57 (0.30-1.08)
5		0.92 (0.29-2.88)

CI = confidence interval.

[a]Quintiles of cardiorespiratory fitness in each age–sex group; [b]p for trend = .001; [c]1 = sedentary, 2 = some activity or walking or jogging 1 to 10 miles per week, 3 = walking or jogging 11 to 20 miles per week, 4 = walking or jogging 21 to 40 miles per week, 5 = walking or jogging >40 miles per week; [d]p for trend = .217; [e]p for trend = .011; [f]categories 4 and 5 combined for the women's analyses because of small numbers of deaths.

Data from J.B. Kampert, S.N. Blair, C.E. Barlow and H.W. Kohl, 1996, "Physical activity, physical fitness, and all-cause and cancer mortality: A prospective study of men and women," *Annals of Epidemiology* 6: 452-457.

to focus on coronary heart disease outcomes. These studies were conducted before it became widely appreciated that coronary heart disease is the leading cause of death in women, as it is in men. Another issue regarding consistency of associations in women and men is that some studies found that physical activity was inversely associated with mortality in men but not in women. One reason for such observations may well be fewer endpoints in women, thus leading to insufficient statistical power to detect a true difference. Indeed, this has been observed in the ACLS as well (see table 9.3), although we consistently have observed similar steep inverse gradients for health outcomes across fitness categories in both women and men. Recent analyses have been conducted among women in the ACLS and are consistent with our earlier data reported among men. Thus, there is little doubt that cardiorespiratory fitness is associated with mortality in women, as it is in men.

A likely explanation for the inconsistent results regarding physical activity and mortality between women and men is that many of the physical activity questionnaires were developed for use in study populations of men and then later applied to women participants. This led to questionnaires that may not have adequately assessed the habitual activity patterns that are common in women, including physical activity around the home and childcare. This would have produced more misclassification

A similar pattern of an inverse graded association exists for cardiorespiratory fitness and physical activity exposures with mortality outcomes, and the association persists after adjustment for potential confounding factors such as age, smoking, and conventional risk predictors. The higher magnitude of association with mortality observed with fitness exposures than with activity exposures is likely attributable to the greater objectivity and lower likelihood of misclassification on exposure with laboratory-based quantification of fitness compared with self-reported recall of past physical activity habits.

of physical activity in women than in men and thus would have been more likely to bias study results in women toward the null hypothesis.

The data on cardiorespiratory fitness and later studies of physical activity with better questionnaires leave little doubt that a sedentary and unfit way of life is indeed hazardous to women. A thorough review of physical activity and health in women was recently published (Oguma et al. 2002).

Age

Numerous investigators have evaluated the associations of physical activity or cardiorespiratory fitness with mortality in various age groups. The results are highly consistent and show a similar inverse gradient among middle-aged and older study participants, up to at least the eighth or ninth decade of life. Results from the ACLS in women and men 60 years of age and older show a steep inverse gradient for mortality across fitness groups, as we observed in younger study participants (figure 9.7). For men, where there were sufficient deaths for subgroup analyses, we observed similar gradients for those in the 60 to 69, 70 to 79, and 80+ years age groups.

> The Harvard Alumni Health Study was one of the first large, prospective, epidemiological studies to demonstrate that physical activity protected against mortality even among individuals of advanced age.

Racial or Ethnic Groups

Most of the studies on physical activity or cardiorespiratory fitness and mortality have been conducted in populations of non-Hispanic whites. However, several recent studies have been undertaken in other racial or ethnic groups, and the results are largely consistent with earlier studies in non-Hispanic white populations. One example is a recent report from the Tokyo Gas Company Study on cardiorespiratory fitness and cancer mortality (Sawada et al. 2003). These investigators followed 9,039 men ranging in age from 19 to 59 years for an average of 16 years, during which time 123 men died of cancer. Cardiorespiratory fitness was assessed by a submaximal cycle ergometer test at baseline. Results show a steep inverse gradient of cancer deaths across quartiles of fitness (figure 9.8), and these findings

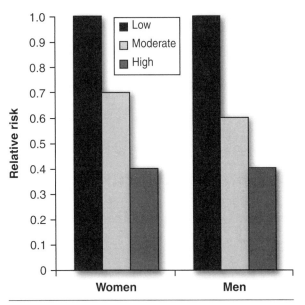

Figure 9.7 Relative risk of all-cause mortality across low, moderate, and high cardiorespiratory fitness groups (see table 9.2 for definitions) in 749 women and 1,758 men 60 years of age or older in the ACLS. Relative risks are adjusted for age, examination year, body mass index, cholesterol, blood pressure, smoking habit, health status, and parental history of cardiovascular disease.

Data from S.N. Blair and M. Wei, 2000, "Sedentary habits, health, and function in older women and men," *American Journal of Health Promotion* 14: 1-8.

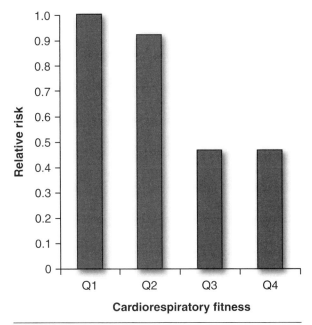

Figure 9.8 Cancer mortality (123 deaths) for quartiles of cardiorespiratory fitness in 9,039 Japanese men. Relative risks are adjusted for age, body mass index, systolic blood pressure, alcohol intake, and smoking habit. p for trend = .005.

Data from S.S. Sawada, T. Muto, H. Tanaka, I.-M. Lee, R.S. Paffenbarger Jr., M. Shindo and S.N. Blair, 2003, "Cardiorespiratory fitness and cancer mortality in Japanese men: A prospective study," *Medicine and Science in Sports and Exercise* 35: 1546-1550.

were similar in smoking and nonsmoking men. These investigators also evaluated other health outcomes in this population, and the results are similar to extant studies.

Investigators from the Corpus Christi Heart Project evaluated changes in physical activity patterns in a group of survivors of myocardial infarction. The study group included women and men as well as Mexican American and non-Hispanic white participants. Compared with patients who remained sedentary, those who increased their physical activity had an 89% lower mortality risk during 7 years of follow-up. The authors conclude that increasing physical activity was beneficial for Mexican American and non-Hispanic white men and women. Thus, it appears that the health benefits of becoming and staying active and fit apply across various demographic groups.

Body Mass Index

Obesity **prevalence** has increased in most countries of the world since the mid- to late 1980s. Because obesity is associated with numerous chronic diseases and health problems, its increasing prevalence constitutes a major public health problem and has received much attention. Physical activity is recognized as having a role in the prevention, treatment, and amelioration of obesity and related health risks. This last point deserves further comment in the context of this chapter. Few if any individuals would argue that obesity is not an important clinical or public health issue, but one equally important point is often overlooked; that is, physical activity confers health benefits for people of all sizes and shapes. Being physically active may not prevent weight gain in modern societies where average daily energy expenditure continues to be driven to lower and lower levels by labor-saving devices at home, on the job, and during leisure time, but physical activity has value, even for those who are obese.

Investigators from the ACLS have published several reports on the independent relationship of cardiorespiratory fitness to mortality in normal weight, overweight, and obese individuals. One of these reports presented data on the association between fitness and all-cause mortality among study participants separated into three categories of body composition (Lee et al. 1999). A total of 21,925 men aged 30 to 83 years at baseline were followed for an average of 8 years for all-cause mortality. There were 428 deaths during 176,742 man-years of observation. We assessed body composition by hydrostatic

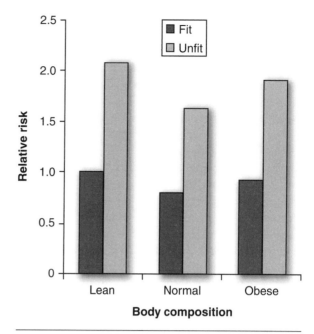

Figure 9.9   Relative risks for all-cause mortality are adjusted for age, examination year, smoking habit, alcohol intake, and parental history of coronary heart disease. Lean men = <16.7% body fat; normal fat men = 16.7 to <25% fat; and obese men = ≥25% fat. Relative risks are shown for fit men and unfit men.

Data from C.D. Lee, S.N. Blair and A.S. Jackson, 1999, "Cardiorespiratory fitness, body composition, and all-cause and cardiovascular disease mortality in men," *American Journal of Clinical Nutrition* 69: 373-380.

weighing, sum of seven skinfolds, or both methods, and the men were sorted into lean, moderate, and obese categories based on percentage of body fatness. Summary results are shown in figure 9.9. It is clear that within each body composition group, unfit men were more likely to die than were their fit peers. Obese men who were fit had less than one half the risk of dying compared with the lean men who were unfit. These results persisted after various adjustments for confounding factors and have recently been observed in ACLS women as well.

An active and fit way of living confers protection against mortality among overweight and obese individuals. In the ACLS, obese men who were fit had less than one half the risk of dying compared with the lean men who were unfit. These results have also been observed in ACLS women, and similar findings have been reported in other studies.

Other investigators have reported similar findings during subgroup analyses on the association between physical activity and body size, quantified most often as body mass index, among men and women. Clearly, a growing evidence base indicates that an active and fit way of living confers health benefits even among individuals who are overweight or obese.

Chronic Disease

Physical activity and moderate to high cardiorespiratory fitness extend longevity in individuals with chronic disease. Several reports indicate increased survival in active and fit individuals who have coronary heart disease, hypertension, or diabetes. A 2004 case–control study in India among women and men with acute myocardial infarction is a good example (Rastogi et al. 2004). The investigators collected data from 350 persons 21 to 74 years of age who had been hospitalized with acute myocardial infarction in New Delhi or Bangalore during 1999. Each case participant was matched with two control participants from noncardiac outpatient clinics or inpatient wards. Physical activity was assessed by a questionnaire validated for the Indian population and included information on occupational, leisure-time, and sedentary activities. Patients in the highest stratum of physical activity, which was equivalent to 36 min of walking per day, had a 55% lower risk (95% confidence interval, 34-69%) of myocardial infarction compared with sedentary persons. This observation held after adjustment for several potential confounding variables. The investigators also observed higher myocardial infarction rates in individuals who reported more hours per day of sedentary activities such as television watching.

Reports from ACLS have shown significantly lower mortality rates among fit compared with unfit men who have existing hypertension (Church et al. 2001) or diabetes (Church et al. 2004). The latter study showed a substantial inverse gradient for all-cause mortality across quartiles of cardiorespiratory fitness among men with documented type 2 diabetes. The investigators followed 2,196 men with diabetes (average age = 49.3 years) for up to 26 years. There were 275 deaths in 32,161 person-years of follow-up. Relative risk of death across fitness categories is shown in table 9.4. Men in the least fit category were more than four times more likely to die than were men in the most fit category, and this gradient persisted after adjustment for several potential confounding variables, including body mass index. In fact, adjusting the fitness data for body mass index made little difference in the mortality association, but conversely adjusting the body mass index data for fitness caused this association to become nonsignificant (p for trend = .22). Data from other investigations provide consistent evidence that physical activity and cardiorespiratory fitness are important considerations for secondary disease prevention.

> Higher levels of physical activity and cardiorespiratory fitness are associated with lower death rates among individuals with existing diseases like coronary artery disease, hypertension, and diabetes. ACLS investigators showed that men with diabetes and in the least fit category were more than four times more likely to die than were men with diabetes and in the most fit category, even after accounting for obesity status.

Table 9.4 Cardiorespiratory Fitness and All-Cause Mortality in Men With Documented Type 2 Diabetes, Aerobics Center Longitudinal Study, 1970 to 1995

Fitness group (maximal METs from the treadmill test)	Adjusted relative risk[a]	95% confidence interval
≤8.82	4.49	2.64-7.64
8.83-10.08	2.77	1.65-4.66
10.09-11.71	1.60	0.93-2.76
>11.71	1.0	Reference group

[a]Adjusted for age, examination year, history of cancer or cardiovascular event, parental history of premature cardiovascular disease, smoking habit, systolic blood pressure, cholesterol, glucose, and body mass index.

Data from T.S. Church, Y.J. Cheng, C.P. Earnest, C.E. Barlow, L.W. Gibbons, E.L. Priest and S.N. Blair, 2004, "Exercise capacity and body composition as predictors of mortality among men with diabetes," *Diabetes Care* 27: 83-88.

Biological Mechanisms

The data reviewed here make a compelling case for physical inactivity and low cardiorespiratory fitness as important independent causes of all-cause and cause-specific mortality in women and men, in middle-aged and older participants, in various racial or ethnic groups, and in both healthy and unhealthy persons. A causal association requires that plausible biological mechanisms can explain, or at least possibly explain, the protection resulting from an active and fit way of life.

In the early years of studies on physical activity and mortality, it was widely assumed that the protection observed in active individuals must be attributable, at least in part, to a beneficial effect of activity on the conventional risk factors such as blood lipid abnormalities and elevated blood pressure. Indeed, regular activity has a beneficial effect on these variables, but the associations between activity or fitness and mortality persist, and in most studies are not substantially reduced, after adjustment for these variables. Thus, there are obviously other perhaps more important mechanisms whereby physical activity provides protection. Skeletal muscle is one of the largest metabolically active organs in the human body, and it seems reasonable to assume that healthy skeletal muscle is important to avoid dis-

orders of carbohydrate and lipid metabolism. After all, skeletal muscle is the major site of insulin action, and controlled laboratory studies show both acute and chronic benefits of exercise on insulin resistance, which likely have important regulatory implications on downstream adipose and hepatic glycemic control and lipid metabolism. Other variables, such as flow-mediated vasodilation, inflammatory responses, clotting mechanisms, and catecholamine regulation, have recently received extensive attention for their role in the development of type 2 diabetes, hypertension, and atherosclerotic and arrhythmogenic heart disease. It may be that physical activity plays a role in these relationships, although research is in the early stages on these issues.

Much of this book is devoted to the acute and chronic effects of physical activity on various physiological, biochemical, and metabolic systems. It appears that physical activity has the potential to affect nearly every system in the body and thus has the potential to play a role in numerous disease processes. As mentioned at the onset of this chapter, physical activity is not a fad; rather, it is a return to our evolutionary way of living—the kind of life for which our bodies are engineered and which facilitates proper function of our biochemistry and physiology. A sedentary existence is an aberration that results in maladaptive changes in our constitution and increases the likelihood of disease and premature death.

> Several mechanisms exist through which physical activity and fitness protect against premature mortality. Skeletal muscle is a huge metabolic furnace with a high oxidative potential for substrate clearance. An active and fit skeletal muscle mass likely has important regulatory implications on downstream adipose and hepatic glycemic control and lipid metabolism. Additional mechanisms may be mediated through changes in vascular compliance, inflammatory responses, clotting mechanisms, and catecholamine regulation, all of which have a role in the development of type 2 diabetes, hypertension, and atherosclerotic and arrhythmiogenic heart disease. A careful reading of the other chapters in this book will provide additional insight into the biological mechanisms whereby physical activity protects against mortality.

> Physical activity has the potential to affect nearly every system in the body. An active lifestyle is not a fad; rather, it is a return to our evolutionary way of living. Sedentary existence is unnatural and results in maladaptive changes in our anatomical, physiological, and biochemical constitution that increase the likelihood of disease and premature death.

Summary

Human beings evolved to live a physically active lifestyle, but modern men and women live in a world where physical activity has largely been engineered out of daily living. There is compelling evidence from many prospective epidemiological studies showing that an unfit and sedentary way of life increases risk of morbidity and mortality from chronic diseases. The associations between inactivity or low fitness and mortality are strong, graded, and temporally consistent, and they remain after adjustment for numerous potentially confounding variables. Sedentary habits and low cardiorespiratory fitness are among the strongest predictors of premature mortality and pose a major public health threat in most countries of the world.

KEY CONCEPTS

cardiorespiratory fitness: See definition on page 47.

confounding factor: Variable that is related to both the exposure and outcome under analysis. Statistical methods are used to control or adjust for confounding factors so that the true or unbiased association between the exposure and outcome may be examined.

dose–response: Relationship wherein a change in the dose of exposure (e.g., physical activity level) is associated with a graded change (increase or decrease) in outcome (e.g., mortality).

exercise energy expenditure: Net transfer of energy required to support skeletal muscle contraction during physical activity. This term is used to quantify the volume or dose of physical activity, computed as the product of frequency, duration, and intensity of a specified physical activity and typically is expressed as kcal/week or MET hour/week.

epidemiology: For definition, see page 47.

exposures: Agents or factors that can affect health. In epidemiological analyses, these factors are examined for their association with study outcomes. Exposure variables are also referred to as independent variables, predictor variables, or explanatory variables.

incidence: Rate of new cases that have developed over time, such as the rate of new heart attacks per year in a specific population. For example, the incidence of cardiovascular disease deaths in a population might be 7/10,000 per year.

muscular fitness: Component of physical fitness that encompasses the expression of maximal skeletal muscle strength and skeletal muscle endurance at submaximal work loads.

outcome: Effect on a health parameter that results from an exposure. Outcome variables are often referred to as dependent variables or response variables.

person-years: Summary of the amount of follow-up in a prospective epidemiological study. One person followed for 1 year would provide 1 person-year of observation, and 10 persons followed for 10 years would provide 100 person-years of observation. This method of determining the amount of exposure in a study is necessary when the length of follow-up is different for different individuals. It is not necessary to use person-years if all study participants are enrolled at the same time and are followed for the same length of time.

physical activity: For definition, see page 19.

physical fitness: For definition, see page 19.

prevalence: Proportion of cases within a population at a particular point in time. For example, in 2003 the prevalence of stroke (individuals who have had a documented stroke) might be 2/100, which would mean that of every 100 persons in the population, two had a stroke. Therefore, the prevalence of stroke within the population at risk in 2003 would be reported as 2%.

prospective study: Study in which the exposure variables are collected at baseline, and the population is followed over time with monitoring for outcomes. For example, deaths occurring in both sedentary and physically active groups would be counted, and incidence for the two groups would be calculated.

rates: Way of expressing exposures and outcomes in epidemiology; rates are derived from a numerator (number of cases or individuals with the characteristic) and denominator (number of individuals in the population from which the cases are derived). Rates in epidemiology may be expressed as a percentage or as a rate per some number in the population, for example, 15 events/10,000 individuals.

STUDY QUESTIONS

1. Describe the general components and study design of an epidemiological investigation of physical activity and mortality.

2. List three epidemiological studies on the association between physical activity (or cardiorespiratory fitness) and mortality.

3. Describe what is meant by the terms *confounding* and *bias*. How are these issues typically addressed in epidemiological studies of exposures and outcomes?

4. Why are associations between cardiorespiratory fitness and mortality apparently stronger than associations between self-reported physical activity and mortality?

5. What is the primary concern regarding changes in an exposure during follow-up for mortality outcomes?

6. True or false: Protection against mortality by physical activity and cardiorespiratory fitness exposures is only seen among men, individuals who are normal weight and overweight, younger and middle-aged individuals, and individuals without other diseases such as diabetes.

7. List four plausible biological mechanisms that might mediate the protection against mortality that is observed among active and fit individuals.

Physical Activity, Fitness, and Cardiac, Vascular, and Pulmonary Morbidities

■ Ian Janssen, PhD

CHAPTER OUTLINE

Physical Inactivity and Low Cardiorespiratory Fitness As Risk Factors for Cardiovascular Morbidities

- Physical Inactivity and Low Cardiorespiratory Fitness As Risk Factors for Specific Cardiovascular Morbidities
- Questions on Physical Inactivity, Low Cardiorespiratory Fitness, and Risk for Cardiovascular Morbidities

Physical Inactivity and Low Cardiorespiratory Fitness As Risk Factors for Pulmonary Morbidities

Biological Mechanisms

Role of Physical Activity in Patients With Cardiac, Vascular, and Pulmonary Morbidities

- Cardiovascular Disease
- Chronic Obstructive Pulmonary Disease and Asthma

Summary

Review Materials

In this chapter we explore the role of physical activity and fitness in preventing and treating **cardiovascular disease** and lung disease. Cardiovascular disease—including **coronary heart disease, stroke, hypertension,** rheumatic fever, congenital heart defects, congestive heart failure, and **peripheral vascular disease**—is the leading cause of death in men and women in industrialized countries. This chapter focuses on four specific forms of cardiovascular disease that are particularly important in the context of physical activity and fitness: coronary heart disease, stroke, hypertension, and peripheral vascular disease. As with cardiovascular disease, lung disease is a leading cause of hospitalization, disability, and death. This chapter focuses on two specific forms of lung disease that are important in the context of physical activity and fitness: **chronic obstructive pulmonary disease and asthma.**

Physical Inactivity and Low Cardiorespiratory Fitness As Risk Factors for Cardiovascular Morbidities

The first part of the chapter examines the role of physical activity and cardiorespiratory fitness in the prevention of cardiac and vascular diseases. The relationship between physical activity and fitness outcomes is discussed, as is the dose–response relationship between physical activity and fitness and cardiovascular disease. The influence of race and gender on these associations is also covered.

Physical Inactivity and Low Cardiorespiratory Fitness As Risk Factors for Specific Cardiovascular Morbidities

We begin by examining the role of physical activity and fitness for each of the major forms of cardiovascular disease. Research conducted over the past 30 years has provided a substantial knowledge base for most of the cardiovascular morbidities.

Coronary Heart Disease and Stroke

A vast scientific literature has examined the role that physical activity (both leisure-time and occu-

pational) and cardiorespiratory fitness play in the risk of coronary heart disease. These studies indicate that physical inactivity and low cardiorespiratory fitness are causally linked to an increased risk of coronary heart disease (Kohl 2001). That is, physically inactive individuals and individuals with a low cardiorespiratory fitness are at greater risk of developing coronary heart disease than physically active and fit individuals. Furthermore, both occupational and leisure-time physical activity levels affect coronary heart disease risk (Kohl 2001). Results from a **meta-analysis** suggest that physically inactive individuals have a 45% greater risk of developing coronary heart disease than physically active individuals (Katzmarzyk and Janssen 2004).

There is a dose–response relationship between physical activity level and coronary heart disease risk (Kohl 2001), as illustrated in figure 10.1. This figure shows an inverse, curvilinear association between total weekly physical activity energy expenditure—whether it be strenuous sports or nonvigorous activities—and the **relative risk** of having a heart attack (Paffenbarger, Wing, and

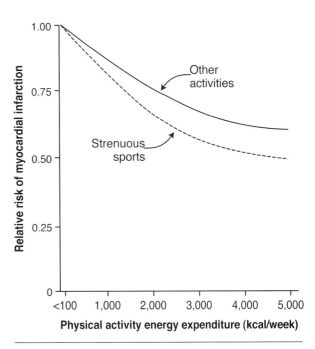

Figure 10.1 Dose–response relationship between physical activity level and coronary heart disease risk.

Reprinted from R.S. Paffenbarger Jr., A.L. Wing and R.T. Hyde, 1978, "Physical activity as an index of heart attack risk in college alumni," *Am J Epidemiol* 108(3): 161-75, by permission of Oxford University Press.

Hyde 1978). Two conclusions can be drawn from this dose–response curve. First, there is an increasing protective effect of physical activity throughout the energy expenditure range such that the most active individuals have, on average, the lowest risk of developing coronary heart disease. Second, the reduction in coronary heart disease risk is greatest at the lower end of the physical activity spectrum, implying that the largest impact on coronary heart disease risk in the population would be achieved by having sedentary individuals perform modest amounts of physical activity.

Physical inactivity and low cardiorespiratory fitness are causally linked to an increased risk of transient ischemic attacks and stroke (Kohl 2001; Lee et al. 2003). The protective effects of physical activity and fitness are apparent for both **ischemic strokes** and **hemorrhagic strokes.** Thus, physically inactive individuals and individuals with a low cardiorespiratory fitness have a greater risk of having a stroke than do physically active and fit individuals. A meta-analysis of the scientific literature indicates that moderately active individuals have a 17% reduction in stroke risk compared with inactive persons, whereas highly active individuals have a 25% reduction in stroke risk compared with inactive persons (Lee et al. 2003). Thus, there appears to be a curvilinear association between physical activity level and stroke risk, which is similar in shape to the dose–response relationship observed between physical activity and coronary heart disease risk (see figure 10.1).

Hypertension

In-depth reviews of the published scientific literature have concluded that physical inactivity and low cardiorespiratory fitness are causally linked to an increased risk of hypertension (Katzmarzyk and Janssen 2004; Pescatello et al. 2004). Thus, physically inactive and unfit individuals are at greater risk of developing hypertension than physically active and fit individuals. A meta-analysis concluded that physically inactive individuals have a 30% greater risk of developing hypertension than physically active individuals (Katzmarzyk and Janssen 2004).

The beneficial effects of physical activity on the risk of developing hypertension may be influenced by race and gender (Pereira et al. 1999; Pescatello et al. 2004). This race- and gender-dependent effect is illustrated in figure 10.2, using 6-year follow-up

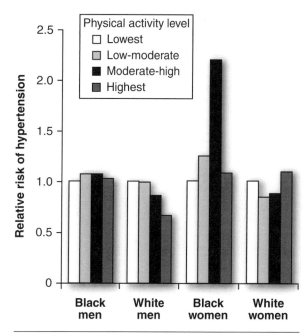

Figure 10.2 Relationship between physical activity level and hypertension risk in black and white men and women.

Data from M.A. Pereira, A.R. Folsom, P.G. McGovern, M. Carpenter, D.K. Arnett, D. Liao, M. Szklo and R.G. Hutchinson, 1999, "Physical activity and incident hypertension in black and white adults: The atherosclerosis risk in communities study," *Prev Med* 28(3): 304-312.

data from the Atherosclerosis Risk in Communities Study, a study of more than 7,000 black and white men and women aged 46 to 65 years. In this study, baseline leisure-time physical activity level was associated with the development of hypertension in white men but not in black men, black women, or white women. In white men, there was evidence of a graded dose–response relationship between physical activity level and hypertension risk, with the lowest risk being in the most active men. The mechanisms that account for this racial and gender difference are unclear but may be genetic.

Peripheral Vascular Disease

The influence of physical activity and fitness on peripheral vascular disease risk has not been thoroughly examined. However, because about 75% of the deaths in individuals with peripheral vascular disease are caused by coronary heart disease or stroke, and because physical inactivity and low fitness are risk factors for coronary heart disease and stroke (as discussed previously), it is reasonable to assume that physical inactivity and

It appears to be particularly important for those with overweight or obesity problems, which increase the chances of developing hypertension and type 2 diabetes, to engage in regular aerobic exercise to reduce their risks of cardiovascular morbidities.

low fitness are also risk factors for peripheral vascular disease.

In one of the few studies that have examined the relationship between physical activity and peripheral vascular disease risk, a sedentary lifestyle was reported to be an independent risk factor for *asymptomatic* peripheral vascular disease—that is, sedentary men and women had a 60% greater risk of developing asymptomatic peripheral vascular disease than physically active men and women (Hooi et al. 2001). However, in that study a sedentary lifestyle was not an independent risk factor for *symptomatic* peripheral vascular disease—that is, physical inactivity was not a risk factor for symptomatic peripheral vascular disease after the researchers considered the effects of other risk factors such as hypertension and diabetes mellitus. However, although physical inactivity did not have an effect on symptomatic peripheral vascular disease that was independent of these other risk factors, physical inactivity is a well-established risk factor for hypertension (described previously) and diabetes (see chapter 12). Thus, participation in physical activity may indirectly protect against the development of symptomatic peripheral vascular disease by protecting against the development of the primary risk factors for this disease, such as hypertension and diabetes.

Questions on Physical Inactivity, Low Cardiorespiratory Fitness, and Risk for Cardiovascular Morbidities

This next section addresses several questions. Is the relationship between physical activity and fitness and cardiovascular morbidities consistent across race and gender? Do both aerobic and resistance exercise have a positive effect on cardiovascular risk? What is the dose–response relationship between physical activity, fitness, and cardiovascular disease?

Consistency of Findings for Physical Activity and Cardiorespiratory Fitness

When we compare the effects of cardiorespiratory fitness and physical activity as risk factors for cardiovascular disease, there is a noticeably stronger and more consistent effect for cardiorespiratory fitness than there is for physical activity (Blair et al. 2001). An example of this comparison is shown in figure 10.3. This figure represents the relative risk for myocardial infarction over a 5-year follow-up in a sample of 1,453 Finnish men aged 42 to 60 years (Lakka et al. 1994). In this study, there was a strong dose–response relationship between fitness and heart attack risk, whereas the effects of physical activity were not as strong or as clear as those for fitness.

The finding that cardiorespiratory fitness is a better predictor of cardiovascular disease morbidity and mortality than is physical activity warrants consideration. Cardiorespiratory fitness and physical activity are closely related in that an individual's fitness level is in large measure determined by his or her physical activity participation over recent weeks or months. Although there is no consensus, most experts believe that the beneficial health effects of cardiorespiratory fitness are mediated by physical activity level. If cardiorespiratory fitness is determined by physical activity, and if the protective effects of fitness are mediated by physical activity, why then is physical activity not a stronger predictor of cardiovascular disease risk than cardiorespiratory fitness? The answer to this question is likely explained by the fact that fitness is measured using precise and objective measures (e.g., $\dot{V}O_2$max exercise test), whereas physical activity is assessed using imprecise and subjective measures (e.g., self-reported questionnaire). The imprecise physical activity measures will therefore be subject

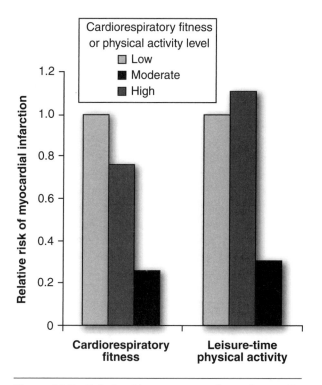

Figure 10.3 Relationships among cardiorespiratory fitness, physical activity level, and risk of myocardial infraction in men.

Data from T.A. Lakka, J.M. Venalainen, R. Rauramaa, R. Salonen, J. Tuomilehto and J.T. Salonen, 1994, "Relation of leisure-time physical activity and cardiorespiratory fitness to the risk of acute myocardial infarction," *N Engl J Med* 330(22): 1549-1554.

to higher rates of misclassification than the precise fitness measures, and the greater misclassification rates for physical activity likely explain the poorer associations with cardiovascular disease outcomes (Blair et al. 2001). This concept was discussed in greater detail for all-cause mortality in chapter 9.

> Cardiorespiratory fitness level is a stronger predictor of cardiovascular disease than is physical activity level. This is explained by the greater degree of measurement error for physical activity.

Consistency of Findings for Aerobic-Based and Resistance-Based Physical Activities

A plethora of scientific studies have examined the influence of aerobic-based physical activity and cardiorespiratory fitness on cardiovascular disease risk. The relationships between physical activity

and fitness and cardiovascular disease risk that have been covered up to this point in this chapter are for the most part based on aerobic-based measures of physical activity and fitness. Less than a handful of scientific studies have examined the influence of strength-based physical activities or musculoskeletal fitness on cardiovascular disease risk. The available information suggests that, as with aerobic-based physical activity, resistance activities protect against cardiovascular disease. For instance, in the Health Professional's Follow-Up Study of more than 40,000 American men, those who participated in 30 min or more of weight training per week had a 23% reduction in coronary heart disease risk compared with those who performed no weight training (Tanasescu et al. 2002). Although the effect of resistance training on the development of hypertension has not been studied, exercise training studies ranging in duration from 6 to 30 weeks have shown that resistance exercise will reduce resting systolic and diastolic blood pressures by an average of 3 mmHg (Kelley and Kelley 2000), further suggesting a protective effect of strength training.

Influence of Age, Gender, Race, and Changes in Physical Activity and Fitness

A major unresolved issue when we consider the effects of physical activity and fitness on coronary heart disease, stroke, hypertension, and peripheral vascular disease is the effects of gender and race on the observed relationships. At present, studies are sparse in women and even more so in non-Caucasian races. The available data for coronary heart disease and stroke indicate that the relationships for physical activity and fitness are relatively constant and independent of gender and race. However, as discussed previously in this chapter, the data for blood pressure suggest that the effects of physical activity may be specific to men and may be race dependent.

The vast majority of published studies used a single baseline measure of physical activity or fitness as it related to the risk for the cardiovascular disease endpoint, which at times was measured more than 25 years after the baseline exam (Kohl 2001). To put this into context, envision measuring physical activity level in a large cohort of research subjects in 1975 and then relating that single physical activity measure to the risk of cardiovascular disease up to the year 2000. Examination of a single baseline time point does not allow the changes in physical activity and other behaviors and risk factors that occurred during the follow-up period to be considered. Thus, in the aforementioned example, some of the subjects who were physically active in 1975 would have become sedentary over the 25-year follow-up, whereas some of the subjects who were sedentary in 1975 would have become physically active over the 25-year follow-up. Consequently, an important question is whether these changes in physical activity influence cardiovascular disease risk.

Limited research data are available on changes in physical activity and fitness as they relate to the development of coronary heart disease, stroke, hypertension, and peripheral vascular disease. Figure 10.4 presents the data from one of the few studies that has examined this issue. In this study of 9,777 men aged 20 to 82 years, cardiorespiratory fitness was assessed at baseline and after an average 5-year follow-up period. Study participants were then classified into one of four groups: (1) those who were unfit at baseline and unfit after follow-up, (2) those who were unfit at baseline but fit after follow-up, (3) those who were fit at baseline but unfit after follow-up, and (4) those who were fit at both baseline and after follow-up. After the second fitness test was performed, participants were then followed for an additional 5 years, on average,

to determine the influence of baseline fitness and changes in fitness on the risk for cardiovascular disease mortality. As shown in figure 10.4, men who maintained a high level of fitness were the least likely to die from cardiovascular disease whereas men who were persistently unfit were the most likely to die (Blair et al. 1995). More important, an increase in fitness in unfit men was associated with a decrease in cardiovascular mortality risk, whereas a decrease in fitness in fit men was associated with an increase in cardiovascular mortality risk. An important public health message can be drawn from this study: Fit individuals should remain physically active, and unfit individuals should become physically active.

Physical Inactivity and Low Cardiorespiratory Fitness As Risk Factors for Pulmonary Morbidities

This section examines the role of physical activity and cardiorespiratory fitness in the primary prevention of asthma and chronic obstructive pulmonary disease. Scientific evidence for a role of physical activity and fitness in the development of lung disease is extremely sparse. Only one research study has examined the role of physical activity or fitness as it pertains to the development of asthma. In this study, 757 asymptomatic children (aged 8-11 years at baseline) were followed for 10 years. During that time, 7% of the children developed asthma. There was a clear, inverse dose–response relationship between cardiorespiratory fitness level at baseline and the risk of developing asthma during adolescence, such that 16% of the least fit subjects developed asthma whereas only 4% of the most fit subjects developed asthma. This effect is illustrated in figure 10.5.

As for the risk of developing chronic obstructive pulmonary disease, the limited information available suggests that physical activity participation does not offer a protective effect. In a 22-year follow-up of 3,686 longshoremen (men who load and unload ships at a seaport), cigarette smoking was a risk factor for death from chronic obstructive pulmonary disease and cardiovascular disease, as would be expected. On the other hand, physical inactivity was a risk factor for cardiovascular disease–related mortality, but not for chronic obstructive pulmonary disease–related mortality (Paffenbarger, Brand, et al. 1978).

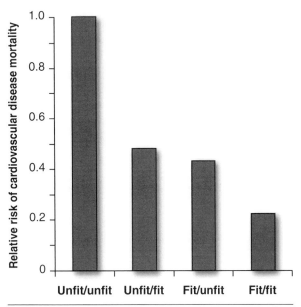

Figure 10.4 Influence of baseline fitness and changes in fitness on cardiovascular disease mortality.

Data from S.N. Blair, H.W. Kohl III, C.E. Barlow, R.S. Paffenbarger Jr., L.W. Gibbons and C.A. Macera, 1995, "Changes in physical fitness and all-cause mortality. A prospective study of healthy and unhealthy men," *JAMA* 273(14): 1093-1098.

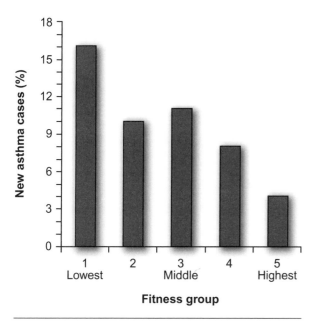

Figure 10.5 Influence of fitness on the development of asthma during adolescence.

Reprinted, by permission, from F. Rasmussen, J. Lambrechtsen, H.C. Siersted, H.S. Hansen and N.C. Hansen, 2000, "Low physical fitness in childhood is associated with the development of asthma in young adulthood: The Odense schoolchild study," *Eur Respir J* 16(5): 866-870.

It is unknown whether physical activity and fitness have comparable effects on pulmonary disease risk, whether the effects are different for aerobic and resistance types of physical activity, and whether gender or race influences the relationships. Additional research is required in the area of physical activity, fitness, and pulmonary disease risk.

Biological Mechanisms

The next part of this chapter briefly reviews the potential biological mechanisms that help interpret the cause-and-effect relationship between physical activity and fitness with various forms of cardiovascular and lung disease.

• *Hypertension.* Blood pressure is determined by the cardiac output (volume of blood pumped by the heart) and the total peripheral resistance of the blood vessels (determined by blood viscosity, length of blood vessels, and radius of blood vessels). Reductions in cardiac output do not occur after exercise training, indicating that a reduction in total peripheral resistance is the primary mechanism by which physical activity reduces resting blood pressure. The changes in total peripheral resistance in turn

are primarily mediated by changes in blood vessel diameter. A number of neural and local changes occur in response to chronic physical activity participation that reduce the vasoconstrictive state of the peripheral vasculature and in so doing decrease total peripheral resistance and blood pressure. These changes include less sympathetic neural influence on the peripheral blood vessels and local vasodilator influences on the blood vessels from molecules such as nitric oxide (Pescatello et al. 2004).

• *Coronary heart disease, stroke, and peripheral vascular disease.* There are several plausible biological mechanisms by which physical activity reduces coronary heart disease, stroke, and peripheral vascular disease risk, and these mechanisms are similar for these three forms of cardiovascular disease. Physical activity reduces blood pressure, improves the blood lipid profile (e.g., decreases triglycerides, increases high-density lipoprotein or "good" cholesterol), and decreases systemic inflammation (e.g., decreases blood levels of C-reactive protein) and in so doing decreases damage and **atherosclerosis** of the cardiac, cerebral, and peripheral blood vessels. Physical activity also improves endothelial function (e.g., improves the vasodilation and vasoconstriction properties of the blood vessels) and has an antithrombotic effect (e.g., reduces blood clotting), which further reduces the risk of adverse cardiac and cerebrovascular events. The body of evidence supporting the role of inflammation, endothelial function, and an antithrombotic effect is considerably smaller than the body of evidence supporting the role of blood pressure and the blood lipid profile.

• *Asthma.* There is no clear and accepted biological mechanism that explains the cause-and-effect relationship between physical activity and fitness and asthma. It has been proposed that high fitness could increase the threshold for respiratory symptoms and increase the level at which respiratory discomfort develops. Furthermore, a lower ventilation rate and volume during more intense aerobic exercise in fit individuals, which would occur consequent to a training-induced increase in the ventilatory threshold (e.g., the exercise intensity at which ventilation starts to increase at a quicker rate with further increases in intensity), would result in a smaller ventilatory stimulus to induce an asthma attack. In short, increased physical activity during childhood may positively influence the lungs and decrease the risk of developing asthma (Rasmussen et al. 2000).

Role of Physical Activity in Patients With Cardiac, Vascular, and Pulmonary Morbidities

In the next section of this chapter, we consider the role of physical activity in reducing morbidity and mortality in patients with cardiac, vascular, and pulmonary morbidities. Patients with preexisting cardiovascular disease (coronary heart disease, stroke, peripheral vascular disease) or lung disease (chronic obstructive pulmonary disease) need to participate in a medical evaluation, including an exercise stress test, before beginning a physical activity program. Furthermore, decisions regarding the degree of supervision and monitoring during the physical activity program and the nature of the physical activity program itself need to be determined by a multidisciplinary team of health care professionals. The physical activity program for cardiovascular disease and lung disease patients also needs to be geared toward the unique problems of the individual, because persons with cardiovascular and lung disease vary greatly in their clinical (e.g., severity of disease, existence of other diseases) and functional (e.g., impairments that limit physical activity) status. Detailed discussions of physical activity prescription for individuals with cardiovascular and lung disease, although an important topic, go beyond the scope of this chapter and are not covered here.

In general, individuals with coronary heart disease, stroke, peripheral vascular disease, and chronic obstructive pulmonary disease have a number of additional illnesses and diseases. For instance, many people who have suffered a stroke have also had a heart attack. Functional status is often severely impaired in individuals with the more severe types of cardiovascular and lung disease, and many of these individuals will have difficulty performing simple activities of daily living (e.g., walking up a flight of stairs, standing up from a chair). The illness, disease, and reduction in function often lead to depression and social isolation. In short, most individuals with cardiac and pulmonary morbidities have a reduced quality of life. Thus, in addition to focusing on the symptoms of the disease, cardiac and pulmonary rehabilitation programs should aim to improve cardiorespiratory and musculoskeletal fitness, decrease functional disability, improve psychological well-being, and improve the overall quality of life.

> Individuals with cardiovascular and lung disease often have many illnesses, diseases, functional problems, and psychosocial issues. Thus, physical activity programs for these patients need to focus on improving overall health and well-being and not merely the symptoms of the specific disease.

Cardiovascular Disease

Physical activity programs are effective treatments for improving exercise tolerance, functional status, cardiovascular risk factor profile, psychological status, and quality of life in individuals with peripheral vascular disease (Regensteiner and Hiatt 2002), coronary heart disease (American College of Sports Medicine 1994), stroke (Gordon et al. 2004), and hypertension (Pescatello et al. 2004). Likewise, physical inactivity and low cardiorespiratory fitness are risk factors for mortality in individuals with cardiovascular disease. An illustration of this effect for all-cause mortality is shown in figure 10.6. In this study of 772 men with established coronary heart disease, various types of

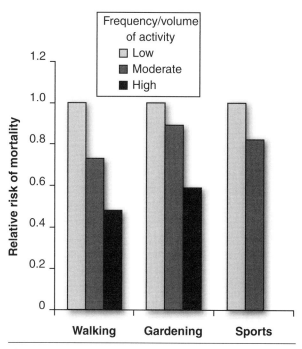

Figure 10.6 Influence of physical activity on all-cause mortality in men with coronary heart disease.

Data from S.G. Wannamethee, A.G. Shaper and M. Walker, 2000, "Physical activity and mortality in older men with diagnosed coronary heart disease," *Circulation* 102(12): 1358-1363.

physical activity were related to all-cause (as shown in figure 10.6) and cardiovascular disease mortality risk over a 5-year period (Wannamethee et al. 2000). The effect of low to moderately intense physical activities, such as walking and gardening, was particularly apparent in this study.

The observation that low to moderately intense activities had the greatest benefit is consistent with current exercise rehabilitation programs, which recommend low to moderately intense physical activity for individuals with cardiovascular disease. More specific physical activity recommendations based on current clinical guidelines for coronary heart disease (American College of Sports Medicine 1994; Pollock et al. 2000), hypertension (Pescatello et al. 2004; Pollock et al. 2000), and stroke (Gordon et al. 2004) are provided in table 10.1.

Chronic Obstructive Pulmonary Disease and Asthma

Pulmonary rehabilitation programs including physical activity reduce disease symptoms while increasing exercise tolerance, functional status, psychological status, and quality of life in individuals with chronic obstructive pulmonary disease (American Thoracic Society 1999). These beneficial changes occur despite the fact that participation in physical activity has little or no effect on the degree of airway obstruction in chronic obstructive pulmonary disease. Given the overlying theme of this chapter, it should not be surprising that cardiorespiratory fitness level is negatively related to mortality risk among individuals with chronic obstructive pulmonary disease. In fact, low cardiorespiratory fitness is a better predictor of mortality in patients with chronic obstructive lung disease than more traditional markers of disease severity such as measures of lung function and smoking (Oga et al. 2003).

Current clinical guidelines recommend that physical activity training be included in the management of patients with moderate to severe chronic obstructive pulmonary disease (American Thoracic Society 1999). Pulmonary rehabilitation programs emphasize aerobic exercise with a supplement of resistance training. More specific details are provided in table 10.1.

Table 10.1 Physical Activity Guidelines for Individuals With Selected Cardiovascular and Pulmonary Diseases

Disease	Exercise type	Frequency (days/week)	Duration and intensity
Coronary heart disease	Aerobic	3-7	20-40 min (continuous or intermittent) at a moderate intensity (40-85% $\dot{V}O_2$max)
	Strength	2-3	1 set of 10-15 repetitions of 8-10 exercises
Hypertension	Aerobic	5-7	30 or more min (continuous or intermittent) at a moderate intensity (40-60% $\dot{V}O_2$max)
	Strength	2-3	1 set of 10-15 repetitions of 8-10 exercises
Stroke	Aerobic	3-7	20-60 min (continuous or intermittent) at a moderate intensity (40-70% $\dot{V}O_2$max)
	Strength	2-3	1-3 sets of 10-15 repetitions of 8-10 exercises
	Flexibility	2-3	Stretches held for 10-30 s
	Neuromuscular	2-3	Coordination and balance exercises
Chronic obstructive pulmonary disease[a]	Aerobic	2-5	20-30 or more min (continuous or intermittent) at a moderate intensity (~60% $\dot{V}O_2$max)
	Strength	2-3	Loads ranging from 50% to 80% of 1-repetition maximum

[a]For chronic obstructive pulmonary disease, the exercise type, frequency, duration, and intensity reflect current practices rather than specific exercise clinical guidelines.

As with more severe forms of lung disease, asthma is characterized by a wide range of severity both within and between patients. In severe cases of asthma, when airflow obstruction is very poor with no reversibility, the physical activity capabilities and exercise rehabilitation of the disease are similar to those for chronic obstructive pulmonary disease (Satta 2000). However, in patients with mild to moderate asthma, the physical activity capabilities are relatively comparable to those of healthy, asthma-free persons. In fact, when free of symptoms, asthmatic and nonasthmatic people have the same physiological response to physical activity. With appropriate training and medication, individuals with mild to moderate asthma can successfully train and compete in high-intensity endurance events. However, regardless of fitness level and the degree of asthma, there remains the possibility of an exercise-induced asthma attack, a transient airway obstruction that sometimes occurs immediately after exercise. Although treatable with medication, repeated exercise-induced asthma attacks can alter psychosocial behaviors of asthmatics, leading to a restriction in exercise and a negative attitude toward physical activity. This may often result in an unnecessarily sedentary lifestyle, even in asthmatics who rarely experience exercise-induced asthma (Satta 2000).

Summary

Cardiovascular and lung diseases are highly prevalent and are leading causes of disability and death. The materials covered in this chapter provide strong support for the notion that physical inactivity and low fitness are leading risk factors for the development of most forms of cardiovascular disease and lung disease. Current public health guidelines are that adults should accumulate 30 to 60 min of moderately intense physical activity (e.g., brisk walking) on most, or preferably all, days of the week to help prevent these diseases (Pate et al. 1995).

It is also clear that participation in physical activity has many beneficial effects on disease symptoms, functional mobility, overall health and quality of life, and mortality risk in individuals with cardiovascular disease and lung disease. Physical activity recommendations for individuals with these diseases will vary depending on the type of disease, severity of disease, and needs and characteristics of the patient.

KEY CONCEPTS

asthma: A chronic disease affecting the airways of the lungs in which the inside walls of the airways are inflamed (swollen), making the airways very sensitive to normal irritants or allergens. When the airways react, they become narrow, causing symptoms like wheezing, coughing, chest tightness, and trouble breathing. Asthma cannot be cured, but for most patients it can be controlled.

atherosclerosis: The process in which deposits of fatty substances, cholesterol, cellular waste products, calcium, and other substances build up in the inner lining of an artery to form plaque.

cardiovascular disease: Dysfunctional conditions of the heart, arteries, and veins that supply oxygen to vital life-sustaining areas of the body like the brain, the heart itself, and other vital organs.

coronary heart disease: A narrowing of the coronary arteries that feed the heart (atherosclerosis), resulting in an insufficient blood supply to the heart muscle and causing angina (chest pain) or a myocardial infarction (heart attack).

chronic obstructive pulmonary disease: A lung disease in which the lung is damaged and the airways are partly obstructed, making it difficult to breathe.

hemorrhagic stroke: Occurs when a blood vessel bursts inside the brain. Bleeding irritates the brain tissue, causing swelling and forming a mass (hematoma), which displaces normal brain tissue.

hypertension: High blood pressure. Typically defined as a resting systolic blood pressure $\geq$140 mmHg or a resting diastolic blood pressure $\geq$90 mmHg. Hypertension is preceded by prehypertension, which is typically defined as a resting systolic blood pressure ranging from 120 to 139 mmHg or a resting diastolic blood pressure ranging from 80 to 89 mmHg. Blood pressure is dependent on output from the heart, blood vessel flexibility and resistance to blood flow, volume of blood, and blood distribution to organs, which are in turn influenced by hormones and the nervous system.

ischemic stroke: Occurs when too little blood reaches an area of the brain, which is usually caused by a clot that has blocked a blood vessel.

meta-analysis: A quantitative systematic review of the scientific literature in which the results from many studies dealing with the same topic are combined. This statistical method of combining the results of a number of studies allows more accurate estimations of effects than can be determined from looking at individual studies.

peripheral vascular disease: Disease of the blood vessels outside the heart and brain. It is often seen as a narrowing of vessels that carry blood to the legs and in rare cases to the arms, stomach, or kidneys.

relative risk: The probability of an event occurring in a comparison group divided by the probability of the event occurring in a referent group. The relative risk value is set at 1.00 in the referent group, and relative risk values >1.00 in the comparison group indicate a higher risk whereas relative risk values <1.00 in the comparison group indicate a lower risk.

stroke: Damage to part of the brain caused by interruption to its blood supply or leakage of blood outside the vessel walls. Sensation, movement, or function controlled by the damaged area is impaired. Strokes are fatal in about one third of cases.

STUDY QUESTIONS

1. What is the difference between ischemic stroke and hemorrhagic stroke? Is physical inactivity a risk factor for ischemic stroke, hemorrhagic stroke, or both?

2. Explain the biological mechanisms by which high physical activity and fitness protect against the development of hypertension.

3. Provide examples of variables that are typically improved with physical activity participation in patients with severe forms of cardiovascular disease or lung disease, other than the symptoms of the disease itself.

4. Does physical activity or cardiorespiratory fitness have a stronger effect on the risk of cardiovascular disease? Why?

5. Explain the dose–response relationship between physical activity level and the risk of coronary heart disease.

6. Are there any racial or gender differences in the protective effects of physical activity on coronary heart disease, hypertension, and stroke risk? If so, what are the differences and what are some possible explanations for these differences?

7. What are the current public health recommendations for physical activity that are aimed at reducing the risk of chronic diseases, such as cardiovascular and lung disease?

8. What frequency, duration, and intensity of aerobic physical activity are recommended for patients who have had a stroke?

Physical Activity, Fitness, and Obesity

■ Robert Ross, PhD ■ Ian Janssen, PhD

CHAPTER OUTLINE

Obesity is a leading risk factor for premature mortality and numerous chronic health conditions that reduce the overall quality of life. The prevalence of obesity has increased to epidemic proportions in both developed and developing countries during the past two to three decades, and this condition affects virtually all ages, sexes, races, and socioeconomic groups. Obesity reflects a continued positive energy balance, which is accompanied by unhealthy weight gain, and is linked to physical inactivity.

In this chapter we explore the role that physical activity plays in preventing and treating obesity. The first part of the chapter provides a definition and assessment system for overweight and obesity, examines prevalences and trends in overweight and obesity from a global perspective, and discusses the impact of obesity on health risk. In the second part, the role of specific fat depots in determining obesity-related health risk is examined. This is followed by an examination of the role of physical activity in the etiology of obesity from a population perspective. The third part of this chapter reviews the relationship between excess weight and physical activity and fitness, and the fourth section identifies the role of physical activity in the prevention and treatment of obesity.

Definition and Problem of Overweight and Obesity

Obesity is a condition of excessive fat accumulation that may impair health. The underlying disease reflects a continued positive energy balance, which is accompanied by undesirable weight gain. However, the degree of excess fat, its distribution within the body, and associated health consequences vary considerably between obese individuals.

Overweight and obesity are commonly assessed in the research and clinical setting using the **body mass index (BMI)**, a simple index of weight for height, calculated as weight in kilograms divided by the square of height in meters (kg/m^2). The globally accepted BMI classification system for adults is shown in table 11.1 (World Health Organization 1998). This classification system is based on the relationships among BMI, mortality, and chronic disease. Note that the BMI values are age-independent and are the same for both men and women. At present, these BMI values can be used for all racial and ethnic groups, with the exception of Asian populations, for whom a BMI of 23 kg/m^2 should be used to denote overweight and a BMI of 27.0 kg/m^2 should be used to define obesity (World Health Organization Expert Consultation 2004). The lower overweight BMI cut point for Asians reflects that body fat and health risk are higher for a given BMI in these subgroups (Deurenberg-Yap and Deurenberg 2003).

In addition to BMI, **waist circumference** can be used as a simple anthropometric index of overweight and obesity. But there is currently no consensus on what waist circumference thresholds should be used to denote increased health risk. The most commonly used cut points in men are ≥94 cm, which denotes a moderately increased risk of obesity-

Table 11.1 Classification of Overweight in Adults According to BMI

Classification	BMI (kg·m⁻²)	Morbidity and mortality risk	
		Low waist (men ≤102 cm, women ≤88 cm)	High waist (men >102 cm, women >88 cm)
Underweight	<18.5	Low[a]	NA
Normal range	18.5-24.9	Low	Increased
Overweight	≥25		
Pre-obese	25-29.9	Increased	High
Obese class I	30-34.9	High	Very high
Obese class II	35-39.9	NA	Very high
Obese class III	≥40	NA	Extremely high

NA = not applicable. All underweight individuals have a low waist circumference and all class II and class III obese individuals have a high waist circumference.

[a]Low risk for obesity-related complications but risk for some other health complications is increased.

Data from National Institutes of Health, National Heart, Lung, and Blood Institute, 1998, "Clinical guidelines on the identification, evaluation, and treatment of overweight and obesity in adults: The evidence report," *Obes Res* 6(S2): S51-210, and World Health Organization, 1998, *Obesity: Preventing and Managing the Global Epidemic* Vol. [WHO/NUT/NCD/98.1.1998]. Geneva: Report of a WHO Consultation on Obesity.

related complications, and ≥102 cm, which denotes a substantially increased risk of obesity-related complications. The corresponding waist circumference thresholds in women are 80 and 88 cm, respectively (World Health Organization 1998).

Ideally, waist circumference should be used in combination with BMI as an indicator of health risk, because waist circumference explains an additional component of obesity-related morbidity and mortality. An example of this effect is illustrated in figure 11.1. This figure illustrates the incidence rate of coronary heart disease according to BMI and waist circumference in a longitudinal follow-up study of 44,702 female American nurses. Within the low, moderate, and high BMI categories, larger waist circumference values were associated with increased risk of coronary heart disease.

Organizations such as the U.S. National Institutes of Health (National Institutes of Health National Heart Lung and Blood Institute 1998) have proposed that waist circumference values of ≥102 cm in men and ≥88 cm in women can be used within the BMI categories listed in table 11.1 to differentiate between those with and without abdominal obesity. For example, an obese class I male with a waist <102 cm would be considered to have a low abdominal fat mass and a "high" obesity-related risk, whereas an obese class I male with a waist ≥102 cm would be considered to have a high abdominal fat mass and a "very high" obesity-related risk.

> Waist circumference can be used alone, or in combination with BMI, as an indicator of obesity-related health risk

In children, BMI changes substantially with age, rising steeply during infancy, falling during the preschool years, and then rising again continuously into adulthood. This effect is illustrated in figure 11.2, which shows the mean BMI by age and gender in six nationally representative datasets. For this reason, overweight and obesity in children and adolescents are determined using age-specific BMI thresholds. There is not the same level of agreement over the classification of overweight and obesity by BMI in youth as there is in adults. A number of countries have produced nationally representative BMI-for-age growth curves, which allow an individual's BMI to be expressed as an age- and sex-specific percentile. Historically, the 85th and 95th percentiles have been used to determine overweight and obesity status, respectively (World Health Organization 1998). International BMI standards for defining overweight and obesity in youth have also been developed by regressing the adult BMI cut points of 25 kg/m² and 30 kg/m² at age 18 back through the growth curve (Cole et al. 2000). This approach also provides age- and sex-specific BMI cut points that, from a global perspective, may be the most appropriate means for defining overweight and obesity in children and adolescents.

Impact of Overweight and Obesity on Health

Overweight and obesity are leading risk factors for premature mortality and numerous chronic health conditions that reduce the overall quality of life. These chronic health conditions include type 2 diabetes, coronary heart disease, hypertension, stroke, certain forms of cancer, gallbladder disease, and osteoarthritis (National Institutes of Health National Heart Lung and Blood Institute 1998; World Health Organization 1998). Table 11.2 summarizes the extent of the morbidity and mortality risk for the average obese individual compared with the average individual with a normal body weight.

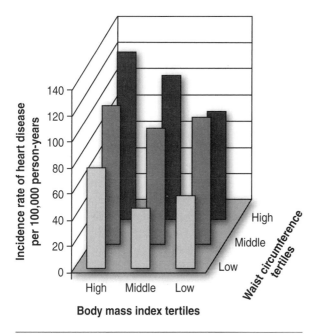

Figure 11.1 The incidence rate of coronary heart disease according to body mass index and waist circumference in a longitudinal study of 44,702 female American nurses.

Adapted from *JAMA*, 1998, 280(21): 1843-1848. *Copyright © 1998, American Medical Association.* All rights reserved.

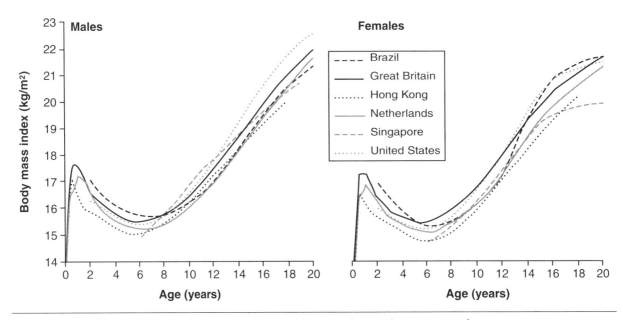

Figure 11.2 Six nationally representative datasets with children aged 0 months to 18 years of age.

Reprinted from T.J. Cole, M.C. Bellizzi, K.M. Flegal and W.H. Dietz, 2000, "Establishing a standard definition for child overweight and obesity worldwide: International survey," *British Medical Journal* 320: 1240-1243, with permission from the BMJ Publishing Group.

Table 11.2 Health Risks Associated With Obesity

Health condition	Extent of increased risk in obese individuals
Premature mortality	↑
Coronary artery disease	↑↑
Stroke	↑
Hypertension	↑↑↑↑
Type 2 diabetes	↑↑↑
Colon cancer	↑
Postmenopausal breast cancer	↑
Gall bladder disease	↑↑↑
Osteoarthritis	↑↑

↑ Risk for condition increased by approximately 25-50%; ↑↑ risk for condition increased by approximately 200%; ↑↑↑ risk for condition increased by approximately 350%; ↑↑↑↑ risk for condition increased by more than 400%.

Data from P.T. Katzmarzyk and I. Janssen, 2004, "The economic costs associated with physical inactivity and obesity in Canada: An update," *Can J Appl Physiol* 29: 90-115.

Global estimates indicate that approximately 58% of diabetes cases, 21% of ischemic heart disease cases, and 8% to 42% of certain cancers are directly attributable to excess body weight (World Health Organization 1998). Conservative estimates indicate that obesity alone, not including overweight, accounts for 2% to 7% of total direct health care costs in developed countries (World Health Organization 1998). Indirect costs, which are far greater than the direct costs, include lost work days, disability pension, impaired quality of life, and premature mortality. Indeed, in industrialized countries obesity is one of the leading causes of death.

Global Prevalence and Secular Trends in Obesity

Overweight and obesity are at epidemic proportions globally, with more than 1 billion overweight adults and at least 300 million of these clinically obese (World Health Organization 1998). Obesity affects virtually all ages, sexes, races, and socioeconomic groups. Even in developing countries, obesity coexists with malnutrition. Current obesity levels range from less than 5% in China, Japan, and many African nations to more than 75% in urban Samoa (World Health Organization 1998). In a World Health Organization–sponsored study of 48 mainly European countries conducted between 1983 and 1986, between 50% and 75% of adults aged 35 to 64 were either overweight or obese as defined by BMI (World Health Organization 1998). Thus, the majority of middle-aged adults in these developed nations were at increased health risk attributable to excess body weight. The following evidence indicates that the situation in the 21st century is likely to be significantly worse.

The distribution of BMI is shifting upward in most child and adult populations, and obesity prevalences have increased at an alarming rate in both developed and developing countries during the past 2 to 3 decades. This effect is shown in figure 11.3, which illustrates the prevalences of adult obesity in the 50 U.S. states as well as the 10 Canadian provinces and three Canadian territories in the years 1990, 1994, 1998, and 2000. During this 10-year time span, there was a dramatic increase in adult obesity prevalences across North America. The sudden increase in obesity prevalences appears to have started during the 1980s, and there is no indication that these trends have slowed in the 21st century. Although there are limited representative population data available, it appears that the prevalence of a high waist circumference is also high and increasing. Data from the United States indicate that the prevalence of a high waist circumference ($\geq$102 cm in men, $\geq$88 cm in women) in 1988 to 1994 was 29.5% in men and 46.7% in women. From 1988-1994 to 1999-2000, these prevalences increased to 36.9% in men and 55.1% in women (Ford et al. 2003). Obesity has become so widespread in North America that it is causing not only the indirect economic costs of reduced productivity due to poor health, but direct costs, such as higher fuel costs for airlines because of increased passenger weight, and ambulance companies having to buy extra large, heavy-duty stretchers with hydraulic lifts because so many patients are now too big to fit on standard stretchers, and too heavy to be lifted safely by emergency workers.

> Overweight and obesity are highly prevalent conditions in both developed and undeveloped nations. There has been a dramatic increase in obesity prevalences in recent years, and there is no indication that these temporal trends are slowing.

Fat Depots

The next section considers the importance of specific fat depots. A vast number of scientific studies conducted during the course of the past half century have provided clear evidence that the distribution of fat is a more important determinant of obesity-related health risk than is the degree of overall adiposity.

Total Fat

The relationship between BMI and obesity-related health risk is explained in large measure by the ability of BMI to predict total fat. Although BMI is a good marker of total fat, the relationship between BMI and adiposity is marked by large interindividual variation, and somewhere on the order of 25% to 50% of the between-individual variation in total fat is not accounted for by BMI (Janssen et al. 2002). Gender, race, age, genetic factors, and fitness level all influence the relationship between BMI and total fat. More direct measures of body fat (e.g., skinfolds) provide more accurate measures of total fat and obesity-related health risk than does BMI (Janssen et al. 2002). Waist circumference is also a good marker of total fat and, interestingly, is as strong a correlate of total fat as is BMI (Janssen et al. 2002).

Abdominal Subcutaneous and Visceral Fat

In addition to the degree of excess fat, the distribution of fat within the body is an important determinant of the health consequences associated with obesity. In particular, the two abdominal fat depots—**abdominal subcutaneous fat** and intraabdominal or **visceral fat**—are associated with the pathogenesis of numerous cardiovascular disease and diabetes risk factors (National Institutes of Health National Heart Lung and Blood Institute 1998; World Health Organization 1998). The accumulation of excess visceral fat is of particular relevance for obesity-related health risk.

Anatomically defined, visceral fat consists of **adipocytes** contained within the visceral peritoneum. The visceral peritoneum is the membrane that covers the abdominal organs of the gastrointestinal tract, and it extends from about the 11th thoracic vertebra to about the 5th lumbar vertebra (figure 11.4 on page 179). This definition excludes perirenal adipocytes that comprise the extraperitoneal fat surrounding the kidneys. However, because the peritoneum cannot be seen on magnetic resonance imaging (MRI) or computed tomography (CT) images, in obesity studies visceral fat is typically defined as the sum of the fat contained within (intraperitoneal, ~80% of total visceral fat) and behind (extraperitoneal, ~20% of total visceral fat) the peritoneum.

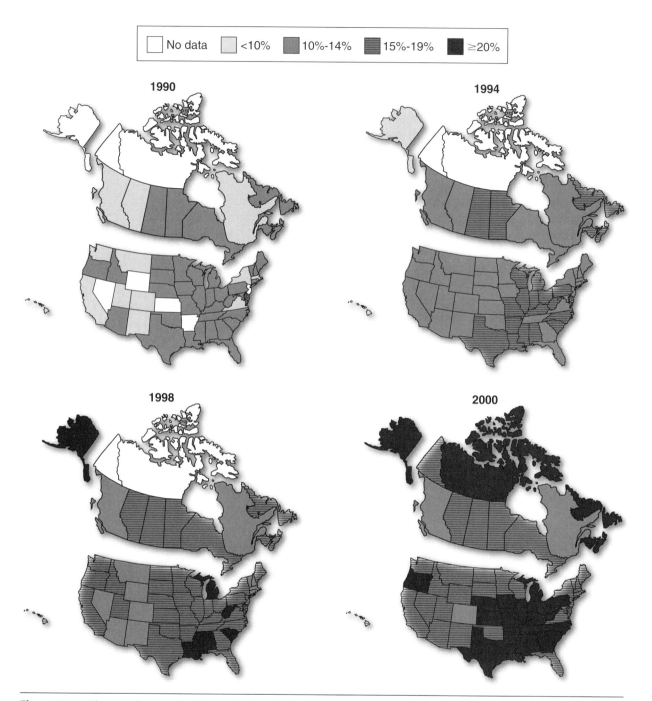

Figure 11.3   The prevalences of adult obesity in the United States and Canada in the years 1990, 1994, 1998, and 2000.

Data from P.T. Katzmarzyk, 2002, *Canadian Medical Association Journal* 166(8): 1039-40, and Mokdad et al., 2001, *JAMA* 286: 1195-2000.

The mechanisms that explain the association between visceral fat and obesity-related health risk are not completely understood. Free fatty acids, glycerol, and hormones released from adipose tissue within the visceral peritoneum are drained into the hepatic portal vein that travels directly to the liver. A common postulate is that sustained exposure of the liver to an increased flux of free fatty acids via the portal circulation is the anteced-

ent to many of the disturbances in glucose and lipid metabolism that are associated with abdominal obesity (Björntorp 1990). This is often referred to as the "portal theory." However, emerging evidence provides strong support for the notion that adipose tissue–derived cytokines may be the mechanism that links visceral fat with metabolic health risk. Adipose tissue releases more than 100 hormones and autocrine–paracrine factors. The majority of these factors are cytokines, and many of these are involved in the pathogenesis of atherosclerosis, hypertension, and insulin resistance. Most of these factors are secreted to a different extent in different fat depots, and for the most part, visceral fat is a more active producer of these factors than is subcutaneous fat (Matsuzawa 2002). Furthermore, the portal drainage of visceral fat through the liver and other visceral organs may amplify the effect of the cytokines released by this fat depot.

Subcutaneous fat is the layer of adipocytes that lies directly between the dermis layer of the skin and extends over the whole body. There are no commonly accepted boundaries that define the proximal and distal borders of abdominal subcutaneous fat. It is well established, however, that there are regional differences in **lipolysis** between subcutaneous adipocytes in the abdomen and those in the leg. Subcutaneous adipocytes in the abdominal region have an increased lipolysis rate and deposit free fatty acids into the systemic circulation at a greater rate than subcutaneous adipocytes in the appendicular regions (Arner et al. 1990). On a metabolic level,

the contribution that abdominal subcutaneous fat makes to obesity-related health risk may result from the increased spillover of free fatty acids into the systemic circulation that are subsequently delivered to other tissues such as the liver, pancreas, and skeletal muscle.

A number of factors influence fat distribution. For a given level of total fat, men have more abdominal and visceral fat than women, older adults have more abdominal and visceral fat than younger adults, Caucasians have more abdominal and visceral fat than African Americans, and physically inactive and unfit individuals have more abdominal and visceral fat than physically active and fit individuals.

Radiological imaging methods such as CT and MRI are the most accurate means available for in vivo quantification of body composition at the tissue level. Although access and cost are obstacles to routine use, these imaging approaches are now used extensively in body composition research and are the methods of choice for measuring abdominal fat. CT and MRI systems generate cross-sectional images of human anatomy, as illustrated in figure

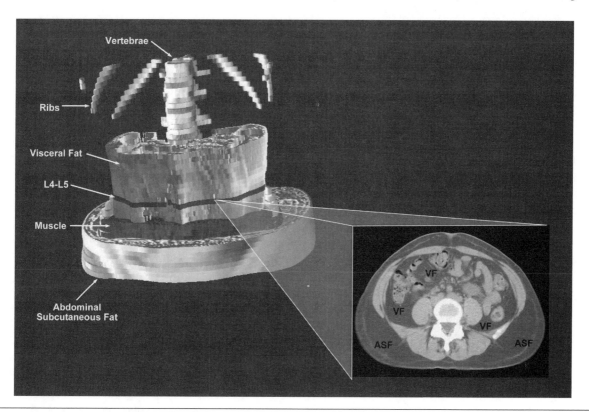

Figure 11.4   Tissue distribution in the abdominal cavity. The two-dimensional image on the lower right represents a cross-sectional computed tomography image obtained at the level of the intervertebral disk between the fourth and fifth lumbar vertebrae. The abdominal subcutaneous fat (ASF) is the dark gray tissue located directly beneath the skin. The visceral fat (VF) is the dark gray tissue located underneath the skeletal muscle and surrounding the visceral organs. The three-dimensional image is a computerized re-creation of the various tissue layers in the abdomen.

11.4. Image analysis programs are used to determine the area (cm²) of abdominal subcutaneous fat and visceral fat in these cross-sectional images, with a greater area representing a greater tissue size. Typically a single image strategy is used; that is, one image is obtained at a predetermined level of the abdomen, usually at the level corresponding to the intervertebral disk between the fourth and fifth lumbar vertebrae. The area measures from the single cross-sectional image are used as an index of tissue size. However, multiple cross-sectional images can also be obtained across the abdominal region, either contiguously or with gaps between two consecutive images, and the area values derived from these multiple images can be used to calculate the volume or mass of the entire abdominal subcutaneous and visceral fat depots.

Unfortunately, CT and MRI are not appropriate techniques for assessing abdominal fat in the clinical setting because of their high cost and limited accessibility. In that regard, waist circumference is a convenient and simple measurement that is an approximate index of abdominal subcutaneous and visceral fat (Janssen et al. 2002). Furthermore, changes in waist circumference reflect changes in abdominal obesity and risk factors for cardiovascular disease (Janssen et al. 2002; Ross et al. 2000). Thus, waist circumference is a useful clinical tool that can be used to identify individuals at increased health risk attributable to abdominal obesity.

> The accumulation of excess visceral fat is of particular concern in overweight and obese individuals. In the research setting visceral fat can be measured using sophisticated imaging techniques, whereas in the clinical setting waist circumference is used as an index of visceral obesity. Although visceral fat tends to increase with increasing BMI, other factors such as age, race, physical activity, gender, and genetics influence visceral fat accumulation

Non–Adipose Tissue Fat

An emerging area of research has focused on the relevance of non–adipose tissue fat, otherwise known as **ectopic fat,** in determining obesity-related health risk. Of primary concern are the ectopic fat depots in skeletal muscle and the liver.

In skeletal muscle, lipids can be stored within the muscle fibers, referred to as intramyocellular lipids, and outside the muscle fibers, referred to as extramyocellular lipids. The intramyocellular lipids are primarily stored in droplets close to the mitochondria, where they serve as a readily available energy source. The extramyocellular lipids accumulate to form adipocytes between the muscle fibers and bundles, where they serve as a long-term energy storage site. When numerous extramyocellular adipocytes accumulate, they are seen as marbled fat within the skeletal muscle.

Obesity is associated with an increase in both intramyocellular lipids and extramyocellular lipids, although the increase in intramyocellular lipids is of greater concern for obesity-related health risk. Most notably, intramyocellular lipids have been linked with insulin resistance, a leading risk factor for type 2 diabetes and cardiovascular disease (Kelley et al. 2002). In fact, the relationship between intramyocellular lipids and insulin resistance is independent of total and abdominal fat content.

There are, however, situations where the accumulation of intramyocellular lipids is not related to insulin resistance. Most notably, a hallmark adaptation to chronic exercise is an increase in lipid oxidation, and, accordingly, intramyocellular lipids levels are increased in aerobically trained athletes. Despite high intramyocellular lipid levels, athletes are characterized by elevations in insulin sensitivity (Goodpaster et al. 2001). The periodic depletion and repletion of intramyocellular lipids that occur with regular exercise may disrupt the mechanistic link between intramyocellular lipids and insulin resistance that occurs in obesity. That is, high levels of intramyocellular lipids do not appear to have adverse metabolic consequences in skeletal muscle that has the capacity for efficient lipid utilization, as in aerobic athletes (Goodpaster et al. 2001). Thus, the role of intramyocellular lipid accumulation in explaining obesity-related health risk is confounded by decreasing levels of physical activity. This suggests the ability to oxidize intramyocellular lipids may be more important than the storage of lipids. In this way physically active obese individuals would be at less risk for a given level of intramyocellular lipid accumulation than their sedentary counterpart.

The second ectopic fat depot of interest is liver fat. In obesity studies, liver fat is usually measured by CT, as illustrated in figure 11.5. The density or attenuation of muscle and liver in CT images is an indication of the lipid content of these tissues—the greater the lipid content, the lower the tissue density on the CT images and the darker the color of the liver on the CT image. Determination of liver fat may provide a mechanistic link between abdominal obesity, in particular visceral fat, and increased obesity-related health risk. It has been hypothesized that the metabolic importance of visceral fat may be mediated by the delivery of free fatty acids into the hepatic portal vein, exerting potent and direct effects on the liver. Consistent with this position, liver fat content is correlated with total and visceral fat content, and reductions in total and visceral fat that are induced by caloric restriction are associated with corresponding reductions in liver fat (Malnick et al. 2003). Evidence indicates that liver fat is related to metabolic risk factors such as plasma triglyceride levels and insulin resistance and that the effects of liver fat are at least in part independent of visceral fat content, suggesting that the amount of fat within the liver carries an independent health risk (Malnick et al. 2003). Preliminary evidence suggests that cardiorespiratory fitness is negatively associated with liver fat.

Ectopic fat depots in skeletal muscle and the liver increase with obesity, and research suggests that these fat depots partially explain the effect of obesity on health risk. More research is needed to clarify the role of obesity and physical activity on ectopic fat content and related health risks.

Relationships Among Excess Weight, Physical Activity, and Fitness

We now consider the role that physical activity plays in the development of obesity. In so doing, we consider the following questions. Do population-wide changes in physical activity levels explain the recent obesity epidemic? Are physically inactive individuals more likely to develop obesity than physically active individuals? Does participating in physical activity increase hunger and energy intake?

Figure 11.5   Measurement of liver fat by computed tomography. The three-dimensional image of the liver on the left was generated using a series of contiguous images. The cross-sectional image on the top right was obtained in an individual with a fatty liver and on the bottom right in an individual with a lean liver. The fatty image is darker in color than the lean liver, which indicates a lower density and, hence, greater lipid accumulation.

Role of Physical Activity in the Etiology of Obesity

Obesity results from a chronic energy imbalance whereby intake exceeds expenditure. Thus, the hypothesis that physical inactivity has contributed to the obesity epidemic is a very reasonable assumption. However, the relative contributions of physical inactivity versus dietary consumption to the obesity epidemic are unclear because there are no accurate and reliable data available on the two components of energy balance in large representative populations. Data on physical activity levels and dietary intake in large representative surveys are based entirely on questionnaire measures.

There are few ecological data on temporal changes in energy intake. In the United States, survey data suggest that energy intake remained relatively stable from 1965 to 1995, despite an average body weight increase of about 10% (Jeffery and Utter 2003). If correct, the dietary data would indicate that the primary cause of the American obesity epidemic has been physical inactivity. However, this contention is not supported by the available data, and there appears to have been a general stability in the percentage of the adult U.S. population meeting physical activity guidelines from the mid-1980s to the mid-1990s (Jeffery and Utter 2003).

There are a number of possibilities as to why population trends in physical activity and dietary intake data are not congruent with the corresponding trends in body weight and obesity. Among these are the methodological and technical problems of reporting physical activity and food intake that may have changed over time. Another factor to consider is the increased public awareness of the health benefits of physical activity and of what "counts" as physical activity. In recent years there has been an increased knowledge that activities such as walking "count" as physical activity, whereas in the past people may have only considered vigorous activities such as running or swimming as physical activity. The most likely explanation, however, is that population trends in physical activity participation have focused solely on leisure-time physical activity and thus are not sensitive to any changes that have occurred in non-leisure-time physical activity levels. This is essential, because leisure-time physical activity accounts for a small proportion of the day (<1 hr) for most individuals in most countries around the world and consequently a small proportion of total daily energy output. Thus, when considering obesity it is also necessary to take into account the energy expenditure during the remainder of the day.

Unlike leisure-time physical activity, non-leisure-time physical activity has decreased substantially in the past 20 to 30 years (Jeffery and Utter 2003; Saris et al. 2003). Physical activity has been engineered out of daily life by increasing mechanization at work and in the home. In the workplace, for example, there are far fewer blue-collar manual labor jobs and far more white-collar desk jobs in the new millennium than there were in the 1970s. There has been a decline in the use of walking and biking as a means of transportation, whereas the use of automobiles has steadily increased. At home, inactive forms of entertainment have been on the rise. Videocassette recorders, digital video disc players, cable and satellite television, and home computers and the Internet are technologies that have only become widely available to the public in the past 2 decades. The accessibility to countless labor-saving devices has also increased in recent years, such as remote controls, scrub-free cleaning products, lighter and more efficient tools, and dish-washing machines, just to name a few. Consistent with these changes in technology, adults in industrialized nations spend approximately 20% less time on housework now than they did 30 years ago (Jeffery and Utter 2003). These changes have fostered sedentary habits and have greatly reduced energy expenditure in the portion of the day not spent on leisure-time physical activity.

What is the implication of a decrease in non-leisure-time physical activity? Because leisure-time physical activity levels have changed minimally, the implication is that total physical activity–induced energy expenditure at the population level has likely decreased during the past 30 years. Even a 10 kcal per day reduction in total daily physical activity would be substantial when added up over time. An individual who is 42 J (10 kcal) per day above his or her weight maintenance energy requirements would gain about 0.5 kg (1 lb) of body fat per year. Coincidentally, the average young to middle-aged adult in North American gains about 0.5 kg (1 lb) per year in body weight (Williamson et al. 1993).

> Population-based data suggest that dietary intake and leisure-time physical activity levels have changed minimally during the recent obesity epidemic. Recent technological advances have engineered non-leisure-time physical activity out of the daily routine of most individuals, and these changes have likely contributed to the obesity epidemic.

Given the potential importance of the decrease in non-leisure-time physical activity it is reasonable to ask, Is the general population going to abstain from using these labor-saving devices to curtail the obesity epidemic? The answer to this question is almost certainly no, because these labor-saving devices make life easier and more enjoyable and they free up additional leisure time. The implication is that the public must be willing to compensate by decreasing caloric intake or increasing leisure-time physical activity levels. This fundamental public health message is not novel. Indeed, Jean Mayer, the nutritionist who founded the School of Nutrition at Tufts University, made the following observation in 1955:

> In many cases adaptations to modern conditions without development of obesity implies that the person will have either to step up his activity or endure mild or acute hunger all his life. . . . If the first alternative, stepping up activity, is difficult, it is well remembered that the second alternative, life-time hunger, is so much more difficult that to rely on it for weight control in cases of sedentary overweight can only continue the fiascos of the past. (Mayer 1955)

Relationships Among Excess Weight, Physical Activity, and Fitness

Observational or cross-sectional data on the relationships between physical activity and cardiorespiratory fitness level and body weight and obesity have, in general, shown that there is an inverse association between these measures. That is, physically active and fit individuals are considerably less likely to be obese than physically inactive and unfit individuals. Population-based longitudinal studies on the relationship between physical activity and weight gain have also shown an inverse relationship between physical activity level and increase in weight and body fat or the prevalence of overweight or obesity. That is, physically active and fit individuals are considerably less likely to develop obesity than physically inactive and unfit individuals. The doses of physical activity required to maintain a normal body weight and to reverse obesity are discussed next.

Interactions Among Physical Activity, Energy Intake, and Body Weight

It is commonly held that physical activity increases hunger and food intake, thereby compromising its utility as a strategy for controlling body weight and obesity. The scientific evidence, however, does not support this contention. To the contrary, adult men and women can tolerate daily energy deficits of up to 4,187 J (1,000 kcal) for 2 weeks when partaking in physical activity programs without influencing hunger or **ad libitum** food intake (Blundell et al. 2003). Thus, the short-term effect of participating in a physical activity program will be weight loss, although the weight loss would be minimal over a 2-week period.

In the long term, about 2 weeks after commencing a physical activity program, food intake begins to increase. As noted in chapter 5, on average the increase in ad libitum energy intake compensates for about 30% of the physical activity–induced energy expenditure. Thus, the average individual will likely continue to lose weight in the weeks and months after commencing a physical activity program. However, there is a large variation in the level of compensation that occurs between subjects. Some individuals do not compensate at all whereas others compensate close to 100% (Blundell et al. 2003). Phrased differently, some individuals show no increase in ad libitum energy intake even weeks after commencing a physical activity program, whereas the ad libitum energy intake in some individuals will increase by an amount comparable to what they expended during their physical activity session. These compensators, who unfortunately cannot be readily identified, must make a concerted effort not to increase food intake if weight loss is a goal.

> Beginning a physical activity program will not result in an increase in food intake in the short term, and over the long term, the increase in caloric intake will on average only compensate for 30% of the physical activity–induced energy expenditure.

In terms of physical activity and diet interactions, it is important to consider the alternative. That is, what happens to energy intake in physically active individuals who become physically inactive? This situation does not induce a compensatory reduction in ad libitum energy intake and leads to a markedly positive energy balance, most of which is stored as fat (Blundell et al. 2003). Thus, if physically active individuals suddenly become inactive, they will in all likelihood gain body weight and fat unless they make a concerted effort to reduce their food and caloric intake.

Role of Physical Activity in Prevention and Treatment of Excess Weight

From the preceding discussion it is clear that a decrease in physical activity contributes to the increased prevalence of obesity worldwide. Accordingly, it is intuitive to suggest that an increase in physical activity levels would be associated with a decrease in obesity. Indeed, evidence from population-based studies with long-term follow-up confirms that age-related weight gain is attenuated in physically active adults compared with sedentary adults (Saris et al. 2003). However, although the experts agree that an increase in physical activity is associated with a lower prevalence of obesity, precisely how much physical activity is required to prevent age-related weight gain is the subject of considerable debate.

Physical activity guidelines for adults were initially formulated to prevent morbidity and mortality. Indeed, there is now a large body of evidence suggesting that the accumulation of 30 min or more of moderate-intensity exercise on most days of the week provides substantial benefits across a broad range of health outcomes. Refer to chapter 2 for a thorough review of these guidelines. Although 30 min of daily physical activity may prevent unhealthy weight gain in some individuals, it is now generally reported that this volume of physical activity may be insufficient to prevent age-related weight gain in many if not most adults. Table 11.3 summarizes reports from various expert groups who have considered how much physical activity is required to prevent weight gain. In general, the expert groups derived their recommendations from analysis of longitudinal population-based studies that related estimates of self-reported physical activity levels over time to corresponding changes in body weight. The principal exception was the report from the Institute of Medicine, which based its recommendations on cross-sectional analysis of data that used the doubly labelled water method to estimate total energy expenditure. Combined with the measurement of basal metabolic rate, the total energy expenditure values derived by doubly labelled water can be used to calculate an individual's **physical activity level (PAL)** (see table 11.4). The PAL is also used in population-based studies as a way to standardize the various approaches used to determine physical activity energy expenditure. In short, the higher the PAL, the higher the level of physical activity performed daily.

The PAL is defined as the ratio of total energy expenditure to 24 hr basal energy expenditure.

Table 11.3 Current Recommendations for Prevention of Weight Gain

Expert group (year)	Recommendation for prevention of weight gain
World Health Organization (1998)[a]	Men and women should achieve a PAL of 1.75.
U.S. Surgeon General (2001)[b]	Adults should get at least 30 min of moderate physical activity on most days of the week. Children should aim for 60 min.
International Obesity Task Force (2002)[c]	Men and women should achieve a PAL of 1.8 to prevent unhealthy weight gain. Vigorous activity is more clearly linked to weight stability.
Institute of Medicine (2002)[d]	All adults should accumulate 60 to 90 min of daily physical activity. This corresponds to a PAL greater than 1.6
Stock Conference (2003)[e]	Men and women should undergo moderate-intensity physical activity for about 45 to 60 min per day or achieve a PAL of 1.7.

PAL = physical activity level.

[a]World Health Organization. 1998. *Obesity: Preventing and managing the global epidemic* (WHO/NUT/NCD/98.1.1998). Geneva: WHO.

[b]U.S. Surgeon General. 2001. *Call to action to prevent and decrease overweight and obesity.* Washington, DC: U.S. Department of Health and Human Services.

[c]Erlichman, J., A.L. Kerbey, and W.P. James. 2002. Physical activity and its impact on health outcomes. Paper 2: Prevention of unhealthy weight gain and obesity by physical activity: An analysis of the evidence. *Obesity Reviews* 3(4): 273-287.

[d]Institute of Medicine. 2002. Dietary reference intake for energy, carbohydrate, fiber, fat, fatty acids, cholesterol, protein and amino acids. Washington, DC: National Academy Press.

[e]Saris, W.H., S.N. Blair, M.A. van Baak, S.B. Eaton, P.S. Davies, L. Di Pietro, M. Fogelholm, A. Rissanen, D. Schoeller, B. Swinburn, A. Tremblay, K.R. Westerterp, and H. Wyatt. 2003. How much physical activity is enough to prevent unhealthy weight gain? Outcome of the IASO 1st Stock Conference and consensus statement. *Obesity Reviews* 4(2): 101-114.

Table 11.4   Physical Activity Level (PAL)

PAL category	PAL value	Walking equivalence, at a pace of 3-4 mph, for individuals weighing		
		44 kg (97 lb)	70 kg (154 lb)	120 kg (264 lb)
Sedentary	1.0-1.39	0	0	0
Low active	1.4-1.59	~2.9 miles 43-58 min	~2.2 miles 33-44 min	~1.5 miles 22-30 min
Active	1.6-1.89	~9.9 miles 148-198 min	~7.3 miles 109-146 min	~5.3 miles 79-106 min
Very active	1.9-2.5	~22.5 miles 337-450 min	~16.7 miles 250-334 min	~12.3 miles 184-246 min

Thus, PAL depends to a certain degree on body size and age, because these variables contribute to basal energy expenditure. Individuals can be placed into one of four activity categories based on their PAL, as shown in table 11.4.

The four PAL categories in the table correspond roughly to quartiles in the population. Thus, the Sedentary category is the lowest 25% of the population, whereas the Very Active category is the highest 25%. The Sedentary category was defined according to basal energy expenditure, the thermic effect of food, and the energy expended in physical activities that are required for independent living. Incorporating about 40 min per day of walking at a speed of 3 to 4 mph, in addition to the activities that are part of daily living, raises the PAL to the Low Active level in the average 70 kg person. To reach a PAL of 1.7 to 1.8—which is currently recommended for the prevention of age-related weight gain—the average 70 kg person must incorporate about 2 hr per day of walking at 3 to 4 mph, in addition to the activities that are part of daily living.

The distances and times required to move to the more active categories vary considerably by body weight. Thus, more walking is required for lean individuals and less for obese individuals. The times can be reduced substantially by walking faster or performing other activities of a more vigorous nature. For example, if the average person walked 30 min per day at 4 mph, cycled moderately for another 25 min, and played tennis for 40 min, the PAL would increase to about 1.75 (Active).

The consensus opinion at present is that the prevention of weight gain in both developed and undeveloped countries is associated with a PAL of about 1.7 to 1.8 (table 11.3). To achieve a PAL of 1.8 would require a physical activity habit equivalent to walking 5 to 7 miles a day at 3 mph in addition to the habitual activity required by a sedentary lifestyle.

For most inactive individuals, this would require adding a minimum of 60 min of physical activity to their daily routine.

However, the guidelines for prevention of weight gain have been derived from, in large measure, population-based cohorts of men and women, and thus the implementation of the guidelines may vary substantially among individuals. In other words, some individuals will maintain body weight by accumulating only 30 min of daily physical activity, whereas others may find it necessary to accumulate 60 to 120 min or more to maintain energy balance and to prevent weight gain. The point is that the quantity of physical activity required to maintain body weight (e.g., energy balance) will vary depending on the individual. Also note that the guidelines for prevention of weight gain by physical activity are derived from studies that used primarily Caucasian adults. Accordingly, the potential influence of race on these guidelines is unknown.

On average a PAL of about 1.75, which is equivalent to about 60 to 90 min of daily leisure-time physical activity, is recommended to prevent age-related weight gain.

Treatment of Obesity

The independent role of physical activity (exercise) as a treatment strategy for obesity has received considerable attention. Early reviews of the literature suggested that the reduction in body weight (1-2 kg) associated with physical activity alone (e.g., no caloric restriction) was marginal and thus physical activity in the absence of caloric restriction was not a particularly useful strategy for the treatment of obesity (National Institutes of Health National

Heart Lung and Blood Institute 1998). Subsequently, a careful inspection of the exercise studies revealed that, for the most part, few of the early studies prescribed an exercise program wherein one would expect meaningful weight loss (Ross and Janssen 2001). On the other hand, for those studies wherein the exercise program prescribed did result in a meaningful negative energy balance, weight loss was substantial (Ross and Janssen 2001). In other words, weight loss is positively related to the volume of physical activity performed. This point is illustrated in figure 11.6, wherein a dose–response relationship is shown between caloric expenditure and the time spent exercising with the corresponding reduction in body weight and total fat. However, daily exercise is not always associated with reductions in body weight or body fat. Some investigators report a resistance to weight loss in response to daily exercise performed for about 30 to 40 min for several months (Donnelly et al. 2003). Nevertheless, the majority of studies suggest that regular physical activity without restricting caloric intake is associated with weight loss and a reduction in total fat in overweight men and women.

From figure 11.6, it is also clear that exercise performed for as little as 200 min per week is associated with weight loss. In fact, weight loss in the order of 0.5 kg (1 lb) per week is achieved in response to exercise performed for between 300 and 400 min

per week or about 50 min per day. This observation is consistent with the position of the American College of Sports Medicine, which recommends that overweight and obese persons seeking weight loss should exercise between 200 and 300 min per week, the equivalent of about 8,374 J (2,000 kcal) per week (American College of Sports Medicine 2001).

Treatment of Abdominal Obesity

Whether an increase in physical activity is associated with a significant reduction in abdominal obesity is an important question. Minor reductions (~2 cm) in waist circumference (a surrogate for abdominal fat) are observed in response to exercise-induced weight loss in the order of 2 to 3 kg (National Institutes of Health National Heart Lung and Blood Institute 1998). In other words, similar to body weight, waist circumference is reduced a small amount in response to a small amount of physical activity. Although it is unclear if a dose–response relationship exists between the amount of physical activity and the reduction in waist circumference, it is evident that larger reductions in waist circumference are observed in response to a significant amount of daily exercise. Indeed, exercise performed for 300 to 400 min per week, or about 60 min per day, is associated with reductions of about 0.5 cm per week. In fact, exercise performed for about 60 min per day for 3 to 4 months is associated with reductions in

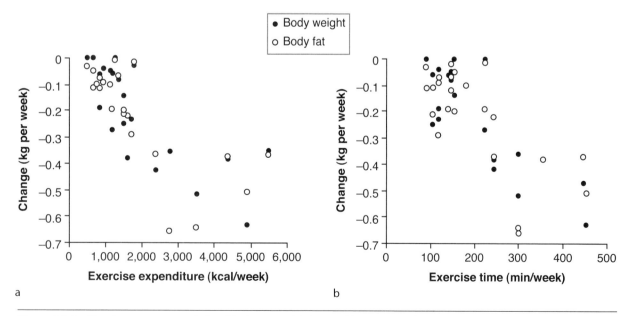

Figure 11.6 The dose–response relationship between *(a)* the weekly exercise caloric expenditure or *(b)* weekly minutes of exercise and the corresponding changes in body weight (black data points) and total fat (gray data points). Each data point represents the mean of one study.

Data from R. Ross and I. Janssen, 2001,"Physical activity, total and regional obesity: dose-response considerations," *Medicine and Science in Sports and Exercise* 33(6 Suppl): S521-527.

waist circumference that approach 5 to 6 cm in both men and women, as illustrated in figure 11.7.

Whether exercise-induced weight loss is associated with corresponding reductions in abdominal subcutaneous and visceral fat has also been considered (Ross and Janssen 2001). It is generally observed that exercise is associated with a substantial reduction in abdominal subcutaneous and visceral fat independent of gender and age. An example of this effect is shown in figure 11.8, where a 10% reduction in body weight is associated with a reduction in abdominal subcutaneous and visceral fat that approximates 25% and 35%, respectively. Figure 11.8 also shows that the greater the exercise level expressed in minutes per week, the greater the reduction in both abdominal subcutaneous and visceral fat.

Because visceral fat is such an important predictor of health risk, practitioners have questioned whether this fat depot is selectively reduced in response to exercise-induced weight loss. The answer to this question depends on how the reduction is presented. That is, for a given weight loss, a greater reduction in abdominal subcutaneous fat is observed if the reduction is expressed in absolute values (e.g., cm² at the L4-L5 image); this is because most adults have more abdominal subcutaneous fat than visceral fat. On the other hand, if the reduction in abdominal subcutaneous and visceral fat is expressed in relative terms (e.g., relative to the initial size of the depot), then the reduction in visceral fat will be greater than the reduction in subcutaneous fat (figure 11.8).

Exercise-Induced Reduction in Obesity Without a Change in Body Weight

Emerging evidence suggests that regular exercise can reduce total and abdominal obesity in the absence of any change in body weight. This is supported by at least two lines of evidence. First, for any given level of BMI between 18 and 35 kg/m², adults who are physically active (e.g., have a higher level of cardiorespiratory fitness) have a lower waist circumference and lower levels of abdominal subcutaneous and visceral fat compared with their sedentary counterparts (who have a lower level of cardiorespiratory fitness) (Janssen et al. 2004). Second, results from well-controlled, randomized trials reveal that obese men and women who participate in exercise programs for 3 to 4 months can experience significant reductions in both waist circumference (figure 11.7) and abdominal subcutaneous and visceral fat despite no change in BMI (Ross et al. 2000, 2004). These observations are important because they suggest that those who seek obesity reduction by increasing physical activity should be educated about the possibility that reductions in waist circumference, total fat, and abdominal fat can occur with or without a corresponding weight loss. On the other hand, it is equally important to note that the reduction in both total and abdominal fat depots is much greater in response to exercise with weight loss than exercise without weight loss (Ross et al. 2000, 2004). These observations highlight the importance of monitoring obesity reduction using both BMI and waist circumference.

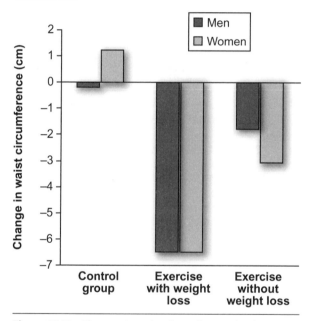

Figure 11.7 Changes in waist circumference in obese men and women after a 12- to 14-week program. The mean values from three treatment groups are shown. The control group did not exercise or change caloric intake. The "exercise with weight loss" group and "exercise without weight loss" group participated in daily exercise that consisted of about 60 min of vigorous walking. The "exercise with weight loss group" did not change caloric intake and lost ~7% of their body weight. The "exercise without weight loss" group increased caloric intake to compensate for the exercise-induced energy expenditure and had no change in body weight.

Data from R. Ross, D. Dagnone, P.J. Jones, H. Smith, A. Paddags, R. Hudson and I. Janssen, 2000, "Reduction in obesity and related comorbid conditions after diet-induced weight loss or exercise-induced weight loss in men: A randomized, controlled trial," *Ann Int Med* 133(2): 92-103, and R. Ross, I. Janssen, J. Dawson, A.M. Kugl, J. Kuk, S. Wong, T.B. Nguyen-Duy, S.J. Lee, K. Kilpatrick and R. Hudson, 2004, "Exercise with or without weight loss is associated with reduction in abdominal and visceral obesity in women: A randomized, controlled trial," *Obesity Research* 12(5): 789-798.

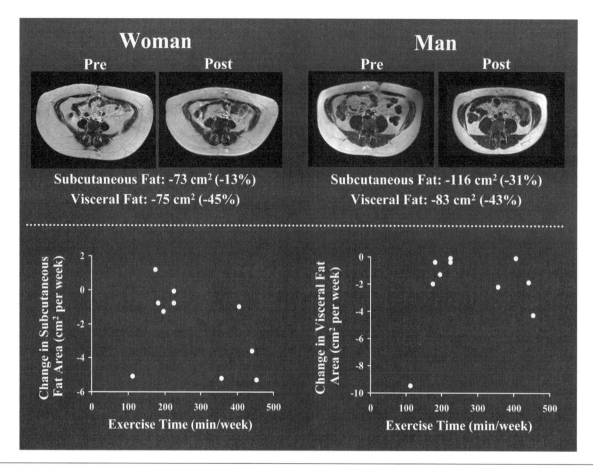

Figure 11.8　Exercise-induced reductions in abdominal subcutaneous and visceral fat. The images on the top are the pre- and posttraining images of a woman (left) and man (right) who lost substantial amounts of abdominal fat in response to an aerobic exercise program that consisted of about 60 min of daily moderate-intensity walking. The absolute reduction (cm²) in subcutaneous fat was greater than the absolute reduction in visceral fat, but the relative (%) reduction in subcutaneous fat was less than the relative reduction in visceral fat. The figures on the bottom illustrate the dose–response relationship between the weekly minutes of exercise and the corresponding reductions in abdominal subcutaneous fat (left) and visceral fat (right). Each data point represents the mean of one study.

Data from R. Ross and I. Janssen, 2001, "Physical activity, total and regional obesity: dose-response considerations," *Medicine and Science in Sport and Exercise* 33(6 Suppl): S521-527.

Exercise in the absence of weight loss is associated with significant reductions in total, abdominal, and visceral fat in obese men and women. These reductions are, however, smaller than those associated with exercise-induced weight loss.

Summary

The prevalence of obesity is already high and is increasing worldwide. This poses a major threat to public health, and innovative, multidisciplinary strategies are required to combat this problem. The materials covered in this chapter provide strong support for the recommendation that physical activity should be an integral component in the strategies developed to both prevent and treat the obesity epidemic. Current guidelines suggest that adults should accumulate about 60 min of moderate-intensity physical activity daily to prevent unhealthy weight gain. Results from shorter obesity treatment studies wherein dietary intake was carefully controlled suggest that 60 min of moderate-intensity exercise without a change in energy intake is associated with substantial reductions in total and abdominal obesity in obese men and women.

Although it is now clear that an increase in daily physical activity is required for most individuals, the challenge that remains is how to engage in and maintain a physically active lifestyle. Increasing physical activity to the levels recommended for obesity prevention and reduction will require a multidisciplinary approach that includes such components as educating allied health care providers about the benefits of physical activity, reestablishing daily physical education programs in our school systems, and working with urban planners to develop environments that encourage physical activity. Although the societal challenge to increase physical activity levels to an appropriate amount to combat the obesity epidemic is immense, the benefits are many, and thus the problem must be approached with vigor, step by step.

KEY CONCEPTS

abdominal subcutaneous fat: The layer of fat that lies directly underneath the skin in the abdominal region.

adipocyte: An adipose tissue or fat cell that stores lipids.

ad libitum: At one's pleasure; as one wishes.

body mass index (BMI): A simple index of weight for height, calculated as weight in kilograms divided by the square of height in meters (kg/m^2), that is commonly used to determine overweight and obesity status in the research and clinical setting.

ectopic fat: Fat that is stored outside of the adipose tissue depots.

lipolysis: The lipid breakdown reaction in adipocytes whereby the triglyceride molecule is hydrolyzed in the cell's cytosol into its component glycerol and three fatty acid molecules.

obesity: A condition of excessive fat accumulation to the extent that health may be impaired.

physical activity level (PAL): Total daily caloric expenditure divided by total calories from resting metabolism. This term is being increasingly used as an overall indicator of energy expenditure.

visceral fat: Internal fat in the abdominal region that surrounds the organs of the gastrointestinal tract. Visceral fat consists of omental and mesenteric adipocytes and is contained within the visceral peritoneum.

waist circumference: A measurement of abdominal circumference commonly obtained at the top of the iliac crest. Waist circumference is used to characterize levels of abdominal obesity in research and clinical settings.

STUDY QUESTIONS

1. What is BMI, how is it calculated, and what cut points are used to define overweight and obesity in adult men and women?

2. List five major chronic diseases that are associated with obesity.

3. Provide two examples of ectopic fat deposition in obesity and explain their relationship to obesity-related disease.

4. Describe how the average dietary intake, average leisure-time physical activity levels, and average total physical activity levels have changed in the past 3 decades and how these changes have contributed to the obesity epidemic.

5. What is PAL, how is it calculated, and what levels are currently recommended for the prevention of unhealthy weight gain?

6. What changes occur in abdominal subcutaneous and visceral fat in obese individuals in response to daily exercise performed for about 60 min at a moderate intensity?

7. Describe the importance of waist circumference when monitoring success in obesity reduction programs.

8. What changes occur in waist circumference in response to exercise with or without weight loss?

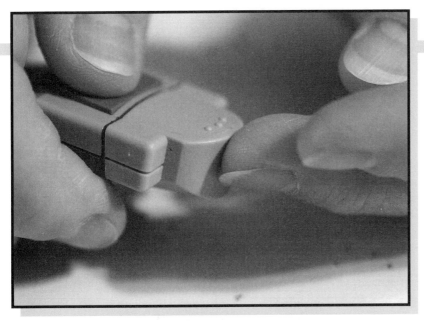

Physical Activity, Fitness, and Diabetes Mellitus

■ Oscar Alcazar, PhD ■ Richard C. Ho, PhD ■ Laurie J. Goodyear, PhD

CHAPTER OUTLINE

Diabetes mellitus is a chronic disease that encompasses a heterogeneous group of disorders, with the predominant form being type 2 (or non-insulin-dependent) diabetes. Even though the symptoms of diabetes mellitus have been known for centuries, only in the last few decades has research begun to unravel the various causes of diabetes. Physical activity provides remarkable health benefits to those with type 2 diabetes. Moderate-intensity regular physical activity has been demonstrated to prevent or delay the onset of diabetes in high-risk subjects. Changes in lifestyle that include introducing physical activity have a very positive impact for public health in developed countries, where diabetes and its long-term complications constitute an enormous economic burden. In this chapter, we discuss several aspects of this endocrine disease and its complications, focusing primarily on type 2 diabetes. We then discuss the basis of this disease at a molecular level and, finally, provide the epidemiological body of evidence that supports the important role of physical activity in preventing and treating type 2 diabetes.

Diabetes: Concept and Prevalence

Diabetes, the most common endocrine disorder, affects multiple organs and body functions, causing serious health complications, such as renal failure, heart disease, nerve damage, stroke, and blindness. The body cannot control the level of circulating blood glucose because of either insufficient **insulin** production or inadequate response by organs to circulating levels of insulin, the major hormone controlling the body's **glucose homeostasis.** Some of the most characteristic symptoms associated with the onset of diabetes include frequent urination, excessive thirst, and fatigue. Diabetes is diagnosed when the level of glucose in the blood is greater than 7.0 mmol/L (fasting). Three major types of diabetes have been defined: type 1 or insulin-dependent diabetes mellitus, type 2 or non-insulin-dependent diabetes mellitus, and gestational diabetes.

In type 1 diabetes, which is an autoimmune disorder, the cells responsible for releasing insulin, the pancreatic β-cells, are mistakenly recognized by the immune system as "foreigners" and selectively destroyed. As a result, pancreatic β-cells are virtually erased from the body, and circulating insulin levels in the blood dramatically decrease or even disappear. Regular insulin injections become nec-

essary to sustain life. This disease is also known as juvenile-onset diabetes because it commonly begins in childhood or adolescence, although not always. This type of diabetes occurs more frequently in populations descended from northern European countries rather than in southern European countries, the Middle East, or Asia. In the United States, approximately 3 people in 1,000 develop type 1 diabetes.

Type 2 diabetes, unlike type 1, is basically a metabolic disorder where insulin levels may be normal, elevated (hyperinsulinemia), or even decreased (hypoinsulinemia), although hyperinsulinemia characterizes most patients. At least in the early stages, most people with type 2 diabetes can produce insulin but cannot use it effectively. Two major components act in this pathology: pancreatic β-cell dysfunction, in which these cells gradually fail to adequately produce insulin in response to circulating levels of glucose in the blood (i.e., impaired insulin secretion), or **insulin resistance,** in which insulin-sensitive tissues such as skeletal muscle, fat, and liver become, to some extent, desensitized to insulin. Insulin resistance appears to precede, and at least in part be responsible for, pancreatic β-cell dysfunction. The onset of type 2 diabetes typically occurs with advancing age, but, as we discuss later in this chapter, it is increasing alarmingly in young individuals.

Type 2 diabetes is frequently found with obesity, which indicates that it is a metabolic disease. The negative consequences of a sedentary lifestyle and obesity may be exacerbated by genetic components in individuals and certain ethnic groups. For example, type 2 diabetes is more common in people of aboriginal, Hispanic, and African American descent than it is in other populations. It is the most common form of diabetes worldwide and in the United States, where approximately 90% of the cases of diabetes are classified as type 2. This disorder typically has a slow onset, usually developing over several years. Treatment usually combines diet with exercise, oral medication (e.g., metformin, sulfonylureas), and sometimes insulin injections. The major features of types 1 and 2 diabetes are highlighted in table 12.1.

The third major type, gestational diabetes, usually develops during the second or third trimester of pregnancy and resolves after the baby is delivered. Treatment of this condition focuses on diet and sometimes requires insulin injections. Women who have diabetes during pregnancy are at higher risk for developing type 2 diabetes later in their lives.

Table 12.1   Characteristics of the Predominant Types of Diabetes

Type of disorder	Type 1 diabetes (IDDM) Autoimmune disorder	Type 2 diabetes (NIDDM) Metabolic disorder
Insulin level	Hypoinsulinemia	Hyperinsulinemia
Age of onset	Predominantly in youth	Predominantly after age 40
Genetic component	Weak	Strong
Other		Frequently linked to obesity

IDDM = insulin-dependent diabetes mellitus; NIDDM = non-insulin-dependent diabetes mellitus.

Epidemiology, Etiology, and Complications of Type 2 Diabetes

The alarming increase in the incidence of Type 2 diabetes worldwide makes it imperative that the health care community understand the dimensions of this increase; the factors that lead to individuals' developing the disease; and the health consequences that will result of the condition is not diagnosed and controlled.

Epidemiology of Type 2 Diabetes

Type 2 diabetes ranks among the world's most common chronic diseases, with a remarkably high economic impact in developed countries. The term *prevalence* indicates the number of people with a particular condition, whereas *incidence* refers to the number of new cases. The prevalence of diabetes is increasing dramatically in the United States and worldwide, causing some health organizations and researchers to consider it an epidemic.

Epidemiological surveys for diabetes are complicated to perform and prone to underestimate the real magnitude of the problem. Recent estimates by the International Diabetes Institute and the World Health Organization suggest that globally the number of persons with diabetes will increase from 151 million in the year 2000 to 221 million by the year 2010 and to 300 million by 2025. This rate of increase is predicted to occur in virtually every country throughout the world. However, the greatest increases for the next decade are expected to occur in developing countries, particularly in Asia.

One of the most important recent population-based studies from the United States (the Third National Health and Nutrition Examination Survey, NHANES III) showed a marked increase in the prevalence of diabetes. As some excellent reviews have discussed in depth (Boyle et al. 2001; Mokdad et al. 2000), epidemiological predictions drawn by NHANES III are consistent with the trend seen in virtually every developed Western country. It is predicted that the number of people diagnosed with type 2 diabetes in the United States will increase by 165% in the next 50 years, rising from 11 million in 2000 to 29 million by the year 2050. The highest increases are expected to happen among people aged 75 years and over (336%) and among African Americans (275%). During 1991-1992, only eight states had a diabetes prevalence exceeding 6%, but by 1997-1998 the number of states with such a prevalence had risen dramatically to 23 (see figure 12.1). In most of the states, the increase was independent of age, gender, ethnic group, or educational level. The greatest morbidity and mortality rates from type 2 diabetes occur in the elderly and minority groups in the United States, a trend that epidemiological studies predict will not change in the near future. Another trend and an enormous challenge for the future is the emergence of an obesity-induced diabetes epidemic in children and adolescents.

Among the major reasons behind the expected dramatic increase in the incidence of type 2 diabetes are the anticipated world population growth, mostly in developing countries, the increase in longevity in most Western countries, and certain environmental factors. In particular, rapidly changing and unhealthy dietary patterns, along with increasingly sedentary lifestyles, lead to obesity, a major risk factor in the development of type 2 diabetes.

Even small changes in lifestyle can be very significant for preventing and treating type 2 diabetes and for determining the incidence of this metabolic disorder in the long term. In contrast, unhealthy diets rich in high saturated fats, together with reduction in physical activity even at early ages, may result in a public health problem even greater than predicted to date by epidemiological studies.

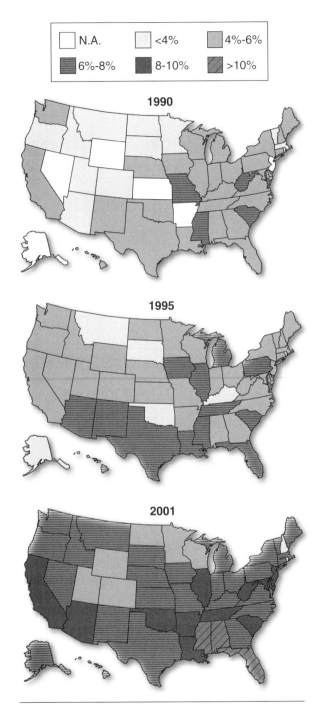

Figure 12.1 Epidemiological data maps developed by the Centers for Disease Control and Prevention, showing trends of incidence of type 2 diabetes throughout several years in the United States.

Data from the Centers for Disease Control and Prevention (CDC).

Mechanisms Leading to Type 2 Diabetes

Type 2 diabetes is caused by a combination of two different defects: (1) impaired insulin secretion by pancreatic β-cells and (2) insulin resistance, a

reduced sensitivity of the body to insulin. Individuals with type 2 diabetes show a marked reduction in early insulin secretion, that is, the phase of insulin secretion that occurs immediately after eating, also referred to as the cephalic phase, a consequence of malfunctioning of the pancreatic β-cells. These cells progressively lose their ability to "sense" and adequately respond to changes in the concentration of blood glucose. The detailed molecular and signaling mechanisms underlying this defect are not yet fully understood. However, it is known that progression in the decay of β-cell function takes a long time, so by the time type 2 diabetes is diagnosed, β-cell dysfunction may have been occurring for as long as 12 years.

The increasing hyperglycemic state, characteristic of this metabolic disorder, is exacerbated in the long term by the cytotoxic effects of high blood glucose concentrations. Animal studies suggest that chronic hyperglycemia is detrimental to insulin secretion and may also induce insulin resistance, and it has been shown that this glucotoxicity may eventually lead to a permanent loss of β-cell function. Thus, improving blood glucose control is essential to prevent further deterioration in β-cell function and consequent progression of diabetes.

It is widely accepted that insulin resistance plays a central role in the pathogenesis of type 2 diabetes. The term *insulin resistance* refers to the subnormal response to a given concentration of insulin by the major insulin-sensitive organs of the body, that is, liver, muscle, and adipose tissue. In other words, an adequate level of insulin is released at mealtimes but its effect (glucose disposal) is impaired. Aside from the influence of some important genetic factors, insulin resistance is a common feature of individuals who are obese—particularly when the excess fat is concentrated in the abdominal region (i.e., central adiposity)—and who are physically inactive. In fact, a primary mechanism causing insulin resistance is weight gain. It has been suggested that insulin resistance may represent a feedback mechanism to prevent further weight gain above a certain threshold.

Figure 12.2 highlights the so-called two-step model leading to type 2 diabetes. This model proposes that insulin resistance precedes and contributes to β-cell failure, which ultimately causes the symptoms of diabetes. The first change from normoglycemia to **impaired glucose tolerance (IGT)** is largely attributable to insulin resistance, whereas the second, from IGT to type 2 diabetes, arises from β-cell dysfunction and subsequent declining insulin secretion. In studies performed on a variety of ethnic groups who developed IGT from normal glucose tolerance, the insulin resistance phenomenon preceded the defect in insulin secretion.

Complications Associated With Type 2 Diabetes

People with type 2 diabetes are at risk for a variety of serious long-term complications (see *Type 2 Diabetes Complications*), attributable to the wide scope of insulin pleiotropic actions on glucose, lipid, and protein metabolism. In some cases, there is also a risk for acute emergency complications such as diabetic hyperglycemic hyperosmolar coma, which is characterized by severe dehydration, decreased consciousness, and extreme hyperglycemia. Usually the kidneys compensate for high glucose levels in the blood by excreting excess glucose in the urine. However, under dehydration, the kidneys conserve fluid and blood glucose levels increase greatly.

The long-term complications of type 2 diabetes are frequently broken down into microvascular and macrovascular complications.

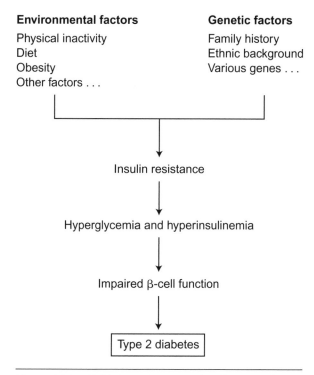

Figure 12.2 Multiple environmental and genetic factors lead to type 2 diabetes. These factors are not fully defined, and there are likely to be more elements involved in the development of this disease.

Data from the Centers for Disease Control and Prevention (CDC).

Type 2 Diabetes Complications

Acute complications
Hyperglycemic coma (unusual)

Chronic complications
Diabetic retinopathy
Diabetic nephropathy
Diabetic neuropathy
Cardiovascular diseases

• *Microvascular complications* are those that damage organs through their effects on small blood vessels. The most common microvascular complications are retinopathy, nephropathy, and neuropathy. Diabetic retinopathy is caused by damage to blood vessels of the retina and is the leading cause of blindness in the United States and in most developed countries. Diabetic nephropathy causes kidney damage and is the most common cause of chronic kidney failure in the United States. The earliest step of this pathology consists of thickening of the renal glomerulus, reducing the kidney's filtration capacity. Increasing numbers of glomeruli are destroyed with time. As a consequence, the kidney allows more albumin than normal in the urine, leading to a condition called microalbuminuria, which heralds the onset of diabetic nephropathy and may result in the need for dialysis or a kidney transplant. Finally, diabetic neuropathy is caused by nerve damage as a result of high blood glucose levels.

• *Macrovascular complications* are those that affect large blood vessels and result in cardiovascular diseases (CVD), such as coronary heart disease and stroke, and peripheral vascular disease. Peripheral vascular disease causes arteriosclerosis of the extremities, characterized by narrowing of the arteries that supply the legs and feet. This decreases blood flow, which can injure nerves and other tissues in the extremities. Type 2 diabetes is a major risk factor for CVD, and statistics predict that about 80% of type 2 diabetic patients will die because of this complication. On average, type 2 diabetic patients will die 5 to 10 years earlier than their nondiabetic counterparts, mainly because of CVD. The treatment of cardiovascular diseases accounts for a large part of the huge health care costs attributed to type 2 diabetes. The fact that some of the complications associated with type 2 diabetes often begin to develop well before the disorder is diagnosed contributes to this economic burden.

Impact of Physical Activity on Insulin and Glucose Metabolism

In previous sections we discussed the concept of diabetes, its associated conditions, and the worldwide problem of its prevalence. We now approach this disorder from a physiological and cellular perspective and focus on the molecular mechanisms in skeletal muscle underlying the beneficial aspects of physical activity. More in-depth information on this topic can be found in previous reviews (Goodyear and Kahn 1998; Tomas et al. 2002).

Glucose Metabolism and Type 2 Diabetes

To maintain whole-body glucose homeostasis, coordination of three different metabolic events is required: adequate secretion of insulin by pancreatic β-cells, suppression of hepatic glucose production, and stimulation of glucose uptake by insulin-sensitive tissues, primarily muscle. During an acute bout of exercise, the increased need for metabolic fuel is met by increases in both carbohydrate and fat utilization in the skeletal muscle. Glucose is taken up from blood into the working skeletal muscles. In people who do not have diabetes, unless the exercise is of extremely long duration, blood glucose concentrations do not decrease appreciably. This is because glucose output by the liver is precisely matched to glucose uptake in the muscle and because insulin secretion by the β-cells of the pancreas is reduced. In contrast, in people with type 2 diabetes who have moderate hyperglycemia, glucose concentrations can be decreased with moderate-intensity exercise, an important health benefit of exercise.

Skeletal muscle is the major organ in the body responsible for glucose disposal and consequently is of prime importance in metabolic disorders such as IGT and type 2 diabetes. Three potential rate-controlling steps and molecules for insulin-stimulated muscle glucose metabolism (i.e., synthesis of glycogen from glucose) have been identified: glucose transporter 4 (GLUT4), hexokinase, and glycogen synthase. Each of these steps is defective in people with type 2 diabetes (see figure 12.3).

The major physiological stimulators of muscle glucose uptake are exercise and insulin. These stimuli enhance glucose transport into the muscle

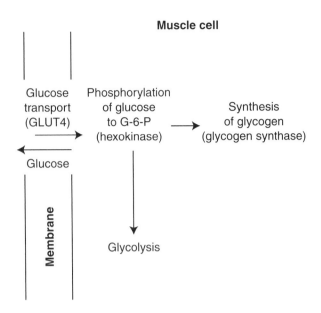

Figure 12.3 Glucose is transported in the skeletal muscle fibers via glucose transporter 4 (GLUT4). Glucose is phosphorylated to glucose-6-phosphate (G6P) via the enzyme hexokinase. Depending on the energy needs of the muscle, G6P can be stored as glycogen via the rate-limiting enzyme glycogen synthase or can be used as substrate for glycolysis.

cells, where it can be used for adenosine triphosphate (ATP) production or stored in the form of glycogen. Glucose transport in skeletal muscle occurs primarily by facilitated diffusion, using glucose transport carrier proteins. In mammalian tissues, glucose transporters constitute a family of structurally related proteins (isoforms) with tissue-specific expression patterns. There are 12 different glucose transporter isoforms, GLUT4 being the most abundant isoform present in skeletal muscle. Glucose transport is the rate-limiting step in muscle glucose utilization. In response to exercise or insulin, GLUT4 moves from an intracellular location to the plasma membranes and tubules (translocation). The amount of GLUT4 in the plasma membrane is tightly regulated by exercise and insulin, exerting a fine control on glucose disposal and metabolism within the muscle.

Glycogen synthesis is the primary pathway for nonoxidative glucose disposal in nondiabetic persons. The rate of glycogen formation in people with diabetes is decreased, representing 40% of the rate typical of control subjects. It is believed that defects in muscle glycogen synthesis play a significant role in the insulin resistance that precedes the development of type 2 diabetes. In addition, impairment of GLUT4 translocation and hexokinase phosphorylation, causing defects in glucose transport, has been shown to be an early factor in the pathogenesis of type 2 diabetes rather than a consequence.

Insulin Signaling in Skeletal Muscle

Understanding type 2 diabetes requires a closer view of the signaling mechanisms triggered by insulin in skeletal muscle (see figure 12.4). The cascade of intracellular signaling events stimulated by insulin involves multiple effector proteins that orchestrate diverse cellular responses. Insulin binds to the extracellular portion (α-subunit) of the transmembrane insulin receptor, and this event leads to activation of the transmembrane β-subunits and further autophosphorylation of the insulin receptor. The signal is next transduced through phosphorylation of a family of closely related proteins, which are referred to as insulin receptor substrates (IRS).

In addition to the IRS, the insulin receptor may phosphorylate and activate the protein Shc, which ultimately results in the activation of mitogen-acti-

vated protein kinase. IRS molecules contain multiple tyrosine phosphorylation sites that, after becoming phosphorylated by insulin stimulation, bind additional downstream signaling molecules. IRS-1 appears to be the predominant isoform mediating signal transduction in skeletal muscle, whereas IRS-2 appears to be important in β-cell development. Both isoforms are important for insulin regulation of glucose metabolism in the liver. Once IRS-1 is activated in muscle, the signal is transduced by activation of the phosphatidylinositol-3-kinase (PI3-K). PI3-K is a heterodimer protein consisting of a regulatory subunit (p85) associated with a catalytic subunit (p110). Interaction of phosphorylated IRS with the p85 subunit of PI3-K ultimately activates the enzyme. The p110 catalytic subunit of PI3-K uses phosphatidylinositol-4,5-bisphosphate as substrate, resulting in the phosphorylated lipid phosphatidylinositol-3,4,5-trisphosphate (PIP_3). PIP_3 is required to activate the membrane-associated enzyme 3-phosphoinositide-dependent protein kinase 1 (PDK1). On activation,

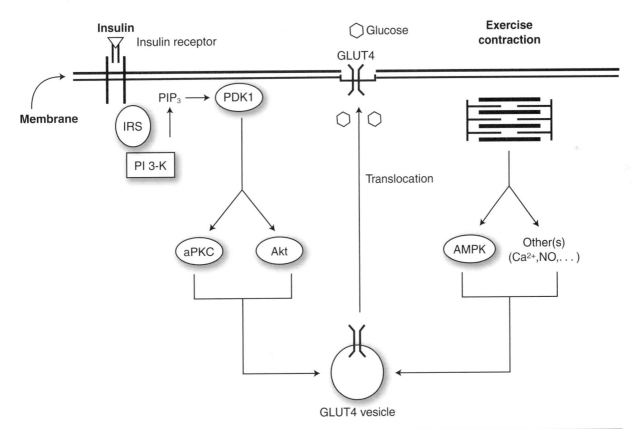

Figure 12.4   Insulin increases glucose transport in skeletal muscle through the translocation of the glucose transporter 4 (GLUT4) protein. Signaling to the GLUT4 vesicle involves insulin binding to its receptor, phosphorylation of insulin receptor substrate proteins (IRS), activation of phosphatidylinositol-3-kinase (PI3-K), generation of phosphatidylinositol-4-phosphate 3 (PIP_3), and activation of atypical protein kinase C (aPKC) and the protein kinase Akt via the 3-phosphoinositide-dependent protein kinase 1 (PDK1) protein. Mechanisms downstream of aPKC and Akt are not fully understood. The signaling for exercise-stimulated glucose transport is poorly understood. Although there is some evidence that adenosine monophosphate (AMP)-activated protein kinase (AMPK) may be involved, activation of this molecule can not fully explain how exercise increases GLUT4 translocation.

PDK1 phosphorylates and activates the protein kinase Akt (also known as protein kinase B) and also the atypical protein kinase C isoforms ζ and λ. Akt and atypical protein kinase Cs represent two different signaling branches that are activated by the same molecule, PDK1. Further downstream events in the insulin-signaling cascade remain to be elucidated. However, AS160 protein has recently emerged as an Akt substrate that may play an important role in transducing the insulin signal to the translocation of GLUT4. The final step is the movement of GLUT4 from intracellular vesicles to the plasma membrane (translocation), resulting in increased glucose transport into the cell.

Exercise Signaling in Skeletal Muscle

As mentioned earlier, exercise and insulin are the major mediators of glucose transport activity in muscle. Both stimuli cause translocation of the GLUT4 glucose transporter from an intracellular location to the plasma membrane in skeletal muscle. In fact, insulin and exercise share many similar biological effects in skeletal muscle, because both stimuli can increase glucose transport, amino acid uptake, and glycogen synthesis. Because of these similarities, it was first believed that insulin and exercise used similar signaling cascades. However, it has been demonstrated that activity of one of the major insulin signaling molecules, PI3-K, is not increased immediately after exercise or muscle contraction. The lack of activation of PI3-K is consistent with findings that insulin and contraction use differential signaling leading to glucose uptake and glycogen synthesis in skeletal muscle, and that contraction-stimulated glucose uptake occurs through a PI3-K-independent mechanism. However, whether insulin and exercise signaling converge further downstream at a point distal to PI3-K is unknown.

Exercise-stimulated glucose disposal is mediated by a mechanism that is independent of PI3-K. In recent years, there has been a flurry of research into the role of the 5'AMP-activated protein kinase (AMPK) in skeletal muscle glucose transport. AMPK is a member of a metabolite-sensing protein kinase family that acts as a fuel gauge monitoring cellular energy levels. When AMPK "senses" decreased energy storage, it switches off ATP-consuming metabolic pathways and switches on alternative pathways for ATP regeneration. AMPK activity is unaffected by insulin but increases in response to muscle contraction or exercise, an event that has been correlated with GLUT4 translocation and glucose transport in skeletal muscle. However, recent evidence suggests

that although AMPK may be important in many insulin-independent mechanisms for increased glucose transport (e.g., hypoxia), AMPK, or at least one of its major isoforms, may not be necessary for exercise-mediated muscle glucose uptake. Instead, additional AMPK-independent signaling mechanisms may contribute to the regulation of glucose uptake in skeletal muscle in response to exercise. Further studies are necessary to unravel more detailed molecular pathways involved in exercise signaling in muscle.

Given that exercise increases glucose metabolism by insulin-independent signaling cascades, activation of this pathway provides an alternative strategy to increase glucose transport in insulin-resistant skeletal muscle of people with diabetes and prediabetes. This is beneficial in the management of metabolic abnormalities associated with type 2 diabetes, given that exercise-induced AMPK activity is not impaired in insulin-resistant skeletal muscle. Lessons learned from the AMPK signaling cascade and other insulin-independent mechanisms in skeletal muscle may become very valuable in developing drugs to treat patients with diabetes.

Insulin Resistance and Physical Activity

Insulin resistance is the inability of peripheral target tissues (especially muscle, adipose tissue, and liver)

Courtesy of Joslin Diabetes Center

Loading gel to analyze gene expression of muscle proteins. Such analysis helps researchers determine the mechanism responsible for the beneficial effects of exercise on the prevention and treatment of diabetes.

to respond properly to normal circulating concentrations of insulin. After a period of compensated insulin resistance, IGT eventually develops despite elevated insulin concentrations. Finally, pancreatic β-cell failure results in decreased insulin secretion. This leads to the onset of overt clinical type 2 diabetes, when insulin resistance and impaired β-cell function occur simultaneously, resulting in fasting hyperglycemia. Remember that insulin resistance can be found in patients more than a decade before diabetes appears and is the best predictor for later development of the disease.

A single bout of exercise can increase skeletal muscle glucose transport and metabolism and can also have profound effects on glycogen metabolism. Rates of glycogenolysis are increased during exercise, followed by a rapid resynthesis of glycogen in the postexercise state. The molecular basis for this phenomenon has not been completely elucidated but appears to be dependent on multiple factors, including muscle glycogen concentrations, humoral factors, and autocrine–paracrine mechanisms. These metabolic changes occur in the skeletal muscle and can improve glucose homeostasis in persons with insulin resistance. In addition, these metabolic changes may also be responsible for the ability of physical activity to prevent or delay the onset of type 2 diabetes.

Insulin sensitivity is related to the degree of physical activity. Exercise training improves glucose tolerance and insulin action in insulin-resistant people and patients with type 2 diabetes. This stems from adaptations in multiple tissues including the pancreas, liver, adipose tissue, and skeletal muscle. In the pancreas, secretion of insulin in response to glucose is reduced with exercise training, although part of this may be attributable to the acute effects of the previous bout of exercise. The effects of chronic exercise training on the liver include decreased hepatic glucose production for a given workload of exercise, most likely attributable to increased ability of muscle to use fatty acids for a cellular fuel source. However, the maximal capacity of liver for hepatic glucose production is increased, a function of increased gluconeogenesis. This enhanced liver capacity allows for increased exercise duration with higher intensity exercise. In adipose cells, training increases the capacity to store and mobilize free fatty acids. Exercise training leads to increased expression of GLUT4 in skeletal muscle, which has been correlated with improved insulin action on glucose metabolism.

These findings are clinically relevant because insulin-stimulated tyrosine phosphorylation of

IRS-1 and activity of PI3-K are reduced in skeletal muscle from patients with type 2 diabetes. Thus, exercise training is an important therapeutic strategy to partially restore insulin sensitivity in people with type 2 diabetes. Indeed, recent epidemiological studies have determined that regular physical exercise can reduce one's risk of developing type 2 diabetes.

Epidemiological Evidence Indicating Benefits of Physical Activity in Preventing Type 2 Diabetes

It is well accepted that a physically active lifestyle plays an important role in preventing a variety of chronic diseases, including type 2 diabetes. In particular, exercise induces metabolic changes that significantly affect both high-risk individuals with IGT and patients with diabetes. As a consequence of the physiological benefits of regular exercise, active individuals show better insulin and glucose profiles. The epidemiological evidence in favor of exercise as a major lifestyle component in preventing or delaying the onset of type 2 diabetes is quite convincing, irrespective of ethnicity, gender, or age group (table 12.2).

In a **cohort study** (Helmrich et al. 1991) performed on approximately 6,000 male participants ranging from 45 to 55 years, physical activity during leisure time was inversely related to the development of type 2 diabetes. In particular, for each 500 kcal/week increment in energy expenditure, the risk for developing type 2 diabetes decreased by 6%. Interestingly, this association was unaltered by other factors, such as obesity, hypertension, and a family history of diabetes. Another interesting conclusion derived from this study was that the beneficial effects of physical activity were strongest in participants at highest risk for type 2 diabetes, that is, those individuals with high **body mass index (BMI),** hypertension, or a family history of diabetes.

Another epidemiological study (Manson et al. 1991) focused on women and examined the association between regular vigorous exercise and the incidence of type 2 diabetes. Participants were approximately 87,000 nondiabetic women aged 34 to 59 years. The follow-up duration of this study was 8 years. Women who engaged in vigorous exercise at least once per week showed a 16% lower relative risk of developing type 2 diabetes. Similarly to the previously cited study (Helmrich et al. 1991), adjustments for age, BMI, family history of diabetes, and other variables did not alter the beneficial effect of exercise.

In another prospective cohort study, Manson and colleagues (1992) evaluated approximately 21,000 male participants aged 40 to 84 years who were initially free of diagnosed type 2 diabetes mellitus. After 5 years of follow-up, the outcome showed an inverse correlation between incidence of type 2 diabetes and the frequency of vigorous exercise. Men who engaged in vigorous physical activity at least once per week had a 29% lower risk of developing diabetes, compared with men who performed no vigorous exercise. In agreement with previous studies (Helmrich et al. 1991), the inverse relationship between exercise and risk of this metabolic disorder was particularly pronounced among overweight men. However, there was still a significant reduction after data adjustment for BMI as well as for age.

In nonrandomized feasibility studies performed in Sweden and China, physical activity was instituted as part of an intervention to prevent the development of diabetes among persons with impaired glucose tolerance. The Swedish feasibility study (Eriksson and Lindgarde 1991) included 41 subjects aged 47 to 49 at the early stage of type 2 diabetes and 181 additional participants with IGT.

The aim was to test the feasibility of long-term intervention with emphasis on changes in lifestyle. The intervention program consisted of dietary treatment or an increase in physical activity. After a 5- to 6-year follow-up period, **oral glucose tolerance tests** showed normalization in more than half of the participants with IGT and more than 50% of the participants with diabetes went into remission. As expected, the improvement in glucose tolerance was correlated with body weight reduction and increased physical activity.

Pan and colleagues (1995) focused on individuals with IGT. Participants aged 25 to 74 were matched according to their BMI and assigned to four different intervention groups: control, diet, exercise, and diet plus exercise. After a 6-year follow-up period, the incidence of diabetes was 8.3 cases per 100 person-years versus 15.7 cases in the control group. In agreement with the previously cited studies, the exercise intervention and the incidence of type 2 diabetes were inversely related.

All of these epidemiological approaches unambiguously indicate that physical activity is a powerful means to prevent or delay type 2 diabetes among high-risk people. As mentioned before, physically active people have better profiles of blood insulin and glucose concentrations than their sedentary counterparts, partly attributable to an exercise-

Table 12.2   Epidemiological Data on the Effects of Exercise on Prevention of Type 2 Diabetes

Study population	Main findings	Reference
Cohort studies		
Males aged 45-55	Physical activity level was inversely related to risk of developing type 2 diabetes especially in men at high risk for type 2 diabetes.	Helmrich et al. (1991)
Females aged 34-59	Eight-year follow-up study with vigorous exercise at least once per week showed a 16% lower risk of type 2 diabetes.	Manson et al. (1991)
Males aged 40-84	Five-year follow-up study with vigorous exercise at least once per week showed a 29% lower risk of type 2 diabetes.	Manson et al. (1992)
Feasibility studies		
IGT men aged 47-49	Five-year follow-up study showed OGTT normalized in >50% of the participants who exercised regularly.	Eriksson and Lindgarde (1991)
IGT men and women aged 25-74	Six-year follow-up study showed 8.3 cases of type 2 diabetes per 100 person-years in the exercised group versus 15.7 cases in the control group.	Pan et al. (1995)

IGT = impaired glucose tolerance; OGTT = oral glucose tolerance test.

induced increase of insulin sensitivity in peripheral tissues. In addition, some of the beneficial effects of physical activity on insulin sensitivity may be an indirect consequence of weight loss or changes in body composition (decreased adiposity).

Summary of Randomized Controlled Trials on the Prevention of Type 2 Diabetes

Randomized controlled trials are used to evaluate the effectiveness of particular interventions on health indicators. Randomized controlled trials are extremely useful tools to evaluate preventive and public health measures, pharmacological treatments, physical and psychological therapies, and more. These trials need to be large enough and of sufficient duration to allow for a proper evaluation of the intervention programs tested. The main advantage of a random allocation is that intervention groups are comparable in terms of all factors that might influence the outcome. Therefore, any differences in outcome can be attributed to a particular intervention program.

In 1997, the Oslo Diet and Exercise Study in Norway (Torjesen et al. 1997) investigated the effect of diet and exercise intervention on insulin resistance for 1 year. Participants were randomly allocated to the following groups: control, diet only, exercise only, and diet plus exercise. The diet intervention consisted of reduced total fat intake, whereas the exercise protocol entailed endurance training three times a week. The exercise intervention group did not notably change insulin resistance, unlike the diet only and diet plus exercise programs. This outcome could have been attributable to the nature of the exercise program, the short duration of the study, or compliance rates of participants with the various treatment arms.

The Da Qing IGT and Diabetes Study in China (Pan et al. 1997), also published in 1997, drew more convincing conclusions. This study focused on individuals with impaired glucose tolerance, a high-risk factor for developing type 2 diabetes. A large number of participants (577) were distributed among the same intervention groups applied in the Oslo study protocol: control, diet only, exercise only, and diet plus exercise. The follow-up evaluation was conducted at 2-year intervals over 6 years, and the risk for developing type 2 diabetes was

assessed. After 6 years, the incidence of diabetes was 67.7% in the control group, 43.8% in the diet group, 41.1% in the exercise group, and 46.0% in the diet plus exercise group. This study demonstrated that lifestyle interventions significantly reduced the risk for developing type 2 diabetes in a high-risk population group.

The Finish Diabetes Prevention Study (Tuomilehto et al. 2001) focused on middle-aged, overweight participants with impaired glucose tolerance. This study analyzed two different groups, a control and an intervention group. Intervention was aimed at reducing body weight and total fat intake while increasing intake of fiber and physical activity. The follow-up duration of the interventional study was 3.2 years. The cumulative incidence of diabetes was 11% in the intervention group and 23% in the control group. In other words, the risk of developing type 2 diabetes was reduced by 58% in the intervention group. This study and the Da Qing study both demonstrate that changes in lifestyle are critical to prevent type 2 diabetes in high-risk persons.

In a more recent study, the Diabetes Prevention Program (Knowler et al. 2002) compared the beneficial effects of diet, physical activity, and weight loss with metformin treatment. Metformin is an antihyperglycemic drug widely used in the management of type 2 diabetes. To date, metformin is the most efficient single pharmacological treatment for this metabolic disorder, acting by increasing the sensitivity of peripheral tissues to insulin. Participants in this study had IGT and thus were a high-risk population group for developing type 2 diabetes. Individuals were randomly assigned to three different groups: placebo (control), metformin administration, and diet plus exercise. This last intervention consisted of reducing body weight by at least 7% and incorporating a moderate-intensity exercise protocol (minimum of 150 min per week). The follow-up evaluation was conducted over an average period of 2.8 years. The physical activity plus diet program was the most effective intervention to prevent the progression of impaired glucose tolerance to type 2 diabetes. The combined exercise, diet, and weight loss intervention reduced the incidence of diabetes by 58%, whereas metformin reduced the risk by 31%.

From most of the major randomized controlled trials aimed at preventing type 2 diabetes, the conclusion can be drawn that lifestyle changes play a major role. In particular, moderate physical activity, alone or in combination with diet, seems to be the

Sporadic bouts of exercise will not provide significant preventive/curative effects for insulin resistance and type 2 diabetes. But consistent aerobic exercise—such as kayaking, jogging, hiking, biking, or vigorous walking—will mitigate or even correct insulin resistance.

most effective intervention to reduce the incidence of diabetes in persons at high risk. Whether this benefit is solely attained through the associated weight loss has not been fully clarified.

Importance of Regular Physical Activity for People With Type 2 Diabetes

The therapeutic benefits of regular exercise in the treatment of type 2 diabetes have long been recognized. As early as 1919, there were reports of exercise lowering blood glucose concentrations in diabetic patients and improving glucose tolerance. In the 1935 edition of *The Treatment of Diabetes Mellitus* by Joslin and colleagues, exercise was recommended in the "everyday treatment of diabetes." From the epidemiological studies discussed in the previous sections, it is clear that regular moderate-intensity exercise can be an important part of a regimen to prevent and treat type 2 diabetes. Regular physical activity potentiates the effects of diet and oral antihyperglycemic therapy (e.g., metformin and sulfonylureas) to lower glucose levels and improve insulin sensitivity in obese people with type 2 diabetes.

Many health benefits that regular physical activity provides in the prevention of chronic metabolic diseases, such as type 2 diabetes, may be attributable to the overlapping actions of individual exercise sessions and long-term adaptations to exercise training. As mentioned elsewhere, acute exercise produces major effects on whole-body glucose disposal and skeletal muscle glucose uptake and metabolism. However, the elevated insulin-stimulated glucose disposal rates, responsible for the improved insulin sensitivity, tend to disappear after about 5 to 7 days of inactivity. Hence, the effects of exercise training in increasing insulin action are transient and require a regular and constant practice of physical activity.

In addition to improving glucose tolerance and insulin resistance, exercise training has other beneficial effects, such as improving cardiovascular fitness, lowering blood pressure, improving blood lipid profiles, promoting weight loss, reducing abdominal and intra-abdominal fat (a major risk factor for insulin resistance), and promoting a sense of well-being, all of which are known to be associated with diabetes. Therefore, regular physical activity improves morbidity and mortality in people with type 2 diabetes. Multiple factors may modulate the response to exercise training in subjects with diabetes, such as the degree of insulin resistance and insulin deficiency, the frequency and intensity of exercise, the adherence to diet, and weight loss. Insulin sensitivity and the rate of glucose disposal are related to cardiorespiratory fitness even in older persons. The additional potential beneficial effects of exercise training to lower cardiovascular risk in people with type 2 diabetes may reduce the risk of macrovascular or atherosclerotic complications typical of diabetes.

The recommendations provided by the Centers for Disease Control and Prevention and the American College of Sports Medicine (Pate et al. 1995) encourage sedentary people to increase their level of physical activity in a moderate and feasible manner. The types of physical activity should be flexible and fit the life demands of individuals regardless of income, race, or other socioeconomic factors. For instance, walking for about 30 min on most days, averaging approximately 150 min of moderate physical activity per week, seems to be sufficient in many individuals. Ideally, these changes in lifestyle should be maintained over the years to prevent the risk of developing type 2 diabetes or to reduce its severity.

Summary

The prevalence of type 2 diabetes has increased dramatically in recent years and is predicted to reach epidemic levels in the years to come. The increase in this disease correlates with increased rates in obesity and decreases in physical activity. There is now strong epidemiological evidence that regular physical exercise can prevent or delay the onset of type 2 diabetes. The mechanism for this beneficial effect is thought to be attributable to the effects of repeated bouts of exercise to improve overall glucose homeostasis. As examples, exercise increases glucose uptake in skeletal muscles and reduces insulin secretion from the pancreas. Future research should focus on understanding how these important metabolic changes occur on the molecular level, which may provide novel tools for developing antidiabetic drugs.

KEY CONCEPTS

body mass index (BMI): For definition, see page 189.

cohort study: A type of epidemiologic study design in which participants are grouped on the basis of their self-selected exposure of interest, in this case, physical activity. They are then followed over time for the development of the disease of interest, in this case, cancer. For example, three groups of participants, inactive, moderately active, and highly active, may be followed for the development of colon cancer. The rates of colon cancer in the three groups are then compared to assess whether higher levels of physical activity are associated with lower rates of colon cancer.

diabetes mellitus: A chronic disorder characterized by a deficiency of insulin secretion or insulin action, which impairs the body's ability to regulate the levels of blood glucose. It is diagnosed when the level of glucose in the blood is greater than 7.0 mmol/L (fasting) or greater than 11.1 mmol/L (random).

glucose homeostasis: Every process involved in maintaining an internal equilibrium of glucose within the organism.

impaired glucose tolerance (IGT): Condition where blood glucose during the oral glucose tolerance test is higher than normal but not high enough for a diagnosis of diabetes. This prediabetic state is associated with insulin resistance. People with IGT are at a greater risk of developing type 2 diabetes.

insulin resistance: Inability to respond properly to normal insulin concentrations present in blood by peripheral target tissues, that is, muscle, fat, and liver.

insulin: A polypeptide hormone secreted by the β-cells of the pancreatic islets. Insulin is one of the most important hormones in maintaining glucose homeostasis and also regulates the metabolism of fats and proteins.

oral glucose tolerance test: Test to measure the body's ability to use or metabolize glucose, used to diagnose diabetes. It consists of drinking a 100 g glucose solution and measuring blood glucose concentrations for up to 3 hr after ingestion.

randomized controlled trial: Trial in which patients are randomly assigned to two groups:

STUDY QUESTIONS

one treatment group and one control group. Assigning patients at random reduces the risk of bias and increases the probability that differences between the groups can be attributed to the treatment.

1. What is the importance of insulin in regulating normal physiology in the body?

2. Explain the major differences between type 1 and type 2 diabetes.

3. Name the major reasons for the worldwide increase in the incidence of type 2 diabetes.

4. Why is the economic burden associated with type 2 diabetes so high? Name at least four health complications associated with this metabolic disorder.

5. Name some of the biological effects that insulin and exercise share in skeletal muscle.

6. Discuss why unraveling the cell signaling events occurring with exercise may be of great relevance in the treatment of diabetes.

7. Identify the metabolic changes that occur in response to physical activity that may prevent the onset of type 2 diabetes.

8. Drawing on epidemiological evidence, discuss whether ethnic group, gender, or age is a factor determining the beneficial effects of physical activity on the prevention of type 2 diabetes.

9. Explain the benefits of using randomized controlled trials to draw epidemiological conclusions.

10. Summarize how regular physical activity can benefit people at high risk of developing type 2 diabetes.

Physical Activity, Fitness, and Cancer

■ I-Min Lee, MBBS, ScD

oday, we have clear evidence that physical activity decreases the risk of developing many chronic diseases. As we look back on the history of epidemiologic studies investigating the health benefits of physical activity, we find that many of the early studies focused on cardiovascular disease (or risk factors for this disease) as the health outcome of interest. This is appropriate because cardiovascular disease is the leading cause of death globally. It was not until the mid-1980s that researchers also began to place emphasis on the question of whether physical activity plays any role in the prevention of **cancer,** the second leading cause of death globally.

Over the subsequent 2 decades, many studies were conducted that addressed the relationship between physical activity or physical fitness and cancer development. The large body of evidence that accumulated led to recent recommendations specifically targeting physical activity as a cancer preventive measure for the first time by two cancer agencies. The American Cancer Society (ACS) periodically publishes guidelines on healthy nutrition to prevent cancer. In 2002, the ACS guidelines included physical activity for the first time, recommending that regular physical activity be undertaken to decrease the risk of developing colon and breast cancers (Byers et al. 2002). That same year, in one of their handbooks on cancer prevention, the International Agency for Research on Cancer (IARC) of the World Health Organization also recognized the importance of physical activity in reducing the risk of colon and breast cancers and, possibly, other cancers (IARC 2002). In this chapter, we review the evidence that led to these recommendations.

Importance of Cancer

In the United States, cancer is a leading cause of mortality and morbidity. In fact, cancer ranks as the second leading cause of death among both U.S. males and females, trailing only deaths from heart disease (figure 13.1) (Jemal et al. 2004). In 2001, 2.4 million persons died in the United States; of these deaths, 29% were attributed to heart disease and 23%—more than half a million deaths—were attributed to cancer. Looked at in a different way, these data indicate that approximately one out of every four deaths in 2001 was caused by cancer. The most common fatal cancers occurring among U.S. men are, in order of frequency, cancers of the lung,

prostate, colorectum, and pancreas and leukemia (table 13.1). Among U.S. women, they are cancers of the lung, breast, colorectum, ovary, and pancreas. Heart disease and cancer are important chronic diseases not only in the United States but also globally; it has been estimated that of the 58 million deaths worldwide in 2005, 30% will be attributable to cardiovascular diseases and 13% to cancer (Strong et al. 2005). (As a comparison, 30% of worldwide deaths in 2005 were estimated to be attributable to communicable diseases, maternal and perinatal conditions, and nutritional deficiencies.)

In addition to deaths occurring from cancer, each year a large number of males and females also are newly diagnosed with cancer. In 2004, the American Cancer Society estimated that 1.4 million new cases of cancer will be diagnosed in the United States (not counting the common basal and squamous cell cancers of the skin) (Jemal et al. 2004). The most common sites of new cancers occurring in men are, in order of frequency, the prostate, lung, colorectum, bladder, and skin melanoma (table 13.1). Among women, they are cancers of the breast, lung, colorectum, uterus, and ovary. You will note that these most common, newly diagnosed cancers are similar, but not identical, to the most common fatal cancers. This is because some cancers have better

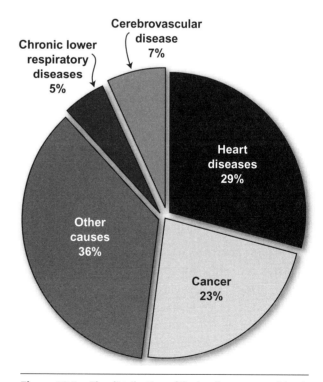

Figure 13.1 The distribution of the leading causes of death in the United States for the year 2001.

Table 13.1   Most Common Cancers in U.S. Men and Women

Most common fatal cancers		Most common newly occurring cancers	
Men	**Women**	**Men**	**Women**
Lung (32%)	Lung (25%)	Prostate (33%)	Breast (32%)
Prostate (10%)	Breast (15%)	Lung (13%)	Lung (12%)
Colon or rectum (10%)	Colon or rectum (10%)	Colon or rectum (11%)	Colon or rectum (11%)
Pancreas (5%)	Ovary (6%)	Bladder (6%)	Uterus (6%)
Leukemia (5%)	Pancreas (6%)	Melanoma (4%)	Ovary (4%)

prognosis than other cancers, partly because of the ability to diagnose the cancers at an earlier (and, hence, more treatable) stage and also partly because some cancers are more amenable to treatment than other cancers. For example, both prostate and breast cancers are more easily diagnosed at an earlier stage than lung cancer. Also, these cancers are more successfully treated than lung cancer.

Because cancer poses a major public health burden, in the United States as well as globally, there has been a great deal of interest in searching not only for a cure for cancer but also for the causes of cancer. Cancer is multifactorial in its origin, with contributions coming from both genetic and environmental components. From a public health perspective, modifiable determinants of cancer are important, because such factors are amenable to change. One determinant, or risk factor, that has received a great deal of attention recently, and which is the focus of this chapter, is sedentary behavior or physical inactivity.

Cancer— A Major Cause of Death

Globally, as well as in the United States, cancer is the second leading cause of death, after cardiovascular disease.

How Physical Activity and Physical Fitness Decrease the Risk of Developing Cancer

Although the exact mechanisms underpinning lower cancer rates among active compared with sedentary

persons are unknown, several broad categories of plausible mechanisms have been proposed to explain why physically active men and women may be at lower risk of developing cancer (Lee and Oguma in press).

Plausible Mechanisms for Reduced Cancer Risk With Physical Activity

- Modulation of reproductive hormone levels
- Decrease in body weight and adiposity
- Change in levels of insulin-like growth factors and their binding proteins
- Decrease in intestinal transit time
- Enhanced immune function

First, sex hormones may play a role. These hormones have powerful mitogenic and proliferative effects and are important in the development of reproductive cancers. Investigators have proposed that physical activity reduces the risk of developing breast cancer through its effects on menstrual function and female sex hormone levels. Girls who participate in physical activity and sports tend to have later age menarche and are more likely to have cycles that are anovulatory. These effects can decrease breast cancer risk because later age at menarche is associated with lower breast cancer rates, whereas anovulatory cycles are associated with lower levels of estrogen. Physical activity also has been associated with changes in female sex steroid hormones in adult women. Among adult pre- and postmenopausal women, higher levels of physical activity have been correlated with lower levels of estrogen and progesterone. Additionally, increased concentrations of sex hormone binding globulin have been observed in women who are physically active. These globulins bind to estrogens in the circulation, leading to lower concentrations of the free, active hormones. Such changes in estrogen levels with physical activity also can be expected to decrease the risk of developing endometrial cancer among physically active women, because higher levels of estrogen strongly predict higher rates of endometrial cancer.

In men, changes in androgen levels with physical activity have been postulated to decrease the risk of prostate cancer. Although androgen levels may be acutely elevated after a session of aerobic exercise, basal levels appear lower, within physiological range, among highly trained men compared with sedentary men. The amount of physical activity associated with lower androgen levels is high (e.g., elite marathon runners), and it is not clear whether more moderate levels of physical activity can lower testosterone levels.

A second major pathway through which physical activity may influence the risk of cancer regards weight and adiposity. Because physical activity is associated with lower body weight and fat, this may reduce the risk of developing several obesity-related cancers, such as postmenopausal breast cancer, endometrial cancer, and colorectal cancer. Among obese postmenopausal women, higher levels of estrogen are present, compared with lean women, because adipose tissue can convert estrogen precursors to estrogen. As discussed previously, higher levels of estrogen increase the risk of female reproductive cancers.

Sedentary persons who take up physical activity may be more likely to lose abdominal fat. Obesity, in particular abdominal obesity, is associated with insulin resistance, hyperinsulinemia, hypertriglyceridemia, and higher levels of insulin-like growth factors (IGF). Insulin and IGF have been implicated in the etiology of several cancers, such as breast, prostate, and colon cancers. A recent study suggested that exercise-associated decreases in IGF also were associated with increases in cellular p53 protein content, which in turn reduced growth and increased apoptosis (cell death) of cultured prostate cancer cells (Leung et al. 2004). Thus, this may represent a third pathway, separate from the sex hormone pathway, through which physical activity has the potential to influence cancer development.

A fourth commonly cited mechanism for lower rates of, specifically, colon cancer among physically active persons relates to change in intestinal transit time. It has been proposed that physical activity speeds up transit time within the colon, decreasing exposure to carcinogens, cocarcinogens, or promoters in the fecal stream. However, although some studies have shown faster transit time among physically active persons, not all studies have supported this.

Finally, physical activity may reduce the risk of developing cancer by enhancing the innate immune system. Because the innate immune system is responsible for regulating susceptibility to cancer development, this may represent another mechanism through which physical activity can protect against cancer development. In general, the available evidence suggests that moderate levels of physical activity can enhance the immune system. However, prolonged and intense exercise (e.g., running a marathon) may have the opposite effect, leading to a temporary period of immunosuppression, lasting perhaps days to as long as 2 weeks.

One in four U.S. persons died from cancer in 2004.

How We Study Whether Physical Activity and Physical Fitness Decrease the Risk of Developing Cancer

Although several plausible mechanisms have been proposed to suggest lower cancer rates among physically active or fit persons, epidemiologic studies are needed to provide direct evidence of a protective effect of physical activity on cancer risk. **Epidemiology** is the study of the distribution and determinants of disease in human populations. Several epidemiologic study designs are available; next we describe three that have the most relevance to this chapter.

Evidence for a Role of Physical Activity in Preventing Cancer

The evidence for a role of physical activity in preventing certain cancers comes primarily from epidemiologic studies. The findings from such studies are supported by plausible biologic mechanisms.

The **randomized clinical trial** generally is considered the gold standard of epidemiologic study designs. In this study design, investigators take a group of eligible participants and randomly assign them to "treatment" groups. With regard to an investigation of physical activity, for example, researchers might assemble a group of eligible participants—such as individuals who are sedentary—for study. The researchers might then randomly assign these participants to exercise at three levels: no activity, moderate activity, and vigorous activity. These participants are then followed over time to assess the outcome of interest.

The reason this study design is considered the gold standard is because the investigator assigns the exposure: Participants do not select their own physical activity. Self-selection of physical activity may lead to **confounding** by other health habits, because healthy behaviors tend to cluster. For example, active persons also are likely to smoke less and to follow more prudent diet patterns. Because of the random assignment by the investigator, this clustering of healthy behaviors is unlikely to occur.

Although this study design is considered the most rigorous study design, it may not always be feasible, or even desirable. A major factor is cost: Randomized clinical trials are by far the most expensive study design. Another factor to consider is compliance with the assigned "treatment." To have valid results from a randomized clinical trial, compliance has to be high. It is not difficult to imagine that in the preceding example, previously sedentary individuals who are assigned to exercise have a high likelihood of dropping out of their exercise program, particularly if the study lasts for many years. However, to examine directly whether physical activity is associated with lower cancer rates, a study of long duration is required because cancer takes years to develop.

Thus, primarily for reasons of cost and compliance, no randomized trials of physical activity and cancer rates have been conducted. Instead, shorter term studies—which cost less and are more likely to engender high compliance from participants—have assessed predictors of cancer (e.g., body fat) (Irwin et al. 2003) rather than cancer occurrence itself. One limitation of randomized clinical trials, whether long or short term, is the characteristics of participants being studied. The criteria for inclusion in a clinical trial tend to be strict; participants often need to be in good health to enter the study, must agree to be randomized to the different "treatment" groups, and must agree to remain committed to the study protocol for the duration of the trial. This leads to selection of individuals who tend not to be representative of the population of interest (e.g., all women). Although study participants of the other study designs described next also may not be representative, the lack of representativeness generally is more pronounced for randomized clinical trials because participants have to agree to be randomized to different treatment groups, as opposed to merely being observed for their usual habits.

The direct epidemiologic evidence that we have on the association of physical activity and cancer risk comes, instead, from two other study designs: cohort and case–control studies. In a **cohort study** design, participants are grouped on the basis of their self-selected exposure. For example, in a cohort study of physical activity and colon cancer, investigators might study three groups of participants who are free of colon cancer: individuals who have chosen to be inactive, moderately active, and highly active. The three groups are then followed over time for the development of the disease of interest, in this example, colon cancer. The rates of colon cancer in the three groups are then compared to assess whether higher levels of physical activity are associated with lower rates of colon cancer.

Because participants in this study design are selecting their physical activity, confounding is a concern. This concern can be minimized by using statistical methods in data analyses, to take into account differences in other health habits that may be associated with physical activity and that independently predict the occurrence of cancer.

In a **case–control study** design, participants are grouped on the basis of the outcome of interest, for example, colon cancer. Participants with cancer are referred to as cases. A comparison group of participants without colon cancer is needed; they are referred to as controls. Both groups are assessed for their physical activity in the past, and case participants are assessed for physical activity that occurred before the onset of cancer. The prevalence of different levels of physical activity in the two groups is then assessed. This will allow the investigators to determine whether people with colon cancer have a lower prevalence of physical activity compared with control participants (i.e., are case participants less active?).

As with the cohort study design, because physical activity is selected by participants and not assigned at random by the investigator, we are concerned with confounding by other risk factors for cancer. Similarly, this concern can be minimized by statistical methods in data analyses that adjust for differences in other risk factors that may be associated with physical activity and that independently predict the occurrence of cancer.

In interpreting the findings from epidemiologic studies, several concepts are important. First, the term **relative risk** (often abbreviated as RR) is a measure of association used in these studies. The relative risk measures the magnitude of association between the exposure (in our example, physical activity) and the disease (in our example, cancer). Relative risk essentially compares the rates of cancer among persons with different levels of physical activity. A relative risk of 0.5 for cancer associated with physical activity, compared with inactivity, indicates that active persons have 0.5 times (or 50%) the risk of developing cancer, compared with those who are inactive (who serve as the reference group).

A second important concept is that of a **confidence interval** (often abbreviated as CI). When we calculate a relative risk, we can also calculate a band of uncertainty, the confidence interval, around our estimate of the relative risk. Typically, 95% confidence intervals are used in epidemiologic studies. For example, if we estimate the relative risk for cancer associated with physical activity, compared with inactivity, to be 0.5 and the associated 95% confidence interval to be 0.3 to 0.8, this means that we are 95% certain that the true estimate of the relative risk lies between 0.3 and 0.8. The narrower the width of the confidence interval, the more precise the estimate of the relative risk. Additionally, if the 95% confidence interval does not span a value of 1.0, this implies that the findings are statistically significant at $p < .05$.

Finally, epidemiologic studies use the term **dose–response** to describe the phenomenon of a graded effect of exercise. In the case of physical activity and cancer, the presence of an inverse dose–response indicates that the higher the level of physical activity, the lower the rates of cancer. The shape of the dose–response curve can take many forms—linear, curvilinear, and nonparametric. Most commonly, epidemiologic studies test for a curve that is linear.

Epidemiologic Study Designs

There are three major epidemiologic study designs:

1. Randomized clinical trial
2. Cohort study
3. Case–control study

Physical Activity, Physical Fitness, and Site-Specific Cancers

Next we discuss findings from epidemiologic studies of physical activity or physical fitness and cancer risk. The studies are all of cohort or case–control study designs. The available data primarily come from studies of physical activity; there have been relatively few studies of physical fitness or markers of physical fitness (e.g., heart rate). Because the evidence indicates that the association of physical activity or physical fitness with cancer differs for different cancers, we discuss the individual cancers separately.

Colorectal Cancer

The large intestine comprises both the colon and the rectum, with the rectum being continuous with the end of the colon. Although they are closely linked anatomically, the epidemiologic literature suggests that physical activity has different associations with colon and rectal cancer. Thus, we discuss colon cancer and rectal cancer separately. Studies that have looked at colorectal cancer as a single outcome will not be considered, because combining the two separate cancers may obscure the findings for each individual cancer site.

Colon Cancer

Some 50 published studies have examined the association between physical activity or physical fitness and the risk of developing colon cancer. These studies have been conducted in several countries in North America, Europe, and Asia and in Australia and New Zealand. Most of the studies have assessed physical activity during middle age and later years. Additionally, a handful of studies have assessed physical fitness, using markers of fitness such as resting pulse rate or heart rate. The evidence from these cohort and case–control studies indicates that physically active individuals have a lower risk of developing colon cancer compared with inactive individuals. The data have been clear for both men and women. Because few studies have included participants from racial or ethnic minority groups, the data are heavily drawn primarily from studies of white persons. Additionally, most of the data relate to physical activity during middle age and later years; there have been few studies of physical activity during childhood and adolescence and almost no data on changes in physical activity over time.

How much of an effect does physical activity have? Although different studies have shown different results, the median or average finding across all studies gives an idea of the magnitude of effect of physical activity. For colon cancer, when we compare most active participants with least active participants, the median relative risk across all studies is 0.7. That is, on average, these studies indicate that the colon cancer rate among active individuals is 0.7 times that among inactive persons, or a 30% reduction in risk. When we examine studies of men and women separately, the data indicate perhaps a somewhat larger effect in women. The median relative risk for men is 0.7 (or a 30% reduction in risk) and for women 0.6 (40% reduction in risk).

Although there are limited data, the available information suggests that some 30 to 45 min per day of moderate-intensity physical activity is needed to reduce risk, as indicated by the findings from men in the Harvard Alumni Health Study (Lee et al. 1991) and women in the Nurses' Health Study (Martinez et al. 1997). In figure 13.2*a*, significantly lower rates of colon cancer were observed among men expending 4,200 kJ (1,000 kcal) or more per week. This amount of energy expenditure can be accomplished by 30 min of moderate-intensity physical activity most days of the week. In figure 13.2*b*, significantly lower rates

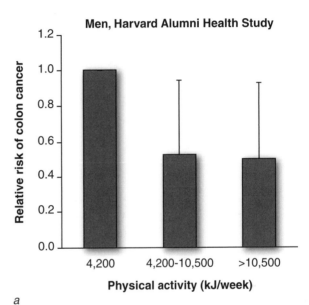

a

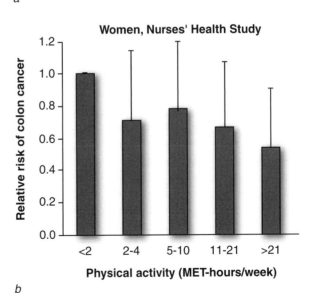

b

Figure 13.2 Relationship between physical activity and colon cancer risk among (a) men in the Harvard Alumni Health Study and (b) women in the Nurses' Health Study.

of colon cancer were seen among women expending >21 MET-hours per week, or approximately 45 min per day of moderate-intensity activity.

Many of the studies of physical activity and colon cancer risk provided some information on a dose–response relationship, because they investigated participants belonging to at least three levels of physical activity. However, because the dose–response relationship was not often studied in detail, the available information suggests only that an inverse dose–response is likely. There are not enough data to provide details on how much additional protection physical activity might confer at increasingly higher levels of activity. That is, it is unclear what shape of the dose–response curve best describes the nature of the relationship between physical activity and colon cancer rates.

Previously, we discussed concerns about confounding arising from cohort and case–control studies. Many of the investigators who studied physical activity and colon cancer risk did adjust for potential confounders in the data analyses, including controlling for differences in age, body mass index, smoking, and diet. Even after these adjustments, physical activity was found to be associated with lower colon cancer risk. Thus, it is unlikely that confounding by other risk factors for colon cancer can explain the inverse relationship seen between physical activity and the risk of developing colon cancer.

Rectal Cancer

About 30 studies have investigated the relationship between physical activity and the risk of developing rectal cancer. As with the colon cancer studies, these studies of rectal cancer were carried out in North America, Europe, Asia, and Australia. Although some studies reported lower rectal cancer rates among active individuals, the data on the whole do not support any relationship between physical activity and rectal cancer risk. Across all studies, the median relative risk comparing most with least active participants is 1.0. That is, on average, the data indicate that the rates of rectal cancer among most and least active persons are the same. This holds for both studies of men and women. The studies of physical activity and rectal cancer risk have been conducted primarily among participants who are white, with few data on minority racial or ethnic groups.

Breast Cancer

Like the studies of colorectal cancer, almost all of the epidemiologic studies of breast cancer have

investigated physical activity rather than physical fitness. Approximately 60 published studies have examined whether physical activity is associated with a lower risk of developing breast cancer in women. These studies have been conducted in various countries in North America, Europe, and Asia and in Australia. Overall, the data have been reasonably consistent in supporting an inverse relationship between physical activity and breast cancer incidence rates.

Although individual studies have indicated that active women may have half, or even less than half, the incidence rates of breast cancer compared with inactive women, the data on the whole suggest a smaller magnitude of association than that observed for colon cancer. Across all studies, the median relative risk for developing breast cancer, comparing most with least active women, is 0.8. That is, the data on average indicate that active women have a 20% lower breast cancer risk. Because risk factors for breast cancer in premenopausal women and postmenopausal women may be different, several studies examined these women separately. Physical activity may have a somewhat larger effect for postmenopausal women: The median relative risk is 0.8 (or a 20% risk reduction) for premenopausal women and 0.7 (30% risk reduction) for those who were postmenopausal.

There are relatively few data on how much physical activity is needed to decrease the risk of breast cancer. It appears that some 4 to 7 hr/week of moderate- to vigorous-intensity physical activity is required, which is similar to that observed for colon cancer. For example, in a Norwegian study, significantly lower rates of breast cancer were observed among women undergoing regular physical activity, defined as exercising to keep fit for at least 4 hr/week or participating in competitive sports (figure 13.3) (Thune et al. 1997). In this study, higher levels of physical activity on the job also were associated with lower risk. Note that the magnitude of reduced risk in this study—on the order of 40% to 50%—is larger than that seen in other studies, such as the Women's Health Study (Lee, Rexrode, et al. 2001) or the Women's Health Initiative (McTiernan et al. 2003), where the degree of risk reduction observed was closer to the overall average of 20%. The data from the Women's Health Initiative are noteworthy, because a sizeable number of participants in this study were women belonging to minority racial and ethnic groups.

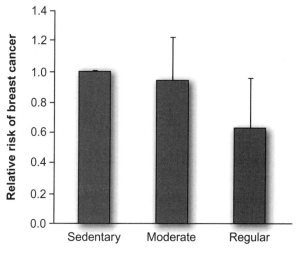

a

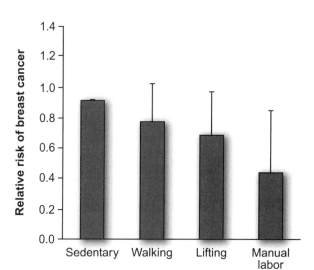

b

Figure 13.3 Relationship between *(a)* leisure-time physical activity or *(b)* occupational physical activity and breast cancer risk in a Norwegian study.

The timing of physical activity, in order for it to decrease breast cancer risk, is unclear. Most studies have assessed physical activity only at one time in relationship to breast cancer rates. However, several studies have tried to estimate physical activity throughout a woman's life, including physical activity carried out during adolescence. As discussed previously, this has biological rationale, because physical activity during adolescence is associated with favorable changes in menstrual function and female sex steroid hormones. Although some studies suggest that physical activity throughout a woman's life is

more strongly protective, other studies have not supported this, indicating instead that physical activity still is inversely related to breast cancer risk, even if carried out during middle age and older ages.

Many of the breast cancer studies examined women belonging to at least three levels of physical activity, allowing for the assessment of a dose–response relationship. As with the colon cancer studies, the dose–response relationship has not been investigated in detail. Thus, the data suggest an inverse dose–response relationship between physical activity and the risk of breast cancer, with higher levels of activity being associated with increasingly lower risks. However, we do not have enough information to indicate how much additional decrease in risk might be observed with increasingly higher levels of activity.

Although there have been no randomized clinical trials, the data that we have for breast cancer are unlikely to be confounded by risk factors that are associated with a physically active lifestyle. Many studies continued to find a lower risk of breast cancer among active women, even after adjustment for other risk factors such as age, body mass index, alcohol intake, use of oral contraceptives and hormone therapy, reproductive variables (ages at menarche and menopause, menopausal status, parity, age at first birth, breast feeding), benign breast disease, and family history of breast cancer.

Prostate Cancer

In contrast to the other site-specific cancers that have been discussed here, the findings for prostate cancer have been less consistent. About 40 studies have been published on the relationship between physical activity or physical fitness and the risk of developing prostate cancer. As with the other studies discussed here, the studies of prostate cancer, which have been conducted in several countries in North America, Europe, and Asia, have primarily assessed physical activity rather than physical fitness.

Although individual studies have indicated that physical activity is associated with a lower risk of prostate cancer, on the whole the data from epidemiologic studies do not provide strong support for an inverse relationship between physical activity and the risk of this cancer. In fact, several studies also have reported an opposite effect—that physical activity is associated with higher rates of prostate cancer. The median relative risk across all studies, comparing most with least active men, is 0.9. This average result is close to a relative risk of 1.0, which indicates no difference in prostate cancer incidence rates between active and inactive men.

It is unclear why the findings have been inconsistent. Apart from age and race, few risk factors for prostate cancer have been established. Thus, it is unlikely that the different findings reflect uncontrolled confounding by risk factors related to both physical activity and prostate cancer risk. It has been postulated that the age at which prostate cancer develops is important: Early-onset prostate cancer may be more likely influenced by genetic factors, with physical activity being less important, and the opposite being true for prostate cancers occurring at older ages. However, studies of physical activity and prostate cancer occurring at different ages have not provided clear evidence supporting this hypothesis. Another factor that may be relevant is the level of physical activity that may be needed to reduce prostate cancer risk. As discussed previously, one postulated mechanism for lowered risk of this cancer is the modulation of male sex steroid hormones. Changes in androgen levels with physical activity have generally been observed in studies where the level of activity is very high. Epidemiologic studies of physical activity and prostate cancer risk have typically been conducted in the general population, where there is a low prevalence of very high levels of physical activity. Perhaps these studies have not been able to document an inverse relationship between physical activity and prostate cancer rates because the level of physical activity among participants is not sufficiently high to effect changes in hormone levels and, hence, subsequent prostate cancer risk.

Finally, the issue of screening for prostate cancer may explain the discrepant findings. Recently, a new test was developed that has the ability to diagnose very early stage prostate cancer. This test, called prostate-specific antigen (PSA) screening, for early detection gained widespread use in the United States in the 1990s. If physically active men also are more health conscious and undergo PSA screening for prostate cancer more frequently, this may result in higher observed rates of prostate cancer among these men, because of increased detection of early stage cancers. This may obscure the protective effect of physical activity, because of the detection of early stage cancers, which would not have been diagnosed clinically, among physically active men. Some support for this argument is provided by two studies conducted as part of the Harvard Alumni

Health Study (Lee, Sesso, and Paffenbarger 2001; Lee, Paffenbarger, and Hsieh 1992). In the first study of physical activity and prostate cancer, the cases of interest were prostate cancer diagnosed in 1988 or earlier. Investigators reported an almost halving of prostate cancer incidence rates among men aged 70 years or older who expended 16,800 kJ (4,000 kcal) or more per week (equivalent to some 5 hr per week of vigorous activity), compared with sedentary men expending <4,200 kJ (1,000 kcal) or more per week. However, in an updated analysis of these men that examined prostate cancer diagnosed after 1988, the earlier observations were not replicated. In this later study, no differences in prostate cancer rates were observed among men with different physical activity levels. These different findings may have been attributable to increased screening for prostate cancer among the most active men in the later study.

Another study of health professionals also appears to support the argument of biased findings attributable to a screening effect. In the Health Professionals' Follow-Up Study, investigators observed an approximate halving of risk of metastatic prostate cancer among very active men compared with sedentary men (Giovannucci et al. 1998). However, when total prostate cancers were examined, no differences in cancer rates were seen among men with different physical activity levels. A plausible explanation is that the diagnosis of metastatic prostate cancers was unaffected by PSA screening. But all prostate cancers included early-stage prostate cancers that may have been picked up more frequently during PSA screening among the more active men.

Although all these factors may have contributed to the inconsistent findings on physical activity or physical fitness and prostate cancer rates, it remains unclear whether the risk of developing this cancer can be influenced by higher levels of physical activity or fitness.

Lung Cancer

In recent years, several studies have suggested an intriguing inverse relationship between physical activity or physical fitness and the risk of developing lung cancer (with the data coming primarily from studies of physical activity). Some 20 studies, conducted in North America and Europe, have examined this hypothesis and observed a median relative risk for lung cancer of 0.8, comparing most with least active participants. That is, on average, the studies indicate that those who are physically active have a 20% lower risk of developing lung cancer compared with those who are sedentary. The data, on the whole, suggest an inverse relationship in both men and women; however, the data are sparse in women. There also are few data on minority racial and ethnic groups for this cancer and few data examining the effect of physical activity carried out at different ages.

Although the data supporting a role of physical activity in lowering lung cancer rates are promising, a major concern, given that the findings are derived only from cohort and case–control studies, is confounding by cigarette smoking. As discussed previously, physically active persons tend to be health conscious and thus they are less likely to smoke cigarettes, a major risk factor for lung cancer. Although almost all studies did adjust for cigarette smoking in the analyses (typically, the number of cigarettes smoked and the duration of smoking), it is difficult to be certain that the effect of cigarette smoking was completely controlled. Residual confounding, as well as confounding by other smoking-related factors (e.g., use of low-tar cigarettes or filter tips, depth of inhalation when smoking, passive smoking), might still be present.

To be certain that the inverse relationship between physical activity and lung cancer rates does not reflect confounding by smoking, a study could be conducted only among never-smokers. However, lung cancer occurs very infrequently among never-smokers, making this a difficult study to conduct. Several studies have been able to examine nonsmokers (never and past smokers) separately, with similar associations observed for nonsmokers and smokers. For example, in the Harvard Alumni Health Study, similar results were observed among all participants as well as among nonsmokers (Lee, Sesso, and Paffenbarger 1999). The magnitude of risk reduction observed among the most active participants compared with the least active was 39% (figure 13.4a). When nonsmokers only were examined separately, the corresponding risk reduction was 46%.

Only one study has been large enough to examine never-smokers as a separate group. In a case–control study conducted in Canada (Mao et al. 2003), an inverse relationship between physical activity and lung cancer risk was observed among all participants. The most active men had a 26% lower risk of developing lung cancer compared with least active men. For women, the corresponding risk reduction was 28%. The Canadian study was large enough to examine never-smokers as a separate group, with

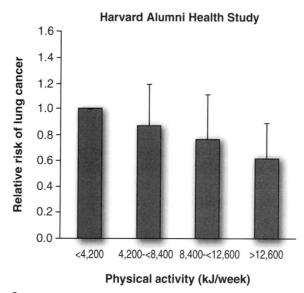

Figure 13.4 Relationship between physical activity and lung cancer risk in *(a)* the Harvard Alumni Health Study and *(b)* in never-smoking participants of a Canadian study.

126 lung cancers occurring among these participants (men and women combined). Although this number of lung cancers may have been too small to reveal statistically significant findings, it is of note that the magnitude of risk reduction among never-smokers expending 34.4 MET-hours per week or more in recreational physical activity was 32%, compared with the least active participants expending <6.1 MET-hours per week (figure 13.4b). This risk reduction was similar to that seen among past smokers (34% reduction) and current smokers (41% reduction) exercising at that level in the same study.

A few studies also have been able to investigate lung cancers of different histological types. Although cigarette smoking increases the risk of all types of lung cancer, the increase in risk is stronger for certain histological types such as squamous cell and small cell carcinomas than for others such as adenocarcinomas. In a Norwegian study, investigators reported significant inverse associations between physical activity and small cell carcinomas and adenocarcinomas of the lung, but not for squamous cell carcinomas, in men. This may provide some indirect evidence that the observed association in this study was not reflecting confounding by smoking. However, in the Canadian study described previously, the strongest inverse relationships were noted between physical activity and squamous cell carcinomas in men but between physical activity and small cell carcinomas in women. Thus, on balance, although the data indicate an inverse relationship between physical activity or physical fitness and lung cancer risk, the findings should be considered preliminary.

Other Cancers

Several other cancer sites have been investigated for a relationship with physical activity or physical fitness. These include endometrial cancer, ovarian cancer, testicular cancer, pancreatic cancer, kidney cancer, bladder cancer, and hematopoietic cancers. The data suggest an inverse relationship between physical activity and endometrial cancer risk, consistent with the hypothesis that physical activity can lower estrogen levels, which are related to increased risk of this cancer. However, many of the studies of endometrial cancer did not control for the use of postmenopausal estrogen therapy, an important risk factor for this cancer. With regard to the other cancers, the data are too sparse to make any conclusions regarding whether physical activity is associated with decreased rates of occurrence.

Summary

Many published studies have examined the role of physical activity or physical fitness in preventing cancer (table 13.2). The data are clearest in supporting physical activity or fitness as a means for preventing colon and breast cancers, as recommended by the American Cancer Society and the International Agency for Research on Cancer (Byers et al. 2002; IARC 2002). It also appears reasonably clear that physical activity or fitness does not influence rectal cancer rates. For prostate cancer, the data have been equivocal, whereas for lung and endometrial cancers, there are suggestive, although not definitive, data to support inverse associations for these cancers.

Thus, this chapter describes yet another important benefit of physical activity—decreasing the risk of developing certain types of cancer. The evidence discussed in this chapter provides yet another compelling argument for why all persons should adopt and maintain a physically active way of life.

In Review

1. Cancer is a leading cause of mortality in the United States as well as worldwide.

2. Biologically, it appears plausible for higher levels of physical activity or physical fitness to result in lower rates of cancer.

3. The data from epidemiologic studies indicate that higher levels of physical activity or physical fitness are associated with lower rates of colon and breast cancers.

Table 13.2   Summary of Epidemiologic Data on the Association of Physical Activity With Risk of Developing Cancer

Cancer site	Epidemiologic data
Colon	Inverse association; likely dose–response
Rectum	No association
Breast	Inverse association; likely dose–response
Prostate	Equivocal data
Lung	Possible inverse
Endometrium	Possible inverse
Other cancers	Sparse data

KEY CONCEPTS

cancer: A chronic disease that is multifactorial in its etiology, with contributions from both environmental factors (the term *environmental factors* is used broadly here and refers not only to factors such as radiation but also to smoking, physical inactivity, diet, viruses) and genetic factors. The disease arises when there is excessive, uncontrolled, and purposeless proliferation of cells, in the absence of physiological stimuli. These cells invade their surrounding tissues and spread by means of the lymphatics and blood vessels to give rise to secondary cancers, or metastases.

case–control study: A type of epidemiologic study design in which participants are grouped on the basis of the disease of interest, in this case, cancer. Participants with cancer are referred to as cases. A comparison group of participants without cancer are referred to as controls. Both groups are assessed for their physical activity in the past (before the onset of cancer in cases), and the prevalence of different levels of physical activity in the two groups is compared. This will allow, for example, the determination of whether participants with colon cancer have a lower prevalence of physical activity compared with control participants.

cohort study: For definition, see page 203.

confidence interval: A band of uncertainty around an estimate of the relative risk. Typically, 95% confidence intervals are used in epidemiologic studies. For example, if we estimate the relative risk for colon cancer associated with physical activity, compared with inactivity, of 0.5 and a 95% confidence interval of 0.3 to 0.8, this means that we are 95% certain that the true estimate of the relative risk lies between 0.3 and 0.8. (See also *relative risk*.)

confounding: A phenomenon that may exist in the data from epidemiologic studies and cloud the interpretation of findings. The simplest way to explain confounding is with an example. In this example, we compare the rates of colon cancer that develop among a group of athletes and among a group of sedentary persons. We find that the rates are lower among the athletes compared with the sedentary group. Can we conclude that physical activity lowers the risk of developing colon cancer? If we look more closely at the groups, we might find, perhaps, that the athletes are much younger than the sedentary group. Cancer rates also increase with age. Thus, the lower rates of colon cancer that we see among the athletes may have nothing to do with physical activity but, rather, reflect the fact that the athletes are younger. This phenomenon is termed *confounding*. Statistical adjustment can be made in data analyses to factor in the age differences. If this is done and we continue to observe lower colon cancer rates among the athletes, it is reasonable to conclude that physical activity is associated with lower colon cancer rates, provided there are no other confounders.

dose–response: For definition, see page 158.

epidemiology: A scientific discipline that involves the study of the distribution and determinants of disease in human populations. In this chapter, epidemiologic studies are used as the basis to directly assess whether physically active individuals have lower rates of cancer than inactive persons (see also *case–control study*, *cohort study*, and *randomized clinical trial*).

randomized clinical trial: A type of epidemiologic study design in which participants are grouped on the basis of the investigator-assigned exposure of interest, in this case, physical activity. For example, among a group of eligible participants, investigators may randomly assign them to exercise at three levels: no activity, moderate activity, and vigorous activity. These participants are then followed over time to assess the outcome of interest, such as change in abdominal fat, for example. This is often considered the "gold standard" of epidemiologic study designs. However, because of the cost and issues regarding compliance with an assigned activity level, it may not always be feasible, or even desirable, to conduct randomized clinical trials.

relative risk: For definition, see page 170.

STUDY QUESTIONS

1. List the leading causes of death in the United States. Where does cancer rank among the causes of death?

2. What are the most common fatal cancers among U.S. men and women? What are the most common newly occurring cancers among them?

3. Describe some of the mechanisms that might explain why physically active persons have lower rates of cancer than inactive persons.

4. What are some of the key differences in the following epidemiologic design strategies: randomized clinical trial, cohort study, case–control study?

5. Describe the meaning of the terms *relative risk* and *confidence interval.*

6. Describe the meaning of the term *confounding*. Why should we be concerned about this in studies of physical activity and cancer?

7. Which cancers that have been shown to occur at lower frequencies among physically active, compared with inactive, persons?

8. How much of an effect does physical activity have in reducing the risk of developing colon cancer?

9. What do the available data show regarding how much physical activity may be needed to reduce the risk of developing colon cancer?

10. Discuss what epidemiologic studies have shown regarding the relationship of physical activity and the risk of developing breast cancer.

© Photodisc

Physical Activity, Fitness, and Joint and Bone Health

■ Jennifer Hootman, PhD, ATC, FACSM

CHAPTER OUTLINE

As you learned in chapter 1, the human body is engineered for movement. The neurological, muscular, and skeletal systems all work together to produce human movement, with forces being produced and transmitted within the musculoskeletal system. Most forces are generated for required daily functioning activities and are well tolerated by the human body. However, unexpected, excessive, or accumulated forces can have unwanted effects. This chapter discusses the prevalence of select musculoskeletal conditions, including musculoskeletal injuries, **osteoarthritis,** and **osteoporosis** and the relationship between different levels of activity and the development of these conditions.

Burden of Selected Musculoskeletal Diseases in the Population

Acute and chronic musculoskeletal conditions, including trauma-related injuries and **arthritis and other rheumatic conditions,** are among the most frequently occurring medical conditions that have a substantial impact on the general health of the population and the health care system. In terms of acute conditions, almost 100,000

Americans and 3.6 million persons globally die each year from unintentional injuries (National Center for Injury Prevention and Control 2002; World Health Organization 2004). Regarding chronic conditions, arthritis and rheumatism are the leading cause of disability among persons aged 15 years and older (Centers for Disease Control and Prevention 2001). In fact, as shown in figure 14.1, 3 of the 10 leading causes of disability in the United States are musculoskeletal conditions: arthritis and rheumatism, back and spine problems, and extremity or limb weakness. These three conditions account for 38.2% of all disabilities. The prevalence of musculoskeletal impairments attributable to all causes exceeds 36 million in the U.S. population, accounting for 488.6 million restricted-activity days and 152.8 million days restricted to bed (Praemer et al. 1999). Internationally, unintentional injury accounts for 51.1 million musculoskeletal conditions and 29.0 million years lost because of disability annually (World Health Organization 2004).

Unintentional Injury

In the United States and worldwide, unintentional injury is the leading cause of death among young persons aged 1 to 34 years and ranks fifth among all ages in the United States. Globally,

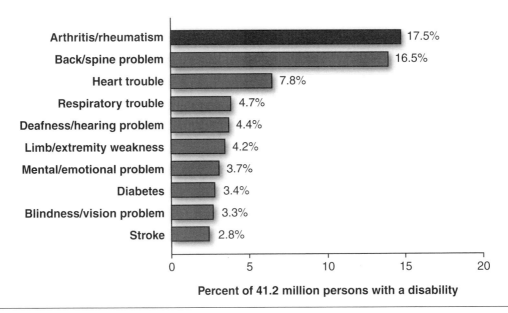

Figure 14.1 Percent of 41.2 million persons with a disability.

Data from Centers for Disease Control and Prevention (CDC), 2001,"Prevalence of disabilities and associated health conditions among adults- United States 1999," *Morbidity and Mortality Weekly Report* 50: 120-125.

220

81% of all injury deaths among persons aged 0 to 29 years are unintentional (National Center for Injury Prevention and Control 2002; World Health Organization 2004). Injury fatalities are only a small part of the problem. In 2001, 1 in 10 Americans experienced a nonfatal injury severe enough to require medical attention. Children, seniors, males, persons of low income, and select ethnic minorities (American Indian–Alaska natives) have the highest risk of unintentional injury. Each year 20% to 25% of all children will require medical attention or miss school because of injury. Fall-related injuries are the most common cause of nonfatal injuries among adults aged 65 and older, resulting in 1.6 million emergency department visits and more than 350,000 hospitalizations (National Center for Injury Prevention and Control 2002). Activity-related injury rates range from <2% to >50% per year depending on the type of activity. Walking has relatively low rates of injury compared with more vigorous activities, such as football or soccer (Hootman et al. 2001; Powell et al. 1998). Fortunately, few of these injuries are fatal or require hospitalization (National Center for Injury Prevention and Control 2002). However, the physical, emotional, and societal effects of injury are extensive. Injury-related medical costs exceed $117 billion, which equates to 10% of all U.S. medical expenditures (Finkelstein et al. 2004). Motor vehicle accidents alone cost more than $500 billion in U.S. dollars worldwide (World Health Organization 2004).

Osteoarthritis

Osteoarthritis, or degenerative joint disease, affects more than 21 million in the United States. It most commonly affects the large weight-bearing joints, such as the knees, hips, and spine. Osteoarthritis is more common among women, Caucasians, older persons, overweight or obese persons, and those with a history of significant joint injuries (Arthritis Foundation 2001; Felson et al. 2000). Osteoarthritis is a disease of the articular cartilage and the subchondral bone. Clinical diagnosis is based on specific radiographic features including joint space narrowing (indicating cartilage degeneration), **sclerosis** of subchondral bone, and **osteophyte** formation. Physical signs and symptoms include pain, swelling, **crepitus,** and restricted range of motion (Arthritis Foundation 2001).

Limitations attributable to osteoarthritis are common. This condition accounts for the major-

ity of reported difficulties with climbing stairs and walking reported by individuals as well as lost work time and early retirement. Almost 640,000 joint replacement surgeries are performed every year because of osteoarthritis. Total knee replacements make up almost 50% of all joint replacement surgeries, representing a 67% increase in recent years (Praemer et al. 1999). These procedures are costly, and many patients require substantial long-term postoperative care. Because of the aging of the population, as well as the obesity epidemic, rates of osteoarthritis and associated disabilities are expected to increase exponentially in the next 2 decades.

Osteoporosis

Osteoporosis is a disease of the skeletal system that is characterized by low bone mass and fragility. It is estimated that more than 30 million people have osteoporosis; most cases occur among post-menopausal white women (Arthritis Foundation 2001; Kahn et al. 2001). Osteoporosis is largely a silent disease, producing no symptoms until it is manifested clinically in the form of overt fractures; common fracture sites include the hip, spine, and wrist. More than 1.3 million osteoporosis-related fractures are treated annually in the United States, resulting in 432,000 hospitalizations and 3.4 million outpatient medical care visits and costing more than $13.8 billion (National Center for Injury Prevention and Control 2002). The lifetime risk of a hip, vertebral, or wrist fracture is 40% for white women, but men, smokers, persons with physically disabilities, and sedentary persons are also at risk.

Osteoporosis-related hip fractures are of special concern among elderly persons. Up to 20% of patients die in the first year after hip fracture, and two thirds do not return to their preinjury level of functioning (Arthritis Foundation 2001). Several recent public health intervention studies examined ways to reduce the morbidity and mortality associated with fall-related injuries among older adults. Promising interventions to prevent falls and hip fractures include programs that integrate improving balance and mobility through exercises, such as tai chi and walking, with patient education programs that address modification of environmental factors that increase the potential for falling (Wolf et al. 2003). (See *Focus on Research* on the next page.)

Focus on Research

Wolf et al. (2003) reported that the National Institutes of Health funded a multiple-center research study to investigate the effects of various types and levels of physical activity for the prevention of falls and fall-related injuries among older adults. The FICSIT trials (Frailty and Injuries: Cooperative Studies of Intervention Techniques) used strategies such as resistance training, balance training, and tai chi and evaluated these programs on their ability to reduce falls and related injuries. At one site (Atlanta, Georgia), 200 community-dwelling adults aged 70 and older were randomized into three arms: a tai chi exercise program, a computerized balance training group, or an education class. All groups met one to two times per week for 15 weeks. Biomedical, functional, and psychological outcomes were measured at baseline, 15 weeks, and 4 months. Falls were monitored monthly for the entire length of the follow-up (range 7-20 months). Overall compliance was 87%. In total, 209 falls were recorded: 56 in the tai chi group, 76 in the balance training group, and 77 in the education group. In time-dependent regression analyses, the tai chi group had a 47% reduced rate of falls compared with the education group. Other significant results reported for the tai chi group pre- to postintervention include lower exercise systolic blood pressure, increased grip strength, and less fear of falling. These data suggest that exercise programs such as tai chi can significantly influence older adults' physical functioning ability and reduce the risk of falls and related injuries.

Total bone strength is determined by both material and structural properties (Kahn et al. 2001). Figure 14.2 depicts the material properties, which are independent of bone size, and the structural features that contribute to skeletal strength. Of these, bone density is the feature most commonly studied in relationship to physical activity and exer-

cise; however, all features may respond to changes in activity level.

In addition to nutritional, environmental, and genetic factors, physical inactivity is an established risk factor for low bone density and osteoporosis. Bone will become brittle and porous without bone-loading stress from weight-bearing activity and muscle contraction (Kahn et al. 2001; Vuori 2001). Bone cells respond to mechanical loading stresses (axial compression and bending forces) that are generated during weight-bearing activity by temporarily deforming, which in turn stimulates bone formation. This process is described by the **mechanostat theory,** which suggests that a minimum effective strain is needed to initiate and maintain osteogenesis (Kahn et al. 2001). Bone's adaptation to mechanical stress and strain is a complex process, but the three basic elements are as follows:

1. An effective stimulus involves dynamic versus static loading.

2. Relatively few loading cycles of short duration are needed to stimulate bone tissue response.

3. Over time, bone cells cease to accommodate to routine loads and, subsequently, a periodic increase in load is necessary to continue to effect bone growth (Arthritis Foundation 2001; Kahn et al. 2001).

Dose–Response Relationships Between Physical Activity and Injury, Osteoarthritis, and Osteoporosis

The total amount and type of physical activity play an important role in the development of a variety of musculoskeletal conditions. This section focuses on the relationship between the dose (frequency, intensity, duration, and type) of physical activity and selected outcomes such as activity-related injuries, osteoarthritis, and osteoporosis.

Physical Activity and the Risk of Injury

Most data available on injury rates come from studies of very specialized populations such as elite athletes (high school, college, Olympic, and professional athletes) or military trainees. Comparatively little information is available on the relationship between

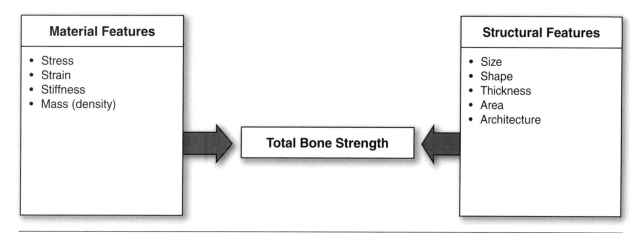

Figure 14.2 Structural and material properties that contribute to total bone strength.

a wide range of physically active lifestyles and the probability of sustaining musculoskeletal injury in the general population. Some facts do seem to be evident, however. First, sedentary persons are not immune to activity-related injuries; in fact, as many as 16% of sedentary adults report activity-related injuries during the previous 12 months. Second, the type of activity influences the risk of injury; walkers seem to have no higher risk of injury than their sedentary peers, but runners and sport participants are 1.5 to 2 times more likely to be injured. Third, frequency, duration, and intensity of activity also play a role in injury risk (figure 14.3) (Hootman et al. 2001; Powell et al. 1998; van Mechelen, 1992).

Runners and walkers have been studied most frequently, and a few risk profiles have been developed. Among runners, the risk of injury increases linearly with increasing dose of running per week; however, this same relationship has not been consistently observed among walkers. Despite the fact that an increasing dose of activity is related to an increased risk of injury, as shown in table 14.1, there are modifiable factors that may mitigate injury risk during activity. Among adults, running or walking more than 20 miles per week almost doubles the risk of a lower extremity injury, and injury risk increases 11% for every hour of total weight-bearing activity per week. Among young girls, each hour of high-impact exercise per week may result in a 5% increased risk for stress fracture; running, gymnastics, and cheerleading appear to be the highest risk activities (Loud et al. 2005). Slower training pace ($\geq$15 min/mile) is associated with a 40% to 50% reduction in injury risk. Among women, engaging in weight training at least twice weekly lowers the risk 44%. Other factors that may help

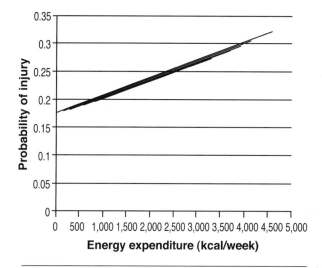

Figure 14.3 Probability of activity-related injury per level of energy expenditure in leisure-time physical activity.
Unpublished data, Aerobics Center Longitudinal Study

lower injury risk associated with physical activity include maintaining normal body weight, limiting competitive racing events, and cross-training in different activities and on different surfaces (Hootman et al. 2001; van Mechelen 1992). Studies in military populations confirm these modifiable risk factors. However, more comprehensive research is needed in these areas, especially randomized, controlled studies of injury prevention strategies.

Physical Activity and the Development of Osteoarthritis

Because the incubation period for the development of degenerative joint disease may be upward of 30 years, it is difficult to connect participation in various activities throughout a lifetime and the

Table 14.1 Modifiable Risk Factors Studied in Relation to Running Injuries

Modifiable risk factor	Risk level	Direction of risk
Body mass index	Body mass above 25 (women only)	↑
Running experience	Less than 3 years of running experience	↑
Running purpose	Competitive running versus recreational	↑ risk with more training per year and number of competitive races
Weekly mileage	>20 miles per week	↑
Frequency	Contradictory data, no specific risk level confirmed	Highly correlated to total weekly mileage
Intensity (pace)	Faster pace (<15 min/mile)	↑; fastest pace associated with competitive running
Duration	Not associated with injury in most studies	Highly correlated to total weekly mileage
Warm-up, stretching	Warm up or stretch before running	Inconsistent data but suggest no association or ↑ risk
Resistance training	Engaging in regular resistance training or weightlifting program	↓
Other sport participation	Cross-training	Inconclusive data; several studies suggest no association with participation in other sports besides running. Others report that excessive hours of weight-bearing exercise per week, including running, may cause accumulated risk.
Psychological factors	Select personality traits	Type A and risk-taking behavior, high motivation = ↑ risk
Training surface, terrain	Hill running, graded roads, hard surfaces	Inconsistent data, but suggest no relationship except running on concrete surface may ↑ risk among women
Training patterns	Time of day, seasons, interval training	No relationship
Cardiorespiratory fitness	Low fitness levels	↑ among military populations; no association among general adult population

long-term risk of osteoarthritis. The important consideration in regard to physical activity is how the actual activity loads the joint surfaces. Three main features likely contribute to the development of osteoarthritis: rate and magnitude of contact forces, extent of joint torsion and twisting, and total dose of exposure to different physical and occupational activities over a lifetime. Sudden single high-impact torsional loadings can significantly damage articular cartilage and subchondral bone (Vuori 2001). Loads that are applied more slowly are tolerated much better because muscles can absorb and disperse the load more efficiently and articular cartilage can deform slowly, thereby safely transmitting forces across the joint.

Articular cartilage will fracture in response to joint contact forces in the range of 25 MPa (megapascals, newtons per square meter). In activities such as running, jumping, and throwing, peak articular contact forces range from 4 to 9 MPa, which is far below the maximum level associated with cartilage injury (Vuori 2001). In animal models, activity-related joint contact stresses are not associated with cartilage damage in healthy joints. In fact, moderate physical activity produces many structural and functional beneficial alterations within the joint. However, in joints with abnormal anatomy or biomechanics related to trauma or injury, even normal joint stress can cause further articular cartilage damage and eventually osteoarthritis. Also, chronic or repetitive articular forces below 25 MPa may eventually lead to articular surface damage, initiating the development of osteoarthritis through a cumulative effect (Vuori 2001). In addition to trauma-related joint injury, the normal physiological changes that occur within the musculoskeletal system attributable to aging may perpetuate the progression of superficial surface damage, leading to disease progression.

Observational data suggest that select types of activities may contribute to the development of osteoarthritis, especially in the knee but also in other large weight-bearing joints such as the hip, ankle, and spine. Activities with high-load joint stresses, especially twisting and torsional types of stress, have been associated with degenerative joint conditions. Soccer, football, tennis, and certain track-and-field events involve twisting and repetitive impact compression forces. These forces, over time, may wear excessively on the hyaline cartilage of the joint, resulting in osteoarthritis—this has been referred to as the **wear and tear theory** of degenerative joint disease (Buckwalter 2003; Felson et al. 2000). Occupational activity can also contribute to joint stress. Persons employed in occupations requiring a lot of knee bending, carrying of heavy loads, and twisting, such as farming, warehouse work, and carpet laying, are at high risk for knee and hip osteoarthritis (Felson et al. 2000). These high-load activities, whether from sports, exercise, or occupational exposure, also have higher risk of major joint injuries, and joint injuries are a known risk factor for osteoarthritis.

No experimental studies have been published examining the dose–response relationship between various activity levels and the development of osteoarthritis. However, data from observational epidemiological studies may help us understand

Sport Participation and Osteoarthritis

Lifelong participation in sports that cause minimal twisting and torsional joint forces (e.g., walking, cycling, swimming) does not increase an individual's risk of developing osteoarthritis. High-impact sports such as football and soccer have higher inherent risks of major joint injuries and, subsequently, are associated with the development of post-traumatic osteoarthritis.

this relationship. When compared categorically, former elite athletes are at higher risk of hip and knee osteoarthritis than are recreational athletes and nonathletes. Also, in terms of total exposure to sports, whether measured in total hours of sport participation or frequency of participation over a lifetime, those in the highest exposure category are two to four times more likely to develop knee or hip osteoarthritis than persons in the lowest exposure groups (Vuori 2001). The combination of high-load occupational activities with high-load sport participation multiplies the risk of osteoarthritis. The dose associated with a significantly increased risk of knee osteoarthritis is 4 hr per day of high-load occupational and leisure-time activity.

Participation in moderate types and amounts of activity, such as running, walking, or cycling, has low, if any, risk of osteoarthritis. In fact, we know that some level of activity is necessary for optimal joint health. Loss of joint motion, either from immobilization postinjury or postsurgery or from spinal cord transection, is associated with muscle atrophy, declines in bone density, slowed tissue metabolism, connective tissue stiffness, and other deleterious effects on the musculoskeletal system.

Several factors should be considered when we attempt to quantify the dose of activity in regard to osteoarthritis. The focus should be primarily on the amplitude and rate of force loading as well as the frequency (number of repetitions and accumulation over a lifetime) and duration of each force loading. The **torsional joint loading** of the activity must also be considered because torsion or twisting forces produce the highest joint forces and tend to be concentrated in small areas of the joint surface

(Buckwalter 2003). For a variety of reasons, most researchers have found it impossible to quantitatively define dose–response relationships between activity level and osteoarthritis that incorporate all these measurements. However, by incorporating data from animal, clinical, and observational studies, we can hypothesize that this relationship is likely nonlinear and is probably J-shaped. Figure 14.4 illustrates this theoretical relationship. However, no direct evidence exists showing that physical activity in the optimal zone prevents the development of osteoarthritis.

Exercise As a Treatment for Osteoarthritis

In the previous section, we discussed the fact that some types of vigorous sport participation may contribute to an increased risk of osteoarthritis. However, once an individual has been diagnosed with osteoarthritis, physical activity can also ameliorate the effects of this condition. Exercise, both aerobic and resistance, has been consistently shown to decrease pain; improve physical function, mental health, and overall quality of life; and delay disability among adults with knee osteoarthritis (Arthritis Foundation 2001; Stuck et al. 1999; Vuori 2001). For example, persons with knee osteoarthritis typically present clinically with severe quadriceps atrophy and weakness and poor cardiorespiratory fitness. Exercise that strengthens the anterior thigh musculature can vastly improve gait performance, mobility, and balance and will help reduce pain in the long term. Stronger muscles help disperse the joint stresses associated with daily movement. Range

of motion and flexibility exercises promote optimal joint lubrication and connective tissue extensibility. Appropriate aerobic exercise improves fitness levels and emotional well-being and is integral to weight loss and maintenance.

What type and amount of exercise are most beneficial for persons with osteoarthritis? The published studies are relatively homogenous in regard to dose of activity. The standard exercise program involves moderate types of exercise such as walking, cycling, or swimming, three to four times per week for 30 to 60 min. Expert panel activity recommendations suggest that average people with arthritis should engage in moderate activity at least 3 days per week for at least 30 min. The sessions can be broken into 10 min bouts (Vuori 2001). Few if any adverse events have been reported with these doses of exercise, and they are generally regarded as safe. However, a small subgroup of persons with severe knee osteoarthritis who also have abnormal joint deformity (varus or valgus angulation) and excessive joint laxity should consult a professional therapist and undergo supervised exercise programs because some types of quadriceps exercises may worsen symptoms. More research is needed to identify both the minimal dose of activity that results in disease-specific benefits and the maximal dose that may exacerbate symptoms or cause injury.

Physical Activity, Bone Development and Maintenance, and Preservation of Function

Physical activity acts on the musculoskeletal system in various ways and can have important effects at various periods during the life cycle. One important issue in bone health is the development of peak bone mass during adolescence and young adulthood and the maintenance of peak mass throughout the adult years. Also important during the later adult years is the contribution of physical activity to maintaining or even recovering lost muscle mass and neuromuscular deficits.

Physical Activity and Bone Development and Maintenance

Physical activity has the potential to affect bone density at three critical time periods during a lifetime (Kahn et al. 2001; Vuori 2001):

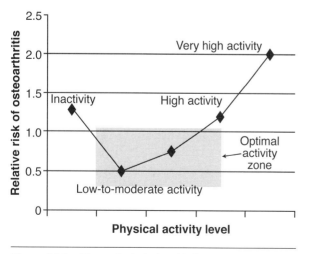

Figure 14.4 Theoretical relationship between activity level and osteoarthritis.

1. Weight-bearing activity in childhood and adolescence helps to develop peak bone mass. Puberty is the most critical time period because the rate of bone mass accrual is highest during this period. In most adolescents, bone mass peaks around age 18. Children who participate in sports that involve running, jumping, and twisting have higher peak bone mass than more inactive children when they reach young adulthood. Targeted bone-loading, school-based interventions have also been successful at increasing bone density as much as 10% above the control group.

2. During the second critical time period, the second through fifth decades, physical activity helps maintain peak bone mass. Longitudinal intervention studies have reported that bone density may be augmented about 1% to 3% in active young adults compared with controls.

3. In later adult life, the third critical time period, physical activity slows the rate of decline in bone mass as much as 1% per year. Because of the hormonal changes associated with menopause and their relation to bone health, postmenopausal women are a high-risk population with an accelerated risk of osteoporosis. However, evidence suggests that physical activity can slow bone loss and even facilitate some gain in bone density among postmenopausal women who already have osteoporosis and are taking hormone replacement therapy.

Bone will respond variably depending on the extent of strain applied. Loads below 50ϕ (microstrain) have no appreciable effect on bone. Loads in the 50 to $200\ \phi$ range are physiological in effect and promote healthy remodeling of bone. Loads ranging from 2,000 to $4,000\ \phi$ will overload bone and stimulate new bone formation. Loads in excess of $4,000\ \phi$ are pathological, resulting in microdamage and initiating the production of unorganized bone cells to assist in the repair process. Cortical bone fractures at compressive loads of about $25,000\ \phi$. Fortunately, loads generated during normal activities, including high-intensity, vigorous sports and exercise, rarely exceed 1,500 to $3,000\ \phi$ (Kahn et al. 2001).

No evidence exists that defines a clear dose–response relationship, in terms of frequency, intensity, and duration, between physical activity and bone health. Despite this, research in both animal and humans suggests several components of exercise that will provide the appropriate stimulus to bone (Kahn et al. 2001; Vuori 2001). Bone-loading activities should consist of forceful and fast mechanical loads that may be generated from either muscular contraction or weight-bearing forces and that load the bone at various angles. Specific bones must be targeted for the best localized effect. For instance, to increase bone density in the tibia, activities must incorporate rhythmic contractions of the large muscles of the lower leg as well as compressive forces on the foot, ankle, and lower leg. The number of repetitions or loading cycles need not be high; strain magnitude, rate, and distribution play the most important role in stimulating bone growth. Also, because bone accommodates to customary loads over time, the magnitude of loads applied needs to increase periodically to continually stimulate bone formation. Exercise programs that incorporate high-impact jump training (50-200 jumps per day, variable height 3-10 in.), high-intensity aerobic training (walking, jogging, stair climbing; 50 min sessions four times per week at 55-75% $\dot{V}O_2max$), or resistance training (12 different exercises, three sets of 8-12 repetitions, 70-80% 1-repetition maximum) have been successful at maintaining or increasing bone density.

Role of Physical Activity in Preserving Function

Evidence suggests that physical activity levels are inversely associated with the development of disability in older adults. In fact, several longitudinal studies report that persons who engage in the highest levels of activity had almost 50% reduced risk of disability, compared to their inactive peers (Singh 2002; Spirduso and Cromin 2001; Stuck et al. 1999). There are a variety of mechanisms by which physical activity and exercise can help prevent incident function loss and disability during the aging process. Directly, exercise may affect physiologic capacity (cardiorespiratory function, muscle strength, flexibility, balance). Indirectly, physical activity may affect psychosocial factors such as self-efficacy and depression, which in turn can influence activity levels. In the context of a general disability prevention framework, four constructs help elucidate the complex relationship between physical activity level and disability:

1. Physical activity may delay various physiological processes associated with aging.

2. Physical activity has been shown to modify risk factors (hypertension, high blood glucose, hypercholesterolemia) for common chronic and disabling diseases.

3. Physical activity may affect the course and sequelae of diseases already present.

4. Physical activity can affect other contributors to disability, such as depression, low self-efficacy, and lack of social support (Singh 2002; Spirduso and Cronin 2001).

To illustrate these concepts, let's look at the case of muscular function and its relationship to the onset of disability. Sedentary individuals lose about 10% of muscle mass every decade of their adult life. This age-related loss of muscle mass is called **sarcopenia,** which is an established risk factor for disability (Singh 2002). Engaging only in aerobic activities does not provide enough stimulus to preserve muscle mass as we age: Muscle must be properly "loaded" through resistance training activities. Progressive strength training programs have demonstrated increases in muscle strength of 40% to 150%; these strength and subsequent functional gains have been reported for adults as old as 80 to 90 years, suggesting that it is never too late to start a strengthening program. Women may particularly benefit from strength training, because they tend to have lower baseline reserves of muscle mass and progress to functional disability an average of 10 years earlier than men (Spirduso and Cronin 2001).

Relatively few studies have investigated primary prevention of disability through exercise interventions, and these published studies have been small in size, have been of short duration, and have used exercise interventions of very low intensity. Subsequently, no significant differences were noted in the rate of disability between exercise and comparison groups, even though strength gains and other beneficial effects were reported (Spirduso and Cronin 2001). There is a significant need for additional research in this area.

On the other hand, the evidence is much stronger regarding secondary and tertiary prevention of disability among adults with existing functional impairments. Physical inactivity is an independent predictor of functional decline and resulting disability. Moderate-intensity activity such as walking seems to show the same favorable effects on function as more strenuous activity programs. Observa-

tional studies suggest a dose–response relationship between activity level and secondary and tertiary prevention of functional disability; however, the ideal dose, and especially the minimum intensity, have not been identified (Singh 2002; Spirduso and Cronin 2001; Stuck et al. 1999).

Despite the lack of tested exercise prescriptions designed to ward off age- and disease-related functional loss and disability, some recommendations have been published. A comprehensive disability prevention activity prescription should, at minimum, contain

- a functional aerobic component (such as walking activities) to promote cardiorespiratory function as well as muscle mass and strength development,

- static and dynamic balance activities to improve movement self-efficacy and reduce fall risk, and

- resistance training to strengthen all major muscle groups (Singh 2002).

Challenges in Defining Exposure Data for Musculoskeletal Outcomes

It is difficult to determine with certainty the exact relationship between physical activity and various musculoskeletal outcomes for two primary reasons. The first is the fact that historically, physical activity has been measured in epidemiological studies using self-report survey instruments. These surveys tend to have problems with recall bias, especially when survey instruments ask participants to recall activity levels over an entire lifetime. In regard to musculoskeletal outcomes, the lifetime accumulation and patterns of activity are the most important exposure elements to capture accurately. Also, because most surveys have been developed for studies of cardiovascular health or other non–musculoskeletal disease outcomes, the types of activities queried to ascertain physical activity level tend to be based on their cardiorespiratory effect and not necessarily their musculoskeletal effect. For example, in terms of cardiorespiratory effect, swimming and running are similar; however, these two activities are very different in terms of their net loading effect on the musculoskeletal system. Many surveys also fail to ask questions regarding light-intensity activities that significantly stress bones or joints (e.g., repetitive squatting and kneeling during household tasks) but do not sufficiently stress the cardiorespiratory system.

The second reason it is difficult to determine the relationship between physical activity and musculoskeletal outcomes is that standardized scoring systems do not assess the dose of activity that is actually delivered to the musculoskeletal system. The most common scoring method uses a standardized set of absolute MET (metabolic equivalent) values for each activity and then applies this information in an equation with frequency and duration of activity to calculate a summary dose of activity. However, MET-based intensity values primarily represent the cardiorespiratory effects of a given activity. Thus, neither the design of the self-report surveys nor the scoring algorithms used in analysis reflect the dose of activity that is delivered to the musculoskeletal system. There is a critical need to develop a standardized scoring system that can be used to weight self-reported physical activity data according to the musculoskeletal loading aspects of the activity.

Summary

Physical activity is as important for the health of the musculoskeletal system as it is for other bodily systems, such as cardiovascular, respiratory, and metabolic. Quantifying physical activity in terms of dosage delivered to the musculoskeletal system may require different approaches and analytic methods than used for other systems and is an important area for future research. Identifying doses of activity that (1) quantify the minimum threshold for musculoskeletal health benefit; (2) result in important benefits for muscles, bones, heart, lungs, and brain; and (3) minimize potential adverse effects such as injury or degenerative joint disease is critical. Evidence suggests that accumulating physical activity in amounts recommended for improving general health is not associated with high rates of adverse musculoskeletal effects and thus should not be considered a barrier to adopting and maintaining a physically active lifestyle.

KEY CONCEPTS

arthritis and other rheumatic conditions: A set of more than 120 medical conditions or diseases that primarily affect the musculoskeletal system. The most common conditions include osteoarthritis, rheumatoid arthritis, gout, fibromyalgia, systemic lupus erythematosus, arteritis, polymyalgia rheumatica, and psoriatic arthritis.

crepitus: A grating, crunching, or crackling sensation under the skin, in the lungs, or around the joints; can be felt and sometimes audible while palpating a joint during movement.

mechanostat theory: Theory of the process by which bone formation is initiated in response to mechanical loading. It posits that a minimum effective strain must be present to promote bone cell proliferation.

osteoarthritis: Condition of the hyaline cartilage and underlying bone in which cartilage fractures and disintegrates, resulting in joint swelling, pain, and deformity; also called degenerative joint disease. Diagnosis is based on specific radiographic features (joint space narrowing, osteophytes, and subchondral bone cysts) and clinical symptoms (pain, swelling, and reduced motion).

osteophyte: Excess bone formation, usually caused by abnormal stress or pressure at joint margins or sites of ligament or tendon attachment.

osteoporosis: Loss of bone mineral density and the inadequate formation of bone, which can lead to an increase in bone fragility and a risk of fracture.

sarcopenia: Loss of lean muscle mass attributable to factors such as aging, poor nutrition, and lack of exercise often leading to weakness, decreased metabolic energy needs, fatigue, and eventually functional limitation and loss of independence.

sclerosis: Stiffening of bone directly beneath the hyaline cartilage of a joint; caused by accelerated bone turnover in response to superficial cartilage degeneration (cracks and fissures).

torsional joint loading: Activities that are characterized by weight bearing and twisting motions, such as tennis, soccer, and basketball. These types of activities can transmit high-impact forces to the large weight-bearing joints and may contribute to the development of osteoarthritis.

wear and tear theory: One mechanism by which osteoarthritis may develop. Suggests that accumulation of joint stresses from high-load activities over a lifetime initiate and promote the degradation of hyaline cartilage in the joints.

STUDY QUESTIONS

1. What three leading causes account for almost 40% of all disabilities?

2. What is the leading cause of death among persons aged 1 to 34 years?

3. Clinical trial research suggests that fall and hip fracture rates can be reduced through exercise programs that focus on which of the following?

   a. Cardiorespiratory fitness

   b. Flexibility

   c. Balance and mobility

   d. Strength and function

4. What is the lifetime risk of osteoporotic fracture among white women?

5. What are the three basic principles of mechanical bone loading?

6. List four modifiable risk factors for activity-related musculoskeletal injuries.

7. What are four ways physical activity can affect the development of disability?

Physical Activity, Muscular Fitness, and Health

Neil McCartney, PhD ■ Stuart Phillips, PhD

CHAPTER OUTLINE

This chapter considers muscular fitness in terms of muscular strength and power and properties of muscle that contribute to its mass and quality. The relationships between muscular fitness and health are reviewed throughout the life span, both in healthy individuals and among groups with diseases and disabilities. We demonstrate that muscular fitness is perhaps as important as aerobic fitness in improving health and maintaining good health. Because progressive overload **resistance training (RT)** is the method of choice to increase muscular fitness, this literature is reviewed.

History of Resistance Training and Its Role in Health

In this chapter we define resistance training as a technique whereby external weights are used to provide progressive overload to skeletal muscles to strengthen and enlarge them. In any given resistance exercise, it is typical to set the load as a percentage of the individual's **1-repetition maximum (1RM,** the maximum load that can be lifted once only throughout a complete range of motion). The dose of resistance training is described by the load lifted, the number of repetitions, and the number of sets of repetitions. It is typically believed that 3 to 5 sets using high loads ($\geq$80% 1RM) and low repetitions (3-6) are best to increase muscle strength and mass, whereas lower loads (50-70%) and higher repetitions (10-20) are best to increase muscular endurance and muscular power (the ability to move an external load rapidly). As discussed later, the evidence for these assertions is not overwhelming.

The formal origins of RT are found in the 1945 publication by Captain Thomas De Lorme, showing that heavy RT restored muscular strength and power in physically disabled war veterans. De Lorme's strategy of using heavy loads and low repetitions to build strength versus low loads and high repetitions to build muscular endurance is still the accepted approach to this day. After that original publication, RT was often recommended as part of a balanced exercise program until the burgeoning literature on the cardioprotective effects of aerobic exercise tended to make this form of exercise preeminent. The 1968 book *Aerobics*, by Kenneth Cooper, catalyzed a decade of publications on the health and fitness benefits of activities such as running, but RT received little attention. Indeed, the first posi-tion statement by the American College of Sports Medicine (ACSM) in 1978 on "The Recommended Quantity and Quality of Exercise for Developing and Maintaining Cardiorespiratory and Muscular Fitness in Healthy Adults" did not include any RT guidelines. By the mid-1980s, publications began to appear demonstrating the benefits of RT in various clinical populations, such as those with coronary artery disease, hypertension, and low bone mineral density, and in 1990, the ACSM revised its position statement to include guidelines for RT. This was followed by recognition of the importance of RT by the American Heart Association and the American Association of Cardiovascular and Pulmonary Rehabilitation. RT is now considered an important component of a balanced exercise program for healthy individuals and many patient groups.

> The well-documented cardioprotective effects of aerobic exercise previously over-shadowed resistance exercise and resistance training; however, RT is now recognized as a critically important part of any formal exercise program.

Fundamental Aspects of Resistance Training

Muscle tissue is very malleable and changes in response to external stimuli. One such stimulus is resistance exercise, which involves performance of high-force, but brief, contractions. The usual adaptation to this stimulus is that the muscle grows larger or undergoes **hypertrophy,** which means that the volume of muscle increases so long as the stimulus is maintained and is frequent enough. The volume of high-force contractions necessary to elicit this response is still not completely clear, but enough data are now available on which to base some general guidelines.

We do know that for hypertrophy to occur there needs to be a shift in the continual turnover of muscle proteins toward a state of net synthesis. In this section, we introduce the concept of skel-etal **muscle protein turnover** and show how this system is affected by resistance exercise and feeding. We also examine other potentially important health effects of the increase in skeletal muscle mass and introduce the concept of skeletal muscle "quality." Differences between men and women with regard to the impact of sex on the ability to accrue muscle proteins will also be examined.

Effects of Resistance Training

Resistance training is fundamentally anabolic. Thus, when regularly performed, RT stimulates skeletal muscle to synthesize new muscle proteins and to preserve existing proteins. Although RT can damage muscle by disrupting muscle ultrastructure and releasing myocellular constituents, such damage is repaired and the muscle becomes stronger and remarkably resistant to subsequent damage. RT can lead to an accumulation of muscle protein, or muscular **hypertrophy,** when performed regularly and with sufficient intensity, usually greater than 60% of a person's 1RM, which is the greatest load a person can lift once only (figure 15.1), and sufficient volume (figure 15.2). Muscular **atrophy,** or loss of muscle protein, results from acute (i.e., immobilization, exposure to microgravity) or prolonged and persistent (i.e., aging, wasting diseases) withdrawal of a sufficiently intense contractile stimulus. Hence, as one might imagine, RT is an attractive therapeutic tool for persons who are losing skeletal muscle mass.

> RT promotes anabolism and thus promotes muscle growth and retention of muscle protein. These characteristics make RT a highly valuable clinical tool in treating declining skeletal muscle mass.

Dose–Response Effects

The effective dose of RT that induces a beneficial outcome depends greatly on the outcome desired and the target population. Where strength is concerned, one can see changes in as few as one or two sessions of RT, reflecting neural adaptations rather than muscle protein accumulation (i.e., hypertrophy). However, the optimal dose or volume of exercise required to maximize strength gains is an understudied area in which information is truly lacking. Some attempts for consensus in this area have been made, but we believe that the evidence behind such recommendations is inadequate (Feigenbaum and Pollock 1999). We have attempted to

summarize a number of published studies in which hypertrophy, not strength gains, was the measured outcome and have put those results in schematic form in figures 15.1 and 15.2. These are still theoretical constructs, and much work still remains to establish true dose–response effects concerning resistance exercise and hypertrophy. We believe that viewing resistance exercise from its ability to increase muscle mass versus simply strength, which almost always increases to a varying degree, is likely more beneficial for reasons outlined subsequently. Most agencies have aimed to develop both muscular strength and power in relatively sedentary, diseased, or aged individuals; their recommendations can be summarized as 1 to 2 sets of 8 to 12 repetitions per set, with 8 to 10 exercises per session and no more than 2 to 3 sessions per week (Feigenbaum and Pollock 1999).

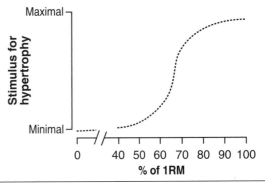

Figure 15.1 Theoretical curve showing the relationship of resistance training intensity to a relative stimulus for hypertrophy. Note that at intensities of ~60% to 80% of 1-repetition maximum, the stimulus for hypertrophy increases sharply but plateaus at higher intensities.

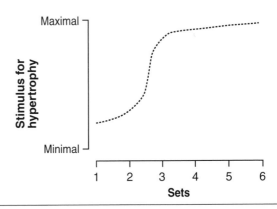

Figure 15.2 Theoretical curve showing the relationship between exercise volume and stimulus for hypertrophy. It appears that beyond 3 sets, so long as voluntary fatigue is induced, the stimulus for muscle mass as well as strength gains is relatively minimal.

Turnover of Muscle Proteins

Figure 15.3 illustrates proteins in muscle "turnover." The process of turnover illustrates the interplay of protein synthesis and breakdown that occurs simultaneously and continuously in skeletal muscle and all body tissues. In figure 15.3, the rate of protein breakdown (B) exceeds that of protein synthesis (S), and the net protein balance (equal to S minus B) is negative, or a net loss of protein. This is the situation observed in the fasted state, which begins approximately 4 to 5 hr after food consumption. Note that there is a net loss of amino acids from muscle (i.e., the outward flux of amino acids is greater than the inward). Consumption of protein results in a fed-state hyperaminoacidemia (i.e., increase in blood amino acids), promoting inward transport of amino acids from the blood into the skeletal muscle pool of free amino acids. Protein synthesis is then stimulated to yield a net positive protein balance (figure 15.4). The stimulation of muscle protein synthesis appears to be mediated by the amino acids themselves and relies on a small amount of insulin being present. As figure 15.5 indicates, the magnitude of the negative net balance when fasted is equal to that of positive net balance when fed; hence, skeletal muscle mass is maintained by feeding. In fact, this happens daily in people who perform no RT but simply eat sufficient protein containing adequate quantities of essential amino acids.

As previously noted, RT is anabolic and shifts the balance of protein turnover toward net anabolism by increasing both the rate of protein synthesis and the rate of protein breakdown. In a fasted state, performance of RT stimulates protein synthesis but insufficient amino acids are available in the circulation to be transported into the muscle to shift a

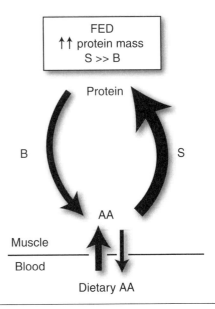

Figure 15.4 Schematic of muscle protein turnover when fed. Muscle protein synthesis (S) is stimulated by amino acid influx into the cell; breakdown (B) may be suppressed somewhat but net balance is positive and equal in magnitude to the fasted losses seen in figure 15.3.

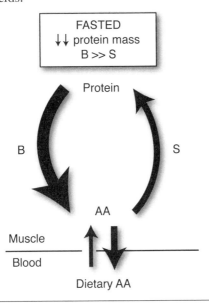

Figure 15.3 Schematic of muscle protein turnover when fasted. Muscle protein breakdown (B) exceeds synthesis (S) so that net muscle balance (S minus B) is negative, resulting in a net loss of muscle protein.

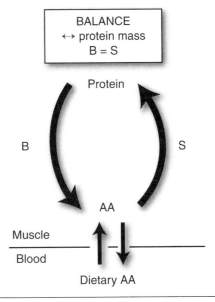

Figure 15.5 If adequate energy and sufficient protein (i.e., containing all essential amino acids) are consumed daily, then muscle protein balance is maintained and muscle mass remains relatively unchanged.

negative net protein balance to a state of net protein gain. Net balance is less negative following RT but is not positive (figure 15.6). Only when protein is consumed following RT does a maximal stimulation of protein synthesis yield a net positive muscle protein balance (figure 15.7). The concepts presented in figures 15.3 through 15.7 are shown dynamically in figure 15.8. The curves presented in this figure show that swings in net protein balance occur with feeding and fasting and that RT results in a reduced loss of protein in the fasted state and a higher gain in the fed state. Although the curves in figure 15.8

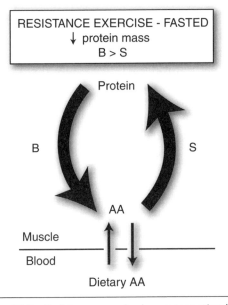

Figure 15.6 Resistance exercise is a potent stimulator of synthesis (S) and also of breakdown (B) to a much smaller degree. Net muscle protein balance is improved (i.e., less negative than that depicted in figure 15.3) attributable to improved intracellular recycling of free amino acids but is not positive.

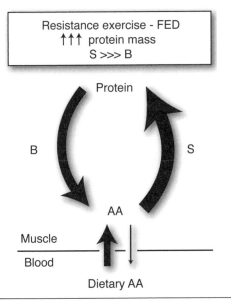

Figure 15.7 After resistance exercise, in the presence of amino acids, synthesis (S) is additively stimulated by feeding and resistance exercise. The response yields a greater net balance than feeding or resistance exercise alone, resulting in muscle protein accretion.

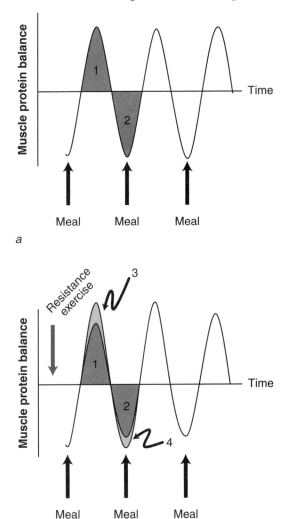

Figure 15.8 *(a)* Normal fed-state gains and fasted-state losses in skeletal muscle protein balance (synthesis minus breakdown). Note that the area under the curve in the fed state (indicated by 1) would be equivalent to the fasted loss area under the curve (indicated by 2); hence, skeletal muscle mass is maintained by feeding. *(b)* Fed-state gains and fasted-state losses in skeletal muscle protein balance with performance of resistance training. In this scenario, fasted state gains are enhanced by an amount equivalent to the stimulation of protein synthesis brought about by exercise (indicated by 3). Additionally, fasted-state losses appear to be less (indicated by 4), attributable to persistent stimulation of protein synthesis in the fasted state.

Reprinted from *Nutrition,* Vol. 20, S.M. Phillips, "Protein requirements and supplementation in strength sports," pp. 689-95, copyright 2004 with permission from Elsevier.

are perfectly shaped and sinusoidal in nature, they are generalized. Greater meal frequency or longer periods of fasting would affect the temporal nature of the response of muscle protein net balance and possibly the rate of rise and fall of the response. It is also likely, but has not yet been shown, that the intensity and volume of exercise would affect the amplitude and duration of the anabolic response.

> The balance of **muscle protein turnover,** which includes both muscle protein synthesis (S) and muscle protein breakdown (B), determines muscle protein gain or loss. When S chronically exceeds B, then muscle protein accretion—hypertrophy—occurs; in the opposite situation (B>S), wasting or atrophy occurs. Both feeding of protein and RT stimulate S and thus shift the balance of muscle protein turnover toward gain of protein mass.

Sex-Based Differences in Hypertrophic Gains

One school of thought is that a large portion of the acute exercise-induced muscle mass gain (see figure 15.8) is attributable to the postexercise increase in anabolic hormones such as growth hormone (GH), insulin-like growth factor-1, and testosterone. What is now evident is that pharmacological doses of GH do not have a marked effect in increasing muscle mass or strength in persons with normal pituitary function, whereas pharmacological doses of testosterone do. Hence, the acute postexercise increase in testosterone has been implicated as being an important factor in determining hypertrophy. The rise in testosterone following intense RT is, however, small (relative to the daily circadian release of the hormone) and remarkably transient (lasting only 30-60 min); occurs in the absence of a significant increase in luteinizing hormone (LH); and tracks closely with shifts in hematocrit. The last points highlight the fact that the postexercise increase in testosterone concentration is simply attributable to increased hemoconcentration of the hormone from plasma shifts and not increased production, which would be accompanied by higher levels of LH. Hence, it is questionable whether the acute increase in testosterone (or of GH) after exercise is important in determining the hypertrophic response.

Women have a 10-fold lower concentration of testosterone than men, yet a number of studies have shown that women who follow an intense progressive RT program have significant strength gains and muscle hypertrophy. In fact, it has been concluded that absolute and relative maximal dynamic strength increase in a similar manner for both sexes. Moreover, data suggest that skeletal muscle adaptations that may contribute to strength gains of the lower extremity are similar for men and women during the early phase of RT. Such RT-induced hypertrophic responses are not related to either transient or chronic changes in circulating testosterone, except possibly in elderly women. Even in elderly women the change in systemic testosterone cannot explain more than 20% to 25% of the variance in strength gains with RT. Instead, local muscle factors potentially acting in an autocrine or a paracrine manner are more important than circulating systemic hormones in determining exercise-induced hypertrophic gains.

Importance of Skeletal Muscle Mass and Quality

Aside from the obvious role that skeletal muscle mass plays in locomotion, why is it important to maintain skeletal muscle mass? Because skeletal muscle has a large working range of adenosine triphosphate turnover rates, it has tremendous potential to consume energy and hence is important in weight maintenance or loss, as discussed later. Because of its mass, skeletal muscle is highly important thermogenic tissue and an important contributor to basal metabolic rate (BMR), which for most people is the largest single contributor to daily energy expenditure; again, this fact highlights the importance of maintaining muscle mass. Because of its oxidative capacity (i.e., mitochondrial content), skeletal muscle is also a large site of lipid oxidation, potentially playing a role in maintaining balance in lipoprotein and triglyceride homeostasis. Skeletal muscle is also, mostly because of its mass, the primary site of glucose disposal in the postprandial state. Hence, maintaining skeletal muscle mass would also reduce the risk for development of type 2 diabetes mellitus. Finally, the decline in maximal aerobic capacity with age and other muscular wasting conditions has also been found to result largely from a decline in skeletal muscle mass (figure 15.9).

Although skeletal muscle mass per se is important, the quality of skeletal muscle must be maintained. The term **muscle quality** has usually been used in the context of the ability of the skeletal muscle to generate force. In most cases, the potential for force generation of a skeletal muscle is directly proportional to its cross-sectional area. However,

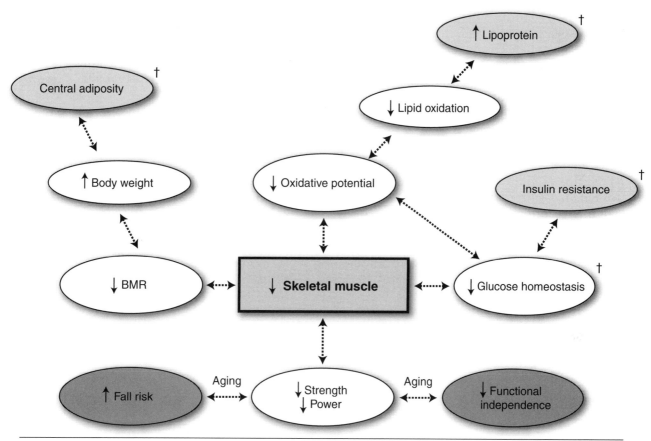

Figure 15.9 Interplay of skeletal muscle mass and its relative metabolic activity with certain factors important in the development of various chronic diseases—heart disease, diabetes, metabolic syndrome, and obesity. In particular, the decline in skeletal muscle mass and the metabolic quality of skeletal muscle can lead to derangements that are considered to be risk factors for metabolic syndrome. The grayed ovals indicate either a process that negatively affects skeletal muscle mass or a negative outcome of reduced skeletal muscle mass.

with aging, and in other neurological and metabolic disorders, force per cross-sectional area declines, which may be a manifestation of reduced skeletal muscle quality. In addition, a relatively inactive skeletal muscle mass would be of low "metabolic quality" because of a lower oxidative capacity, capillary supply, and transport capacity for fatty acids and glucose. Although aerobic exercise is associated with increased mitochondrial capacity, capillarization, and transporters for glucose and fatty acids, most moderately intense RT programs, particularly in sedentary people, would probably enhance many of these parameters also.

Resistance Training Throughout the Life Span

There is no doubt that RT can benefit health, but what about effects on different age groups? Are

Beyond the obvious important role that skeletal muscle plays in locomotion, it is also a critically important tissue in maintaining health. Skeletal muscle is active tissue. It burns a tremendous amount of lipid, stores the majority of ingested glucose, and is a significant contributor to basal metabolic rate (BMR). As such, one can easily appreciate how important it is to maintain a large and metabolically active skeletal muscle mass. Doing so would likely decrease the risk for numerous health problems including obesity, diabetes, and frailty in the elderly.

they similar among young and aged individuals, or is there an age-related loss in the capacity for strength or hypertrophic gain? Should volume and intensity be different? We attempt to answer these questions in the following sections.

Resistance Training in Children and Adolescents

Resistance training programs have proven to be a safe and effective method of "conditioning" for children. A caveat concerning this statement is that exercise guidelines need to be rigorously followed. In fact, participation in well-designed and appropriately supervised programs of strength and conditioning by children and adolescents is supported by the American Academy of Pediatrics and the ACSM. These position stands are in accordance with such reports as the Surgeon General's report, *Physical Activity and Health*, which aims to increase the number of children who participate regularly in physical activities that enhance and maintain muscular strength and muscular endurance. Many of the same health-related benefits from RT seen in adults hold true for this age group as well (i.e., improvements in aerobic fitness, reduced risk of osteoporosis and increased bone mass, prevention of obesity and hypertension, increased high-density lipoprotein [HDL] cholesterol, and improved psychological health indexes). Also, data support the thesis that physically active children are more likely to become physically active adults. Hence, encouraging children who already enjoy free play to participate in RT programs would be advantageous and beneficial.

Although some early reports cast doubt on whether children or adolescents could actually gain strength as a result of RT, subsequent reports clearly showed that RT does result in strength gains, enhanced motor performance skills, and possible benefits to selected anatomic characteristics such as bone density and architecture and lean mass. Furthermore, RT enhances self-efficacy for exercise and other tasks. Significantly, participating in programs of RT appears to reduce injuries in sports and recreational activities. For example, strength-trained athletes (13-19 years old) had a lower injury rate and required less time for rehabilitation than their teammates who did not strength train. Strength training has been shown to reduce the incidence of shoulder pain in 13- to 18-year-old swimmers and to decrease the number and severity of knee injuries in high school football players.

The most efficacious training frequency and intensity rates for children are difficult issues to resolve, because excessive frequency and intensity of resistance activities could lessen adherence and enthusiasm for participation and potentially be inju-

rious. It appears, from the few studies in this area, that a training frequency of two sessions per week results in improved strength gains versus just one session per week. In addition, in terms of enhancing at least upper-body strength and local muscle endurance of untrained children, the prescription of higher repetition (13-15 repetitions per set) training protocols, at least during the initial adaptation period, would be optimal.

> It is a common misconception that properly designed RT programs harm children or adolescents. In fact, numerous reports show that RT can increase bone density, enhance skeletal muscle growth, and reduce the incidence of sport-related injuries in children and adolescents.

Resistance Training in Middle-Aged and Elderly People

Beginning in midlife, aging is associated with a time-dependent loss of muscle, coined **sarcopenia**. This loss is commonly thought of as a consequence of old age itself, although chronic illness, poor diet, and inactivity all accelerate its progression. Sarcopenia, because of the associated loss of muscle strength and stamina, is a major cause of disability, frailty, loss of independence, and increased risk of falling and fractures in the elderly. Muscle wasting in the elderly is associated with a 50% loss of muscle fibers between 20 and 80 years and a loss of fiber area from those fibers remaining, particularly type II. The increase in longevity has forced physiologists to reconsider exactly where middle age actually lies in a person's lifetime. It is clear that sarcopenia is not a precipitous phenomenon, because muscle mass is progressively lost at a rate of ~1% to 2% per annum beginning at ~40 years of age. Hence, unlike osteoporosis, sarcopenia is insipid and is not associated with a sudden change in hormonal status. The decline in muscle mass must come about because of an imbalance between muscle protein synthesis and breakdown (figure 15.3); however, exactly which process, synthesis, breakdown, or combination of these alters with age to bring about sarcopenia is a matter of great debate. What is clear, however, is that older individuals can accrue muscle mass as a result of RT in the same way as younger subjects. The best evidence suggests that muscle maintains its plasticity and capacity to hypertrophy even into the 10th decade of life.

Projections are that life expectancy of North Americans will increase by 3 to 4 years from 1980 levels by the year 2020, but for more than 40% of people, the years of added life will be spent in a full-time care facility. Data show that 43% of males and 37% of females between the ages of 65 and 74 have a disability or disease that limits the kind or amount of activity they engage in at home, work, or school or in other activities such as travel, sport, or leisure. The most common disabling condition for all elderly is arthritis (44% males and 56% females) and the coincident muscle weakness, which may be etiologically linked. RT is an effective intervention to increase total muscle mass, muscle quality (force per cross-sectional area), strength, and power. Because of the lowered functional status of the majority of elderly people, probably no other population segment would benefit more from RT to combat age-related muscle wasting.

The consequences of this age-related reduction in skeletal muscle mass are numerous, including reduced muscle strength and power (Reinsch et al. 1992), reduced BMR, reduced capacity for lipid oxidation, increased abdominal adiposity and insulin resistance, and increased risk for falls (figure 15.9). Ultimately, given the decline in skeletal muscle mass and reduction in strength and power in elderly people, sarcopenia contributes substantially to morbidity and reduced mobility in the elderly and thus to significant health care costs (Tseng et al. 1995). According to Tseng and colleagues (1995), sarcopenia is "a progressive neuromuscular syndrome that will lower the quality of life in the elderly by (1) decreasing the ability to lift loads (progressing to difficulty arising from a chair), and (2) decreasing endurance (leading to an inability to perform the activities of daily living, which increases health care costs)" (p. 113). Numerous studies have shown that RT programs for the elderly can increase muscle mass, strength (maximal force generating capacity), and power (Campbell et al. 1999). It has also been shown that RT programs in elderly men and women can cause substantial fiber hypertrophy (Campbell et al. 1999). All of these RT-induced changes are potent countermeasures to sarcopenia, and hence RT is highly effective in reducing disability in the elderly.

Clearly RT can directly affect sarcopenia, but what about other chronic diseases? As described elegantly by Fiatarone Singh (2001), RT presents a unique opportunity to treat **disease clusters** associated with advancing age. For example, it would not be uncommon for an average 75-year-old man to be overweight and to have hypertension, vascular disease, hyperlipidemia, impaired fasting glucose tolerance (a precursor to diabetes), and some degree of movement impairment. We propose that, for a number of reasons, a regular program of resistance training might be a useful adjunct to any pharmacological therapy such an individual might be receiving. It is likely that this man would take several different medications each day, but regular RT may substantially reduce this man's reliance on pharmacologic therapy. Evidence is now emerging that RT may also have important spillover effects, aside from those directly related to prevention of disability and movement impairment, such as reducing insomnia, improving poor appetite, improving postural instability, improving muscle weakness, possibly lessening the risk of depression, and delaying the threshold for dependency by promoting increased muscle strength and the ability to perform activities of daily living (Fiatarone Singh 2001) (figure 15.9).

> Given the low functional status and often multitudinous health problems of many elderly persons, there is likely no other population segment that could benefit more from a program of RT. In an elderly person suffering from sarcopenia and health conditions such as central adiposity, vascular disease, impaired glucose tolerance, and osteoarthritis, a program of RT could address all of these diseases directly and not simply treat symptoms as many medicines would.

Resistance Training in Disease and Disability

Numerous studies have demonstrated the effectiveness of RT in rehabilitating patients after myocardial infarction. Substantial evidence is now also accumulating that RT is effective in controlling weight, treating depression, improving glucose tolerance, and altering blood lipid profile. It can be used not only in a rehabilitative role but also as a primary treatment for numerous diseases.

Resistance Training, Weight Control, and Obesity

Although it is clear that aerobic exercise is associated with much greater energy expenditure during the exercise bout than RT, numerous studies have shown that regular RT combined with dietary energy restriction is remarkably effective in promoting weight loss. However, in certain populations such as obese or elderly persons, a program of aerobic training may have only a mild effect in favorably altering body weight or body composition. This is because the low aerobic fitness levels prevent significant energy expenditures associated with even longer periods of endurance-based exercise. For example, a 55 kg elderly woman with a $\dot{V}O_2$max of 24 ml·kg^{-1}·min^{-1}, working at 50% of her peak for 30 min, would expend only ~4,200 kJ (~100 kcal), likely with very little residual increase in resting postexercise energy expenditure. In addition, an obese person who was energy-restricted to lose weight and was also performing aerobic exercise would not have an anabolic stimulus to preserve lean mass from the aerobic exercise. The anabolic nature of RT aids retaining skeletal muscle mass during energy-restricted periods so that although people may lose less weight when they are energy restricted and performing RT, more of the weight they lose is fat mass and not skeletal muscle. Because skeletal muscle is so metabolically active, this pattern of weight loss would be highly advantageous.

Adults who underwent a 12-week program of RT showed an increase in their requirements (~15%) for energy to maintain body weight. The increase was not accounted for solely by the energy cost of the exercise itself, which is relatively low, but instead was attributable to an increase in BMR. In addition, participating in a program of RT promoted an increase in free-living physical activity in the elderly. These findings are important because elderly and obese persons do not expend much energy during endurance exercise because of their low fitness levels; hence, RT with the ability to increase BMR and promote greater free-living physical activity is an attractive alternative or critical adjunct to weight loss programs.

Many of these same arguments can be applied to those with type 2 diabetes mellitus, who are frequently overweight. Another potential advantage of having diabetic patients perform RT along with restricting energy intake is that RT, like aerobic exercise, has been shown to increase the amount and insulin responsiveness of the insulin- and contrac-tion-sensitive muscle glucose transporter, GLUT4. Because muscle accounts for more than 80% of the disposal of an oral glucose load, any weight loss strategy that promotes increased skeletal muscle GLUT4 would be of greater benefit than a strategy simply promoting weight loss alone (i.e., dieting). Hence, it is possible that RT can delay the need for insulin injections or lower the dosage of insulin required for glucose homeostasis.

In 1997, the World Health Organization recognized obesity as a worldwide disease that poses a global challenge to public health and health care systems (WHO 1997). Overweight or obese people have substantially increased risk for morbidity and mortality from numerous chronic disorders, such as diabetes, hypertension, and cardiovascular disease; in fact, excess fat deposited in the abdominal region is a strong predictor of cardiovascular disease and type 2 diabetes mellitus. This may be partially explained by excess accumulation of visceral fat, an independent correlate of insulin resistance and poor blood lipid profile. These observations highlight the need to identify appropriate treatment strategies to prevent and reduce obesity and suggest that the effectiveness of these treatments would be enhanced if abdominal obesity, particularly visceral fat, were substantially reduced.

Interventions to treat obesity usually include a combination of diet and pharmacological strategies such as thermogenic agents, appetite suppressants, gastrointestinal lipase blockers, or synthetic fat derivatives. In addition, a number of studies have looked at the inclusion of exercise, either alone or combined with diet or drugs, as a strategy to combat obesity. Because of its large energy expenditure, aerobic exercise has often been the intervention of choice for obese persons. Several studies, however, used RT as an adjunctive therapy in obese persons trying to lose weight. These studies showed that in direct head-to-head comparisons, obese men and women lose the same amount of weight on hypocaloric diets with either RT or aerobic exercise. This is despite an almost four- to fivefold difference in exercising energy expenditure between the two exercise modes (Rice et al. 1999). Given the relatively low aerobic fitness of many obese persons, programs of aerobic exercise may not be the best exercise intervention for weight loss. Hence, a general recommendation might be that exercise-based interventions to promote weight loss, or at least a beneficial shift of lean:fat mass, in obese persons incorporate RT as the main form of exercise, or at least that RT accompany aerobic and dietary interventions.

Despite the observation that aerobic exercise is associated with a larger energy expenditure than resistance exercise, RT has been shown to be an effective tool in promoting weight loss in obese persons. In addition, RT promotes the retention of skeletal muscle mass during periods of energy restriction, making it the logical choice to prevent loss of metabolically valuable skeletal muscle mass during caloric restriction in obese persons undergoing weight loss.

Resistance Training in Coronary Artery Disease

Now considered a standard of rehabilitation for cardiac patients, RT offers substantial benefits. In the following section, we review evidence in this area.

Safety

The first reports of RT in treating coronary artery disease (CAD) appeared in the literature during the mid-1980s (McCartney 1999). Up until that time, RT was contraindicated in this patient group because of untoward symptoms, electrocardiographic changes, and left ventricular wall motion abnormalities that had been observed in patients (with active myocardial ischemia) during sustained isometric handgrip exercise.

Despite these findings in isometric exercise, a later series of studies using intra-arterial catheterization to measure the arterial pressures of patients during RT demonstrated that the responses were no greater than during aerobic exercise and were within clinically acceptable levels (McCartney 1999). Other investigations using two-dimensional echocardiography to assess left ventricular function during RT with intra-arterial pressure measurements reported good preservation of stroke volume and enhanced myocardial contractility, even in patients with documented heart failure. Several investigators reported that RT did not provoke signs or symptoms of myocardial ischemia, most likely because the heart rate was lower and the diastolic pressure was higher than during aerobic exercise, which are conditions favoring coronary artery perfusion. Moreover, myocardial supply-to-demand balance was demonstrated to be more favorable during RT exercises with progressively heavier loads than it was during incremental exercise testing. RT could

no longer be contraindicated based on the earlier studies of isometric handgrip exercise in patients with active myocardial ischemia.

Efficacy

In addition to research on the acute circulatory and left ventricular responses, a proliferation of studies in the past 2 decades examined the efficacy of RT in patients with CAD (see McCartney 1998 for references in this section). Most studies evaluated patients with relatively good left ventricular function who were already participants in community-based maintenance programs, and these studies usually included an aerobic training control group. Training interventions were typically 12 weeks or less and used 1 or 2 sets of upper- and lower-body exercises and 10 to 15 repetitions per set. Lifting intensities ranged from 30% to 80% of the 1RM, with most studies using moderate loads of 60% or less, and the frequency was most often three times per week. All studies demonstrated increases in the 1RM ranging from 3% to greater than 50%. In one study, patients did as many lifts as possible with their pretraining 1RM to yield a measure of lifting endurance. The 1RM measured at baseline was lifted an average of 14 times before fatigue, indicating that strength-related activities of daily living that required almost maximum effort at baseline could be done with ease after training and were potentially safer.

In addition to showing improvements in dynamic strength, patients have demonstrated increased endurance of 12% during a standard Bruce treadmill test and a 15% gain in maximum power output during progressive cycle ergometry. One notable observation during cycle ergometry testing was a marked reduction in perceived leg effort at power outputs above 50% of the pretraining maximum, perhaps suggesting that the improved power and endurance were linked to the reduced symptoms of effort from stronger leg musculature. No such increases in cycling power occurred among control participants who had undergone additional aerobic training, further suggesting that the locus of improvement was in the stronger leg muscles.

Many patients who have experienced a myocardial infarction are limited more by their own perception that they cannot do certain activities than by any real physical limitation, so an intervention that can favorably alter this perception may significantly affect psychological well-being and quality of life. Such effects may be possible with RT. One study demonstrated that 10 weeks of RT resulted in increased self-efficacy for tasks demanding significant arm or leg strength (self-efficacy defined as the level of certainty of an individual that he or she can successfully complete a given task or assume a given behavior), whereas there was no change in aerobically trained control participants. In another study, 38 patients added high-intensity RT (up to 80% of the 1RM) to their usual exercise prescription and demonstrated similar improvements in self-efficacy for strength-related tasks and also for jogging. Another interesting finding was the improvement in quality of life parameters such as total mood disturbance, depression or dejection, fatigue or inertia, and emotional health domains scores (McCartney 1998). Although these findings suggest that RT may improve the quality of life of patients with CAD, more research in this area seems warranted.

A number of investigations have focused on the effects of RT on coronary risk factors, but because the findings are largely equivocal (McCartney 1998) they are only reviewed briefly here. Although there are some reports of decreases in low-density lipoprotein cholesterol and increases in HDL cholesterol after short periods of RT, there are many confounders. These include lack of a control group, only a single blood sample before and after the training, no regulation of diet or account of changes in body mass and composition, and normal pre-intervention lipid profiles.

The results of individual studies on blood pressure have been variable, with some demonstrating decreases and others showing no change. A recent meta-analysis of 320 normotensive and hypertensive males and females demonstrated a small but significant reduction in resting systolic and diastolic pressures of 3 mmHg. This reduction was also clinically significant, because it would theoretically reduce CAD by 5% to 9% (Pescatello et al. 2004).

As mentioned previously, the decrease in muscle mass with advancing age may significantly contribute to the impaired glucose tolerance that is so prevalent among middle-aged and elderly people. It is possible that this impairment may be at least partially reversed by increasing muscle mass with RT, because there are encouraging reports of increased glucose tolerance and insulin sensitivity independent of changes in body fat or aerobic capacity. It is likely that more studies in this area will be forthcoming.

Resistance Training in Chronic Heart Failure and After Heart Transplantation

Patients with chronic heart failure (CHF) and heart transplantation (HT) suffer from muscular atrophy and weakness and could theoretically benefit from RT. In CHF, there is a relatively modest association between maximum exercise capacity and left ventricular dysfunction as measured by ejection fraction. Much of the reduced exercise capacity is seemingly attributable to intrinsic abnormalities within peripheral muscles, independent of any reductions in peripheral blood flow. These abnormalities include selective atrophy of fatigue-resistant oxidative muscle fibers (type I), reduced mitochondrial volume and density, and decreased concentration and activity of mitochondrial oxidative enzymes. The overall cross-sectional area of the thigh muscles may decrease by 15% or more, resulting in muscular weakness that in principle could be partially reversed by RT. Nevertheless, despite the hypothetical basis for RT in CHF, there are few reports of RT in this patient group and consequently no published guidelines by health and exercise agencies.

The largest study done so far was the Exercise Rehabilitation Trial (EXERT) from Canada (McKelvie et al. 2002), and the results were inconclusive. In it, 181 patients participated in a 12-month randomized, controlled, single-blind trial of supervised (3 months) and home-based (9 months) aerobic and RT exercise with blinded evaluation of patients on a range of clinically useful outcomes. After the initial period of supervised exercise training, the exercise group demonstrated significant increases in $\dot{V}O_2$max and 1RM for arm and leg strength compared with control participants, but these differences diminished and became nonsignificant after the 9 months of home-based training. The reduced adherence during home-based training may have been a confounder in the study. One encouraging finding was that there were no adverse effects on cardiac function or any greater number of clinical events among the exercising patients. More studies of RT in CHF are needed before evidence-based recommendations can be forthcoming.

HT may ameliorate many of the symptoms of CHF, but HT recipients typically have markedly reduced exercise capacity and a $\dot{V}O_2$max of approximately 55% to 60% of predicted values. Similar relative reductions are seen in quadriceps muscle strength, with a strong correlation between quadriceps strength and $\dot{V}O_2$max in HT patients. Moreover, individuals with CHF also manifest peripheral muscle myopathy and osteoporosis as a consequence of immunosuppression with gluco-corticoids. Because trabecular bone is lost more rapidly than cortical bone, HT patients are particularly susceptible to loss of bone mineral from the lumbar vertebra and suffer a very high incidence of vertebral compression fractures. Because of the inherently anabolic nature of RT, it might be a useful intervention to help prevent or reverse these musculoskeletal changes.

There is limited published information describing the effects of RT on muscle myopathy after HT. Nevertheless, one study (Braith 1998) reported that a 6-month program of RT successfully reversed the glucocorticoid-induced muscle atrophy in seven exercising patients versus seven control participants. Once again, more work in this area is warranted before definite conclusions can be drawn.

A growing body of literature demonstrates that RT may stabilize the loss of bone mineral density (BMD) in fracture-prone populations, but the evidence for increasing BMD is more equivocal. Braith (1998) conducted a randomized controlled trial to evaluate the effects of RT on the losses of BMD that occur after HT and noted that just 2 months after the surgical procedure, the control and RT groups had lost 12.2% and 14.9% of lumbar BMD, respectively. After the 2-month measurement, the experimental group began RT on 2 days per week using a single-set, 10 to 15 RM program, and the control group took part in usual care activities. After 6 months of RT, the BMD of the training group was similar to pre-HT levels, whereas there was

Courtesy of Stuart M. Phillips

Early fears of risks for cardiac transplant patients performing RT have now been laid to rest. These patients clearly benefit from RT in recovery. Even after a major surgery, impacts on strength, endurance, and quality of life are enormous.

no meaningful recovery of BMD among control participants. This preliminary evidence indicates that RT could be a useful intervention to ameliorate losses of BMD after HT, but more studies are needed in this area.

Inclusion and Exclusion Criteria

Recent guidelines from the American Heart Association (Pollock et al. 2000) define the following contraindications to resistance training: unstable angina, uncontrolled hypertension (systolic pressure 160 mmHg or diastolic pressure 100 mmHg), uncontrolled dysrhythmias, a recent history of congestive heart failure without evaluation and effective treatment, severe stenotic or regurgitant valvular disease, and hypertrophic cardiomyopathy. Preferred inclusion criteria are moderate to good left ventricular function and an exercise capacity of >5 METs.

Exercise Prescription

Patients should take part in 2 to 4 weeks of aerobic training before doing resistance training so they can be observed in a supervised setting. Pretraining instruction should emphasize correct lifting and breathing techniques. Resistance training should be done twice weekly and include 1 set of 10 to 15 repetitions of 8 to 10 exercises designed to train all major muscle groups. If the 1RM is determined, patients should begin training with loads equivalent to 30% to 40% of the 1RM for upper-body exercises and 50% to 60% of the 1RM for lower-body exercises. Older and frail individuals may start training at lower intensities and progress more slowly. Determination of the 1RM is not strictly necessary; patients can begin using light loads that result in moderate levels of fatigue by the end of a set of lifting. Once patients can complete their final lift with ease, the weights can be increased; added loads of 2 to 5 lb (0.9-2.3 kg) per week for the arms and 5 to 10 lb (2.3-4.5 kg) per week for the legs are adequate in most cases. Slower progression may be necessary in older patients. There is no need to rush; patients should determine their own pace based on their levels of fatigue and perceptions of effort.

Resistance Training in Arthritis

There are more than 100 different types of arthritis, which can be divided into three broad classifications: osteoarthritis (the most common form), inflammatory conditions, and rheumatism. Arthritis is the leading cause of disability among Americans,

> RT was once thought to be too intense and dangerous for patients with coronary artery disease and for those who had undergone HT. However, the substantial benefits of RT, including increased strength and endurance, improved self-efficacy, lowered blood pressure, and reduced incidence of depression, are now well documented and recognized. RT can be used safely in patients with CAD, CHF, and HT providing safety measures are followed.

affecting 50 million individuals of all ages, not just older people. The common symptom of arthritis is pain of joints or soft tissues, leading to restricted joint range of motion, sedentary behavior, and concomitant reductions in physical fitness and muscular strength. Indeed, the reductions in fitness and strength may be a leading cause of disability among arthritis sufferers, so RT may be a useful strategy to improve function.

The 1990s witnessed a proliferation of studies that evaluated the role of RT in arthritis (van Baar et al. 1999). The weight of evidence suggests that RT improves muscle strength, balance, and coordination; reduces pain; and increases functional capacity and health-related quality of life for people with arthritis. These adaptations should decrease disability and dependency and also improve an individual's risk profile for diseases such as CAD. The mechanisms responsible for the improvements with RT are not fully understood but are likely a combination of physiological (e.g., improved joint stability resulting from increased strength) and psychological (e.g., greater self-efficacy, mastery, and control) factors.

There are no evidence-based guidelines for RT in arthritis, so caregivers who prescribe exercise must consider the individual differences and comorbid conditions that are associated with different types of arthritis. For example, inflammatory conditions of soft tissue are volatile, and RT should be avoided during times of flare-up. Increased symptoms of general fatigue may signal exacerbation of comorbid conditions, and ankylosing conditions make the back vulnerable to forced flexion, extension, and rotation. Because of these and other considerations, the exercise therapist prescribing RT must be aware of each patient's condition and develop an appropriate exercise program.

More research on RT and arthritis is warranted, but the majority of evidence thus far suggests that RT may be an important therapy in this population.

Resistance Training in Osteoporosis

The National Institutes of Health Consensus Conference has modified the original definition of osteoporosis to include "a skeletal disorder characterized by compromised bone strength, predisposing a person to an increased risk of fracture. Bone strength reflects the integration of two main features, bone density and bone quality" (Hellekson 2002, p. 161). This definition acknowledges that a decrease in BMD is not the only pathological feature of osteoporosis, and the term bone quality refers to microarchitectural elements that contribute to bone strength. Nevertheless, the conventional diagnosis of osteoporosis is a BMD that is 2.5 standard deviations or more below the level of healthy young adults of the same gender. It is estimated that 10 million Americans over the age of 65 have osteoporosis, with another 18 million exhibiting osteopenia, or low bone mass. There are at least 700,000 fractures of the spine, 300,000 hip fractures, and 250,000 fractures of the wrist each year in the United States attributable to osteoporosis. The health costs associated with these fractures are staggering, and yet osteoporosis is viewed as largely preventable.

The theoretical basis for RT in osteoporosis is based on the concept of "minimal essential strain" (when a bone is subjected to forces representing at least 10% of the level that would fracture it), which is believed to be the threshold level required for new bone formation. If the muscular contractions associated with RT can repeatedly load the skeleton above the threshold level, then bone mass should increase. Animal studies have demonstrated that within 2 to 3 months of external loading at appropriate levels, osteoblasts deposit collagen in the bone matrix, and mineralization follows over the next 3 months. This evidence indicates that RT programs of longer than 6 months would be needed to produce any measurable effect on BMD.

The literature on RT and BMD could be viewed as equivocal, because some studies reported improvements and others reported no change. Nevertheless, there were differences between the two groups of studies that may explain the conflicting results. Most of the investigations with positive results were randomized

> RT has been shown to be an effective therapeutic intervention in arthritis and osteoporosis. Not surprisingly, RT has been shown to be as effective as many drug interventions in these diseases.

controlled trials that focused on adult females up to 75 years of age and lasted for a year or more. RT was performed two to three times each week and included 3 sets of up to 12 different exercises, using 8 repetitions of high-intensity loading up to 85% of the 1RM. The increases in BMD in these studies were significant but generally less than 3%. Moreover, increases were most evident in the axial skeleton, which has more trabecular bone than the appendicular skeleton. Studies that showed no increases in BMD over a 1-year period differed from the previous investigations in one important aspect: The intensity of loading during RT was moderate. Although the gains in 1RM were comparable, the data suggest that bone loading during RT may be the most important variable for increasing BMD.

In summary, RT has a sound theoretical basis in the prevention and treatment of osteoporosis and should be administered as a supplement to conventional treatment, not as a stand-alone modality. It appears that RT programs of greater than 1 year using high-intensity loading are required to increase BMD, but more research in this area is needed.

Summary

RT has been used by athletes successfully for many years to improve performance, to increase muscle strength and mass, and to reduce injury, but the benefits of RT for the general population and those with disease and disability have only recently been appreciated. Research has demonstrated that RT promotes the following adaptations that foster and maintain good health: increases in muscle mass and quality, large increases in dynamic strength and endurance, enhanced exercise and functional capacity, improved balance, decreased falls, reductions in body fat, small but significant reductions in systemic arterial pressure, improved blood lipid and lipoprotein profile, improved disposal of blood glucose, increases in BMD, and increases in self-efficacy and health-related quality of life. RT has been used successfully with obese people, frail elderly people, and various patient groups including those with CAD, arthritis, osteoporosis, and type 2 diabetes mellitus. Limited information indicates that RT may be useful in other cohorts, such as those with neuromuscular disorders, kidney disease, chronic obstructive pulmonary disease, and low back pain, but more research is needed. Clearly, RT should be an integral part of a well-rounded exercise program to develop and maintain good health.

KEY TERMS

atrophy: Decrease in the cross-sectional area of skeletal muscle fibers and eventually the muscle itself, occurring when muscle protein breakdown exceeds synthesis.

disease cluster: Linked series of adverse health conditions present in a single individual or group of individuals such as type 2 diabetes mellitus, high blood pressure, and obesity in the metabolic syndrome.

hypertrophy: Increase in the cross-sectional area of skeletal muscle fibers and eventually the muscle itself. For this to occur, muscle protein synthesis must exceed breakdown.

muscle protein turnover: Rates of both muscle protein synthesis and breakdown and ultimately the net flux between these two processes.

muscle quality: Traditionally, the capacity to generate muscle force as a ratio of the muscle's cross-sectional area; here, however, we use it to refer to the metabolic quality of the muscle, that is, the muscle's capacity for oxidative metabolism.

repetition maximum (1RM): The single highest load that a person can lift once (i.e., that results in instant fatigue) is the single repetition maximum or 1RM.

resistance training (RT): Training that uses either mechanical or free moving loads to create a systematic condition of progressive overload on the skeletal muscle. Loads typically range from 50% to 90% of the 1RM.

sarcopenia: For definition, see page 229.

STUDY QUESTIONS

1. Describe how feeding and resistance exercise interact to increase skeletal muscle fiber size (i.e., hypertrophy).

2. Name five reasons why maintenance of muscle mass is important to long-term health in elderly people.

3. Describe why RT would be as effective or potentially more effective than aerobically based exercise in the treatment of persons who are overweight or obese.

4. Define a disease cluster and give an example of how RT might be able to treat such a cluster and why this might be beneficial as opposed to pharmacological treatment of the same disease cluster.

5. Is RT contraindicated in children and adolescents? What benefits might RT offer children and adolescents?

6. Aside from the expected increases in strength with RT in persons with CAD, CHF, and HT, what other benefits have these patients been shown to receive as a result of undergoing a program of RT?

7. What are the main inclusion and exclusion criteria for patients with CAD, CHF, or HT to participation in a program of RT? What might a typical exercise prescription for the same group of patients look like?

8. Describe a limitation to the use of RT in the treatment of arthritis.

9. Can RT effectively reduce the risk of osteoporosis? Why or why not?

© Jack Raglin

Exercise and Its Effects on Mental Health

■ John S. Raglin, PhD ■ Gregory S. Wilson, PED ■ Dan Galper, PhD

CHAPTER OUTLINE

Throughout the ages physical activity has been advocated as a means of attaining optimal health and preventing disease. More than 2,000 years ago Hippocrates stated, "Eating alone will not keep a man well, he must also take exercise." Today, overwhelming scientific research supports the benefits of exercise for physical health, and empirically based exercise regimens have been established for a myriad of **disorders** across the life span.

Exercise has also been promoted as a means to enhance various aspects of mental health, including emotional disorders and cognitive processes such as memory. However, there is far less research on the mental benefits of exercise compared with its physical consequences. In large part, this is because exercise has been regarded as an unorthodox treatment by mental health workers, many of whom continue to discount its benefits or even believe that physical exercise may actually worsen some psychological conditions. Moreover, although research has consistently found that physical exercise improves measures of mental health in both healthy and clinical samples, because the underlying mechanisms responsible for these changes have not been identified, it is uncertain if this relationship is causal or if exercise is merely associated with psychological improvements and a nonspecific mechanism (e.g., socialization, distraction) is at work. Additionally, study of the psychological effects of exercise has been hampered because it typically requires the cooperation of both physiologists and psychologists, a model for research still rarely used in the exercise sciences.

Despite these challenges, a growing scientific literature indicates that physical activity can indeed provide psychological benefits for healthy individuals as well as those suffering from mild to moderate emotional illnesses. There is also research indicating that the extent to which physical activity actually benefits mental health may be influenced by both the mode and the intensity of exercise. However, researchers have also found that exercise is not always beneficial for emotional health and that in some circumstances it is associated with detrimental psychological outcomes.

> Physical activity can improve and maintain mental health, but the optimal mode, intensity, frequency, and duration of exercise have not been determined.

This chapter summarizes the major findings of the exercise and mental health literature with the aim of describing benefits and limitations of exercise in regard to mental health. Particular emphasis is given to the effects of exercise on the major emotional disorders of anxiety and depression. Some discussion is given to mechanisms that are believed to be responsible for the psychological effects of exercise, along with findings that extreme exercise loads can have detrimental consequences on mental health.

Models of Exercise and Mental Health Research

Research on the psychological effects of exercise has generally examined either the changes associated with bouts of exercise (i.e., acute exercise) or changes that occur only after long-term participation in physical activity for a period of weeks or months (i.e., chronic exercise). Acute exercise research has involved different modes of activity ranging in duration from exercise bouts as brief as 5 min to 9 hr or more, such as in the case of athletes completing a triathlon (Petruzzello et al. 1991).

Acute and Chronic Exercise Research

The psychological responses to either acute or chronic exercise are typically measured by having participants respond to self-report questionnaires that measure various aspects of mental health such as anxiety or mood state. In the case of acute exercise, researchers use psychological questionnaires that assess psychological "states." States are transitory and fleeting emotions that can change in intensity in a manner of minutes or even seconds. Because of the labile nature of psychological states, it is preferable to measure them more than once following an exercise session. A single postexercise assessment can be misleading and would not provide information about the duration of changes in psychological states. For example, high-intensity exercise often results in transient elevations in anxiety and mood disturbance immediately postexercise, but mood assessments taken as little as 10 min later reveal significant improvements compared with pre-exercise values. Research using multiple postexercise assessments indicates that the psychological benefits of a single exercise bout typically persist for 2 to 4 hr before mood state gradually returns to baseline values (Raglin 1997).

In contrast, chronic exercise research does not usually assess psychological states but instead examines more stable psychological measures referred to as "traits." Traits are stable emotions that reflect how a person feels "in general" and thus are not altered by single bouts of exercise or brief psychological interventions. Because of their stable nature, exercise programs lasting in duration from weeks to months are used to test the benefits of physical activity on traits, and this includes clinical conditions such as major depression and anxiety disorders.

> Reported psychological benefits of exercise include reductions in anxiety and depression, with improvements in self-esteem and overall well-being.

Quantifying the psychological outcomes of acute or chronic exercise has largely been achieved through the use of standardized psychological inventories of mood states (e.g., the Profile of Mood States; McNair et al. 1992) or emotions such as anxiety (e.g., the State–Trait Anxiety Inventory; Spielberger et al. 1983) developed for use across a range of settings with various populations. It has been less common for researchers to use questionnaires specifically for diagnosing emotional disorders such as depression. In recent years, specialized questionnaires have been developed to examine psychological responses in the specific context of exercise. However, the majority of these measures have not been adequately validated, and there is no compelling evidence that they provide either greater sensitivity or additional information compared with general measures. In some cases, relevant biological variables (e.g., stress hormones, electroencephalographic results; Petruzzello et al. 1991) have been assessed in tandem with psychological variables to study the mechanisms that underlie exercise-associated benefits to mental health or to determine the effects of exercise on variables associated with stress reactivity (e.g., blood pressure).

Research Samples

One of the fundamental issues for exercise and mental health researchers has been to quantify the extent to which exercise can benefit persons with emotional illnesses such as anxiety or depression. Unfortunately, the findings of this research have long been criticized for commonly relying on samples of psychologically healthy individuals instead of samples of people with psychological disorders. Only a minority of studies have assessed individuals possessing elevated scores in anxiety or depression but not clinically diagnosed as depressed using standardized criteria. Access to clinical samples has been hampered by a long-standing reluctance of clinicians to subject their patients to nontraditional and unproven remedies, but in recent years researchers have more successful in testing the effects of physical activity on patients who have been diagnosed with major depression and anxiety.

Not all exercise and mental health research is focused on establishing the benefits of exercise for clinical conditions. Some researchers are specifically interested in the effects of exercise on persons with average or even above-average mental health. Findings from this work are important for determining the efficacy of physical activity in maintaining normal emotional health and preventing the development of mental illness. Moreover, identifying forms of exercise that provide optimal psychological benefits may help fitness professionals keep participants involved in exercise programs. This is particularly important because recent estimates suggest that only 20% of the adults in the United States meet the minimum daily requirements for aerobic fitness, and of those who begin an exercise program approximately 50% will quit.

Although not typically a target population for exercise and mental health research, athletes consistently possess mood state profiles that are superior to those of nonathletes (Morgan 1997). These findings provide another line of evidence that a physically active lifestyle can benefit both physical and mental health. This same research, however, indicates that when athletes undergo periods of physically demanding training, their emotional health worsens as a direct consequence of exercise, a phenomenon that is described later in the section on detrimental consequences of physical activity.

Exercise and Anxiety

Anxiety disorders are the most common form of emotional illness. Large-scale studies indicate that during the course of a single year, 17% of adults in the United States will develop an anxiety disorder that warrants professional treatment, costing the economy an estimated $45 billion. There are several major anxiety disorders: generalized anxiety disorder, phobias, panic disorder, obsessive–compulsive disorder, and posttraumatic stress disorder (American Psychiatric Association 1994). Despite the far-reaching scope of anxiety disorders, relatively few studies have been conducted testing the efficacy of exercise as a potential treatment compared with the research on exercise and depression (O'Connor et al. 2000).

In the first meta-analyses to be conducted of the exercise and mental health literature, Petruzzello and colleagues (1991) examined the consequences of both acute and chronic physical activity on anxiety. The authors found moderate but significant effect sizes (acute exercise, 0.24; chronic exercise, 0.34) and concluded that the benefits of exercise were greater than those associated with control conditions.

At the time of the Petruzzello study, only limited information was available on the effects of specific forms of exercise or its benefits for clinical anxiety disorders, and so the results were collapsed across clinical and nonclinical samples. Relatively few studies compared exercise with other forms of treatment, it being more common to have either no comparison group or a nontreatment control. Although the inclusion of control conditions is a standard aspect of experimental design, these studies have been criticized for not comparing exercise with conditions that have documented benefits. In the first published study to compare exercise to an established treatment, Bahrke and Morgan (1978) found that 25 min sessions of vigorous walking or Bensonian relaxation—a technique with empirically demonstrated antianxiety properties—resulted in similar decrements in **state anxiety.** Unexpectedly, the control condition of quiet rest seated in a sound-dampened room was associated with an anxiety reduction equal to exercise and relaxation. This led to the hypothesis that exercise reduces anxiety because of mental diversion—features shared by both Bensonian relaxation and the control condition—rather than by physiological changes attributable to exertion (e.g., changes in neurohormone production, elevated temperature), an explanation referred to as the distraction hypothesis.

Programmatic Factors

In a comprehensive review of the exercise and mental health literature, Buckworth and Dishman (2002) found that studies conducted before 1993 that examined the changes in anxiety following exercise typically did not quantify exercise intensity based on individual levels of aerobic fitness, nor did they compare differing levels of intensity within the same individual. Moreover, studies comparing different exercise modes were even less common. Empirically established information on the influence of programmatic factors such as intensity is necessary to provide scientifically based exercise prescriptions for mental health.

Information on the psychological consequences of exercise completed at differing intensities is also important because it has long been contended that high-intensity aerobic exercise is ineffective for reducing anxiety in healthy individuals and may even trigger panic attacks in anxiety patients. However, carefully conducted research on this topic does not support this supposition. Studies in which the exercise intensity is precisely controlled and anxiety measured repeatedly following exercise indicate that aerobic activity as high as 100% $\dot{V}O_2$max is associated with significant reductions in state anxiety. But unlike milder activity, high-intensity activity usually results in short-lived elevations in state anxiety immediately postexercise; following another 10 to 15 min state anxiety decreases to a value equal to that found with milder exercise (e.g., >70% $\dot{V}O_2$max). In the case of anaerobic forms of exercise such as strength training completed at mild to moderate intensities, research indicates that state anxiety reductions either are delayed by 90 to 120 min following the cessation of exercise or do not occur at all (Raglin 1997).

Chronic Exercise and Anxiety

Much less research has been conducted on the effects of chronic exercise programs on anxiety, but reviews of the literature indicate that modest reductions in **trait anxiety** are associated with long-term participation in physical activity programs (Petruzzello et al. 1991; Raglin 1997). Unlike acute exercise, both aerobic and anaerobic regimens appear to be associated with similar benefits, although direct comparisons between exercise modes within stud-

Research suggests that aerobic exercise leads to improvements only in anxiety; however, aerobic exercise as well as anaerobic exercise are associated with lowered levels of depression.

ies are lacking. Research involving aerobic exercise performed at different intensities has yielded mixed results, and it is unclear if anxiety changes are influenced by the intensity at which exercise is performed.

Clinical Samples

The long-standing reluctance to test the potential benefits of exercise with individuals with anxiety disorders, particularly for persons with panic disorder, has resulted in a relative paucity of information on its benefit for clinical conditions. Much of this is the result of a widely cited study by Pitts and McClure (1967), who found that infusions of sodium lactate precipitated panic attacks in patients with anxiety disorders. This work has since been interpreted to indicate that vigorous exercise that results in an accumulation of blood lactate would also be likely to trigger panic attacks, stifling research on the topic. But in a study that examined the ability of anxiety patients to tolerate high-intensity exercise, Martinsen and colleagues (1989) had 35 individuals diagnosed with panic disorder complete bouts of bicycle ergometry at 120% $\dot{V}O_2$max. Only one of the 35 participants (4%) experienced a panic attack during exercise. Even this individual was able to successfully complete the exercise protocol, leading the authors to conclude that even high-intensity exercise resulting in high levels of blood lactate (mean = 10.7 mmol/L) does not present an undue risk of panic.

In perhaps the first published study to test the consequences of exercise in patients with clinically diagnosed anxiety, Martinsen and colleagues (1998) found that 8 weeks of walk–run exercise or a mild stretching program significantly reduced anxiety symptoms. Equally important, none of the patients experienced a panic attack while exercising and the majority of individuals (89%) successfully completed the exercise programs. Improvements in symptoms were not associated with fitness gains, indicating that a relatively light exercise prescription could be as effective as more vigorous programs.

Only one study has compared the efficacy of a walk–run exercise regimen with antianxiety medication (clomipramine) and a placebo in anxiety patients diagnosed with panic disorder (Broocks et al. 1998). Participants were randomly assigned to one of three conditions, and symptoms were assessed with a variety of clinical measures. At the end of the 10-week intervention period, the authors found that the exercise program was significantly more effective than a placebo in reducing anxiety but less effective than medication. Compared with medication, exercise took longer before symptoms were relieved. No cases of panic attacks occurred during exercise participation, and the 31% dropout rate was less than the 50% average noted in the general literature.

Exercise and Depression

Depressive disorders will afflict one out of three persons in the course of their lives. Studies in several countries indicate that at any given time, approximately 4% of men and 8% of women are clinically depressed. Major subtypes of depression include major depressive disorder, dysthymia, and bipolar disorder (American Psychiatric Association 1994). Unfortunately, as is the case with other mental health disorders, many people suffering from depression do not seek treatment. Estimates suggest that less than one of three depressed individuals will seek some form of treatment and of those who do, 90% receive less than adequate treatment (Buckworth and Dishman 2002).

Research on the Efficacy of Exercise As a Treatment for Depression

Over the past 4 decades, more than 1,200 studies have evaluated the potential for exercise to treat depression, and the majority of this work has focused on major depressive disorder. Cross-sectional epidemiological studies indicate significantly fewer depressive symptoms among physically active persons, even when adjustments were made for a wide range of potentially confounding variables (Buckworth and Dishman 2002). Studies have reported similar effects among adults and adolescents, and most existing studies have found similar effects among both men and women. Moreover, several recent prospective community-based studies have found a reduced incidence of clinical depression among middle-aged and older adults who remain physically active or increase their activity during the follow-up period, and at least one prospective study (Motl et al., 2004) showed reduced depressive symptoms among physically active adolescents.

In the case of persons who score within the normal range on standardized measures of depression, exercise programs appear to result in little or no reduction in depression scores. In the first controlled study of its type, Morgan and colleagues (1970) found that persons randomly assigned to one of several different exercise modes did not experience significant reductions in depression following 6 weeks of participation. However, depression scores decreased significantly among individuals who possessed values that were elevated above the population norm at baseline. In work involving patients diagnosed with mild to moderate depression, it has been found that the psychological benefits of exercise equal and sometimes exceed those associated with the more conventional forms of psychotherapy. Available evidence also suggests that aerobic and anaerobic forms of activity provide similar benefits (Buckworth and Dishman 2002).

Limited work has directly compared the benefits of exercise to antidepressant medication. In the most important study of its type, Blumenthal and colleagues (1999) had 156 older adults who were clinically diagnosed with moderate depression randomly assigned to 16 weeks of a walk–run program, antidepressant medication, or both exercise and medication. Significant reductions in symptoms were observed in each condition, and the results are presented in figure 16.1. It was also found that the magnitude of reduction did not differ across conditions, although medication was reported to

alleviate symptoms more rapidly, particularly in patients with the most severe depression. A follow-up study of the sample 6 months after the cessation of the treatments revealed that the relapse rate was lowest for the exercise condition.

All these research results indicate that exercise can provide important benefits to patients suffering from mild to moderate depression or anxiety. In the case of depression, the benefits of exercise appear to equal those associated with medication, whereas available evidence indicates that medication provides greater relief for anxiety patients with panic disorder. The modest exercise prescription required to alleviate the symptoms of moderate anxiety or depression is well tolerated, resulting in adherence rates that either equal or exceed those found with individuals free from mental disorders.

Seeking Definitive Dosage Recommendations for Depression

As is the case with anxiety disorders, definitive recommendations for exercise intensity, mode, frequency, and duration have yet to be established for depression (Buckworth and Dishman 2002). Moreover, individuals suffering from depression typically have lower fitness levels than normal populations, further making the prescription of exercise problematic. Although it is logical to assume that the mental health benefits of physical activity would be tied to increases in cardiorespiratory fitness, there is conflicting evidence for such a relationship. Increases in cardiovascular capacity also do not appear to be required for fitness programs to reduce depression, and correlations between increased aerobic capacity and reduced symptoms are modest at best.

The minimal exercise dose necessary to provide effective treatment was examined in a landmark study recently published by Dunn and colleagues (2005). These researchers compared the benefits of different exercise regimens on depression in an effort to identify an effective minimal dosage. Eighty adults diagnosed with major depressive disorder were randomly assigned to 12-week programs of either a placebo condition (flexibility training three times a week) or one of four exercise conditions involving four aerobic activity programs that varied by energy expenditure (low, 7.0 $kcal \cdot kg^{-1} \cdot week^{-1}$; moderate, 17.5 $kcal \cdot kg^{-1} \cdot week^{-1}$) and frequency (three or five days a week). The moderate energy expenditure condition was derived from public health recommendations. At the end

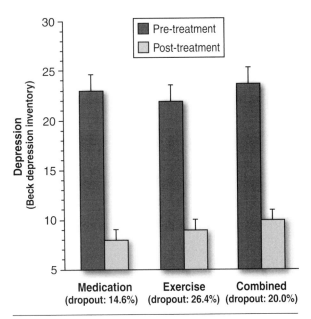

Figure 16.1 Changes in depression following 16 weeks of treatment with exercise, medication, or exercise and medication.

of the intervention, the reduction in depression was significantly greater for the moderate energy expenditure condition compared with the low expenditure or placebo control. The frequency of exercise (either 3 or 5 days a week) had no effect on depression. These results led the authors of the study to conclude that major depressive disorder can be effectively treated with an exercise prescription based on public health recommendations.

Proposed Explanations for the Psychological Benefits of Exercise

Although an increasing amount of research in recent years has indicated that exercise is associated with improvements in mental health, a causal link has not been established, and the underlying mechanism for this relationship remains unknown. This section presents a brief overview of the current physiological and cognitive mechanisms that have been advanced as explanations for the psychological benefits of exercise.

> Both physiological and cognitive explanations have been used in an attempt to explain improvements in mental health as a result of exercise participation. However, the exact mechanism responsible for this relationship is unknown.

Neuroscientific studies provide a number of plausible mechanisms through which exercise may improve mood and reduce symptoms of depression and anxiety. Early neurochemical hypotheses implicated endogenous opioids such as endorphin in the mental health effects of exercise. Endorphins, the most prominent of which are β-endorphin and enkephalin, are chemical substances produced by the brain, pituitary glands, and other bodily tissues. Endorphins act as natural opiates that produce analgesia by binding to opiate receptor sites involved in perceptions of pain, and they have been implicated in reward mechanisms and positive emotion. Physical stressors such as exercise can stimulate the production of endorphins, and this has led to the belief that endorphins are responsible for the popular but unsubstantiated phenomenon known as the "runner's high." Research also has not found endorphin levels following exercise to be correlated with mood change, and the use of pharmacological blocking agents such as naloxone has provided only mixed support for the mood-altering effects of endorphin. Because of the lack of direct evidence, most investigators now favor alternative hypotheses.

Other hormonal and physiological pathways have been proposed to be responsible for the affective benefits of exercise. The monoamine hypothesis is based on the knowledge that neurotransmitters such as norepinephrine (NE), dopamine, and serotonin (5-HT) play a role in both depression and various forms of schizophrenia. It is believed that physical activity may alter mood state through its effect on one or all three of these brain monoamines. Unfortunately, research testing the monoamine hypothesis has been constrained because of the difficulty of accurately measuring levels of hormones such as NE in the brain. Animal studies have shown that increases in central and peripheral levels of monoamine hormones occur following exercise, and other work has found alterations in receptor sensitivity of monoamine neurons in the brain. This work, although suggestive, remains tentative until definitive studies can be done with humans (Buckworth and Dishman 2002).

The thermogenic hypothesis postulates that the increase in body temperature that occurs with vigorous exercise can stimulate neurological changes that are associated with improved mood. Vigorous physical exercise may raise body temperature by several degrees for a period of hours. However, studies examining the efficacy of this hypothesis have not definitively linked temperature to the mood changes that occur following exercise.

It also has been proposed that cognitive or behavioral factors such as social interaction and support, feelings of achievement, self-mastery and self-efficacy, and distraction or diversion of attention may be the initial cause of improved mood following exercise. For instance, the distraction hypothesis posits that exercise results in psychological benefits because it typically is performed in a setting removed from the workplace or other stressful environments, resulting in a form of pleasant diversion. Another hypothesized cognitive explanation suggests that feelings of achievement or personal mastery that result from successfully completing an exercise bout or adhering to a long-term exercise program may contribute to improved mood and mental health.

Tests of these and other hypothetical mechanisms for the psychological benefits of exercise have failed to provide unequivocal support for any single explanation. Given the complexity of physical activity behavior and the various psychological responses to exercise, it is possible that no particular mental health effect can be adequately explained by a single process. Multiple mechanisms may well interact to affect both short-tem and long-term psychological functioning. Table 16.1 summarizes the mechanisms proposed to date. Information concerning how these mechanisms affect moods and traits would

be useful for establishing more effective exercise prescriptions for enhancing mental health, either as a primary treatment or as an adjunct with other therapies.

Detrimental Psychological Responses to Exercise

Although exercise is viewed as having a positive effect on mental health, there is also evidence that

Table 16.1 Proposed Biological and Psychosocial Mechanisms for the Psychological Benefits of Exercise

Mechanism	Hypothesis	Comment
Biological		
	Thermogenic	This hypothesis proposes that corresponding changes in body temperature that occur during exercise are associated with increased central and peripheral neuron activity in the brain, as well as decreased muscle tension. These physiological changes enhance mood state (i.e., reduce state anxiety).
	Monoamine	Neurotransmitters such as norepinephrine, dopamine, and serotonin play a role in both depression and various forms of schizophrenia. Monoamine oxidase inhibitors have been used for several decades to treat depression, although it is not clearly understood how these inhibitors function. Physical activity may alter mood states through its effect on one or all three of these neurotransmitters. Unfortunately, it is impossible to accurately measure levels of these hormones in the brain.
	Endorphin	Endorphins, the most prominent of which are β-endorphin and enkephalin, are chemical substances produced by the brain, pituitary glands, and other bodily tissues. Endorphins act as natural opiates by binding to opiate receptor sites involved in perceptions of pain, and they have been implicated in reward mechanisms and positive emotion. Physical stressors such as exercise can stimulate the production of endorphins, and this has led to the belief that endorphins are responsible for the popular but unsubstantiated phenomenon known as the "runner's high." Research has not found endorphin levels following exercise to be correlated with mood change, and the use of pharmacological blocking agents such as naloxone has provided only mixed support for the mood-altering effects of endorphins. Because of the lack of direct evidence, most investigators now favor other hypotheses.
Psychosocial		
	Distraction	Psychological benefits associated with exercise may not be related to underlying physiological mechanisms but rather to the simple fact that participation in physical activity typically occurs in a setting removed from the workplace or other stressful environments. Hence, a person is distracted from potential stressors and provided with a pleasant diversion.
	Mastery	Feelings of achievement or personal mastery result from the successful completion of task. In the case of exercise, the successful completion of an acute bout of exercise or a long-term exercise regimen improves mood and mental health through enhancing an individual's self-efficacy.

in certain situations exercise can create negative psychological consequences. For instance, an exception to the benefits of physical activity is sometimes seen in highly trained athletes who undergo periods of intense training. Prolonged periods of high-volume training have been shown to lead to mood disturbances in some athletes. Furthermore, evidence suggests that for some individuals, exercise can become a compulsion. In this case, exercise takes on an excessive priority in a person's life and may even precede commitments to family and work.

Athletic Conditioning and the Staleness Syndrome

Detrimental consequences of physical activity have been observed in competitive athletes who undergo intensive training to maximize performance. Studies of sports such as swimming and distance running have consistently found that intensive endurance training is associated with mood disturbances in both men and women athletes, even though these athletes possessed better than average mental health values before hard training (Raglin and Wilson 2000). This work also indicates that training loads are closely associated with the degree of mood disturbance in a dose–response relationship. Increases in either the volume or intensity of training result in corresponding elevations in negative moods such as anxiety, depression, and anger. At the peak of training, mood disturbances are also at their highest with scores that typically exceed the population norm. Reductions in training (i.e., tapers) result in corresponding reductions in mood disturbances and increases in vigor. By the end of the training season when training volumes are low, the mood profiles of most athletes resemble desirable preseason values.

> Athletes engaged in intense training can experience a condition known as staleness in which they often experience bouts of clinical depression and other mood disturbances.

The majority of athletes can tolerate the stress of overtraining, but from 5% to 20% respond adversely and exhibit mood disturbances that exceed those of their healthy teammates during periods of overtraining and suffer from substantial performance decrements that persist for weeks or even months. This condition is commonly referred to staleness or the overtraining syndrome and it is a concern for athletes in endurance sports as well as nonendurance sports that require a high level of physical conditioning. Clinical depression is the most common psychological manifestation of

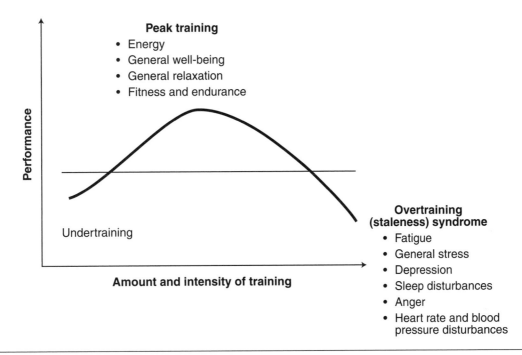

Figure 16.2 Although most athletes can cope with high levels of training, a minority of them will experience the negative symptoms of the overtraining syndrome if intense training is prolonged.

Courtesy of Nina Laidlaw.

staleness, and many athletes are affected to a degree that requires therapy or antidepressant medication. It is unclear why some athletes appear prone to developing the staleness syndrome whereas other athletes rarely or never develop the disorder, but researchers agree that its primary cause is the physical and mental stress of intensive training.

Although the consequences of overtraining and the staleness syndrome are rarely considered by researchers who study the psychological benefits of physical activity, this work has potential relevance. Overtraining research indicates that physical activity should be regarded as a complex phenomenon that, although typically associated with beneficial outcomes that rival conventional forms of treatment, can also result in detrimental psychological and physical consequences. The difference in these outcomes is largely dictated by the exercise dosage, but it is also evident that individual differences in the capacity of athletes to adapt to overtraining play a role.

Exercise Addiction

Although ample evidence exists to suggest a positive influence of exercise to mental health, case studies of compulsive participation in exercise have also been documented (Morgan 1979). In this study, Morgan reported eight individual episodes of runners whom he described as suffering from "running addiction." For these individuals, the need to run exceeded commitments to family, work, social relationships, and even medical advice.

Similar results were described by Dishman (1985), who suggested that such exaggerated emphasis on the role of exercise in one's life may also reflect a predisposition of insecurity and low self-concept. More recently, Phillips and colleagues (1993) proposed a body image disorder they have labeled *muscle dysmorphia*. Muscle dysmorphia is considered a pathological preoccupation with muscularity and can lead to excessive exercise par-

> For some individuals, exercise can become an addiction that can result in negative psychological, medical, and behavioral effects.

ticipation in the attempt to create greater muscle hypertrophy than needed for fitness.

Summary

Over the past few decades, a growing body of research has documented what has been long presumed about physical exercise, namely that it is as beneficial for the mind as it is for the body. For persons who are free from emotional disorders, the primary psychological benefits occur following acute exercise; brief bouts of aerobic exercise are consistently associated with reductions in state anxiety as well as significant mood improvements that last for several hours. The extant research indicates that sessions of anaerobic activities such as strength training are less frequently associated with these benefits. Studies on the psychological effects of chronic exercise have found it to be particularly effective for persons with elevations in anxiety or depression, whereas individuals scoring in the normal range on these variables may not exhibit significant long-term improvements. Notably, there is growing evidence that the benefits of exercise can approach or equal those of psychopharmacological medication and therapy. Mild to moderate exercise programs that are well tolerated and result in high levels of adherence appear to be effective in treating psychological disorders, even if they do not improve cardiovascular fitness. However, more research is needed, not only to replicate these results but also to study the potential for exercise to benefit other forms of psychological illness. Studies are also needed to uncover the mechanisms responsible for the mental benefits of exercise.

Finally, although research has shown exercise to be an effective intervention strategy in the treatment of anxiety and depression, it is not completely accepted by mental health professionals. Further information is needed concerning exercise prescription (e.g., dosage, frequency, mode) before definitive statements can be made. As rates of both anxiety and depression continue to increase, this line of research has important implications for a modern society that is increasingly sedentary in nature.

KEY CONCEPTS

disorder: A chronic condition, whereas an episode is a single acute instance.

state anxiety: An unpleasant emotional arousal in response to perceived danger or threatening demands. A cognitive appraisal of threat is a prerequisite for the experience of this emotion.

trait anxiety: Reflects an individual's tendency to anticipate threatening situations and to respond with state anxiety. Affective or mood symptoms include depressed mood and feelings of worthlessness or guilt. Behavioral symptoms include social withdrawal and agitation. Cognitive symptoms, or problems in thinking, include difficulty with concentration or making decisions. Somatic or physical symptoms include insomnia or hypersomnia.

STUDY QUESTIONS

1. Explain why the relationship between exercise and mental health is considered an associational as opposed to a causal one.

2. Discuss the changes in anxiety and depression following acute versus chronic bouts of exercise.

3. List the leading theories that are used to explain the affective benefits of exercise. Why is it difficult to prove empirically the validity of these theories?

4. Explain the qualitative and quantitative differences found in stress reduction in research examining the effects of exercise and quiet rest.

5. Describe what is meant by the dose–response relationship between exercise volume or intensity and mental health.

Physical Activity, Fitness, and Children

■ Thomas W. Rowland, MD

CHAPTER OUTLINE

Cardiovascular Health
- Serum Lipids
- Hypertension
- Obesity
- Type 2 Diabetes Mellitus
- Synthesis

Bone Health and Osteoporosis

Strategies for Promoting Physical Activity in Youth
- Strategy 1: Suppressing the Development of Long-Term Pathological Processes
- Strategy 2: Introducing Activity to Create Lifelong Habits
- Strategy 3: Shifting the Physical Activity Curve

Summary

Review Materials

Improving and maintaining **physical activity** and **physical fitness** in children may provide both immediate and long-term health benefits. Although the link between exercise and health has been established in adults, much less scientific documentation for such a relationship exists in youth. Still, regular physical activity in children and adolescents can be expected to have long-term salutary outcomes because the adult diseases influenced by activity (atherosclerosis, osteoporosis, obesity) often have their origins in the **pediatric years.** The means by which activity habits can be instilled and sustained in sedentary youth need to be developed.

The accumulated evidence that regular physical activity provides health benefits for adults is incontrovertible. As outlined in the pages of this book, compelling data indicate that the active individual can expect a protective effect from a broad array of disease outcomes, including coronary artery disease, hypertension, obesity, osteoporosis, and type 2 diabetes mellitus. This activity–health link has provided the impetus for aggressive public health and medical initiatives designed to improve the welfare of the general adult population as well as high-risk groups.

At first consideration, the idea that children should expect a similar salutary effect of physical activity is not altogether obvious. Children do not suffer from the previously listed disease outcomes for which activity provides benefits for adults. Indeed, as Blair and colleagues (1989) emphasized, morbidity and mortality in the pediatric age group result principally from accidents, infections, hematologic malignancies, and congenital malformations, conditions for which no beneficial effect of physical activity should be expected. Moreover, the habitual physical activity patterns of children, consisting of frequent short bursts of exercise, are different from those of adults, and children's cognitive and physical immaturity makes exercise interventions more problematic.

These observations notwithstanding, a strong rationale has been developed for the promotion of physical activity and fitness in children for both present and future health. Much of this is based on the recognition that the clinical markers of chronic disease in adults—atherosclerosis, hypertension, obesity, osteoporosis—are expressions of lifelong processes that begin during childhood and adolescence. Other positive outcomes, such as mental well-being and academic performance, are more immediate (Armstrong and van Mechelen 2000). Promotion of physical activity in youth has therefore gained acceptance as a sound strategy for improving health, again both in the general population and in risk-specific individuals.

This effort has been fueled by a concern that the amount of habitual physical activity of children—who are surrounded by an increasingly technological society—is on the decline. There are, in fact, no scientific data on which to base that idea (largely because of the difficulties in accurately assessing physical activity levels in populations). Still, this trend is suggested by indirect evidence: the rising frequency of obesity among children, data suggesting a secular decline in field endurance performance, and increases in television watching time and other **sedentary pursuits** (Tomkinson et al. 2003).

In this chapter, the scientific basis for the pediatric activity–health rationale is addressed as it relates particularly to cardiovascular health and bone development (i.e., prevention of osteoporosis), two outcomes for which sufficiently valid experimental and observational data exist. This discussion presents only a "slice" of a general argument for promoting physical activity in youth, because involvement in regular activity and sports may offer additional social, psychological, and cognitive benefits (Rowland 1990) (figure 17.1). The role of exercise as a therapeutic intervention for specific disease states also is not discussed in this chapter. Evidence exists that improving physical activity in youth may prove beneficial for emotional, cardiopulmonary, and musculoskeletal disorders, but research data are fragmentary (Bar-Or and Rowland 2004).

In this discussion, it is important to separate the health outcomes of physical activity in youth from those related to physical fitness. Different factors influence physical activity, which is a behavior, and physical fitness, which describes the ability to perform a motor task. The potential health benefits from activity and fitness may not be the same, and each calls for a different interventional strategy (i.e., behavioral modification for improving activity, a period of exercise training for increasing fitness).

In adults, an individual's level of regular physical activity is often considered a surrogate marker of physical fitness (and vice versa), but this does not appear to be true in children. Somewhat surprisingly, most studies have indicated little relationship between habitual physical activity and physical fitness (at least as defined by maximal aerobic power) in children (Morrow and Freedson 1994). Moreover, aerobic training programs in prepubertal children

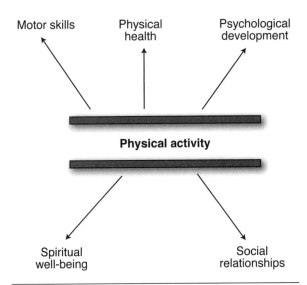

Physical activity

Motor skills

Physical health

Psychological development

Spiritual well-being

Social relationships

Figure 17.1 Benefits of physical activity for youth.

conducted according to the standard criteria for frequency, duration, and intensity cause only small increases in maximal aerobic power (about 5%). From a health-outcomes standpoint, then, activity and fitness may need to be considered separately, at least in the growing years.

Certain issues have proven troublesome for those wishing to scientifically document the rationale for promoting physical activity for health in children. Quantifying levels of activity is particularly difficult, given the recurrent short-burst types of activity characteristic of this age and the inability of young children to accurately report activity levels. In addition, the effectiveness of improving activity habits on well-being in children is often unclear, because potential adverse health outcomes by which to gauge success will not surface clinically for decades to come.

Levels of physical activity decline quite precipitously during the growing years, which is largely a biological phenomenon (Rowland 1998). Consequently, differentiating normal developmental changes in activity from those reflecting environmental, psychosocial, and other mediating influences can be difficult. These challenges notwithstanding, the accumulated evidence outlined in this chapter provides a convincing basis for a pediatric approach to physical activity for health.

Cardiovascular Health

In adults, **cardiovascular health** is defined by the clinical markers of atherosclerosis (coronary artery

disease, CAD) and hypertension, and the effects of physical activity can be readily measured by assessing these disease outcomes. Such an approach in children would require examining the impact of activity levels or intervention over 30 years. As noted previously, such a study has not been performed and, considering the logistical difficulties, probably will not be in the future.

> Obesity decreases aerobic or endurance fitness through the effects of excessive body fat rather than depression of cardiopulmonary function.

The rationale for a pediatric approach to prevent "adult" cardiovascular disease (CVD) is based on observations that the clinical outcomes of high blood pressure and atherosclerotic CAD reflect pathologic processes that begin early in life. Postmortem studies have demonstrated that fat deposition in major blood vessels can be demonstrated even in early childhood, and raised fibrous lesions in the coronary arteries are common by late adolescence. Essential hypertension is highly familial and is often manifest initially in the teen years. Other risk factors (obesity, hyperlipidemia, insulin resistance) are common in youth, often cluster together in a given individual, and tend to track, or persist, into the adult years. Adult CVD has consequently been considered a pediatric problem.

It follows that to have their greatest effect, measures known to ameliorate risk for CAD and hypertension in adults should be introduced as early in these pathologic processes as possible. Although proof of this approach is not at hand, promotion of activity in youth should therefore be expected to reduce lifelong risk of adult cardiovascular disease. To scientifically support this construct, one is left to examine the effect of activity and fitness in children on surrogate markers of future cardiac disease risk, including serum lipid profile, blood pressure, and body fat content.

Serum Lipids

Adults with high levels of physical activity and aerobic fitness often have a more favorable serum lipid profile, especially higher levels of high-density lipoprotein (HDL) cholesterol. The potential influence of activity or fitness on serum lipid profiles in youth has been addressed in both cross-sectional studies of athletes and nonathletes and relatively short (2- to 3-month) aerobic training interventions. Little long-term longitudinal data are available, however, to address this question.

A number of early cross-sectional studies indicated that children who were highly active or more physically fit had higher HDL cholesterol or lower total cholesterol levels compared with their inactive or unfit peers (Armstrong and Simons-Morton 1994). It was surprising, then, that a series of subsequent endurance training programs in children, most lasting from 8 to 12 weeks, generally failed to demonstrate any favorable changes in serum lipid concentrations. Among 12 such studies, HDL cholesterol concentration remained unchanged with training in 7 studies, increased (by 9-20%) in 4, and decreased in 1 (Tolfrey et al. 2000).

Two possible explanations might account for this inconsistency in findings between cross-sectional and interventional studies. First, the training studies were all relatively short (only two lasted for greater than 15 weeks). This may be an inadequate exercise volume dose, because studies in adults suggest that a training intervention of more than 12 weeks is more likely to alter lipid levels. On the other hand, it seems clear that many of the cross-sectional studies were weakened by failure to adjust findings for confounding variables, particularly dietary fat intake and body composition, which could affect lipid levels. For instance, Hager and colleagues (1995) found that a significant relationship between 1-mile performance and HDL cholesterol level in 262 children ($p = .04$) disappeared when body fat content was taken into account.

All of the studies discussed in the previous paragraph were performed in children and adolescents who had serum lipid levels within the normal population range. Youth with an unfavorable lipid profile associated with obesity do demonstrate favorable changes with exercise training (elevations in HDL cholesterol), which occur in association with weight loss (Armstrong and Simons-Morton 1994). The effect of increased activity or fitness training in nonobese youth with familial or acquired hyperlipidemia has not yet been studied.

Long-term longitudinal studies of physical activity level and serum lipid levels have been conducted in American, Finnish, and Dutch youth. These have provided mixed findings, two showing no temporal relationship between changes in activity and lipids and the third indicating a correlation between activity and HDL cholesterol.

Hypertension

More active, fit adults can expect to have lower blood pressure than sedentary individuals. A period of exercise training in adults has been shown to lower systolic and diastolic pressure by an average of 8 and 6 mmHg, respectively, in normotensive individuals and by about half this much in those who are hypertensive.

There is no evidence that either improved physical activity habits or endurance training will alter resting blood pressure in youth who have normal levels to start with. In cross-sectional studies, inverse associations between resting blood pressure and fitness and activity have been demonstrated, but the statistical significance of this relationship typically disappears once body fat is considered (Alpert and Wilmore 1994). Long-term longitudinal studies of 3 to 13 years have found no relationship between development of physical activity and changes in systolic and diastolic blood pressure. In these investigations, no association was seen between habitual activity in the early teen years and adult blood pressure.

> Obesity is associated with increased systemic blood pressure, which is reflected in higher blood pressure values during exercise testing.

On the other hand, all four studies that have investigated the effect of endurance training (of 12-32 weeks) on blood pressure in youth with mild essential hypertension have demonstrated a significant decrease (typically systolic 4-6 mmHg). To maintain this effect, however, training levels of activity must persist, because blood pressure level has been demonstrated to rise again when a training program ends.

All these studies used standard criteria for aerobic training programs (at least three sessions weekly, over 30 min each, at a moderately high intensity). Whether lesser levels of physical activity will influence blood pressure in hypertensive youth is not known.

Obesity

In cross-sectional studies, children with exaggerated body fat content are characterized by unfavorable serum lipid profiles as well as elevations in both systolic and diastolic blood pressure. The principal risk factor associated with childhood obesity, however, is adult obesity. The overweight child is highly likely to carry his or her adiposity into the adult years, with an associated chance for adverse cardiovascular outcomes. That a dramatic increase in the frequency of childhood obesity is being witnessed throughout the world adds additional concerns about future trends for adult cardiovascular morbidity and mortality.

The etiology of obesity—and, in particular, the epidemic of childhood obesity—remains uncertain. In the ultimate analysis, however, increased body fat must represent an imbalance of energy consumption and expenditure. Level of physical activity, as the major volitional means of increasing the "energy out" side of the equation, may be assumed to play a substantial role. Subsequently, efforts to improve activity should prove useful in both prevention and management of childhood obesity.

It is difficult, however, to assess the effect of physical activity on body fat, because in this case the variable in question (i.e., body fat) itself can be expected to decrease both activity and fitness levels. Suppose a study indicates that 3-day physical activity levels in a group of obese 10-year-old boys are significantly less than in a control group of lean children. Does this mean that lower levels of physical activity tip the energy balance and cause accumulation of body fat? Or do the findings imply instead that obese children, because of physical discomfort of moving body bulk, are less prone to exercise? The answer is not clear.

In addition, physical activity and energy expenditure must be considered separately, because they represent different measures. The obese child may move around less, but expend more energy doing so, compared with the nonobese youngster. Moreover, when we analyze these types of data, the means of expressing measurement variables in relationship to body size (relative to lean body mass? body surface area? body mass? or allometric exponents?) are often problematic.

There appears to be no effect of physical activity on body fat content of nonobese youth. Short-term endurance training programs in such youth show no alterations in body fat, and longitudinal observational studies have not demonstrated a relationship over time between level of habitual activity and body composition (Bar-Or and Baranowski 1994).

As would be intuitively expected, however, exercise has proven to be effective as part of a therapeutic program for overweight children and adolescents. Indeed, including exercise as part of a dietary and behavioral modification program carries certain advantages, such as maintenance of or increase in lean body mass and improvements in physical fitness (figure 17.2). The extent to which a program of regular activity or training in obese children can lower body fat content is relatively small, however (reduction in percent overweight by 5-10%). The extent to which such effects can be expected to persist, particularly when structured programs end, remains to be studied.

> Exercise testing protocols may need to be modified for testing of obese, poorly fit youth.

The type of activity intervention may have a bearing on its effectiveness. Some studies have suggested that increasing daily lifestyle activities (i.e., unstructured interventions) may result in more persistent effects on obese children than a regimented structured activity prescription (i.e., exercise training).

Type 2 Diabetes Mellitus

The frequency of type 2 diabetes (also known as non-insulin-dependent diabetes mellitus or NIDDM) has escalated dramatically in the pediatric age group in parallel with the increase in childhood obesity (Pinhas-Hamiel et al. 1996). This condition is characterized by insulin resistance, hyperinsulinism, elevated glucose responses to feeding, and—at least in adults—an increased risk for atherosclerotic vascular disease.

Adults with type 2 diabetes are often characterized as well by a clustering of coronary risk factors, particularly systemic hypertension, hypertriglyceridemia, and depressed levels of HDL cholesterol, defining the so-called metabolic syndrome. Not surprisingly, the coexistence of these factors greatly increases the chances for accelerated atherosclerotic vascular disease.

The metabolic syndrome is frequently observed in youth with type 2 diabetes and obesity as well. Gutin and colleagues (1997) described a significant clustering of all these risk factors with type 2 diabetes in a group of 7- to 13-year-old obese children. Similarly, significant relationships between markers

✓ Decreased percent body fat
✓ Increased lean body mass
✓ Potentialed dietary thermogenesis
✓ Reduced blood pressure
✓ Improved cardiovascular fitness
✓ Benefits to psychosocial health
✓ Prevention of obesity (unconfirmed)

Figure 17.2 Potential beneficial effects of exercise for obese children.

Reprinted, by permission, from T.W. Rowland, 1990, *Exercise and children's health* (Champaign, IL: Human Kinetics), 142.

of insulin resistance and cardiovascular risk factors in children were described by Steinberger and Rocchini (1991) and by Burke and colleagues (1986) in the Bogalusa Heart Study.

Exercise interventions in obese youth with type 2 diabetes may favorably alter insulin resistance. Kahle and colleagues (1996) found that a 15-week exercise program in seven obese children resulted in a 15% decrease in fasting glucose and 51% decline in peak insulin response to a meal.

Synthesis

From the preceding information, several observations emerge that may influence the way we think about physical activity and fitness and health in children. First, improving physical activity levels—even to the extent of exercise training—is unlikely to alter cardiovascular risk factors in normotensive, lean youth with normal serum lipid levels. When levels of blood pressure, body fat, and serum lipids are within the expected range in the population (this creating the definition of "normal"), these surrogate markers of cardiovascular health are generally not altered by repetitive motor activity in youth.

This observation has been described as "discouraging" by those promoting physical activity and fitness for health in children, because it appears to weaken the premise for early exercise interventions. The pediatric rationale, however, may be salvaged by an alternative viewpoint—improving physical activity appears to ameliorate cardiovascular risk factors in those with abnormal values to start with. In fact, this finding has been particularly consistent in the research literature: In every risk factor that has been addressed, abnormal risk factor levels

decrease with exercise interventions. Increasing physical activity has been demonstrated to lower blood pressure in hypertensive youth, improve body composition in obese persons, and favorably influence blood lipids in overweight children with hyperlipidemia.

It can be argued that the question in the pediatric age group is not whether physical activity or fitness will lower cardiovascular risk factors. Instead, the critical issue may be whether regular motor activity will retard the development of risk factors with increasing age. This research question is particularly difficult to address. But the fact that augmented physical activity can lessen abnormal levels of blood pressure, body fat, and serum lipids in children and adolescents strongly implies that sufficient levels of activity can deter the development of abnormal levels of cardiovascular risk factors in the growing years.

Another pertinent observation is the frequency with which obesity serves as a mediating factor between physical activity and fitness and other risk factors. Resting blood pressure levels are sometimes higher in studies of young subjects who are more sedentary—but the relationship seems to be accounted for by their greater body fat content. Exercise training improves abnormal serum lipid profiles in obese persons, when changes in lipid concentrations are related to decreases in body fat content. These observations speak to the central role of obesity as a cardiovascular risk factor.

Bone Health and Osteoporosis

The concept that improved physical activity habits during childhood and adolescence can optimize long-term well-being extends to **bone health** as well. Indeed, stimulating bone development is perhaps the most obvious reason for promoting physical activity during the growing years. Bone mineral density increases progressively during the first decades of life, reaching a peak at about age 20 to 30 years. After that, bone density declines, with a particularly sharp drop-off in women after menopause. This decrease in bone density leaves elderly individuals, particularly females, at risk for exaggerated bone thinning, or osteoporosis, with subsequent disability and death from bone fractures.

Peak bone density in the early adult years is a good predictor of bone health in elderly individuals. It has been suggested that as much as half of the variability in bone mass in older adults can be accounted for by degree of earlier bone mineralization. It follows, then, that interventions to maximize bone development in the growing years can provide long-term dividends by slowing the decay of bone density in later adulthood (figure 17.3). This issue may be of particular importance among girls and women.

Experimental evidence in humans and animals has indicated that two factors—calcium intake and weight-bearing physical activity—are critical to bone development in the early years of life. All this information, then, points to an obvious preventive health intervention: Optimizing diet and promoting weight-bearing physical activity in the pediatric years, especially in girls, should reduce the life-long risk of osteoporosis.

No single study has verified this premise. Still, a large body of research information clearly supports the role of motor activity in the development of bone mineralization in the pediatric years. A literature review by Bailey and Martin (1994) indicated that sports such as gymnastics, hockey, soccer, and volleyball can improve bone mineralization in youth, whereas swimming, a non-weight-bearing sport, is less likely to show such effects. In that review, studies in nonathletes generally demonstrated only modest correlations between regular physical activity and bone mineral density.

Longitudinal studies have indicated that weight-bearing activities can have a significant effect on change in bone mineral density over time. For instance, the extent of weight-bearing activity during early adolescence in the Amsterdam Growth and Health Study was related to degree of bone mineral density at age 17.

Evidence indicates that physical activity enhances bone development independent of genetic and metabolic factors. Such insights are gained from studies showing that favorable bone mineral density is seen in limbs of youth that undergo mechanical stress. Thus, greater density is observed in the bones of the dominant arms of young baseball and tennis players as well in the healthy limb of children who have unilateral bone disease.

Studies investigating changes in bone density with a period of physical training have provided mixed results. Several have shown acceleration of bone development (after training of up to 12 months), whereas others have failed to demonstrate this effect. Bailey and Martin (1994) suggested that such differences in outcome might indicate that exercise must be intense and long term to enhance bone health.

Motor activity stimulates bone growth by triggering stress-responsive osteoblastic actions. As might be expected, then, bone mineralization should be enhanced as a result of short-burst, explosive activities such as skipping, stair climbing, and jumping (Kemper 2000). Non-weight-bearing activities that are designed to improve cardiovascular fitness (i.e., swimming, cycling) should not, therefore, be expected to optimize bone development. These observations have led to recommendations that any program of physical activity in children should be well-rounded in its nature, including weight-bearing activities for promoting bone development.

Strategies for Promoting Physical Activity in Youth

Definitive experimental proof that enhancing physical activity in children will provide health benefits is not at hand. However, the bulk of experimental and observational evidence outlined in this chapter provides convincing (albeit indirect) evidence of the benefits of activity interventions. How best to enhance physical activity in children is not known. From one standpoint, the direction of such interventions depends on what one sees as the most obvious salutary effect of activity in the pediatric years.

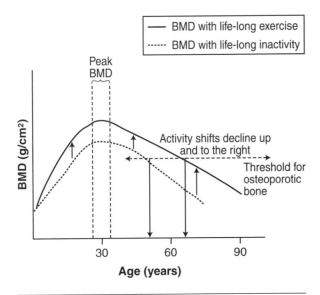

Figure 17.3 Weight-bearing exercise during childhood and adolescence may increase development of bone mineral density (BMD), shifting the curve of bone growth and limiting osteoporosis at older ages.

Reprinted, by permission, from H. Kemper, 2000, "Skeletal development during childhood and adolescence," *Pediatric exercise science* 12: 198-216.

One viewpoint is that retarding the development of atherosclerosis and hypertension is the most important effect of activity in children; thus, threshold amounts of activity need to be prescribed, based on data indicating a dose–effect relationship between amounts of activity and health risk factors in this age group. By an alternative viewpoint, physical activity should be promoted in children with the principal aim of establishing a lifestyle of regular exercise. Health benefits will then be gained as an adult by accrual of a lifetime of activity. Such an approach, which relies on behavioral modification rather than defining threshold amounts of activity for health, is contingent on the expectation that activity habits instituted early in life will persist, or track, to the adult years. Another possible strategy is to identify and modify extrinsic (i.e., environmental) factors that affect the downward slope of the biologically determined **physical activity curve** in children and adolescents. Each of these approaches is discussed next, recognizing that they are by no means mutually exclusive.

Strategy 1: Suppressing the Development of Long-Term Pathological Processes

The major impetus for promoting physical activity in youth rests on the assumption—intuitively attractive and indirectly supported by the data reviewed previously—that a physically active lifestyle in childhood and adolescence is likely to retard the development of lifelong pathological processes (atherosclerosis, obesity, hypertension, osteoporosis) and lessen the risk of their clinical expression in the adult years (figure 17.4). That is, physical activity during the adult years clearly diminishes risks of these disease outcomes, so the earlier in life that regular activity is instituted, the better.

How much physical activity do children need to gain these beneficial effects? According to strategy 1, there should exist a certain threshold of activity in children and adolescents that is necessary to retard pathological disease processes such as atherosclerosis and hypertension. The answer to this question lies in recognizing dose–response relationships between activity instituted and health outcomes measured. Unfortunately, there are virtually no such research data available by which one might establish an accurate necessary "dose" of exercise for youth.

There are other obstacles to identifying a particular exercise prescription for youth as well:

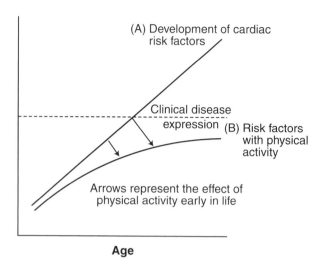

Figure 17.4 The expected development of cardiac risk factors *(a)* may be retarded by physical activity in the early years of life *(b)*, thereby limiting clinical disease expression in the adult years.

Reprinted, by permission, from O. Bar-Or and T.W. Rowland, 2004, *Pediatric exercise medicine* (Champaign, IL: Human Kinetics), 122.

- There is no means of actually assessing the effect of different levels of activity on these disease processes (i.e., pending a means of comfortably visualizing the coronary arteries, one cannot establish to what extent activity retards coronary atherosclerosis in youth). Likewise, the clinical outcomes of these pathological processes are dozens of years down the road and equally inaccessible to the researcher.

- The means by which activity provides benefits is presumably different for each of the various disease outcomes (e.g., caloric consumption for obesity, osteoblastic stimulation for bone health). It is not be unreasonable to assume that dose–effect relationships would vary for these different processes.

- Level of habitual activity declines steadily during the pediatric years. A single recommended "dose" of daily exercise for all age groups thus becomes problematic. Suggesting that a child perform moderate activity for 30 min a day might represent a 25% increase for a 5-year-old but a 50% increase in activity for a teenager.

- Subsequent surveillance studies will rely on guidelines as an indicator of children who are "sufficiently active." Any activity recommendation must be therefore be measurable.

These challenges notwithstanding, it has been deemed important from a public health promotion standpoint to take the best available data and provide population-wide guidelines for threshold amounts of physical activity in youth. Such recommendations have been published by several groups. Besides defining a minimal dose of physical activity, these guidelines have considered different aspects of activity behavior.

• *Amount, intensity, and frequency.* The International Consensus Conference on Physical Activity Guidelines for Adolescents (Sallis and Patrick 1994) recommended that all adolescents should be physically active daily, or nearly every day. It was also concluded that adolescents should engage in three or more sessions per week of activities that last 20 min or more at a time and that require moderate to vigorous levels of exertion. The guidelines from the Health Education Authority (HEA) in the United Kingdom recommended that all young people participate in physical activity of at least moderate intensity for 1 hr per day (Cavill et al. 2001). In Canada, considering that any current level of activity was insufficient, authorities called for inactive youth to increase the amount of time they currently spend being physically active by at least 30 min more per day (Health Canada 2004). Such an increase should consist of a combination of moderate activity (such as brisk walking, skating, and bike riding) and vigorous activity (such as running and playing soccer). Guidelines from the National Association for Sport and Physical Education (NASPE) called for children to accumulate at least 60 min, and up to several hours, of age-appropriate physical activity on all or most days of the week (Corbin and Pangrazi 2004).

• *Accumulation versus single dose.* There are insufficient scientific data to determine if activity-related health outcomes in youth are satisfied by a single dose of activity during the day or by accumulations of multiple shorter episodes. This question bears considerable relevance to children, who presumably are more likely to participate in brief rather than sustained periods of activity. As noted previously, the NASPE guidelines allowed for accumulation of the recommended amount of daily activity. The Canadian guidelines state that inactive children and youth accumulate the increase in daily activity in periods of at least 5 to 10 min each.

• *Progression.* In formulating guidelines, experts have recognized that it is unlikely that a markedly sedentary child is going to abruptly (or enthusiastically) increase activity level to suggested recommendations. The HEA guidelines accommodate this by providing the scaled-down recommendation that young people who currently do little activity should participate in physical activity of at least moderate intensity for at least half an hour a day.

• *Types of activities.* The various health outcomes associated with physical activity reflect responses to different forms of motor activity. Enhancing bone health requires weight-bearing activities. Lowering blood pressure occurs in the setting of endurance training programs. Reducing body fat in obese youth is probably best accomplished by sustained levels of low to moderate activity. For this reason, most of the guidelines specify that activities in youth should be well rounded. The HEA recommendations, for example, state that at least twice a week, some of these activities should help enhance and maintain muscular strength and flexibility and bone health.

• *Decreasing sedentary time.* Intuitively, reducing time a child spends watching television or at the computer should be a reasonable means of improving activity levels. Indeed, some research data indicate a link between time in sedentary pursuits, obesity, and physical activity. Other studies, however, have failed to incriminate sedentary time as affecting overall level of activity. There is no guarantee that a child forced away from the television set will replace this time with physical activity. The current evidence does suggest, in fact, that physical activity and inactivity may represent two separate behavioral constructs. These conflicting data notwithstanding, some guidelines have included a recommendation for limiting sedentary time. The NASPE guidelines indicate that extended periods (2 hr or more) of inactivity are discouraged for children, particularly during the daytime hours. The Canadian recommendations call for a decrease by at least 90 min per day in the amount of time spent on nonactive activities such as watching videos and sitting at a computer.

Strategy 2: Introducing Activity to Create Lifelong Habits

Children are the most active segment of the population, and most youngsters are probably engaging in sufficient physical activity for their health (Blair et al. 1989). However, a "significant minority" of children—presumed to be increasing—do

Limiting sedentary time in front of the television, video game, or computer may be an effective independent strategy for improving overall daily physical activity of children.

not regularly participate in adequate amounts of physical activity. Introducing these young people to activity as a means of creating a lifestyle pattern of regular activity that will carry over to the adult years makes for an attractive preventive health strategy.

This approach does not presuppose any specific desired amount of daily physical activity. Instead, persistence rather than target amounts of activity holds the key to achieving positive health outcomes. The challenge is to identify aspects of activity interventions for this at-risk group that will create an intrinsic desire to continue in a behavior that previously been "vigorously" avoided. Certain aspects of activity seem important in this regard—interventions are unlikely to be successful if they are com-

petitive or are not fun, and peer and family support, self-selection of activity, and charismatic counselors are important factors (figure 17.5). Whether more formal psychosocial constructs used for exercise promotion in adults (such as social cognitive theory and the transtheoretical model) are applicable to youth of different ages remains to be clarified.

The validity of this strategy would be strengthened if one could assume that the activity pattern of a child was, in fact, predictive of how active he or she will be at 40 years old. The few longitudinal studies examining the **tracking of activity levels** between late childhood and the early adult years have not been impressive. These studies have been hampered, however, by difficulties in accurately assessing activity levels at different ages.

Figure 17.5 Keys to motivating children to improve physical activity.

Reprinted, by permission, from T.W. Rowland, 1990, *Exercise and children's health* (Champaign, IL: Human Kinetics), 267.

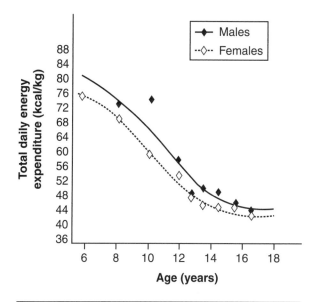

Figure 17.6 Physical activity levels decline with age during childhood.

Reprinted, by permission, from T.W. Rowland, 1990, *Exercise and children's health* (Champaign, IL: Human Kinetics), 35.

As would be expected, tracking correlation coefficients in these reports decrease as the length of follow-up increases (Malina 1996). Short-term studies (i.e., over 3-5 years) show low to moderate correlations (r = .30-.60), whereas longer investigations of the persistence of activity from adolescence to young adulthood indicate lower correlations (.05-.17). There is some evidence that sedentary habits track more closely over time compared with levels of physical activity.

These studies do not actually address the pertinent question: If a previously sedentary child is enticed to participate in regular physical activity, is it likely that this habit will persist over time? Such a study has yet to be performed.

Strategy 3: Shifting the Physical Activity Curve

Levels of daily physical activity decline as children age. This decrease in regular activity is not minor: Between the ages of 5 and 15 years, total daily energy expenditure related to body size is almost halved (figure 17.6). Although some have viewed this as an alarming indication of juvenile sloth, this pattern in fact reflects a normal biological process observed in animals and accompanied by parallel decreases in energy intake and basal metabolic rate.

Can this age–activity curve be shifted upward? Population-based epidemiologic data immediately suggest that this curve has at least some plasticity, because a number of environmental factors are associated with the amount of children's participation in physical activity (i.e., access to recreational facilities,

modeling by parents). Recent studies, as well, have identified influences that will alter activity over time among adolescents, shifting the age-activity curve upward—specifically, physical fitness, ethnicity, and cigarette smoking.

By this approach, then, the idea is not so much to increase activity levels of youth but rather to retard the normal physiological decay in activity levels over time. The goal is to identify extrinsic factors that will shift this falling curve upward and to the right. It may be important, too, to determine if there are certain critical ages when this curve is most susceptible to "bending."

Summary

The rationale for promoting physical activity in youth for present and future health is basically sound. There clearly exist, however, some major gaps in the scientific data to support such an approach. For instance, long-term longitudinal data and specific dose–response information on activity–risk factor outcomes are almost totally lacking. The ultimate study demonstrating that an active, physically fit child is less likely to suffer long-term complications of obesity (coronary artery disease, osteoporosis, and hypertension) as an adult will probably never be performed.

It can be argued that the data outlined here are sufficiently compelling and that absolute (and probably unattainable) proof is not necessary to justify promoting physical activity for health in children and adolescents. Certainly other public health initiatives have been launched with no stronger evidence base. As emphasized by Riddoch (1998) in assessing the activity–health link in youth, "It is likely that, at least in the foreseeable future, we must rely at least as much on theory, common sense, observation, and expert opinion as on hard evidence" (p. 38).

Having accepted such a rationale for activity in children, we face new challenges. By what means can children be motivated to improve their activity habits? Should interventions be population-based or focused on high-risk, sedentary youth? How can exercise interventions be formulated for age-specific groups? What is the role of organized sports—particularly for the nonathletic child—in these efforts?

There are a number of venues on which efforts to enhance activity levels in youth can be focused—home, school physical education class, physicians' offices, community recreation programs. The advantages of one over the other have not been well studied, and some approaches, such as exercise prescription by the pediatrician, remain full of promise but with little research attention.

KEY CONCEPTS

bone health: Health and density of bone. The rate of bone mineralization is highest in the early years of life. Optimization of bone development through weight-bearing activities may diminish loss of bone mineral density and subsequent osteoporosis in later life.

cardiovascular health: Health of the cardiovascular system. Because clinical outcomes of coronary artery disease, hypertension, and obesity are not manifest during childhood, cardiovascular health is defined in the younger age group by surrogate markers of health risk, such as blood pressure, body fat content, and serum lipid profile.

pediatric years: The period of growth and maturation to the adult state, which is divided into childhood (through age 12 years) and adolescence (12-18 years).

physical activity: For definition, see page 19.

physical activity curve: Curve relating physical activity to years of age. The level of daily energy expenditure through physical activity steadily declines during the growing years. This is primarily a biological process. The extent that external factors can alter the shape of this curve needs to be established.

physical fitness: For definition, see page 19.

sedentary pursuits: Time spent in inactivity, particularly watching television, playing video games, and working at the computer; these may or may not detract from total amount of daily physical activity.

tracking of activity levels: Persistence of activity habits. The rationale for promoting activity and fitness for health in children would be supported if tracking, or persistence, of activity habits as children grow to be adults could be documented. Present research data are inadequate to answer this question.

STUDY QUESTIONS

1. Discuss the advantages and disadvantages of school physical education, community recreation programs, family activity, and physician input in creating strategies for promoting physical activity in youth.

2. Identify extrinsic factors that are most likely to alter the biological decline in physical activity during childhood.

3. Define the separate and combined effects of physical activity and physical fitness in children and adolescence on health outcomes.

4. How might the age of the child or adolescent affect the type of intervention that would be effective in altering physical activity habits?

5. Differentiate the potential immediate versus long-term effects of activity and fitness on health in children.

6. Discuss the merits of focusing exercise interventions on high-risk, sedentary youth versus a population-wide approach for all children and adolescents.

Physical Activity, Fitness, and Aging

Loretta DiPietro, PhD, MPH

CHAPTER OUTLINE

Older age traditionally has been viewed as a time of inevitable disease and frailty. However, the current view of aging distinguishes true aging-related decline in function from decline that is secondary to other factors known to decline in older age—especially physical activity. Ample data now exist demonstrating that even the frailest members of the older population can respond favorably to exercise. Therefore, physical activity and fitness remain vitally important in older age with regard to maintaining a functional and independent lifestyle. Understanding the role of physical activity and fitness in modifying aging-related changes in health and function has important public health implications for meeting the needs of the ever-growing population of older adults.

The Aging Process

As advances in public health (sanitation, immunizations, improved nutrition) and health care are keeping people alive longer, the population worldwide is growing older. This aging trend has substantial political, social, medical, and economic implications. Therefore, we need understand the many ways in which the aging process alters human health and function and to distinguish between alterations in function that are reversible and those that are not.

Demographics of Aging

The population aged 65 years and older living in the United States numbers about 38 million and comprises approximately 14% of the population. Because of decreased mortality in older age groups, these **demographic trends** in aging will continue, with the population of older people expected to approach 71.5 million (20% of the total) by the year 2030 (Administration on Aging, U.S. Department of Health and Human Services [USDHHS] 2004). Perhaps of greatest interest in aging research is the increase in the "oldest old" segment of the population—those people 85 years and older. Since 1930, this oldest segment of the U.S. population has doubled in number every 30 years, and it is projected to be the fastest growing sector of the older population well into this century. For example, in 2002 there were approximately 4.6 million persons aged 85 years or older living in the United States, and this number will increase to approximately 9.6 million by the year 2030 (Administration on Aging, USDHHS 2004). The impact of these demographic changes on public health is substantial, particularly because emphasis has begun to shift from tertiary care toward health promotion and disease prevention.

Accompanying these demographic trends in aging is the increasing prevalence of chronic disease and consequent functional impairment. Indeed, more than 80% of older people have at least one chronic health problem such as cardiovascular disease, cancer, diabetes, osteoporosis, **sarcopenia,** or arthritis. Chronic health problems in older adults exact a markedly disproportionate toll on the U.S. economy. For example, despite accounting for about 14% of the population, Americans over age 65 years account for more than 30% of health care expenditures (Federal Interagency Forum on Aging Related Statistics 2004). Clearly, the public health benefits would be enormous if the onset of disease and functional limitations could be postponed or eliminated altogether. By delaying the onset of these chronic conditions, the maintenance of physical function can be extended to a time closer to the life expectancy. This compression of morbidity will undoubtedly improve quality of life and preserve autonomy for older people as well as reduce health care costs to the individual and society. Table 18.1 contains a list of prevalent chronic diseases and their risk factors among older persons that can be ameliorated with a regular exercise program.

> The compression of morbidity refers to the delay of chronic disease and frailty until the end of life or as close to the end of life as possible. Thus, this postponement of chronic disease would maintain physical function, autonomy, and quality of life among older people for a greater proportion of their life span.

Mandatory Versus Facultative Aging

To further accelerate this compression of morbidity, we need to improve our ability to distinguish between the aging-related decline that is mandatory and that which is facultative. Mandatory aging is that over which we have no control. In the absence of disease or injury, biological cells, systems, and organs undergo a process of irreversible decline. The underlying basis for mandatory (i.e., biological) aging has been debated for decades, and there are two general classes of hypotheses that attempt to define it, as shown in figure 18.1. Briefly, the first of these hypotheses proposes that random environmental events such as oxygen free-radical damage, somatic cell gene mutation, or cross-linkage among macromolecules render the cell incapable of functioning normally. Normal function would require the cell to transfer information from DNA to RNA to the synthesis of protein, thereby allowing the cell to contribute to tissue homeostasis. The consequence of this error buildup is that the genetic foundation of the cell is altered and the expression of essential protein either is limited or cannot proceed at all. Individual cell loss is not catastrophic to tissue or organ function until a significant complement of cells in the tissue or organ fail. Aging is therefore the consequence of a progressive accumulation of errors in the makeup of the cell attributable to the inability of the cell's repair processes to keep up. The second of the hypotheses proposes that the aging process is actively programmed by the cell's genetic machinery. In the case of programmed cell death, there is some valid evidence of the death of certain cell lines during development and maturation; however, the

Table 18.1 Chronic Diseases and Risk Factors in Older People That Can Be Ameliorated With Physical Activity

Chronic disease	Risk factor
Cardiovascular disease	Hypertension, dyslipidemia, obesity
Type 2 diabetes	Insulin resistance, glucose intolerance, dyslipidemia, obesity
Cancer	Obesity, bowel immotility, sex hormone profile
Osteoporosis	Low bone density
Physical disability	Sarcopenia, musculoskeletal weakness, poor balance, neuromuscular defects, arthritis

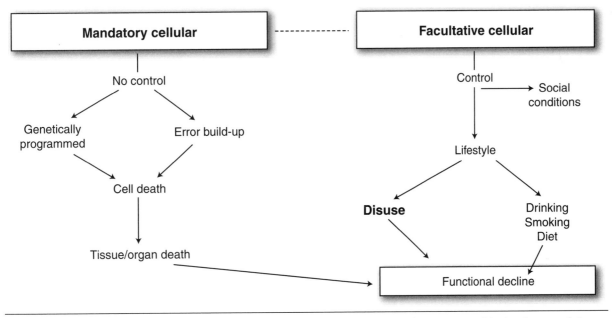

Figure 18.1 Hypothesized models of aging. Mandatory aging is that over which we have no control. In the absence of disease or injury, biological cells, systems, and organs undergo a process of irreversible decline. In contrast, facultative aging is that over which we do have control and comprises factors at the community, as well as at the individual, level.

relevance of this process to the aging of the organism has not been established.

Facultative aging, on the other hand, is that over which we do have control and comprises factors at the community (e.g., quality of health care) as well as individual (e.g., lifestyle) level. Physiological function and resiliency decline with aging—even among the most robust sectors of the older adult population. The degree to which this decline is attributable to true biological aging and the degree to which it is related to changes in social or lifestyle factors that also accompany older age—particularly physical activity or **disuse**—is a primary focus of this chapter.

In sum, the aging process traditionally has been viewed as an inevitable decline in health and function. Although many physiological functions are known to decline with age, the emerging view of the aging process distinguishes the decline in function and resiliency attributable to biological aging from that attributable to disuse. Bortz (1982) was an early proponent of the theory that inactivity causes much of the functional loss attributed to aging, from the cellular and molecular to tissue and organ systems. He noted that many of the physiological changes commonly ascribed to aging are similar to those induced by enforced inactivity, such as during prolonged bed rest or during space flight. He also proposed that the decline in function attributable to disuse could be attenuated, and perhaps reversed, by exercise and stated that this prospect holds much promise for what we now have termed successful aging. This attenuation in functional decline with exercise in older age is extremely important, because closing the "fitness gap" between active and inactive older people can prolong the time to the disability threshold—often independent of the actual improvements made in muscle strength, balance, or bone strength (figure 18.2).

> Successful aging refers to a resilient, disease-free, and highly functional state in older age. Bortz proposes that much of the functional decline commonly attributed to aging per se could be attenuated and even reversed by the reinstatement of regular physical activity.

Physiological Changes Occurring With Aging

As mentioned previously, normal physiological function begins to decline in advancing age as various systems become less pliant and less resilient to environmental stressors. One of the most noticeable and clinically relevant changes occurring with aging is the loss of muscle mass. Longitudinal evidence suggests that during older age, muscle mass decreases about 3% to 6% per decade. This loss of muscle mass, with the accompanied increase in the

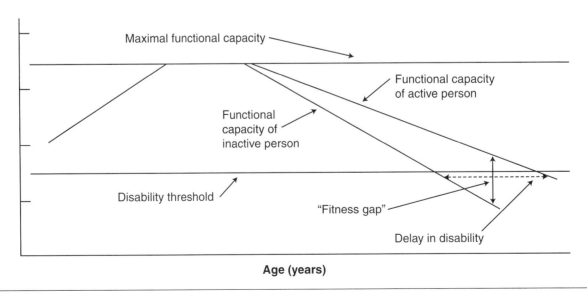

Figure 18.2 The fitness gap and disability threshold. The attenuation in functional decline with exercise in older age would contribute to closing the "fitness gap" between active and inactive older people, thereby prolonging the time to the disability threshold.

proportion of body fat, has negative consequences for maintaining resting metabolic rate and metabolic resiliency as well as for maintaining reaction time, strength, flexibility, and balance—all important variables for maintaining an active and independent lifestyle in older age. Another important functional change accompanying older age is the loss of cardiovascular **plasticity** resulting in a decline in maximal heart rate, stroke volume, cardiac output, and arteriovenous oxygen difference (American College of Sports Medicine 1998; Lakatta 1993). These decrements in cardiorespiratory function (along with the loss of lean body mass) contribute substantially to the age-associated decline in maximal aerobic capacity—an important physiologic indicator of functional capacity in older age. Indeed, maximal oxygen consumption ($\dot{V}O_2$max) declines between 5% and 15% per decade after age 25 (Dempsey and Seals 1995; Heath et al. 1981).

Accompanying the decline in the aforementioned physical capabilities is a decline in sensory function. Defects in vision and hearing may increase the psychological stress associated with common daily interactions with the physical and social environment (e.g., walking to the store, going to the senior center). As older people feel less confident in their abilities to venture safely outside the home, these interactions with the environment will decrease, thereby compounding the cycle of disuse and functional decline.

Although a genetic predisposition toward the loss of muscle tissue and cardiovascular plasticity is important, disuse is also a significant contributor. For instance, many of these physiological changes are similar to transient responses observed after long periods of bed rest or space flight. The common element among these three conditions (aging, bed rest, and spaceflight) is microgravity, and disuse (sedentary behavior) is a form of microgravity, if not hypogravity. However, bed rest and space flight are of short duration relative to the life span of humans, and in the case of these two conditions, decrements in function are usually reversible within short periods of time. This is not the case with older people who may have been inactive for the majority of their middle age and older age. Indeed, the effect of disuse on function will vary by age, with greater impairment in older age attributable to less resiliency and longer duration of disuse. This has important implications with regard to the doses of exercise necessary to reverse existing functional defects (e.g., insulin resistance) in older people.

Nonetheless, ample data have demonstrated the benefits of physical activity in delaying and substantially attenuating aging-related functional decline, even among the least robust members of the older population. Available evidence (see ACSM 1998; DiPietro 2001 for reviews) indicates that older people (even those in their 9th and 10th decades) can respond favorably to both endurance and strength training. Endurance training can maintain and even improve several indicators of cardiorespiratory function—most notably, submaximal work performance, which accounts for the majority of day-to-day physical activity. Strength training is very effective in modulating the loss of muscle mass and its accompanying decline in muscle strength and metabolic function. In addition, both endurance and strength training have demonstrated their effectiveness in improving bone health, postural stability, flexibility, and, in some cases, depressive symptoms and cognitive function in older people (ACSM 1998). Such changes have important consequences for delaying or reducing the risk associated with heart disease, diabetes, osteoporosis, and falling and for increasing life expectancy and quality of life in older age. Thus, it appears that it is never too late in life to achieve the benefits of increased physical activity.

Methodological Considerations in Aging Research

The ability to observe an etiologic relationship between physical activity and human function in aging is dependent on our ability to measure these factors with **accuracy.** This section describes several of the methodological problems and concerns inherent in studying physical activity behaviors in older populations.

Assessment Issues

As noted previously, exercise and other forms of physical activity are known to provide a myriad of specific physiological and psychosocial benefits to older people. Although data from intervention studies demonstrate the effect of more moderate and vigorous aerobic or strength training on improvements in physiological function in older people, the benefits of lower-intensity activity, such as that performed as part of an active lifestyle, are less clear. This is attributable, in part, to the difficulty inherent in assessing habitual activity in older people. Physical activity in older age tends to be unstructured, of low intensity, and highly variable. In addition, the issue of recall of such activity patterns in older people leads to less than accurate estimates. Problems in the definition and measurement of physical activity limit the ability to assess it properly and therefore to determine the health consequences associated with an active lifestyle.

Because many activity behaviors common to very old people are the same behaviors used to assess mobility (walking, housework, climbing stairs), often the assessments of behavior and performance are confused. For instance, questions of "How often *do* you walk half a mile?" become confused with "*Can* you walk half a mile?" Because these activities relate to ordinary life, they are measured in units of activities of daily living (ADLs), instrumental activities of daily living (IADLs), and advanced activities of daily living (AADLs). Measuring performance specifically using the standard fitness tests designed for younger populations (e.g., $\dot{V}O_2$max) may not be appropriate in the very old and so, more recently, objective, performance-based measures have been developed that are very effective in discriminating among a wide range of physical function abilities in older people. The combination of these objective *performance-based* measures with self-reported measures of functional ability (i.e., ADLs, IADLs, and mobility) and self-reported *activity behaviors* (typical of the older person's lifestyle) offers the best opportunity to capture the physical activity behaviors that best relate to health and function in aging (figure 18.3).

> Functional ability refers to the ability to perform activities of daily living (ADLs: bathing, grooming, dressing, eating), instrumental activities of daily living (IADLs: shopping, cooking, housework), and advanced activities of daily living (AADLs: volunteer work, recreational activity), and mobility (stair climbing, walking). In large population-based studies, functional ability typically is assessed by self-report.

Future physical activity assessment for older people needs to focus on separate age categories within older age—that is, age groupings of 65 to 75 years, 75 to 85 years, and the fastest growing segment of the U.S. population, those people ≥85 years. Given the high prevalence of walking and other even less structured lower-intensity activities in older people, we need better validation measures for establishing the internal validity of physical activity assessment survey questions. The combination of field-based with laboratory-based validation measures is necessary to advance epidemiological research in this area. Also necessary in all physical activity assessments in older populations is the inclusion of physical function items, both self-reported measures of functional ability (such as ADLs, IADLs, AADLs, and mobility) and objective performance-based tasks (timed walk, chair stands, balance). Finally, multiple assessments of physical activity, as well as the health indicator of interest, are necessary to determine the true etiological relationship between being active and being healthy

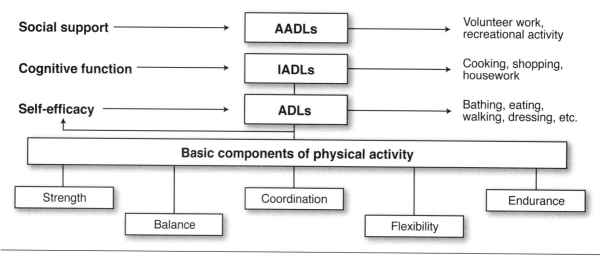

Figure 18.3   Relationship between physical activity and physical function.

over the life span. Newer, longitudinal statistical techniques are now available to model, for example, the influence of changes in physical activity patterns on the trajectory of change in physical function or body weight through older age.

Study Design Issues

Both cross-sectional and longitudinal experimental approaches have been used to study the influence of aging on human function at rest and in response to acute and chronic exercise. Cross-sectional approaches are most common and involve comparisons of different physiological variables among different age groups (say, 20-25 years vs. 40-45 years vs. 65+ years). Longitudinal studies, which make serial measurements within the same subjects over time (e.g., 10-20 years), are less common but give more valuable information with regard to aging-related changes. For instance, in cross-sectional studies, physiological variables among older trained (master) athletes are often compared with those of older sedentary control subjects, whereas in longitudinal designs, functional adaptations to a period of exercise training are determined in older sedentary subjects. In general, the observed positive associations between exercise and physiologic function are much stronger in cross-sectional than in longitudinal studies. Indeed, previously sedentary older people who undergo training are unable to achieve the same high level of physiological adaptation as their athletic counterparts. This is presumably attributable to the genetic advantages of the older athlete as well as the athlete's ability to exercise regularly at much greater intensities and durations than is possible for the untrained older person, who has a much lower physical working capacity. Thus, data from cross-sectional studies should be viewed cautiously, because the influence of regular activity on health and function may be exaggerated based on such data.

Selective survival refers to a select sample of older people who have survived the putative effects of various risk factors in middle age and thus present a more robust physiological profile compared with the general population. Issues of selective survival often affect our ability to generalize results from specific aging studies to the general population of older people or to the population at large.

Finally, selective survival may influence the variability of adaptation to exercise training both within older populations and between younger and older adults. The individuals most susceptible to putative risk factors and who may benefit most from training die at earlier ages. Thus, research results may be biased because they are derived from studies of the effects of exercise training on the more robust surviving cohort.

Demographics of Physical Activity Among Older Adults

Physical activity behavior is not uniform across all age sectors of the population, nor is it particularly stable over time. This section describes the

- activity patterns and choices of older people,
- age differences in activity patterns between younger and older people,
- cross-sectional trends in activity patterns, and
- longitudinal trends in physical activity patterns.

Cross-Sectional Patterns

The *Healthy People 2010* objectives state the goal "to increase to at least 30% the proportion of people aged 18 and older who engage in moderate physical activity for at least 30 min/day." Approximately 15% of the U.S. adult population met this goal in 1999. Cross-sectional studies consistently report that the prevalence of reported inactivity increases with age and is especially evident among older women. Behavioral Risk Factor Surveillance System (BRFSS) survey data suggest that more than 40% of U.S. women aged 65 years and older reported no leisure-time physical activity in 1992 (Centers for Disease Control 1995); this prevalence was 51% in 1999. Indeed, most population-based studies in the United States as well as in other countries report an inverse relationship between weekly energy expenditure from physical activity and age.

Interestingly, some national surveys in the United States report a higher prevalence of vigorous activity among older age groups. This presumably is attributable to the manner in which vigorous activity is defined. The BRFSS survey defines vigorous activity as that which is performed at >60% of maximal aerobic capacity. Because $\dot{V}O_2$max decreases with age, activities of moderate intensity for middle-aged persons (say, brisk walking) may be performed at a greater percentage of V̇O₂max in older people and, therefore, will meet the criterion for "vigorous." These differences in physical activity classification need to be carefully considered in comparing results from different surveys.

Walking is the most prevalent activity reported among adults of all socioeconomic strata in the United States, Canada, Latin America, and Europe (figure 18.4). The most prevalent activities among older people tend to be lower-intensity but sustained activities, such as walking, gardening or yard work, bicycling, and golf. Data from the 1991-1992 National Health Interview Survey–Health Promotion Disease Prevention Supplement (NHIS–HPDP) show cross-sectional patterns in strengthening and stretching activity by age group (Caspersen et al. 2000). Although the prevalence of strengthening activity was markedly lower among older compared with the youngest age group, the prevalence of stretching activities remained fairly stable across age groups. These data have important clinical relevance for older people among whom the loss of muscle strength and flexibility contributes significantly to the decline in physical functioning and mobility.

Cross-Sectional Trends

Cross-sectional trend data rely on information from different representative populations over two or more time intervals. As such, they provide valuable information on temporal trends in physical activity patterns. These data are often referred to as **surveillance data.** Surveillance data from the United States suggest that the prevalence of inactivity (i.e., no reported leisure-time physical activities in the past month) decreased between 1986 and 1990, although this decline was not apparent among nonwhite adults or adults of lower educational attainment. When the data were stratified further by age groups, surprisingly they showed that this overall decline in reported inactivity was explained primarily by reduced inactivity among people more than 55 years old. More recent data from the NHIS–HPDP show a small yet consistent increase in reported physical activity of any kind among older persons living in the United States between 1985 and 1995 (National Center for Health Statistics 2003). These data also show a general increase in reported walking over these 10 years and stable trends in reported gardening activity (figure 18.5). Although these surveillance data are important for understanding temporal changes, they do not give

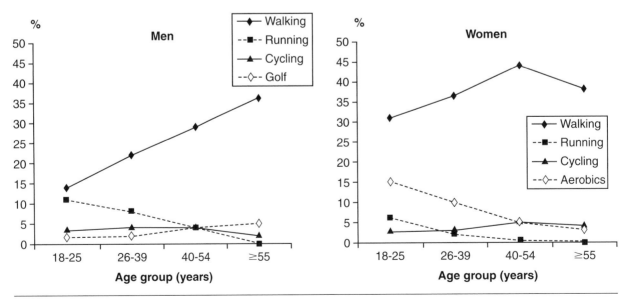

Figure 18.4 Prevalence of popular leisure-time activities by age and sex among persons trying to lose weight.
From Centers for Disease Control and Prevention (CDC), 2003.

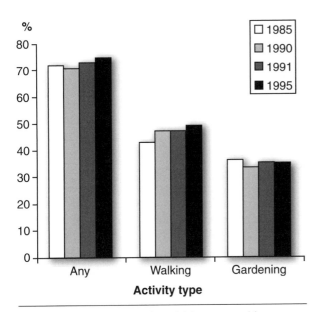

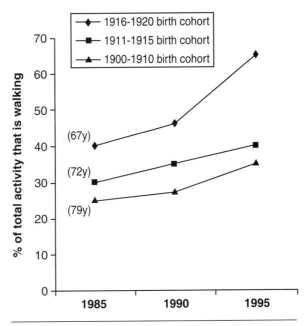

Figure 18.5 Participation in activities among older persons living in the United States. Data are from the National Health Interview Survey–Health Promotion Disease Prevention supplement, 1985-1995.

From Centers for Disease Control and Prevention (CDC), 2003.

Figure 18.6 Age and cohort trends in walking among older men: The Zutphen Elderly Study, 1985-1995.

Reprinted from F.C.H. Bijnen, E.J.M. Feskens, C.J. Caspersen, W.L. Mosterd, D. Kromhout, 1998, "Age, period, and cohort effects on physical activity among elderly men during 10 years of follow-up: The Zutphen Elderly Study," *J Gerontol Series A: Biol Med Sci.* 53: M235-M241. Copyright © The Gerontological Society of America. Reproduced by permission of the publisher.

information about aging-related changes among a single population over time. Thus, multiple assessments in the same population of older people over many years have added tremendously to the study of successful aging.

Longitudinal Trends

Longitudinal trend data make serial measurements on a population and thus can provide information on aging- and time-related changes in activity patterns as well as differences in these trends by birth cohort. Longitudinal data from several countries suggest that although total activity and common activities of gardening and bicycling decline with aging among older people, time spent walking remains relatively stable. For instance, in the Zutphen Elderly Study, walking became an increasing proportion of total activity over 10 years, and this was especially evident in the youngest birth cohort (born between 1916 and 1920) between 1990 and 1995 (Bijnen et al. 1998) (figure 18.6).

Retirement is an important time with regard to potential changes in the physical activity pattern. Not surprisingly, data from the Atherosclerosis Risk in Communities Study suggested that the 6-year odds of adopting or maintaining physical activity were significantly higher among older people who retired at follow-up relative to those who continued

working (Evenson et al. 2002). Moreover, these data suggested a significant effect modification in these odds with regard to race and sex. Older black persons who were already active had a higher odds of maintaining that activity compared with older white persons, and this was especially so among older black women. Older black men had the highest odds of adopting an activity on retirement, whereas older black women had the lowest odds, which was no different than their counterparts who were still working (figure 18.7). Favorite activities adopted on retirement among older people of both races and sexes were walking briskly, walking for pleasure, gardening or yard work, floor exercises, and exercise cycling.

In sum, there is encouraging although limited evidence from cross-sectional surveillance data that the prevalence of reported inactivity is decreasing over time among some sectors (namely, older adults) of the general U.S population, with an increased prevalence of regular leisure-time activity. However, leisure-time physical activity comprises only a portion of total daily activity. Other components involve work or household activity and transportation. Although few surveillance data are available for these

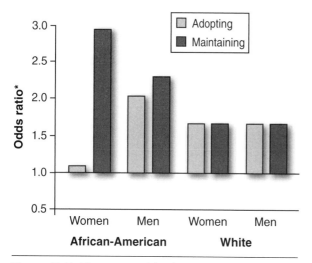

Figure 18.7 The 6-year odds of maintaining or adopting activity among workers who retired at follow-up: The Atherosclerosis Risk in Communities Study.

Adapted from K.R. Evenson, W.D. Rosamond, J. Cai, A.V. Diez-Roux, F.L. Brancati, 2002, "Influence of retirement on leisure-time physical activity." *American Journal of Epidemiology* 155: 692-699, by permission of Oxford University Press.

other components, one can reasonably assume that energy spent in work and household tasks, as well as in transportation, has progressively declined over the years with increasing automation and labor-saving devices and increased use of the Internet. This time-related decline in daily energy expenditure is further accelerated by aging-related declines in physical activity. It therefore is possible that overall total daily energy expenditure from physical activity has declined substantially among our older population despite small increases in their leisure-time activity. Nonetheless, both cross-sectional and longitudinal trend data demonstrate that even though daily energy expenditure from physical activity declines with aging, walking remains an important activity in older age. Because walking is weight bearing, uses large muscle groups, can be sustained, and, if performed regularly, can improve maximal and submaximal physical functioning, its merits should be promoted among older populations.

Dimensions of Physical Activity and Their Relationship to Health and Function in Aging

The relation between physical activity and health is well established. It is important to understand, however, the different aspects of physical activity (and of disuse) that relate in a specific manner to health and function, and how the relative importance of these characteristics changes with age.

Consequences of Sedentary Behavior

Disuse and a sedentary lifestyle are especially detrimental to older people. For example, low levels of physical activity and cardiorespiratory fitness are primary determinants of the decline in metabolic and functional reserve observed in older age. Sedentary behavior has consistently demonstrated a relationship with the development of chronic disease, premature mortality, poor quality of life, and loss of function and independence with aging. Sedentary behavior accumulated over a lifetime may be the risk condition with the biggest global public health impact faced by older people. Indeed, population risk estimates for heart disease attributed to low activity or low fitness levels are comparable, or even greater, in middle-aged and older people than are other well-known predictors of morbidity and mortality, such as hypertension, smoking, dyslipidemia, and obesity.

Given the high prevalence of sedentary behavior, especially among older individuals, and the strong association of low activity or low fitness to many chronic diseases of aging, the direct costs of a lifetime of inactivity among Americans may be as high as $24 billion per year (Colditz 1999). This figure, however, does not account for the economic costs associated with frailty and loss of physical function among older people, and so it actually may underestimate the true economic burden of a sedentary lifestyle.

Physical activity and fitness have been associated with a lower incidence of morbidity and mortality from a number of major chronic diseases affecting older people, namely, coronary heart disease, cancer, and type 2 diabetes, as well as with a lower incidence of most of the risk factors preceding these diseases, such as hypertension, dyslipidemia, and insulin resistance. Physical activity has also demonstrated a significant protective effect on the risk of bone loss, hip fracture, and factors associated with falls as well as on the rate of functional decline so common with aging. There is evidence to suggest that recent or current activity is more protective than past activity; however, cumulative, lifetime activity patterns may be a more influential factor for most of these diseases, especially those with a long developmental (latency) period, such as cancer, bone integrity, or weight maintenance. These etiological relationships are discussed in greater detail in other chapters.

© Marlene Karas Cornman

These older athletes, although they run a greater risk of strains and sprains, are significantly less likely than their sedentary peers to be subject to coronary heart disease, cancer, type 2 diabetes, and osteoporosis.

The inverse association between physical activity and disease or functional decline is consistently strong, graded, and independent and is biologically plausible and specific, thereby meeting most criteria for inferring a causal relationship. The challenge often encountered in epidemiological research in aging—especially when relying on self-reported measures of physical activity—is that of establishing temporal sequencing (i.e., does sedentary behavior truly precede the onset of functional decline?).

Characteristics of Physical Activity Affecting Health

It is difficult to determine the characteristics of physical activity related most specifically to different aspects of health, because there are several dimensions of physical activity behaviors. These dimensions, which are exclusive neither to any one type of activity nor to each other, include aerobic intensity, energy expenditure, weight bearing, flexibility, and muscle strength. Certainly, these dimensions are interrelated because, for example,

activities that increase aerobic capacity also require energy expenditure. Any sustained weight-bearing activity will also expend energy and if done vigorously enough will increase aerobic capacity.

The relative importance and specificity of each of these dimensions to health change with age. For example, among adolescent girls and younger women, dimensions of physical activity related to muscle and bone growth may be of primary interest, whereas among middle-aged people, the influence of energy expenditure on weight regulation or of aerobic intensity on cardiovascular health becomes more important. Among older adults, the influence of weight-bearing, strength, and flexibility aspects of activity on bone and lean mass preservation and balance assumes highest priority with regard to maintaining functional ability and independence.

In the past, it was proposed that exercise of sufficient frequency, intensity, and duration to improve physical fitness was necessary to promote resistance to disease. Although the effects of physical activity on health status may be mediated primarily by the physiological changes that accompany increased fitness, recent data suggest that such changes may have an independent positive impact on several health indicators in aging. A growing epidemiological literature shows significant relationships between low- and moderate-intensity activity and overall health and longevity in aging (see ACSM 1998; DiPietro 2001 for reviews).

The exercise prescription or plan (i.e., frequency, duration, intensity) optimal in achieving one type of health outcome (e.g., weight loss) may be quite different from that necessary to achieve another (increased bone mineral content or muscular strength). Again, age may be an important variable in modifying the effects of a given exercise stimulus on health and functioning. That is, the amount of exercise related to disease-specific morbidity in middle age may be very different than that related to successful aging and overall longevity. Disease status may also be an important effect modifier in that the volume (frequency × duration × intensity) of physical activity necessary to prevent functional decline or to maintain health and function may be lower than the amount needed to reverse an already-established condition such as sarcopenia or insulin resistance (table 18.2). Thus, physical activity may be far more effective with regard to prevention than with regard to therapy. Nonetheless, physical activity and exercise have demonstrated effectiveness in improving function across the aging spectrum—from the very healthy to the most frail—and should be encouraged among all older people.

In summary, there is limited information on the amount of physical activity needed to promote optimal health and function in older age. Prescribing levels of exercise necessary to achieve aerobic fitness may no longer be appropriate in older age; rather, public health policy should focus on sufficient levels when promoting physical activity among more sedentary older or frailer sectors of the general population. Given that health benefits may accrue independently of the fitness effects achieved through sustained vigorous activity, a new lifestyle approach should be emphasized from childhood through older adulthood. This approach calls for incorporating at least 30 min of any activity of moderate intensity (sustained or accumulated) into the daily schedule (USDHHS 1996). Thus, regular participation in activities of moderate intensity (such as walking, climbing stairs, biking, yard work or gardening), which increase accumulated daily or weekly energy expenditure and maintain muscular strength, but which may not be sufficient intensity for improving aerobic fitness, should be encouraged in the community. Because the greatest gains in protection from disease and functional decline continue to be observed with increases in moderate-intensity activity or fitness levels, public health efforts geared toward the most sedentary members of the middle-aged community would most likely offer the most benefits with regard to the prevention of premature morbidity and mortality in later years.

Programmatic Issues in Promoting Physical Activity in Older Populations

Translating scientific knowledge into practical interventions that benefit the health of the community is difficult. A number of factors must be identified and managed before the public health potential of physical activity for older persons can be fulfilled.

Determinants

Older people report a number of reasons for not exercising, including lack of time, fear of injury, caregiving duties, lack of energy, lack of a safe place to exercise, and concerns about appearance.

Table 18.2 Public Health Strategy for Preventing Functional and Metabolic Decline

Level	State	Intervention
Primary	Normal function	30 min or more of moderate intensity activity on most days
Secondary	Compromised function	?
Tertiary	Diseased state	??

? = unknown.

Reprinted, by permission, from J. Dziura and L. DiPietro, 2003, The importance of body weight management in successful aging. In *Obesity, etiology, assessment, treatment and prevention*, edited by R.E. Anderson (Champaign, IL: Human Kinetics), 141-153.

Thus, an array of physiological, psychosocial, and environmental factors may determine physical activity behavior throughout the life span, and these factors become even more important in older age. Physiological factors include aerobic capacity, speed, strength, balance, and flexibility. Because these physiological attributes tend to decline with aging, they become strong determinants of activity level and choice of activity, because older people will participate in activities at which they feel competent and in which they feel safe. Psychosocial determinants include personality, knowledge and beliefs about the health effects of exercise, and social support from family and friends. Safety and accessibility are the two most important environmental factors affecting activity participation across the life span. Although these environmental factors have not been studied extensively, they are becoming quite important as public health efforts are shifting toward modifying the built environment.

Many of these determinants, particularly some of the psychosocial and environmental factors, are particularly amenable to change and should be the focus of community intervention efforts. Strategies for increasing physical activity among the older sectors of the community include

- increased public education about the health effects of moderate-intensity physical activity;
- increased senior center and community center programs that are supervised and provide social support and other incentives for exercise; and
- increased community availability and accessibility of safe physical activity and recreational facilities such as hiking, biking, and fitness trails; public swimming pools; and acres of park space.

Setting

Preferences for where to exercise may vary in older people, but there is evidence that individual home-based programs are preferable to group- or class-based formats. Indeed, data from the Stanford–Sunnyvale Health Improvement Project show significantly greater adherence over 2 years to a higher-intensity home-based program compared with a lower-intensity home-based and a higher-intensity group-based program (King et al. 1997). Although differences in adherence between the two

home-based programs became apparent only in the second year, adherence was substantially higher in both home-based programs compared with the group-based from the start of the project. These data suggest that programs that individuals can perform on their own, whenever they want, may be the most effective types of exercise programs to promote for older people. However, social support for the home-based groups in the Stanford–Sunnyvale project was provided through telephone calls from the research staff, and it is not clear how well home-based exercise works for older people who are not part of a formal research study and don't receive such support.

Programs offered through community settings can be appealing to older people who want to "get out" but may not be willing to leave their familiar surroundings to participate in activity or who may live in environments that are not safe for exercise. Churches, shopping malls, neighborhood recreation or senior centers, and retirement homes are examples of community settings in which physical activity programs can be implemented. Data from community-based interventions suggest that walking programs are the most popular with older adult populations; however, tai chi, yoga, and strength programs are gaining in popularity. Given the multiple components involved in maintaining physical function and mobility in older age, programs that combine the basic elements of aerobic endurance, strength, flexibility, balance, and coordination through a variety of activities (e.g., walking, tai chi, Thera-Bands, juggling, yoga) may be the most efficacious programs for promoting and maintaining the level of activity both necessary and sufficient for achieving the health benefits related to mobility and functional ability in older age. Research that integrates basic and applied physiological science with the social sciences is necessary to understand how improvements in basic components of function (e.g., strength, coordination, balance, endurance, and flexibility) translate into higher-order functions of IADLs (shopping, cooking, housework) and advanced activities of daily living (AADLs) (volunteer work and recreational activity).

Environmental Versus Individual Approaches to Promoting Physical Activity

Environmental interventions that remove barriers to activity at the community level have a greater public health impact than attempts to change barriers at the individual level. Many of the most innovative and successful health interventions have little to do with health per se but rather with altering the barriers that prevent people from gaining control over the determinants of their own health. Public policy that ensures equal and safe access to public spaces for walking, biking, and other forms of recreation is an example of such an environmental strategy. As with any environmental action, an informed and active public takes the lead in influencing the environmental structural changes. Similarly, a number of community agencies can work together to ensure a health-promoting environment for older people.

In addition to involving physicians, the fitness industry, and medical agencies like the visiting nurses associations, collaboration can come from governmental bodies such as state school boards and local health departments, parks and recreation departments, and such nontraditional partners as departments of urban planning and transportation. These groups should cooperate in setting physical activity standards for the entire community (figure 18.8).

> Environmental interventions are those performed at the community level rather than at the individual level. These types of intervention have the ability to affect everyone equally and target risk conditions rather than individual risk factors or behaviors.

Summary

Overall physiological function is known to decline with aging; however, the emerging view of the aging process distinguishes the decline in function and resiliency attributable to biological aging from that attributable to disuse. Low levels of physical activity and fitness are primary determinants of the decline

Physical Activity Promotion

Figure 18.8   Interagency involvement in physical activity promotion.

in metabolic and functional reserve observed in older age, and therefore accumulated lifetime sedentary behavior among older adults is a risk condition of substantial public health importance. There is now substantial evidence demonstrating the benefits of physical activity in attenuating aging-related functional decline, even among older people with established chronic disease and frailty. Therefore, because health benefits can accrue independently of the fitness effects achieved through sustained vigorous activity, a lifestyle approach of incorporating at least 30 min of any moderate-intensity activity into the daily schedule should be encouraged throughout the life span. Population-based data from several countries demonstrate that walking is an important activity through older age, and therefore its merits should be promoted among older people. Healthy public policy that incorporates education and access to safe opportunities for exercise at home and in the community should be a priority among community agencies working together to ensure a healthy lifestyle for all.

KEY TERMS

accuracy: The combination of validity (hitting the mark) and precision (low variability) in measuring a study variable.

demographic trend: A consistent pattern in a given population characteristic.

disuse: Sedentary behavior or physical inactivity.

facultative aging: Components of the aging process over which we have control and that are potentially reversible.

plasticity: Flexibility in function.

sarcopenia: For definition, see page 229.

surveillance data: Regularly collected data that report temporal changes in population characteristics.

STUDY QUESTIONS

1. Describe several differences between mandatory and facultative aging.

2. Plot a curve describing an age-associated decline in maximal oxygen consumption and discuss briefly the implications of this decline on health and function in older age.

3. List three problems in assessing physical activity accurately in older people.

4. Describe how selective survival in an older study sample may limit the ability to observe large improvements in function with exercise training.

5. Differentiate between temporal trends and aging-related trends in walking behaviors.

6. Describe the dimensions of physical activity and their relative importance in an older population.

7. List several determinants of physical activity among older people and identify those that are most amenable to change at the individual or community level.

8. Define and describe an environmental intervention to promote physical activity in older people.

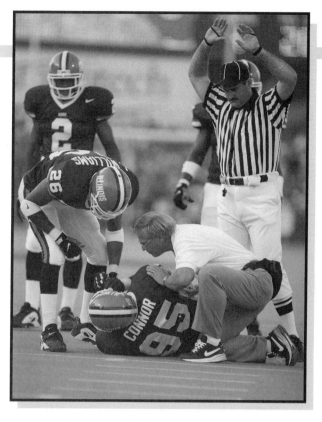

Risks of Physical Activity

■ Evert A.L.M. Verhagen, PhD ■ Esther M.F. van Sluijs, PhD ■ Willem van Mechelen MD, PhD

CHAPTER OUTLINE

Including daily physical activity in one's life provides a number of health benefits, but such participation, especially in vigorous exercise or sports, can result in injury, disability, or death. This chapter focuses on the risk of sudden cardiac death and the female triad, which are important risks in terms of consequences, and on musculoskeletal injuries and asthma, which are important in terms of frequency of occurrence. After discussing the risks and causes, we consider strategies for preventing these risks. After all, the benefits of physical activity need to outweigh the risks in order to maintain joyful and healthy physical activity behavior.

Studies performed over the past 25 years confirm the health benefits of regular physical activity, a concept with foundations in antiquity. The effects of physical activity on certain individual health conditions, the precise dose (intensity or amount) of activity that is required for specific benefits, and the elucidation of biological pathways through which physical activity contributes to health are topics discussed elsewhere in this text. Although numerous details of these topics remain to be clarified by research, it is now clear that regular physical activity reduces the risk of morbidity and mortality from several chronic diseases. It also increases physical fitness, which improves function, physical independence, and quality of life.

In addition to enhancing health, participation in physical activity and, more precisely, vigorous exercise or sports also carries significant risks. These risks can be biomechanical (e.g., injury to various tissues or organs), cardiovascular (ranging from discomfort or pain such as angina pectoris to transient risk of sudden cardiac death), respiratory (asthma or **anaphylaxis**), heat related (heat stroke), or combined (e.g., the female triad, which is the interrelationship among eating disorders, **amenorrhea**, and **osteoporosis** in the female athlete). The occurrence of any serious health problem is very low in the general population who exercise at moderate intensity and in amounts intended to improve health and physical performance.

In a population of athletes or others who exercise very vigorously, the chances of an injury increase with increasing intensity and amount of exercise. Sudden cardiac death, for instance, has a low incidence of one cardiac arrest per 20,000 exercisers per year in the general population. However, the risk of sudden cardiac death during vigorous exercise is 5 to 56 times greater than during usual activities of daily living. The actual prevalence of the female triad is unknown in both the general and the athletic populations. Yet data on eating disorders in the female athlete population suggest the existence of a significant medical problem. The prevalence of anorexia nervosa and bulimia nervosa is as high as 4% to 39% in a female athlete population, whereas it is about 0.5% to 5% in the general population. Taking the risks of physical activity into account, individuals might adopt a "decision-balance" approach in deciding whether it is worthwhile to continue with the same activity level or to become less or more active. Risk is therefore a perception that may partially guide an individual's physical activity behavior.

> The female triad is a combination of three interrelated conditions that can be associated with athletic training and competition and result in a significant health risk to female athletes: disordered eating, amenorrhea, and osteoporosis.

Risks of Physical Activity and Sport Participation

In the new public health move toward greater physical activity, low- to moderate-intensity physical activities are promoted to reduce the health risks in otherwise sedentary people. Just as the health benefits from physical activity seem to increase with an increase in physical activity amount and intensity, so do the risks, as depicted in figure 19.1. When talking about physical activity and health benefits, one often means moderate-intensity daily activities (e.g., gardening, brisk walking, or cycling). It should be obvious that the risks of such low- to moderate-intensity physical activities are relatively low. However, more vigorous physical activity and, in particular, sport participation present a greater risk whether the individual participant is an elite athlete or a recreational athlete.

Where people are encouraged to engage in physical activity for its many health benefits, attempts to reduce the inherent risks of more vigorous activities

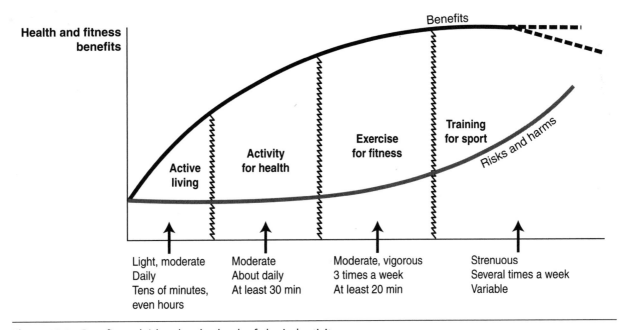

Figure 19.1   Benefits and risks related to levels of physical activity.

From van Sluijs, Verhagen, van der Beek, van Poppel, van Mechelen, 2003, Risks of physical activity. In *Perspectives on health and exercise*, edited by McKenna and Riddoch (United Kingdom: Palgrave MacMillan), 112. Reproduced with permission from Palgrave MacMillan; used with permission from Ilkka Vuori.

need to be considered. The transient risk of sudden cardiac arrest is one of the most hazardous, direct complications of vigorous exercise. Although highly trained athletes are at increased risk during vigorous exercise, those at greatest risk are the physically untrained who infrequently engage in vigorous physical activity. In some cases, vigorous exercise or sport participation may even cause chronic health problems, such as the female triad. Finally, physical activity and sport injuries can cause significant discomfort and disability, can reduce productivity, and are responsible for substantial medical expense and work absenteeism. Before providing suggestions for risk prevention, we address the magnitude of selected physical activity risks.

Sudden Cardiac Arrest

One of the most serious hazards of vigorous exercise is the transient risk of sudden cardiac death during or soon after the exercise, which raises concerns regarding the safety of vigorous exercise for many middle-aged or older persons. The number of sudden deaths during moderate-intensity exercise is small, 0.35% to 0.5% of all sudden deaths in autopsy materials and less than one death per one million exercise hours in middle-aged men (Vuori 1995). The incidence rates of sudden death during vigorous-intensity exercise are estimated at one

cardiac arrest per 20,000 to 45,000 exercisers per year (Mosterd 1999). These figures indicate that the absolute risk is low among apparently healthy middle-aged and older adults. But every death occurring during exercise or sport participation is one too many. Thus, this risk associated with physical activity should be decreased whenever possible.

> Sudden cardiac death is usually caused by an abrupt loss of electrical stability of the heart, causing it to beat rapidly and inefficiently or stop beating altogether (cardiac arrest). Or, a myocardial infarction can occur attributable to a sudden reduction in coronary blood flow causing rapid death to heart muscle and, in some cases, sudden death.

The age-specific mortality rate is lower among physically active people than among inactive people. At the same time, there is an increased risk for sudden cardiac death or myocardial infarction during physical activity compared with inactivity. The mechanisms behind this counterintuitive phenomenon are complex and not fully known. The cause of sudden cardiac arrest during exercise cannot be attributed to just one mechanism.

Pathophysiological evidence suggests that exercise, by increasing the oxygen consumption of the heart muscle and at the same time shortening diastole and coronary perfusion time, may evoke a transient oxygen deficiency at the subendocardial level, which can be worsened by abrupt cessation of activity. A shortage of blood in the heart muscle (myocardial **ischemia)** can alter depolarization, repolarization, and conduction velocity, triggering serious ventricular arrhythmia, which in extreme cases may be the forerunner of ventricular tachycardia or fibrillation. Another cause of sudden death may be the rupture of an atherosclerotic plaque located in a coronary artery, causing a localized blood clot to form and block blood flow to the myocardium. The reasons why strenuous activity may cause such a plaque rupture are not well understood (Burke et al. 1999). The biological mechanisms responsible for exercise-related sudden cardiac death differ with the age of the athlete. The majority of deaths in young athletes (<35 years) are congenital and cardiovascular in nature, the most common being hypertrophic cardiomyopathy (46%), followed by coronary artery anomalies (19%). In contrast, the majority of sudden deaths during exercise in older athletes or nonathletes (>35 years) are attributable to myocardial infarction as a result of underlying coronary artery disease.

> Plaque rupture occurs when an atherosclerotic plaque forms in an artery wall located in the intima just below the endothelium (a one-cell thick inner lining of the artery). The plaque remains separated from the blood flowing through the artery by the endothelium and additional cells and material that slowly form a "cap" over the plaque. If this cap breaks open (ruptures), material in the plaque can cause the blood in the artery to clot and block blood flow to the heart.

When considering the risk of vigorous physical activity, one must be aware of the intensity and duration of the activity and the health status of the person engaged in it. To determine whether vigorous exercise is worth the risk, Siscovick and colleagues (1984) studied the incidence of sudden death during vigorous exercise, paying special attention to the initial level of habitual physical activity. They showed that the relative risk of cardiac arrest

among men with low levels of habitual activity was 56 times greater during vigorous exercise (mainly jogging) compared with other times in their lives. Among men with the highest levels of habitual physical activity, this risk was also elevated but only by a factor of 5. Siscovick and colleagues (1984) (figure 19.2) also studied the overall risk of sudden death (during and not during vigorous physical activity) and showed that men with high levels of habitual physical activity had a risk of sudden cardiac death that was only 40% of the risk of habitually sedentary men. These results support the hypothesis that physical activity both protects against and provokes cardiovascular events: Over the short term, it can provoke clinical cardiac events in those with underlying disease, whereas over the long term, it provides protection.

Female Triad

Competitive athletes frequently are under intense pressure to perform and succeed. Many female athletes experience pressure from their coaches, peers, family, and public to have a low percentage of body fat. They believe that a low level of body fat increases performance and improves their appearance. This is especially true in dancers, distance runners, and gymnasts. Some athletes are driven people, willing to make extreme personal sacrifices to accomplish their goals. But this willingness can drive the athlete to

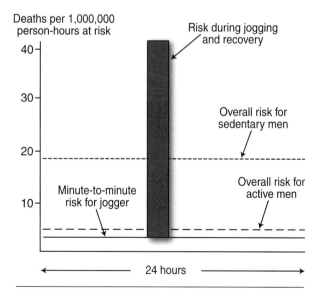

Figure 19.2 The relationship of vigorous activity to sudden cardiac arrest.

Data from D.S. Siscovick, N.S. Weiss, R.H. Fletcher and T. Lasky, 1984, "The incidence of primary cardiac arrest during vigorous exercise," *The New England Journal of Medicine* 311(14): 874-877.

unhealthy eating and exercise behaviors. The results of such behaviors can lead to what is referred to as the female triad. The female triad is a combination of three interrelated conditions that can be associated with athletic training and competition: disordered eating, amenorrhea, and osteoporosis.

The term *disordered eating* refers to a wide spectrum of abnormal patterns of eating that range in severity and include restricting food intake; using diet pills, diuretics, or laxatives; having periods of binge eating and purging; and having anorexia nervosa or bulimia nervosa at the extreme end of the spectrum. The athlete may start simply by monitoring her food intake and then can progress to restricting food such as fats or red meats, limiting food (and calorie) intake, and finally engaging in voluntary starvation. The self-reported prevalence of eating disorders in the general population is in the range of 0.5% to 1.0% for anorexia nervosa and 1% to 5% for bulimia nervosa, although these are most likely underestimations. In athletes, the prevalence of disordered eating has been reported to be between 4% and 39% for meeting the DSM-IV criteria of anorexia nervosa and bulimia nervosa. These prevalence rates are determined through self-report, which may not be a valid way to measure the existence of this disordered behavior: Self-report very likely results in an underestimation of the prevalence of these disorders.

> Disordered eating consists of a wide spectrum of abnormal patterns of eating, ranging in severity from restricting food intake to using diet pills, diuretics, or laxatives; having periods of binge eating and purging; or developing anorexia nervosa or bulimia nervosa.

The second aspect of the female athlete triad, amenorrhea, refers to the delayed onset or the absence of menstrual bleeding. The inability to initiate a menses (menarche) before the age of 16 years is called primary amenorrhea, whereas the cessation of the menstrual cycle function after menarche has occurred is termed secondary amenorrhea. Both primary and secondary amenorrhea have a higher prevalence in female athletes than in the general female population. The prevalence in the general population ranges from 2% to 5%, whereas in the female athlete population prevalence rates have

been reported ranging from 3.4% to 66%. The highest frequencies of amenorrhea have been found among dancers and long-distance runners, but all female athletes training at high intensities and under mental stress are at risk (a list of the athletes at risk, identified by the American College of Sports Medicine, is given under the heading *High-Risk Sports for Development of the Female Triad* below). The pathophysiology of exercise-induced amenorrhea is complex, with varied contributions of a lowered percentage of body fat, body weight loss, and emotional and physical stress. Although amenorrhea is more prevalent in a population with a coexisting eating disorder, Loucks and Horvath (1985) showed that there is no specific body fat percentage below which regular menses ceases. Some athletes with amenorrhea regain their menses after a period of rest, even without regaining body weight or body fat. These findings suggest that amenorrhea is not

High-Risk Sports for Development of the Female Triad

According to the American College of Sports Medicine, participation in the following sports puts female athletes at high risk for developing the female triad, or parts thereof:

- Sports in which performance is subjectively scored (e.g., dance, figure skating, and gymnastics)
- Endurance sports favoring participants with a low body weight (e.g., distance running, cycling, and cross-country skiing)
- Sports in which body contour–revealing clothing is worn for competition (e.g., volleyball, swimming, diving, and running)
- Sports using weight categories for participation (e.g., horse racing, martial arts, and rowing)
- Sports in which prepubertal body habitus favors success (e.g., figure skating, gymnastics, and diving)

Adapted from C.L. Otis, B. Drinkwater, M. Johnson, A. Loucks, and J. Wilmore, 1997, "The female athlete triad," *Medicine & Science in Sports & Exercise* 29(5): i-ix.

caused solely by low body weight or body fat and that other important factors must be considered.

Osteoporosis is the final component of the female athlete triad. Osteoporosis is defined as the loss of bone mineral density (BMD) and the inadequate formation of bone, which can lead to increased bone fragility and an increased risk of fracture. Premature osteoporosis puts the female athlete at risk of stress fractures as well as more devastating fractures of the hip or vertebral column. Amenorrheic athletes have been shown to have low bone mass. However, the prevalence of osteoporosis among athletes is unknown. In women, significant bone loss usually starts at menopause, when they experience a 3% per year bone loss for an average of 10 years, after which bone loss returns to normal levels (0.3-0.5% per year). Osteoporosis in young athletes can result in bone mass loss of 2% to 6% per year, with the total loss during adolescence reaching up to 25%. A young athlete may find herself with the bone mass of a 60-year-old, with a subsequent threefold risk of stress fractures. This accelerated bone loss is the result of estrogen deficiency and subsequent bone resorption. The concern is that the bone loss during early age is partly irreversible, although research shows that regaining the menses can result in increases in BMD. A significant increase in BMD is found in women who decrease their training intensity and regain their menses. However, these athletes are at risk of reaching a level of bone density that is far below normal bone mass for their age and may never reach a normal level again. A low peak bone density in early life is a major risk factor for osteoporosis and increased bone fractures in old age.

Injuries Related to Physical Activity and Sports

As opposed to sudden cardiac death and the female triad, musculoskeletal injury is a risk for all who engage in physical activity, regardless of the level and type of activity. The risk of injury associated with many of the recommended health-promoting physical activities has not been systematically evaluated. Although it is well established that there is an increased risk of injury at the higher end of the physical activity intensity scale, the prevalence of injuries during low-level physical activities (e.g., gardening and walking) is highly variable and not well established. In general, the risks for injury in such low-level physical activities are considered equal to the risks of the activities required for daily living. For instance, when one walks for half an hour a day, there is little or no increased risk for acute or chronic musculoskeletal problems.

One way of looking at the injury problem is to examine the absolute number of physical activity and sports-related injuries that occur in a specific population over a defined period of time. According to the most recent count in the Netherlands, a population of about 16 million, a total of 2.3 million physical activity and sports-related injuries were registered in 1997-1998 (Schmikli 2002). Of all these injuries, 0.9 million injuries required some form of medical attention. Activity-specific numbers are given in table 19.1. Absolute numbers, however, do not precisely represent the injury risk for a person performing a specific activity. Given the popularity of soccer in the Netherlands and the large number of participants in this sport, it is not a surprising to find that in absolute numbers soccer is the most "dangerous" activity. However, a better way of looking at injuries is by calculating the number of injuries per 1,000 hr of participation, that is, the injury incidence. Injury incidence numbers give a more precise estimate of the actual risk of engaging in a particular activity. The injury incidence numbers in table 19.1 show that skiing is the activity with the highest injury risk in the Netherlands. Furthermore, it can be seen that ice-skating, with an absolute number of only 68,000 injuries, has an injury risk equal to soccer.

Each sport or activity has its own injury types and causes. Thus, injury prevention in each activity should focus on the risk factors and injuries inherent to that activity (Conn et al. 2003). For instance, soccer players make hard cuts, sharp turns off a planted foot, and intense contact with the ball and other players. This makes them more vulnerable to acute lower-extremity injuries, especially to the knee and ankle. Acute lower-extremity injuries are also the most common injuries in volleyball. There is general agreement that these acute volleyball injuries result from frequent jumping and landing as well as striking the leg on the floor during defensive maneuvers. The upper extremity is particularly susceptible to injury (acute and chronic) in tennis, because of the use of the racket and the stress it places on the dominant arm and shoulder. Finally, running- and jogging-related injuries are primarily chronic injuries to the lower extremities (e.g., stress fractures). It is beyond the scope of this chapter to discuss risk factors and injury types for each activity

Table 19.1   Sport-Specific Injury Numbers[a]

Type of sport	Number of injuries	Number of medically treated injuries	Number of injuries / 1,000 hr
Soccer	620,000	272,800	2.0
Volleyball	142,000	49,700	2.4
Gymnastics	141,000	43,710	1.6
Indoor soccer	109,000	33,790	6.3
Field hockey	101,000	25,250	2.1
Swimming	92,000	28,520	0.6
Tennis	90,000	35,100	0.4
Skiing	79,000	32,390	10.1
Horse riding	77,000	36,190	0.9
Ice-skating	68,000	21,080	2.1

[a]Estimated total number of injuries, absolute number of injuries medically treated, and injury incidence (number of injuries per 1,000 hr of sports participation).

Table 19.2   Various Sports and Their Main Injury Risk Factors

Soccer	Previous injury
	Player's position
	Playing surface
Volleyball	Jumping technique
	Position
	Playing surface
Tennis	Shoulder strength
	Equipment
	Playing surface
	Muscle imbalance
Hockey	Physical characteristics
	Aggressive play
	Equipment
Gymnastics	Previous injury
	High lumbar curvature
	Protection
Karate or taekwondo	Physical characteristics
	Technique
	Equipment
	Opponent
	Skill level

in full detail (for comprehensive information, refer to Renström, 1993, 1994). Table 19.2 presents the most common risk factors for selected sports.

Asthma and Airway Hyperresponsiveness

A high prevalence of asthma and **airway hyperresponsiveness (AHR)** has been reported in athlete populations. In the general population, the prevalence of asthma is about 5% to 10%, whereas in elite endurance athletes the prevalence ranges from 10% to 50%. The prevalence of asthma and AHR is particularly high among swimmers and athletes exercising in cold air environments. In contrast to elite athletes, amateur endurance athletes do not seem to have an elevated risk for asthma or AHR.

Langdeau and Boulet (2001) stated that although moderate exercise has been shown to be beneficial for patients with asthma, repeated and high-intensity exercise could contribute to the development of asthma. High-intensity physical activity can trigger asthma symptoms in athletes who already have asthma. But high-level exercise performed on a regular basis might also contribute to the development of asthma in previously unaffected athletes. In the general population, the development of asthma is of multifactoral origin. A genetic component can be recognized in addition to environmental factors such as exposure to inflammatory substances. Among athletes, it has been suggested that both prolonged hyperventilation and the quality of the inhaled air during exercise could be contributing factors. Athletes may have an increased exposure to allergens and pollutants attributable to prolonged hyperventilation during and following intensive exercise. This increased exposure could lead to an inflammatory process that might contribute to the development of AHR. The air temperature may be a factor, in that exposure to cold air could induce a bronchoconstrictive response. Whether this is the effect of the low temperature or the low water content in cold air is uncertain. Also, cold air could cause epithelial damage and inflammation and thereby influence airway function. Another factor explaining the high prevalence of asthma and AHR among athletes may be a degree of immunosuppression. Athletes are at increased risk for upper respiratory infection (e.g., common cold, sore throat) during periods of intense training, which may increase their susceptibility for developing asthma.

Minimizing Risk and Maximizing Benefits

Knowing the risks associated with moderate- and vigorous-intensity exercise and, in particular, sport participation raises the question of whether performing vigorous exercise or participating in sports is as healthy as many scientists want the public to believe. It is, but to lead a healthy, physically active life, one must minimize the risks in order to maximize the benefits. The next sections present strategies to minimize the risks of sudden cardiac death, the female triad, and musculoskeletal injury. Prevention of asthma and AHR is not discussed in this section. Preventive measures against asthma and AHR depend mostly on the environment that a person exercises in and will vary from using ventilated indoor swimming pools to using medications.

Preventing Sudden Cardiac Death During Exercise

Although the absolute numbers of sudden cardiac death during exercise are low, screening and providing information about the possible risks and how to reduce them are an important public health endeavor. The number of sudden deaths in exercise can be decreased by identifying susceptible individuals and adjusting their exercise plan based on their medical condition and fitness; it is also necessary to educate athletes, patients, and the general public about exercising in a low-risk manner.

People of all ages with congenital, acquired, or degenerative heart disease can be identified using medical histories, general health examinations, and preparticipation evaluations. Although known heart defects can be important predictors for sudden death during exercise, a large portion of these deaths occur in people with latent or subclinical heart disease. Unfortunately, many of these conditions are not detected by the typical medical evaluation. Furthermore, it is known that the sensitivity of preparticipation examination among apparently healthy persons is low for sudden death during exercise. For example, in one survey the predictive value of a positive exercise test for an exercise-related cardiac death was only 4%. This limited ability to accurately predict an exercise-induced sudden cardiac death is attributable, at least in part, to the rarity of such events. An added difficulty in detecting a pathological cardiac condition that might contribute to the risk of a sudden cardiac death in athletes is the **myocardial hypertrophy**, dilation, and bradycardia produced by vigorous exercise training. These nonpathological changes in cardiac structure and function could mask pathological hypertrophy, dilation, or bradycardia caused by disease.

Providing information about the nature of exercise-related risks and the safest way to exercise is of great importance. A key point is to detect and point out the importance of effort-related symptoms, unexplained tiredness, and febrile infections. Furthermore, attention needs to be paid to the intensity of the activity. Two thirds of sudden deaths with exercise occur during vigorous physical activity, even though most population data indicate that only a minority of exercising people practice vigorous physical activity. As described earlier, the risk of sudden death during vigorous exercise such as jogging is 10 times greater for subjects with a low level of fitness than for subjects with a high level of fitness. Participating in vigorous-intensity exercise when one is not used to it substantially increases the risk for sudden cardiac death. In health promotion activities, people should be cautioned to exercise in moderation in relation to their own exercise capacity and health status. They should be reminded that engaging in moderate-intensity physical activity provides numerous health benefits while keeping risk low. Sedentary people who are changing their physical activity behavior should be discouraged from starting with vigorous-intensity physical activity but rather should initiate a program of moderate-intensity activity and slowly increase intensity as they become more fit.

Preventing the Female Triad

It is not uncommon for female athletes to train at very high intensities, to have an unhealthy diet, and to perform under a great deal of mental and physical stress. But why do they tend to continue these unhealthy practices after a competition is finished? A lack of education about the risks for female athletes from such compulsive behavior might help explain this. Therefore, to identify and prevent the female triad, it is crucial for an athlete's caretaker to provide adequate information about its causes and consequences and to detect the triad early in its evolution.

When providing education on this topic, one must realize that educating only the female athlete will not solve the problem. Education also needs to be directed at the coach and, especially among adolescents, at the parents. Topics to be addressed should include dispelling myths regarding body weight and body fat (e.g., "thinner is better" and "every sport has an ideal body weight") and their relationship to performance; providing nutritional education (e.g., adequate calories from healthy foods should be consumed to meet the energy need); and dealing with other issues of personal wellness (e.g., issues related to sexuality, time and stress management, drug and alcohol use). Female athletes need to be made aware of the long-term consequences of the female triad, such as the possible health consequences regarding fertility and osteoporosis. An effort must be taken to persuade female athletes to change their unhealthy behavior. This is not an easy task, and little is known about the specific approaches that will be effective in this population.

Early detection of the female triad (or separate parts of it) may prevent female athletes from experiencing irreversible consequences of their behavior. In North America, the preparticipation physical evaluation may be the ideal time to screen for the triad, specifically for disordered eating and amenorrhea. However, in Europe such physical exams are not mandatory for athletes, and the triad may therefore remain undetected for a long period. A sports physician or family physician may also screen during office visits for injuries, weight change, amenorrhea, or disordered eating. The physician should screen for signs of disordered eating by asking the athlete about her past eating habits and by asking for a list of "forbidden foods." The patient's highest and lowest body weights should be ascertained, as well as whether she is happy with her current body weight. A history of amenorrhea is another easy way to detect the triad in its earliest stages. The physician should be aware that there is no specific body fat percentage below which regular menses cease. Also, physicians should not discount amenorrhea as a benign consequence of intensive athletic training. This will only reinforce the female athlete in her unhealthy behavior.

Important aspects of treating the female triad include decreasing training intensity, increasing periods of rest, and regaining body weight. Although the female athlete may be unwilling to cooperate, the physician needs to convince her of the importance of these lifestyle changes. In female athletes with signs of the female triad, attention should be paid to the possibility of osteoporosis. The physician should educate the amenorrheic patient about the possible consequences of long-term amenorrhea and the risks of irreplaceable bone loss. To obtain accurate information about her bone mineral density, a dual-energy x-ray absorptiometry (DEXA) evaluation should be considered. This information can be used in considering whether to start hormone therapy with the female athlete to reduce the decrease in bone mineral density.

Preventing Injuries Related to Vigorous Exercise and Sport Participation

Epidemiological data should serve as a basis for prevention programs designed to reduce the injury risk associated with vigorous exercise and sport participation. As postulated by van Mechelen (1992), measures to prevent injuries during vigorous exercise or sport participation do not stand by themselves; they form part of what might be called a sequence of prevention (figure 19.3):

1. The first step in the sequence of prevention is to define the sports injury problem in terms of incidence and severity. This descriptive information provides insight into the magnitude of the problem. It also shows which type of injury is the most common across all sports or in a certain type of sport and which sports are more risky in terms of injury frequency and severity. Furthermore, information on

the severity of sports injuries can help to focus on specific preventive measures.

2. Once it is known where preventive measures are warranted, the etiological risk factors and mechanisms underlying the occurrence of the targeted injury need to be identified.

3. With this information regarding underlying risk factors, preventive measures that are likely to work can be developed and introduced.

4. Finally, the cost-effectiveness of these preventive measures should be evaluated by repeating the first step (time-trend analysis) or preferably by performing a randomized controlled trial (RCT).

As seen in table 19.2, numerous factors that contribute to injuries during vigorous exercise and sport participation have been identified in previous research. However, discussing the third and fourth steps of the sequence of prevention is more difficult. A review by Parkkari and colleagues (2001) indicated that only 16 RCTs on the prevention of sports injuries were conducted over the last 3 decades. Thus, although many risk factors have been

identified, little evidence exists on the effectiveness of specific preventive measures. Because it is impossible to discuss the prevention of each injury in detail here, only some general principles of acute and chronic injury prevention applicable to vigorous exercise and sport participation are given. They are divided into three main categories: athlete, sport or exercise, and risk behavior.

Preparation by the Athlete

The use of stretching and warm-up exercises to promote suppleness and flexibility is historically believed to prevent strain injuries to muscles and tendons. However, there still is much controversy on stretching as a preventive means. For example, no preventive effect of stretching was found in a number of studies (Pope et al. 2000; Shrier 2000; Thacker et al. 2004). General and sport-specific conditioning programs, preferably incorporated in the regular warm-up and cool-down, for athletes are necessary to attain successful performance and avoid injury. Avoiding overuse injuries is a critical part of designing and monitoring the conditioning program, particularly in endurance sports such as running. A common rule to help avoid overuse injuries is not to increase more than one exercise parameter (i.e., intensity, frequency, or duration) at a time. The use of braces and taping to stabilize weak or unstable joints during the return to play or rehabilitation phase is widely practiced. In addition to this secondary preventive application, tape and braces also have a primary preventive function. However, significant controversy exists about the benefit of such procedures. Taping and bracing may function more by improving **proprioception** and thereby stimulating earlier recruitment of supportive muscles rather than by the actual mechanical restriction or support of a joint. Neither tape nor brace applications can guarantee protection from new injury or from exacerbation of a preexisting trauma.

Exercise- or Sport-Specific Measures

The equipment used in sport has the potential for contributing to injury prevention. For instance, facial and head injuries in American football may be prevented through the use of properly fitted helmets and padded chin straps, which eliminate helmet rotation. The effect of headgear in decreasing injuries has also been shown in wrestling and ice hockey. Also, the use of proper cleats or running shoes will

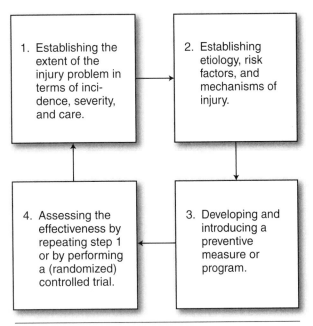

Figure 19.3 Sequence of prevention of sports injuries.

Reprinted, by permission, from W. van Mechelen, H. Hlobil and H.C. Kemper, 1992,"Incidence, severity, aetiology and prevention of sports injuries: A review of concepts," *Sports Med.* 14(2):82-99.

reduce lower-extremity injuries. Preventive equipment can be individually applied depending on sport, position, and bodily dimension. However, equipment should always fit the conditions for which it is designed; for example, basketball shoes should not be used for long-distance running.

In team sports, an important role in injury prevention lies with the referee. The fact that only 8% of all ice hockey injuries are associated with a penalty suggests that referees may allow dangerous play. Many injuries in soccer, for instance, occur during tackling and contact with the opposing player. A good referee needs to keep the game under tight control and try to prevent dangerous behavior by the participants. Coaching within the spirit as well as within the letter of sport rules and regulations should be emphasized, because aggressive actions often lead to injuries in team sports. For instance, only half of teenage ice hockey players understand the seriousness of checking another player from behind. The coach has the power to emphasize the serious magnitude of the injuries that this move can inflict.

Risk Behavior

In contact sports, a debate has arisen about the introduction of preventive measures because changes in injury patterns and mechanisms have occurred hand in hand with the introduction of protective equipment. An example of this phenomenon is found in ice hockey, where the introduction of mandatory helmet use reduced the incidence and severity of head injuries. However, evidence suggests that in ice hockey, neck injuries may have become more frequent since the introduction of mandatory facial and head protection. In the past, the use of shoulder padding in American football was found to be primarily "an offensive weapon" and not within the "spirit" of the game. Shoulder padding is a preventative measure . . . if it is used correctly. It is noticeable that players in a variety of contact sports become more reckless after the introduction of protective measures. This phenomenon is described in the literature as "risk homeostasis."

The theory of risk homeostasis states that individuals maintain their risk behavior at a level they

The referee is a crucial factor in injury prevention during athletic contests. If the official does not penalize dangerous behavior, the likelihood of injuries is significantly increased.

perceive as acceptable and safe. In this viewpoint, individuals adapt their behavior to a certain level of risk they consider acceptable ("target level of risk"). This explains why the manipulation of risk factors and the introduction of preventive measures frequently do not eliminate or even significantly reduce the risk of injuries. In other words, individuals compensate their risk-taking behavior under the influence of preventive measures. According to Bouter (1986), motivating individuals to decrease their target level of risk is the only measure that can decrease the risk for injuries.

> Risk homeostasis is a theory stating that athletes maintain their risk behavior at a level they perceive as acceptable and safe. Such individuals adapt their behavior regarding risk to achieve a "target level of risk."

Recommendations for Future Research

Although a great deal is known about the risks associated with exercise and sport participation, a variety of issues are still undergoing research or need further clarification. Some of the key issues are summarized here.

Sudden Cardiac Death

To reduce the number of sudden cardiac deaths during vigorous exercise or sport participation, a major advance can be made by increasing our understanding of the pathophysiological mechanisms that underlie such events. Questions to be addressed include, "Which persons are at greatest risk?" and "When during their training or competition is their risk the highest?" Such data would enable caregivers to more effectively screen and inform people accurately about their risks associated with exercising vigorously. However, a significant improvement in this effort may be limited by the very low incidence of sudden death during exercise.

Female Triad

The prevalence of the female triad in athletes is substantially higher than in the general population, but estimations from various athlete populations are highly diverse. Researchers should aim to obtain more accurate information about the prevalence of the female triad and its components. Research should consider improving methods for diagnosing and screening for the triad and enlarging our knowledge about the irreversibility of its detrimental effects. Attention should be given to determining safe training volumes and intensities that will not decrease performance during athletic competition. Finally, theory-based health behavior change programs need to be developed and evaluated to assist these competitors in changing their unhealthy behaviors.

Musculoskeletal Injuries

Numerous questions regarding how to more effectively prevent musculoskeletal injuries during exercise of all types and intensities remain to be answered. Procedures to prevent musculoskeletal injuries during exercise do not stand by themselves, because they form part of what is called the "sequence of prevention." A variety of preventive measures need to be evaluated using RCTs after activity-specific risk factors have been identified. Well-designed randomized studies are needed on preventive procedures and devices currently in common use, such as warming up, proprioceptive training, protective equipment, and educational interventions.

Summary

An increase in physical activity, especially vigorous-intensity exercise or sport participation, carries a variety of risks as well as a number of well-documented health benefits. Serious health risks associated with increases in moderate-intensity activities, the core of most public health recommendations, are quite rare but still need to be considered when implementing a physical activity plan designed to maximize benefit while minimizing risk. Major examples of risks associated with vigorous exercise include sudden cardiac death, the female triad, asthma and related pulmonary disorders, and musculoskeletal injuries.

Factors contributing to the likelihood of an exercise-induced medical complication include the medical status of the participant (underlying disease, prior injury, nutritional deficiency, obesity); current physical activity status (sedentary or active, fit or unfit); type (weight bearing, contact sport), intensity (relative to the person's capacity), duration, and frequency of exercise; approach to the exercise session (stretching, warm-up, cool-down), use of appropriate equipment (shoes, protective gear); and the environment (type of surface, air pollution, temperature, humidity). Each of these factors should be considered whether implementing a physical activity plan to enhance health or an exercise training program in preparation for athletic competition.

Comprehensive programs to minimize risk include medical screening when indicated, education of participants about their risks during various activities and how to minimize these risks, professional guidance regarding exercise selection and performance in high-risk persons, and the provision of safe environments for exercise. Although additional research is needed on the pathobiological basis for some health risks, on the most effective medical screening procedures for selected populations, and on intervention programs designed to minimize risk, broad-scale implementation of risk reduction components based on current knowledge can keep overall risks low while maximizing positive health and performance outcomes.

KEY TERMS

airway hyperresponsiveness (AHR): An abnormal condition where the airways (especially the bronchi in the lungs) respond to a stimulus such as cold air during exercise by narrowing and restricting air flow.

amenorrhea: Delayed onset or the cessation of menstruation for 6 or more months associated with high-level exercise training or athletic competition.

anaphylaxis: A sudden, severe, potentially fatal, systemic allergic reaction that can involve various areas of the body, such as the skin, respiratory tract, gastrointestinal tract, and cardiovascular system. Anaphylactic reactions can be mild to life threatening. The annual incidence of anaphylactic reactions is about 30 per 100,000 persons, and individuals with asthma, eczema, or hay fever are at greater relative risk of experiencing anaphylaxis at rest or during exercise.

ischemia: Condition where the oxygen-rich blood flow to a part of the body is not adequate to meet oxygen demands. Cardiac or myocardial ischemia refers to lack of blood flow and oxygen to the heart muscle and happens when sudden exercise is performed without warm-up or when an artery becomes narrowed or blocked for a short time, preventing oxygen-rich blood from reaching the heart. If ischemia is severe or lasts too long, it can cause a heart attack (myocardial infarction) and can lead to heart tissue death. A temporary blood shortage to the heart can cause pressure or pain (angina pectoris). However, in other cases (40-50%) there is no pain, which is called *silent ischemia*.

myocardial hypertrophy: An increase in the muscle mass of the heart attributable to increased thickness of the walls of the heart chambers. This increase in heart mass can be caused by pathological conditions like hypertensive or valvular disease that cause the heart to work harder or to nonpathological conditions such long-term vigorous exercise training.

osteoporosis: For definition, see page 229.

proprioception: Process by which the body can vary muscle contraction in immediate response to incoming information regarding external forces by using stretch receptors in the muscles to keep track of the joint position in the body.

STUDY QUESTIONS

1. Briefly explain how a bout of vigorous exercise might trigger a sudden cardiac death in a 55-year-old man. What are some personal characteristics that might contribute to his risk of having such an event?

2. What are the three major components of the female athlete triad? Briefly explain how a coach, athletic trainer, or physician might identify these features in a 22-year-old highly competitive distance runner.

3. What are the main cardiovascular abnormalities or pathologies that contribute to the risk of sudden cardiac death during vigorous exercise in men under age 35 compared with men over age 35?

4. What is meant by the "sequence of prevention"? Briefly describe the four components of this sequence.

5. If you were advising a group of novice coaches from a youth soccer league about how to reduce the risk of injury in their players, what are four or five issues you would address?

6. A 58-year-old with a body mass index of 33 who has not been physically active since he was 36 years old now wants to begin a physical activity program to increase his cardiovascular endurance and muscle strength. How would you advise him to reduce his risk of musculoskeletal injury?

7. Explain how it is possible that even though the risk of sudden cardiac death is substantially increased during vigorous exercise in middle-aged and older adults compared with periods when they are inactive, these physically active people have an overall lower risk of cardiac death than their sedentary counterparts.

8. What advice would you give a mother regarding participation in cross-country skiing by her 13-year-old son who has exercise-induced airway hyperresponsiveness?

IV

Recommended Activity Amounts and Delivery Modes

An important question frequently asked by physical activity and health professionals, as well the general public, relates to the amount and type of activity necessary to produce specific health outcomes. Part IV of the book deals with a variety of issues raised by this question. A sound scientific basis is needed for recommending a specific activity regimen to improve a person's health. In parts II and III, several biological and clinical responses to changes in chronic physical activity status were presented. Given the numerous health outcomes influenced by physical activity and the range of possible activities, determining the optimal amount and type of activity is still under investigation. In chapter 20, the evolving concepts regarding exercise dose and health response, referred to as "dose–response," are presented with a discussion of the interplay between physical fitness and health. Chapter 21 compares structured and unstructured exercise programs as well as programs for specific groups of people. You should conclude this section with a clear idea of the scope of activities available and the issues involving appropriate levels of activity.

Dose–Response Issues in Physical Activity, Fitness, and Health

■ William L. Haskell, PhD

There is good documentation that moderate amounts of physical activity added to the routine activities of daily living or moderate levels of physical fitness provide meaningful health benefits. Also, there is evidence that more activity and higher fitness levels generally confer somewhat greater benefits. What is still at issue is the more precise dose–response relationship for a number of specific health outcomes, taking into consideration the activity profile (type, intensity, frequency, duration, and amount), a number of characteristics of the person (gender, age, health history, prior exercise training status), and the major health outcomes desired. Such data are critical to design and implement scientifically based recommendations to achieve desired outcomes in specific populations safely and cost-effectively. In this chapter, we address some of the key issues regarding the concept of dose–response and present data that support current physical activity guidelines for disease prevention and health promotion published by major health organizations.

Many people have questions about using physical activity to enhance their health or physical performance. How much activity is enough? How little activity can I get away with? How physically fit do I need to be to be healthy or to help prevent a specific disease? To address such questions, exercise scientists and practitioners have attempted to determine the characteristics of physical activity or level of fitness required to produce meaningful changes in specific health and performance outcomes. *Dose* is the term frequently used to represent the characteristics of the activity (such as type, intensity, frequency, and amount) that are important in producing improvements in health or performance. *Response or **health response*** is used to represent the various changes that occur when a specific dose of activity is performed. Thus, for physical activity and health, *dose–response* refers to the relationship between the characteristics of the activity performed and the nature of the health-related changes produced. For physical fitness and health, dose–response typically refers to the relationship between levels of fitness as defined by a designated measure (e.g., MET capacity on a treadmill or 1-repetition maximum on bench press) and a specific health outcome, such as the development of type 2 diabetes or death attributable to coronary heart disease (CHD).

Principles Guiding the Body's Response to Activity

To better appreciate a number of dose–response issues, it is useful to understand several exercise training principles. These principles form much of the scientific basis for the development of activity plans or exercise prescriptions to improve health and performance. First is the principle of **overload,** which refers to the fact that when a tissue, organ, or system is stressed by an increase in physical activity, it typically responds by increasing its capacity or efficiency. To detect this response, a person may have to perform the exercise bout for a number of days or weeks. But if the activity is greater in intensity or amount than typically performed by this person, the body slowly adapts. For example, in response to an increase in walking speed or distance, components of the oxygen transport and utilization systems will slowly increase their capacity (e.g., increase in stroke volume and skeletal muscle capillary density) or efficiency (decrease in heart rate and systolic blood pressure). This overload principle applies to many of the tissues that are involved in supporting the increase in energy expenditure required for sustained activity or are exposed to the mechanical forces produced by muscle contractions or gravity.

> The principles of overload, progression, and specificity are major determinants of how the body will respond to a dose of physical activity.

A second training principle that is important in establishing a health-promoting dose of activity is **progression.** For overload to produce health benefits rather than injury, it needs to be increased in small amounts or at a slow rate of progression. If the overload progresses too rapidly, it can produce strain or injury and create a negative health outcome. Thus, people who have been very inactive should begin to increase their activity levels by walking daily at increasing speeds over weeks or months rather than immediately starting a jogging program. However, if the overload is not increased and there is no progression in intensity or amount, then very little or no improvement will take place.

Progression does not need to continue when the person has reached an appropriate goal of activity or fitness, such as 30 min of brisk walking on most days. Issues involving progression generally are addressed in detailed activity plans but not in the more general recommendations made to the public. Understanding progression scientifically and clinically is necessary to design and implement activity programs that maximize benefit but minimize injury and chronic fatigue.

The last of the three exercise training principles is **specificity.** The specific types of changes that take place in the body in response to an increase in activity depend substantially on the characteristics of the activity. That is, the changes that occur are specific to the nature and degree of the demands produced by the exercise. The most frequently recognized application of specificity is the prescribing of resistance exercise to increase muscle size and strength and endurance exercise to increase the capacity of the oxygen transport and utilization systems. However, specificity has other applications as well, including the need to exercise a specific muscle to maximize its strength or endurance (arm vs. leg exercise) or stretching specific muscles and ligaments to increase the flexibility or range of motion of a particular area of the body.

Components of the Physical Activity Dose

When exercise dose in relationship to health-related outcomes is discussed, the exercise is characterized by its type, intensity, session duration, session frequency, and total amount per day or week. Recently the terms *activity profile*, *activity volume*, and ***accumulation*** have been included in these discussions. In this section, key features or issues related to each of these characteristics are discussed as they pertain to defining the health-related response to exercise.

Type of Activity

Because muscle contractions have both mechanical and metabolic properties, exercise has been classified by both of these characteristics, a situation that tends to cause some confusion. Mechanical classification stresses whether the muscle contraction produces movement of the limb or not: isometric (same length) or static exercise if there is no movement of the limb, or isotonic (same tension) or dynamic

exercise if there is movement of the limb. A muscle contraction can be either concentric (shortening the muscle fibers) or eccentric (lengthening the muscle fibers). The metabolic classification primarily involves the availability of oxygen for the contraction process and includes aerobic (oxygen available) or anaerobic (without oxygen) processes. Whether an activity is aerobic or anaerobic depends primarily on its intensity relative to the participant's capacity for that type of exercise. Most activities involve both static and dynamic contractions as well as aerobic and anaerobic metabolism. Thus, activities tend to be classified based on their dominant metabolic or mechanical characteristics. Other classifications of exercise include endurance (aerobic) versus strength (resistance), upper versus lower body (arms vs. legs), and classification by purpose (occupation, household chores, self-care, transportation, recreational, leisure time, and physical conditioning).

> The major components used to define a dose of physical activity include activity type, intensity, session frequency, and session duration. The amount or volume of activity is determined by intensity × frequency × duration.

A key issue about exercise type in terms of desired changes is the specificity of the response to a given type of exercise. Specificity refers to the fact that the biological changes that occur in response to exercise are dependent on (1) which tissues or systems are activated or stressed by the exercise and (2) the nature of the stress placed on these tissues or systems. If this activation or stress is of appropriate intensity and amount, these tissues or systems respond favorably by increasing their capacity or efficiency (training effect). If the activation or stress is too little (not more than frequently experienced by the tissues or systems), then there is very little or no training effect or adaptation. If the activation or stress is too great, overuse injury or chronic fatigue can occur. A good example of this specificity is the increase in muscle and bone mass seen in the dominant arm versus the nondominant arm of a professional tennis player as a result of playing tennis. The response of the dominant arm to hitting thousands of tennis balls is not transferred to the nondominant arm, which performs limited exercise just tossing the ball into the air. Vigorous endurance-type exercise such as running provides a major stimulus to the oxygen transport system, substrate (glycogen and fat) processing systems, and oxidative processes in leg skeletal muscle fibers, especially in slow-twitch fibers. These systems increase in efficiency or capacity in response to running and other endurance-type exercise. In contrast are the demands made by heavy resistance exercises, such as powerlifting, that primarily activate fast-twitch muscle fibers and stress the body's support structures, such as bones and connective tissue. Thus, when we describe the response to a dose of activity exercise, the type of activity needs to be accurately defined.

Physical Activity Intensity

Intensity is a key factor when considering the dose of physical activity required to achieve specific health outcomes. Not only does an increase in activity intensity play a major role in producing many favorable adaptations, but it also has a key role in the various health risks resulting from activity (see chapter 19). The intensity of an activity can be described in both *absolute* and *relative* terms. In absolute terms, intensity is either the magnitude of the increase in energy required to perform the activity (endurance exercise) or the force produced by the muscle contraction (resistance or strength exercise). During endurance exercise, the increase in energy is usually determined by measuring or estimating the increase in oxygen uptake, which is expressed in units of oxygen (liters or METs) or converted to an equivalent measure of heat produced (kilocalories) or energy expended (kilojoules). The intensity can also be expressed as a walking or running speed or the work rate (power) on a cycle ergometer (watts). The force of the muscle contraction is measured by how much weight is moved (dynamic contraction) or the force exerted against an immovable object (isometric contraction) and usually is expressed as "work" in kilograms or pounds. If the work performed is expressed per unit of time—such as an oxygen uptake of 1.5 L/min—then the correct term is *power* and not *work*.

In relative terms, the intensity of the activity is expressed in relation to the capacity of the person performing the activity. For energy expenditure, the relative intensity usually is expressed as a percentage of the person's aerobic power, also termed maximal oxygen uptake (%$\dot{V}O_2$max). Because there is a nearly linear relationship between the increase in heart rate and the increase in oxygen uptake during dynamic

exercise, "percentage of maximum heart rate" or "heart rate reserve" (maximum heart rate minus resting heart rate) is also used as an expression of exercise intensity relative to the person's capacity. One advantage of using percentage of maximum heart rate when assigning exercise intensity to a participant is that the recommended value (i.e., 70-85%) can stay the same as the person becomes more fit (but the intensity of the activity will have to increase to reach this heart rate). This is not the case if an absolute intensity is assigned, such as starting at a brisk walk at 4.0 mph. After some weeks at this intensity, the person will have to increase the walking speed to provide a stimulus of a similar magnitude to the cardiorespiratory system. For muscle force, the relative intensity of the contraction is expressed as a percentage of the maximal force that can be generated for that activity (percentage of maximum voluntary contraction or percentage of a 1-repetition maximum).

> When we consider the intensity component of an activity dose, we need to understand the differences between relative and absolute intensity. Relative intensity takes into consideration the exercise capacity of the person, whereas absolute intensity only considers the demands of the activity.

In most experimental studies evaluating the effects of increased activity on various fitness and health outcomes, intensity is expressed relative to each person's capacity (e.g., 60-75% of $\dot{V}O_2$max). However, in most, if not all, large prospective observational studies, exercise intensity is expressed in absolute terms (no adjustment made for each person's exercise capacity). These differences in methodology prevent direct comparison of dose–response data from these two major sources of information. For an activity of a given intensity, such as walking at 5.6 km/hr (3.5 mph, 3.8 METs), the relative intensity varies inversely to the aerobic capacity of the individual. As shown in figure 20.1, for highly fit people with an aerobic capacity of 14 METs, walking at 5.6 km/hr (3.5 mph) has a relative intensity of 27% (light intensity), but for people with only a 6-MET capacity, the relative intensity is at 66% (hard intensity). An example of activity classification by both absolute and relative intensity is provided in chapter 1, table 1.3 on page 13. Standardization of activity intensity classification

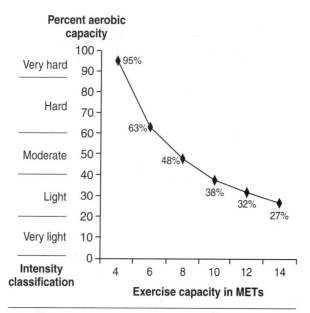

Figure 20.1   At a walking speed of 5.6 km/hr (3.5 mph), the average energy cost for an adult is 3.8 METs. Here we see that for highly fit adults (aerobic capacity of 14 METs), walking at this speed on a flat surface requires only 27% of their aerobic capacity (light activity) whereas for adults with an exercise capacity of only 6 METs, such walking requires 63% of their capacity (hard intensity).

is essential for accurately establishing the relationship between intensity and health or performance outcomes.

Over the past century, the intensity of numerous physical activities was defined by the energy expenditure required to perform the activity. In table 20.1, the energy requirements are listed for selected activities performed during home and self-care, transportation, occupation, leisure time, and physical conditioning (Ainsworth et al. 2000). There can be substantial interindividual variation in the energy cost of performing some activities, especially when the activity has a large skill component such as swimming. However, for many activities like walking or jogging, the between-person variation in energy expenditure when adjusted for body weight is quite small, and thus the average values for such activities listed in table 20.1 apply to most adults.

For some outcomes, it appears that the total amount of exercise performed is more important than the intensity (above some minimal level) at which the exercise is performed (Manson et al. 2002). Establishing which outcomes can be achieved at lower activity intensities is important when using a public health model to promote physical activity

because risks of orthopedic injury or cardiac arrest are closely linked to higher intensities of activity.

Activity Session Duration and Frequency

Discrete recommendations for session duration and frequency traditionally have been provided when prescribing exercise that focuses on performing a defined bout for a designated number of times per week, such as 20 min or more of vigorous exercise three or more times per week. The newer recommendations that focus on adding more activity back into the routine activities of daily living (the lifestyle approach) emphasize activity accumulation throughout the day: accumulate at least 30 min of activity on 5 or more days of the week. When performing resistance exercise, the duration or amount usually is not measured in time but by the number exercises performed, the number of sets per exercise, and the repetitions per set.

When sedentary adults start a program of increased activity, it appears that even quite small amounts of moderate- or vigorous-intensity activity will improve both aerobic capacity and strength. For example, a program of jogging for 10 min per day, three times per week for 12 weeks improved measures of cardiorespiratory fitness in middle-aged men (Wilmore et al. 1970) as did a program of brisk walking 45 min per day two times per week for 20 weeks (Pollock et al. 1977). The results of other studies evaluating the effects of varying exercise session duration or frequency on health and performance outcomes generally support the concept that the greater the increase in the total amount of exercise (duration of session × frequency of the session × intensity) the greater the improvement.

Data are available to support this dose–response concept as long as the exercise intensity is at least moderate relative to the person's capacity and the total energy expended per session is in the range of 100 to 1,000 kcal. On the low side, there are very limited scientific data supporting health or performance benefits from performing just light-intensity exercise (<40% of $\dot{V}O_2$max reserve) or expending less than 100 kcal per day during planned activity. On the high end, few data are available on the level of health benefits achieved when people consistently exercise at very hard intensities (≥85% of $\dot{V}O_2$max reserve) or expend >1,000 kcal a day during vigorous physical activity.

Table 20.1 Estimated Energy Cost (METs) of Selected Physical Activities

Activity	METs
Household and self-care	
Clearing table and washing dishes (with some walking)	2.5
Ironing	2.5
Cleaning house—general	3.0
Vacuuming	3.5
Scrubbing floors on hands and knees	3.8
Sweeping garage, sidewalk, other outdoor locations	4.0
Automobile repair at home	3.0
Carpentry, finishing or refinishing furniture	4.5
Carrying, stacking, or unloading wood	5.0
Mowing lawn—walking with power mower	5.5
Occupational	
Data entry, typing on computer while sitting	1.5
Machine tooling—welding	3.0
Masonry—concrete	7.0
Forestry—general	8.0
Shoveling—heavy, more than 16 lb/min	9.0
Transportation	
Bicycling—leisurely <16 km/hr (<10 mph)	4.0
Bicycling—vigorously 22.4-25.4 km/hr (14.0-15.9 mph)	10.0
Walking briskly—6.8 km/hr (3.5 mph)	3.8
Recreational, leisure time, sports	
Playing piano	2.5
Dancing (ballroom slow—waltz, foxtrot)	3.0
Dancing (ballroom fast—disco, line dancing, folk)	4.5
Jogging—9.6 km/hr (6 mph)	10.0
Running—12.8 km/hr (8 mph)	13.5
Tennis—singles	8.0
Skateboarding—general	5.0
Conditioning	
Stretching (Hatha yoga)	2.5
Calisthenics (light to moderate effort)	3.5
Calisthenics (heavy, push-ups, sit-ups, jumping jacks)	8.0
Aerobic dancing (with 6 to 8 in. step)	8.5
Weightlifting (light weights)	3.0
Weightlifting (heavy weights—vigorous effort)	6.0
Circuit training with some aerobic activities	8.0

Reprinted, by permission, from B.E. Ainsworth, W.L. Haskell, M.C. Whitt, M.I. Irvin, A.S. Schwartz, S.J. Straith, W.L. O'Brian, D.R. Basset, K.H. Smitz, P.O. Emplaincourt, D.R. Jacobs and A.S. Leon, 2000, "Compendium of physical activities: An update of activity codes and MET intensities," *Medicine and Science in Sports and Exercise* 32: S498-519.

Factors Determining Optimal Activity Dose

The interplay of the many factors that must be considered when determining the optimal activity dose for a specific individual can be extremely complex. In this section we consider issues that must be taken into account when determining the optimal activity dose for oneself, a client, or a patient.

Accumulation of Physical Activity Throughout the Day

The concept of accumulation of physical activity by performing multiple short bouts (8-10 min each) throughout the day was first included in major guidelines in 1995 (Pate et al. 1995). These recommendations were based on (a) indirect data from prospective observational studies relating amount of activity to CHD, cardiovascular disease, and all-cause mortality and (b) results of several experimental studies that evaluated the effects of a single longer bout per day versus three to four shorter bouts per day (total exercise performed held constant) on aerobic capacity, plasma high-density lipoprotein (HDL) cholesterol concentration, and body weight. Data from these studies indicated that performing multiple short bouts of moderate- to vigorous-intensity activity produced results not too dissimilar from those produced by the longer bouts. Additional evidence has been reported that supports these initial results (Hardman 2001). There are very limited data from experimental studies evaluating bouts of activity less than 8 to 10 min in duration. These types of data are needed to better determine what the minimum duration of an activity bout needs to be for inclusion in "daily accumulated activity."

Because it now appears that the accumulation concept is valid, we must consider the role of the amount or volume of exercise when defining dose–response. Volume is the product of exercise intensity, duration, and frequency, usually expressed in some units of energy expenditure (e.g., kcal/day, MET-minutes/day). If the key stimulus for certain health benefits is activity volume above some minimal intensity threshold, then a wide variety of exercise profiles can be encouraged for promoting health. Figure 20.2 depicts four different exercise profiles that provide a similar exercise volume expressed in total energy expended (1,800 kcal/ week). These various profiles of exercise, if reason-

ably equal in effectiveness, provide a wide range of opportunities for achieving an adequate volume of activity to maintain or improve health.

Unresolved dose–response research issues regarding the role of session duration and frequency include the following:

- How short can a bout of endurance-type exercise be and still contribute to specific health outcomes (e.g., does stair climbing for 1 min per session 20 times per day provide benefits similar to a single session per day of 20 min)?

- Do sessions of long duration (90 min) on 2 days per week provide similar benefits as 30 min sessions performed 6 days per week (180 min of exercise per week in both cases)?

- Do multiple short bouts of exercise spread throughout the day provide more benefit than a single longer bout of equal amount resulting from repeated acute effects of exercise on specific health outcomes?

Minimal Versus Maximal Dose

Much of the public would like to know the minimal amount or intensity of physical activity needed to produce significant health benefits. From a public health perspective, this presents a problem because most likely the answer to the question depends in a significant way on a person's current level of physical activity or fitness. For example, a person who has been at bed rest for more than several days will benefit significantly by just sitting upright in a chair or standing and moving about slowly for a few minutes several times a day. Also, sedentary older persons appear to benefit from even short bouts of slow walking (<3 mph) or light-intensity activity such as tai chi. Some data from epidemiological studies indicate that there are favorable health outcomes from leisure-time activity that requires an energy expenditure of no more than 75 kcal per day. Although small increases in activity appear to provide some benefit, especially in the very inactive or unfit, there are very few data on this issue in real-life situations for the general public.

Also, at the high end of the **physical activity dose** continuum, we still lack data that define the point where an increase in activity provides no additional health benefits. This would be the point of maximum benefit for a specific health outcome. Documenting such an activity dose would help answer the question, "How much activity do I need

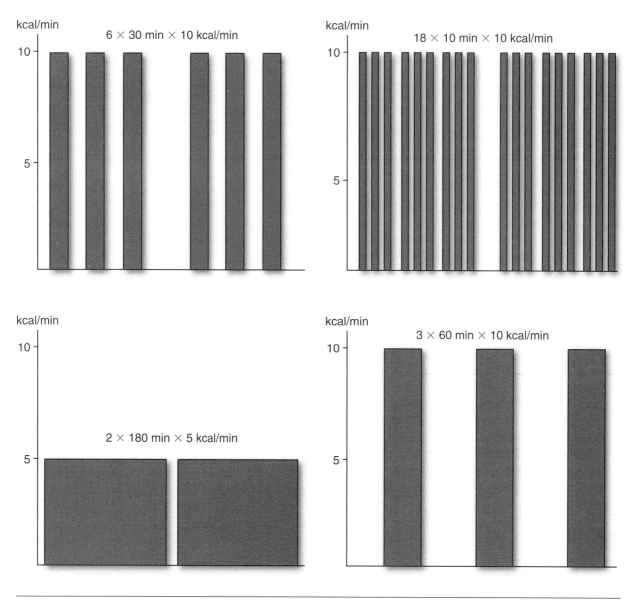

Figure 20.2 Examples of different physical activity regimens that would result in expending approximately 1,800 kcal per week. The examples range from performing three sessions per day (10 kcal/min) for 10 min each, 6 days per week, to one session per day (5 kcal/min) 180 min per day on just 2 days of the week.

to perform to obtain all of the health benefits provided by being physically active?" For some people this amount of activity appears to be quite high. For example, there is a dose–response relationship between distance run per week and cardiovascular disease risk factors in male runners even up to 80 km (48 miles) per week (Williams 1997). Figure 20.3 shows that there is a positive dose–response between distance run per week up to 80 km per week and the percentage of men with clinically defined low levels of HDL cholesterol considered at increased risk for developing CHD. Also, there is a negative dose–response between distance run

and percentage of men with clinically high levels of HDL cholesterol considered to be at decreased risk for developing CHD.

Somewhere between the minimal and maximal doses of activity that provide health benefits, there is likely an optimal dose. This is the dose of activity that would provide the greatest net health benefits for the least amount of time and effort spent in performing the activity. The optimal net health benefit is reached when the difference between positive and negative health outcomes is maximized. Based on what we currently know about dose–response, an optimal dose will vary from one health benefit

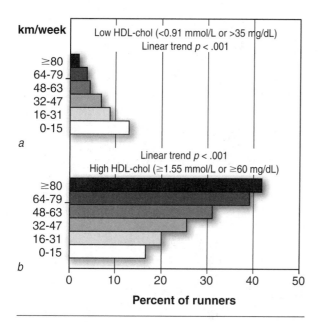

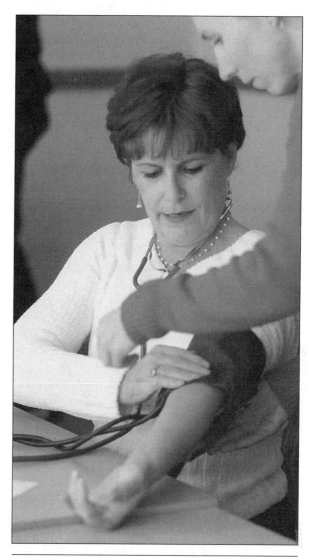

Figure 20.3 Dose–response relationships between the distance run per week and the prevalence of low (bad) values *(a)* and high (good) values *(b)* of high-density lipoprotein (HDL) cholesterol in men who are regular runners. As the distance run increases, the percentage of men with low levels of HDL cholesterol decreases whereas the percentage of runners with high levels significantly increases (Williams 1997).

to another and from person to person. This interindividual variation will have a genetic component and could be influenced by factors such as gender, age, exercise capacity, nutritional status, and overall health.

Use of Threshold Values for Dose–Response Recommendations

Studies addressing dose–response that have sufficient data to evaluate various amounts of activity generally indicate that the relationship between activity and positive health outcomes is more of a continuous relationship than a threshold effect (Manson et al. 2002). In some studies, only a few categories of activity amount are provided, leading to the appearance of a threshold effect (Leon et al. 1987). Also, the way that some guidelines or recommendations have been written makes it appear that a minimum threshold of activity needs to be achieved before any benefit occurs or that above some amount of activity no further benefit takes place. This situation is not surprising, because it is very similar to guidelines addressing the relationship between other major chronic disease risk fac-

It is well established that the relationship is continuous between your blood pressure and health outcomes like heart attacks and strokes; the higher your blood pressure the greater the risk. Thus it is useful for those with extremely high blood pressure to learn to monitor themselves daily.

tors (e.g., blood pressure, cholesterol, obesity) and health outcomes. For example, it is well established that the relationship between level of blood pressure and stroke risk is continuous from very low levels of blood pressure to very high levels.

Yet in blood pressure guidelines from major health organizations, values such as 120/80 mmHg are presented in a way that makes it appear that this is a threshold for benefit. Thus, the use of specific target values, although appropriate for providing easy-to-understand guidelines for the general public, distorts the true relationship between activity or fitness and specific health outcomes. This creates problems when we attempt to develop brief statements on how much activity at what intensity needs to be performed when there is a very wide range in physical activity habits and exercise capacity in the target population. More data are needed on the shape of the relationship between the activity dose (amount and intensity) and specific health outcomes. Such research will depend on having very accurate measures of activity as well as the change in health outcomes.

Multiple Biological Changes May Produce a Specific Clinical Benefit

For some clinical conditions such as CHD, it appears that a number of activity-induced biological changes contribute to a reduction in morbidity and mortality. These changes include but are not limited to such items as improvements in the plasma lipoprotein profile, a decrease in insulin resistance, reductions in blood pressure, increases in coronary blood flow, and decreases in myocardial oxygen demand. It is likely that the dose–response relationship varies among these biological changes, and the dose–response relationship for CHD clinical events is some integration of the dose–response relationship for all of the biological mechanisms. Thus, the dose–response relationship for any one biological variable does not necessarily represent the dose–response relationship for a reduction in CHD clinical events. It would be valuable to know how the dose–response relationships for specific biological risk factors relate to the dose–response relationship for clinical outcomes such as myocardial infarction, stroke, type 2 diabetes, and site-specific cancers. All of these major clinical entities are likely to benefit through a variety of activity-induced biological changes.

Different Doses Required for Different Clinical Benefits

When we try to determine what dose is needed for better health, the specific outcome of interest has to be taken into account because of the specificity of the physical activity effect. Substantial data exist demonstrating that the dose–response relationship varies widely for different health benefits. For example, the stimuli for changes in fat and carbohydrate metabolism (decreasing risk of CHD and type 2 diabetes) are most likely to be responsive to increases in total energy expenditure above some

intensity threshold, whereas changes in bone density are likely attributable primarily to the stress placed on bone by forces working against gravity or by vigorous muscular contraction. To prevent weight gain or to produce weight loss, the most important component of the activity dose is the total amount of activity performed expressed as energy expenditure. To increase aerobic capacity an increase in intensity appears to be critical, but in the prevention or treatment of hypertension, moderate-intensity exercise appears to be at least as effective as vigorous-intensity activity, if not more so. Most current public health physical activity guidelines try to take these differences, as best we know them, into account by promoting a variety of activities.

Activity Dose for Maintaining a Healthy Body Weight

The major physical activity and health recommendations of ≥30 min of moderate-intensity activity on most, preferably all, days of the week are based primarily on chronic disease outcomes, such as decreased rates of heart attack, stroke, type 2 diabetes, colon cancer, and osteoporosis. Although this amount of activity performed over months and years is likely to contribute to the maintenance of a healthy body weight, recent recommendations based on data from various sources have indicated that greater amounts of activity may be needed by many people to prevent unhealthy weight gain, lose weight, and prevent weight regain. For example, the International Association for the Study of Obesity concluded that for adults 45 to 60 min per day is required to prevent the transition to overweight or obese and that prevention of weight regain in formerly obese individuals requires 60 to 90 min of moderate-intensity activity or less amounts of vigorous activity (Saris et al. 2003). Similar recommendations were made in the Dietary Guidelines for Americans 2005 and include the following statements (U.S. Department of Health and Human Services and U.S. Department of Agriculture 2005).

- To reduce the risk of chronic disease in adulthood, engage in at least 30 min of moderate-intensity physical activity, above usual activity, at work or home on most days of the week.
- For most people, greater health benefits can be obtained by engaging in physical activity of more vigorous intensity or longer duration.

- To help manage body weight and prevent gradual, unhealthy body weight gain in adulthood, engage in approximately 60 min of moderate- to vigorous-intensity activity on most days of the week while not exceeding caloric intake requirements.

- To sustain weight loss in adulthood, participate in at least 60 to 90 min of daily moderate-intensity physical activity while not exceeding caloric intake requirements.

Whether a person is in energy balance and either gains or loses body mass, especially adipose tissue, is determined both by the number of calories consumed and by the number expended through physical activity. Thus, any physical activity dose recommendations intended to prevent weight gain or regain or produce weight loss need to consider

© Corbis

Although it is essential to control what and how much one eats to maintain a healthy body weight, exercise is critically important as well. A healthy weight is best achieved by controlling both calories eaten and calories expended during activity.

what is happening to caloric intake. The reader is referred to chapter 11 on physical activity, physical fitness, and obesity for further discussion of this topic.

Acute Versus Chronic Training Responses

When we attempt to define the dose–response for the health benefits of activity, it is useful to consider that some effects may occur as an acute response to a single or several bouts of exercise, some only as a training response, and some as the result of an interaction between acute and chronic training responses. An example of an acute response is that after older men and women with moderate hypertension performed a 30 min bout of stationary cycling at either 40% or 70% of $\dot{V}O_2$max, there were significant decreases in blood pressure for 8 to 12 hr postexercise (Pescetallo et al. 1991). Also, it was demonstrated that in men with high plasma triglyceride concentrations, fasting levels the morning after a 45 min bout of exercise at approximately 75% of aerobic capacity were lower than when such exercise was not performed (Gyntleberg et al. 1977). Over a 5-day period, if exercise was performed every day, fasting triglyceride concentration decreased further the following day. Thus, this acute response was augmented by repeated bouts of exercise. The reduced triglyceride concentration was rapidly reversed if the exercise was not performed for several days. This could be called an augmented acute response in that there was no further decrease even after weeks of regular activity. It is likely that as a person's physical working capacity increases and the absolute intensity of activity performed during a session is increased (relative intensity stays the same), the acute responses of various biological reactions will be enhanced. A more fit person can expend more energy during a set period of time. This would be true for any benefit directly tied to the magnitude of energy expended during the exercise session. For a more detailed discussion of acute responses to exercise, see chapter 5.

Maximizing Health Benefits While Minimizing Health Risks

With regard to health status, physical activity can present a two-edged sword. As the intensity or amount of activity performed is increased, the greater is the risk that injury will occur, especially musculoskeletal problems for many individuals and cardiovascular complications in those people already with underlying disease. Of particular concern when attempting to establish the optimal dose of activity for health outcomes is intensity because it is the major contributor to activity-induced medical complications. Thus, the dose–response assessment needs to consider not only what dose induces the greatest health benefit but also the risk profile for that dose. It may be that high-intensity activity (running) will provide greater benefit for a specific biological outcome but that a moderate-intensity activity (brisk walking) will provide the best overall health benefit in at-risk populations because of its lower risk profile. It will be valuable to establish the risk profile for various activity regimens in different populations, especially those at an increased risk of injury, such as obese and older persons. See chapter 19 for more information on medical risks associated with increased activity and how to reduce these risks.

Physical Activity Versus Fitness When Considering Dose for Health Benefits

In addition to the large volume of published data that link higher levels of activity to reduced chronic disease and all-cause mortality rates, a number of studies have investigated the relationship between level of cardiorespiratory fitness and similar clinical outcomes (see chapters 9, 10, and 13). In general, studies report that the least fit adults have higher morbidity and mortality rates than those who are more fit, with the lowest clinical event rates usually occurring in the most fit men and women. In these studies, fitness levels have been determined using submaximal as well as maximal exercise tests on a motor-driven treadmill or cycle ergometer. Measures of fitness have included heart rate measured at a standard submaximal workload, aerobic capacity (maximum METs) estimated from peak work performed, and $\dot{V}O_2$max measured by analyzing samples of expired air.

> To maximize the health outcomes of an activity dose, the benefits need to be optimized while the medical risks are kept to a minimum.

As is discussed elsewhere in this book (chapter 22), a person's fitness and health are determined by a complex set of interactions between his or her genes and environment, particularly behavior. The major environmental factor that influences cardiorespiratory fitness of most healthy adults is their habitual performance of endurance-type physical activities. Thus, measures of cardiorespiratory fitness indicate the amount and intensity of endurance-type activity performed by an individual over the past weeks or months. In some prospective observational studies evaluating predictors of chronic disease morbidity and mortality or all-cause mortality, a measure of cardiorespiratory fitness has been used as a surrogate for, or an indicator of, habitual physical activity. In studies with a larger sample size and a substantial number of clinical events, it has been possible to establish a dose–response relationship between level of fitness and mortality.

Figure 20.4 shows the dose–response between maximum METs determined during symptom-limited treadmill testing and all-cause mortality in men with and without evidence of cardiovascular disease at time of testing and followed for an average of 6.2 ± 3.7 years (Myers et al. 2002). Mortality is expressed as "relative risk" of death, with the risk for the most fit quintile expressed as 1.0. These data show a strong inverse dose–response gradient across all levels of fitness, with the least fit men having a risk of dying during the follow-up period that was four times greater than their most fit counterparts. The MET capacity at each quintile of fitness was higher for the participants free of cardiovascular disease at baseline compared with those with disease. It is possible that the patients with cardiovascular disease were less active, but their overall lower exercise capacity may have been attributable to their cardiovascular disease.

The data from Myers and colleagues (2002) in figure 20.4 are quite similar to a number of other studies that have evaluated the dose–response relationship between cardiorespiratory fitness and all-cause mortality. They all have shown that the least fit men and women fare the worst. Men and women who have the highest levels of fitness have the lowest age-specific all-cause mortality rates.

Data relating all-cause mortality to levels of reported physical activity and to measures of cardiorespiratory fitness from selected prospective observational studies are plotted in figure 20.5. Fitness levels were categorized by the authors in either quartiles (Laukkenen et al. 2001) or quintiles (Blair

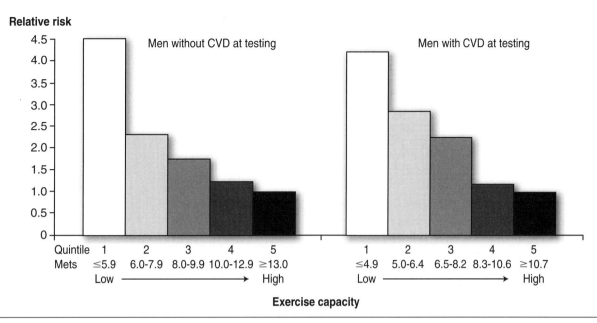

Figure 20.4 Age-adjusted relative risk for all-cause mortality according to quintile of exercise capacity (METs) for men with and without cardiovascular disease (CVD) at the time of exercise testing. Average duration of follow-up was 6.2 ± 3.7 years. The reference was set at 1.0 for the highest quintile of fitness.

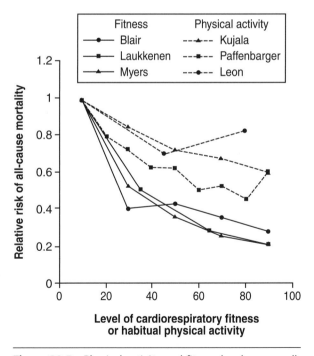

Figure 20.5 Physical activity and fitness levels versus all-cause mortality. The figure shows the dose–response for all-cause mortality by level of either cardiorespiratory fitness determined by exercise testing or reported habitual physical activity in men from six studies reported in the scientific literature. The relative risk for the least fit or active men was set at 1.0.

et al. 1989; Myers et al. 2002), whereas activity levels were categorized in tertiles (Leon et al. 1987), in quintiles (Kujala et al. 1998), or by eight categories of energy expenditure (Paffenbarger et al. 1986). The horizontal axis is an approximate percentile ranking for either fitness or activity for subjects in each study. The relative risk is consistently lower for men in the higher fitness categories compared with the men reporting higher levels of physical activity. Mortality data are presented as relative risks with the risk of death for the least active or fit group in each study designated as 1.0. (Note that these data are displayed with different reference groups—in figure 20.4 the highest fit group is at the reference risk of 1.0, and in figure 20.5 the reference category is the least fit group.) The pattern of results is similar, but the relative risks are different because of the different reference categories. This figure shows that for these studies, the gradient across fitness levels is steeper than the gradient across physical activity levels. This stronger association between fitness and death than between activity and death is likely because fitness is a more accurate and reliable measure than physical activity, and fitness is influenced by factors other than physical activity that can affect health and mortality.

Because of the complex nature of habitual physical activity, highly accurate and reliable measures that allow correct activity classification of each person in a large population are difficult to obtain. Whenever participants are misclassified as to activity level (e.g., they should have been in the first quintile but were placed in the second or third quintile because of overreporting), the strength of the observed relationship between activity and mortality is less than the true relationship. If the measure of fitness results in a more accurate classification of individuals than the measure of physical activity, this alone can account for a stronger association between fitness and mortality than for physical activity and mortality. Also, this stronger relationship between fitness and mortality may be attributable, at least in part, to factors other than physical activity that improve fitness and also increase longevity. Both cardiorespiratory fitness and all-cause mortality have significant genetic as well environmental determinants (e.g., cigarette smoking, poor nutrition) that could enhance the relationship between fitness and decreased mortality beyond that contributed by level of physical activity.

Although the magnitude of the dose–response relationship between fitness and mortality appears greater than that for activity and mortality, data indicate that levels of activity associated with lower mortality rates also will produce meaningful increases in cardiorespiratory fitness. This is consistent with the notion that the lower mortality rates seen in more fit persons are in part attributable to higher levels of activity. For example, in the study by Myers and colleagues (2002), a single MET higher capacity was associated with a 12% lower all-cause mortality rate. For many sedentary adults, following the recommendation of adding 30 min or more of moderate-intensity activity to their routine activities of daily living on 5 days per week

> A stronger dose–response relationship exists between measured cardiorespiratory fitness and mortality than between reported physical activity and mortality. This difference is likely attributable to factors other than physical activity that influence fitness and mortality, including heredity, and the more accurate classification of people into fitness categories than activity categories.

will increase their capacity by at least 2 METs over 6 to 12 months.

Summary and Conclusions

It is now well established that most people who, in addition to the routine activities of daily living, perform moderate-intensity physical activity for at least 30 min on 5 or more days per week significantly reduce their risk of various chronic diseases. Added health benefits frequently occur at a greater activity dose by increasing either intensity (from moderate to vigorous but not exhausting) or duration. Some benefits, like significant weight loss, may have a higher threshold in terms of activity amount than other benefits, such as reductions in moderately elevated blood pressure. For selected health-related benefits, an adequate dose can be obtained by accumulating moderate-intensity activity throughout the day in bouts of 8 to 10 min each. The characteristics of the dose of activity needed to produce a positive health outcome will vary from person to person depending on health and fitness status, genetic makeup, and specific health outcome desired.

KEY TERMS AND CONCEPTS

accumulation: Process of adding together the amount of time spent in performing physical activity (usually in short bouts) throughout the day. The concept of accumulating rather than performing activity all in a single bout has been added to public health recommendations but is still under investigation.

health response: A change (usually an improvement) in a specific measure of health (broadly defined as in chapter 1) or a health-related biological or performance measure. Examples of health-related biological measures include HDL cholesterol, blood pressure, and bone density, whereas examples of health-related performance include changes in aerobic capacity, joint flexibility, and muscle strength.

overload: The increased demands placed on the body by a dose of physical activity characterized by type, intensity, duration, and amount. When tissues, organs, or systems are exposed to an appropriate overload, they typically respond by increasing their capacity or efficiency.

physical activity dose: Characteristics of the physical activity performed by an individual over a defined period of time, which include type, intensity, bout frequency, bout duration, and total amount or volume of activity performed.

progression: Rate at which the physical activity dose (intensity, duration, or amount) is increased over days, weeks, or months. For overload to produce health benefits rather than injury it needs to be increased in small amounts or at a slow rate of progression.

specificity: The concept that changes in the body that occur attributable to an increase in physical activity are specific to the nature of the demands produced by that activity. Specificity applies to the type of activity (endurance vs. strength or arms vs. legs), intensity, and amount.

STUDY QUESTIONS

1. Briefly describe the three major principles that are the basis for the way the body responds to a habitual increase in physical activity.

2. Name the four main components of a physical activity plan that need to be considered when defining a dose of physical activity.

3. What is meant by an acute response to a bout of physical activity, and how does this differ from a training response?

4. Which component of the exercise dose is most closely linked to the risk of orthopedic injury or cardiac arrest? How can the activity plan be changed to reduce this risk?

5. Describe the dose–response relationship between cardiorespiratory fitness and all-cause mortality. Explain how the dose–response relationship differs for physical activity versus cardiorespiratory fitness and all-cause mortality.

6. Describe the concept of activity accumulation and the characteristics of the activity that contribute to this accumulation.

© Dennis Light

Physical Activity and Exercise Programs

■ Adrian Bauman, MB, BS, MPH, PhD, FAFPHM

CHAPTER OUTLINE

What Are Physical Activity and Exercise Programs?

- Physical Activity
- Original Recommendations for Exercise Programs
- Changes to Public Health Recommendations Based on Epidemiological Evidence
- Determining the Success of Programs or Interventions

Settings for Physical Activity and Exercise Programs

Strengths of Structured Versus Unstructured Programs

- Structured Exercise Programs
- Unstructured (Lifestyle) Programs

Physical Activity and Exercise Programs for Subpopulations and Groups

- Physical Activity and Exercise Programs for Minority and Disadvantaged Populations
- Physical Activity and Exercise Programs for Women
- Physical Activity and Exercise for Older Adults
- Physical Activity and Exercise for Children and Adolescents

Summary of the Effectiveness of Physical Activity Interventions

Review Materials

319

The previous chapters identified the health benefits of physical activity for a range of conditions. The World Health Organization has conducted a Global Burden of Disease project which further identifies the disability, morbidity, and mortality attributable to physical inactivity. Especially in developing countries, physical activity, nutrition, and tobacco control are the most important parts of overall efforts at disease prevention. Despite its major contributions to health, physical activity promotion typically receives relatively little of the prevention budget. This is an important area for physical activity research and practice, because physical activity has been often described as "the best buy in public health."

This chapter focuses on what we can do to increase physical activity and exercise in different groups, populations, and settings and at the regional or national level. Different outcomes are sought for different levels of **intervention** and different approaches to the design, objectives, and outcomes for these diverse physical activity and **exercise programs.** This chapter describes the range of interventions from population-wide efforts to increase activity to clinical efforts to increase cardiorespiratory fitness, strength, or other physiological outcomes. The evidence for structured exercise programs compared with unstructured programs is reviewed. Finally, evidence for best practice in physical activity interventions in subpopulations is appraised, focusing especially on **physical activity programs** for disadvantaged and minority groups, children and youth, women, and older adults.

What Are Physical Activity and Exercise Programs?

This section defines physical activity and exercise and examines what comprises a program or intervention to increase or influence them.

Physical Activity

The first issue is to define what is meant by physical activity and how it differs from physical fitness and exercise. This has implications for efforts to change physical activity or exercise habits. The central com-

ponent of physical activity is that it implies any large muscle (body) movement. Intensity, duration, and frequency are attributes used to characterize physical activity. Other attributes, such as the setting for physical activity, describe the context in which the activity occurs. The diverse settings for physical activity include household and domestic physical activity, including vigorous housework and gardening, which may confer some health benefits. Work is also a setting for physical activity, although many occupations have become increasingly sedentary in recent decades. Other settings include active commuting, such as walking or cycling to work or even walking to the bus or train station and then using public transportation. Finally, organized sports and active leisure and recreation are also settings for physical activity.

Original Recommendations for Exercise Programs

One subgroup of physical activity has the purpose of increasing fitness levels; this has been described as "exercise," defined as planned and repetitive structured movement with the aim of increasing or maintaining fitness levels. The kinds of fitness mostly referred to here are those that enhance health (see chapter 1). Some attributes of fitness are related to morphological attributes or muscular power or strength, but the term *fitness* in the context of exercise programs has most often been used to imply cardiorespiratory fitness. The original health-related recommendations described exercise program goals of achieving an exercise frequency of around three to five times per week, for around 20 to 30 minutes each time, and at 65% to 80% of the maximal heart rate. This was described as "aerobic" exercise, and achieving this kind of increase in exercise volume or duration at this level of intensity was expected to improve health-related fitness.

Changes to Public Health Recommendations Based on Epidemiological Evidence

During the 1980s and 1990s, increasing epidemiological evidence led to a change in the physical activity recommendations for populations. This information was based on controlled trials in clinical settings as well as observational epidemiological studies of large cohorts of adults that were followed for decades. Disease risks were assessed in those who were active or fit compared with those who were less active, sedentary, or unfit. The new public health recommendation of half an hour of at least moderate physical activity on most days of the week was developed though various scientific consensus meetings and promulgated through a report from ACSM and the Centers for Disease Control and Prevention (Pate et al. 1995). Epidemiologic studies repeatedly demonstrated that moderate levels of physical activity had health benefits, especially with respect to cardiovascular disease, diabetes prevention, and hypertension control. Most of these studies assessed leisure-time physical activity as the exposure measure rather than other domains of physical activity and types of energy expenditure. It was noted that participation in more vigorous or more sustained activity (formerly referred to as "aerobic" levels of activity) would confer additional health benefits, especially for weight control and cancer prevention. Finally, there was increasing recognition that programs designed to increase muscle strength, often through programs of progressive resistance training, also had health benefits independent of other forms of physical activity or exercise.

Determining the Success of Programs or Interventions

Physical activity or exercise programs are usually planned efforts to influence individuals, groups, or populations to change, improve, or increase their physical activity or exercise levels, with the ultimate goal of producing positive health outcomes. Such programs can be subjected to rigorous scientific evaluation to assess their impact or effects on a range of health and other outcomes. The evidence, and whether it is "causal" or caused by the intervention, is central to any policy decisions regarding program effects.

An intervention program may aim to increase physical activity or may focus on more physiological changes in fitness or muscle strength as a causal consequence of participation in the program. This is illustrated in figure 21.1. It is thought that these changes, if sustained, will eventually lead to health improvements, such as reduced risk of developing coronary heart disease, improved mood or quality of life, or improved lipid profiles and blood pressure.

To judge the effectiveness of a program or intervention, the best possible scientific evidence should be considered to conclude whether the changes that were observed were caused by exposure to the program. This requires a detailed consideration of evidence generation and the quality of causal inference, which is how we might conclude that one program works and another program doesn't. The epidemiological and scientific evidence requires the best possible research design, preferably a randomized controlled trial (RCT). In public health and community interventions, we often recommend quasi-experimental designs. This is where

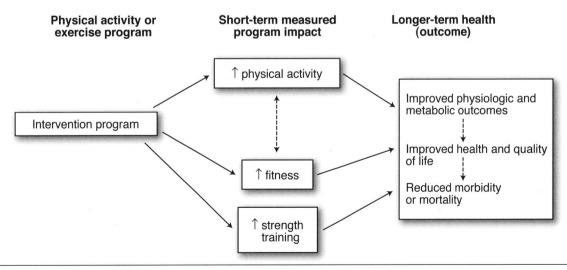

Figure 21.1   Range of outcomes of an intervention program, including both the behavior of being physically active and more physiologic outcomes such as fitness and strength training; these are the short-term effects of programs with subsequent health outcomes observed in the short term (metabolic changes) and long term (mortality).

whole communities or units are compared with comparison or control communities, for example, small communities or worksite programs (Bauman 1998). Less rigorous designs may be required to evaluate large-scale or national programs, where the possibility of a control or comparison region may not exist. Examples of the range of research designs used to assess the effectiveness of physical activity interventions are shown in table 21.1.

The quality of measurement is another issue. Are the outcomes measured in ways that are reliable and valid? How can we distinguish the quality of self-reported physical activity measures, which may be based on questionnaires or interviews, and which are better in terms of measurement properties? A range of more objective measures might be the outcomes of physical activity or exercise interventions; these include measures of fitness, oxygen uptake, adiposity, strength, and flexibility or pedometer or accelerometer counts.

Other research factors might influence the results of an intervention. One issue is whether program attendees are more motivated, more enthusiastic, or more educated than the general community who might subsequently be invited to attend the program. In other words, is the magnitude of the results seen attributable to selection effects, expressed as differences between program participants and those who did not participate? This can cause problems in assessing the **generalizability** of the program, which is how well it would work if offered to a wider group of people. For example, in a worksite program, if only educated executives came to a gym-based program, they might improve their fitness, lipid levels, and blood pressure, but would these results be of the same magnitude if the program was offered to all the blue-collar workers in that workplace?

Finally, statistical methods in approaches to analysis may be required to adjust for differ-

Table 21.1   Hierarchy of Research Designs for Physical Activity Interventions

"Quality of evidence" rating	Research design	Physical activity examples
Excellent	Meta-analysis	Pooled findings from many studies to provide an average effect size (e.g., for the average effect of a worksite PA program; Dishman et al. 1998).
	Cluster randomized controlled trial	Random allocation by community or organizational unit to I and C groups (e.g., CATCH randomly assigned schools to the program or to controls).
	Randomized trial	Random allocation of people to I and C groups; examples abound in exercise programs in clinical settings.
Good	Quasi-experimental: time series design with control group	Nonrandom allocation to I and C groups; time series design adds measurement at multiple time points in both groups; possible matching of I and C groups for baseline similarity (by physical activity level, social disadvantage); for example, monthly fitness tests for a year in workers, serial pre- and postintervention assessments, in four I worksites and four matched C worksites.
Fair	Quasi-experimental: nonrandom selection of I and C groups	Nonrandom allocation of I and C participants or communities; for example, one physicians' practice or one school as the I site and another is chosen as C.
Poor	Single-group design	A single-group before–after study provides relatively weak evidence that any observed changes were attributable to the program; this may be all that is feasible in assessing the effects of a national mass media campaign, but it is preferable to improve the design by having several preintervention measurements (time series design).

PA = physical activity; I = intervention; C = control; CATCH = Child and Adolescent Trial for Cardiovascular Health.

Epidemiology and Causality

These epidemiological texts and one paper might be useful for researchers and practitioners from an exercise science background without public health training. In understanding the "causal" nature of evidence, from interventions to epidemiological studies of health benefits of physical activity, we must consider criteria for causality. These are discussed in standard epidemiological texts, such as the following.

Bauman, A.E., J.F. Sallis, D.A. Dzewaltowski, and N. Owen. 2002. Towards a better understanding of the influences on physical activity: The role of determinants, correlates, causal variables, mediators, moderators and confounders. *American Journal of Preventive Medicine* 23(2 Suppl): 5-14.

Brownson, R.C., and D.B. Petitti, editors. 2006. *Applied epidemiology: Theory to practice.* 2nd ed. New York: Oxford University Press.

Gordis, L. 2004. *Epidemiology.* 3rd ed. Philadelphia: Elsevier Saunders.

Rothman, K.J., and S. Greenland, editors. 1998. *Modern epidemiology.* 2nd ed. Philadelphia: Lippincott-Raven.

There is a difference between the evidence one can generate from a small trial compared with that from a large-scale community-wide program. The small trial should use a randomized controlled design and can produce the best scientific evidence, as if in a laboratory setting (in fact, many will be clinic- or lab-based studies); this is known as an **efficacy** study, where the best case–control designs can be maintained and researchers can ensure sure that participants adhere to the exercise or training protocol, attend all sessions, and perform the exercises at the recommended method and intensity. However, if this study were then conducted in the community, the research to test program impact would be known as an **effectiveness** study, which determines whether the intervention would produce the same effects in field or free-living conditions. The research designs for effectiveness studies may be randomized trials; for example, randomize 20 schools, 10 to receive the intervention program and 10 control schools. This would then be analyzed at the school level as a cluster randomized trial. Alternative designs, such as nonrandom comparison groups, may be used, but careful attention to matching and comparability is required (Bauman 1998).

ences between intervention and control groups. Adjustment for other external factors that might contribute to the outcomes of interest (potential confounders) is the subject of much epidemiological literature, because it influences whether the effects observed are truly attributable to the intervention. Further research methods and epidemiological texts are recommended for more details here (see *Epidemiology and Causality* above), but it is sufficient to say that research design, measurement, selection, and appropriate analyses are essential components of judging the quality of evidence on whether a program works. If similar effects are observed in diverse populations and samples and positive effects are replicated across studies, then the causal evidence is considered of high quality.

Settings for Physical Activity and Exercise Programs

There are numerous settings for physical activity and exercise programs, ranging from individual-based programs to national and international efforts. These are shown in figure 21.2. Individual programs may target an individual who is otherwise well, or one with a specific set of risk factors or chronic conditions, and offer him or her a tailored individualized exercise prescription and plan. This will often be linked to a facility for exercise, such as a health club, gymnasium, or sporting complex. The next category of settings are at the group level and can include facility-based group programs, which may

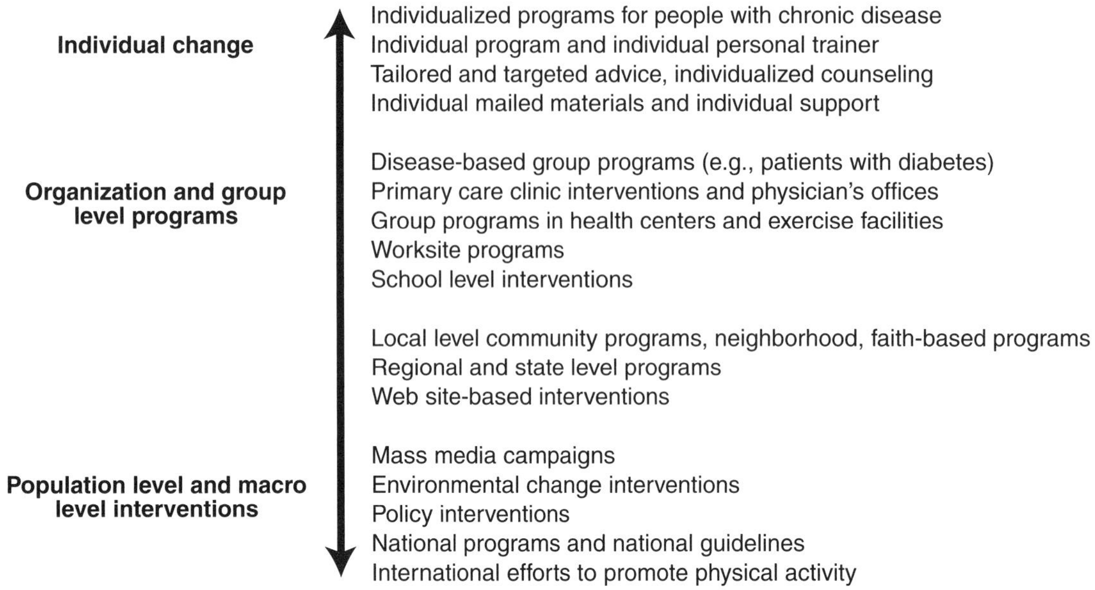

Individual change
- Individualized programs for people with chronic disease
- Individual program and individual personal trainer
- Tailored and targeted advice, individualized counseling
- Individual mailed materials and individual support

Organization and group level programs
- Disease-based group programs (e.g., patients with diabetes)
- Primary care clinic interventions and physician's offices
- Group programs in health centers and exercise facilities
- Worksite programs
- School level interventions

- Local level community programs, neighborhood, faith-based programs
- Regional and state level programs
- Web site-based interventions

Population level and macro level interventions
- Mass media campaigns
- Environmental change interventions
- Policy interventions
- National programs and national guidelines
- International efforts to promote physical activity

Figure 21.2 Range of physical activity interventions on a continuum from individual-level programs to public health and population-wide interventions.

be multiple-session, theoretically driven interventions with the aims of using both group process and social support as well as behavior change theory in a structured setting to increase and maintain exercise levels. Many programs at these levels can focus on a range of biological and physiological outcome measures, and they use randomized controlled trial designs.

Many theories are used to explain how people initially change their behavior and then how they maintain it. This is important in the context of physical activity and exercise programs, because more effective approaches are based on *tailoring* the interventions; this means developing programs matched to people's level of motivation to change. More effective programs are based on principles of adult learning, social learning theories, and various combinations of beliefs and perceptions about the benefits and outcomes of being active (Glanz et al. 2002). Programs that focus on reinforcement and on building long-term skills may be more likely to result in maintained behavior, in this case being active, beyond the duration of an 8- or 16-week clinic- or facility-based program.

Group programs include larger settings, such as worksites, primary care clinics or physicians' offices, schools, and local community groupings such as faith-based programs. There is substantial published literature regarding programs in these areas, but many articles fail to identify the selection effects in those who volunteer for the study. Issues of measurement are also important, because many studies rely on self-reported physical activity changes; the best measures available should be used. Many worksite programs target managers, executives, or senior staff and use existing corporate gyms or exercise facilities. These programs fail to demonstrate convincing effects, especially when one considers the many workers who don't attend the programs, particularly blue-collar workers, women, and minority groups (Dishman et al. 1998).

Effects for primary care and physician settings are more promising in terms of short-term increases in self-reported physical activity but are seldom seen to produce longer term effects. These findings are not universal, leading some researchers to question the effectiveness of this setting for increasing physical activity (Eden et al. 2002). However, the potential remains large, if one could persuade all primary care physicians to provide brief advice about exercise to all sedentary patients. Even with small effect sizes, the reach of the intervention would have important population-level effects. Therefore, the next generation of physical activity and exercise studies in primary care may target providers, or health care organizations, rather than individual patients. The key challenge is to disseminate the need for exercise advice to all health care providers, so that they recommend exercise to the sedentary as often as they recommend smoking cessation to regular tobacco users.

At the population-health end of the settings continuum (figure 21.2), interventions target large groups or whole populations. Although more than half of the adult population fails to meet the modest U.S. Surgeon General's 1996 recommendations of half an hour of physical activity on most days, the undifferentiated nature of the population means that there are many sedentary people not even interested in becoming active (U.S. Department of Health and Human Services 1996). Thus, the outcomes at this level of intervention may include proximal measures, such as awareness of the physical activity recommendations, understanding of local facilities and resources, or intention or level of confidence in one's capacity to be more active. At the national level, qualitative measures of such national guidelines or physical activity policies may be sufficient to indicate that the initiative was a success in showing the difference in outcomes between individual-level and macro-level programs.

It may be possible to use routine serial population health surveys to assess the impact of a regional intervention, if there is sufficient sample size for that region to allow tracking with enough precision to show differences to other regions or national data (see figure 21.3). Examples of this kind of serial surveys are the annual Behavioral Risk Factor Surveillance System (BRFSS) surveys in U.S. states and similar surveys such as the Finbalt surveys in Finland and Baltic states that have been tracking risk factors since 1979.

In summary, there is a range of settings for physical activity or exercise programs. For each setting, the size of the target group, costs of the program, required skills of staffing, and time required to implement and deliver the program should be considered. The resources needed will vary enormously, as will the skills of the staff in exercise science, measurement, behavior change, or even town planning and urban transport (if, e.g., they were elements of a national program). In addition, the intervention

Hypothetical physical activity trends over 6 years

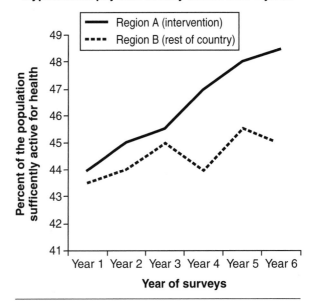

Figure 21.3 In this example, region A had a community-wide physical activity intervention, and the remainder of the country, labeled region B, had no specific intervention. Both regions were monitored across years. The prevalence of activity changed more in the intervention region (region A).

itself will vary, from educational resources and self-help materials to changes to organizations and the physical and social environment. The complexity of developing the right mix of settings, populations, and program approaches is illustrated in figure 21.4. A series of options are possible, even in combination for multistrategy programs addressing inactivity in large populations.

Figure 21.4 shows a model for conceptualizing physical activity and exercise programs from the perspective of health promotion and disease prevention. Much of this thinking is derived from the Ottawa Charter for health promotion and applied to physical activity (Bauman and Bellew 1999). The dimensions are level (or type) of intervention, target population group, and organization or person who is going to deliver the intervention (figure 21.4). The type of intervention ranges from individual-based education to group exercise classes to broader population-level and environmental change interventions. The target population should be specified and the type of intervention considered in light of the number of people who potentially will be influenced by the intervention. Finally, will the intervention be delivered in person, by mail or the Internet, by nongovernment organizations (not for profit organizations), by government agencies, or

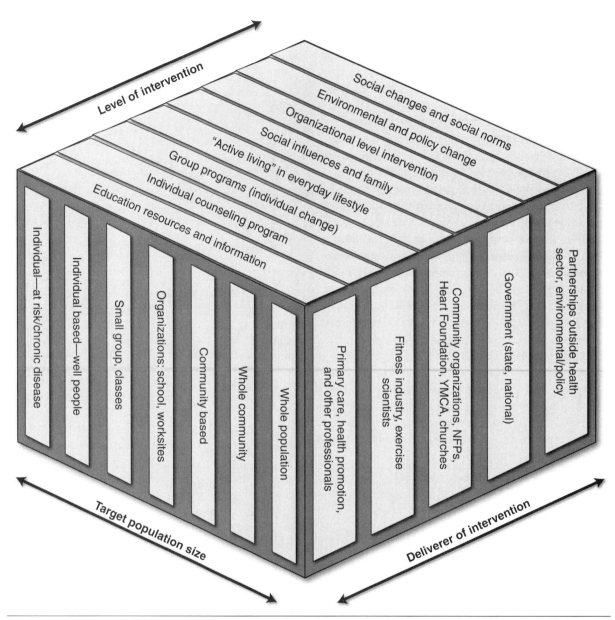

Figure 21.4 The complex interplay between the target group, levels of intervention, and agents or organizations that will deliver the program. This is the challenge for multicomponent and comprehensive programs.

by broad coalitions of stakeholders and agencies in partnership (Bauman and Bellew 1999)?

Strengths of Structured Versus Unstructured Programs

An important issue is the controversy regarding optimal ways of maximizing individual physical activity change. The scope of efforts to engender environmental and policy change is central to public

health approaches to increase physical activity and has been summarized elsewhere (Sallis et al. 1998). Similarly, at the macropolicy level, partnerships are required to organize the resources and deliver programs to influence whole populations.

Recent attention has focused on ways of inducing and maintaining individual-level change. Do individuals achieve more change following supervised structured exercise programs, or can people do as well in home-based physical activity and exercise programs in their discretionary time? This is important in terms of assessing initial program effects as well as long-term maintenance and sustainability of physical activity behaviors.

Two relevant concepts need to be defined here: The first is "structured exercise," defined as predetermined supervised programs consisting of continuous aerobic exercise. Structured exercise programs are usually performed at reasonably high intensity (between 60% and 70% of maximal levels), involve frequent attendance (typically three to four sessions per week), and are held in a facility or exercise center. The duration of such programs varies, but participants typically enroll in programs lasting from 8 to 20 weeks and participate in classes of 40 to 60 min duration. The exercise sessions usually consist of aerobic activities, such as cycling, treadmill jogging, fast walking, or swimming, as well as activities designed to increase muscle strength (Boule et al. 2003).

This is contrasted with the concept of "lifestyle activity," which is the much broader concept of expending energy in a range of ways—around the house, through active commuting, and through utilitarian walking. This concept encompasses

active living through participation in all of the physical activities that might be re-engineered back into daily living. Examples of both structured and lifestyle physical activity are detailed in table 21.2.

Structured Exercise Programs

Part of the argument for **structured exercise programs (SEPs)** relates to their organized and settings-based nature. SEPs can develop and extend the intensity and duration of exercise, tailored for each individual, to maximize fitness enhancement. This may be a time-efficient way, in a safe environment, to provide the maximum health benefits. Because programs are supervised, safety is also a benefit. It is not even essential for SEPs to provide aerobic-level activity; many programs, especially for sedentary older adults, could initially provide lower-intensity programs of walking and stretching, which would confer health benefits. Exercise classes can be tailored to suit the needs of different age and ethnic groups and to provide tertiary prevention through classes for people with established disease; examples are programs for cardiac rehabilitation patients and for people with diabetes or arthritis.

Table 21.2   Examples of Structured Exercise Programs and of Incidental (Daily) Physical Activity

Structured exercise programs	Daily incidental physical activities
Exercise prescription of vigorous activity (50-85% maximal) for at least three sessions per week of supervised sessions with an exercise leader in an exercise facility (Dunn et al. 1999)	Accumulating at least 30 min of physical activity on most days; supported by cognitive–behavioral classes to discuss lifestyle activity options (Dunn et al. 1999)
Structured strength training supervised classes, three times per week, 40 min duration, for 16 weeks, upper-body strength training-resistance exercise with hand weights, dumbbells	Regularly using stairs, incidental walking for chores and local shopping, walking the dog regularly, walking for transport or to and from public transit or bus stations; creating short walking opportunities, and always walking at a moderate to brisk pace
Organized sport and team participation; structured training exercise sessions with a team trainer	Undergoing physical activity through domestic tasks and chores such as energetic gardening, home maintenance tasks around the home, and vigorous continuous housework
	Local walking or cycling for transport for short utilitarian (purposive) trips (up to 1 mile or more for walking and up to 5 miles for cycling)
	Engaging in cultural expression that is physically active, including dance and movement and other cultural and traditional activities that are moderately and continuously active

In addition, SEPs have the potential to add several components to a physical activity regimen, including stretching, aerobic activity, and muscle strengthening. Given the complexities of these modes of activity, initial training in a supervised setting may be beneficial for some individuals who would not otherwise start and maintain an exercise program. This supervision may also reduce the risks of injury and other adverse events, which are higher in untrained, inactive, middle-aged populations. Boule et al. (2003) suggested that at least 80% of participants complete SEPs; this is much higher than adherence to regular physical activity regimens that are self-initiated in the community. Program adherence is important to continuation of the behavior, and repeated, structured training assists this.

Tudor-Locke and colleagues (2002) showed that participation in SEPs led to a "profound increase in walking, even on the same days as the SEP" (p. 354), questioning the previously held hypothesis of compensatory sedentary behavior. This hypothesis was that people who went to structured classes believed that they did not need any more daily activity and would be less likely to walk or perform incidental activity. Tudor-Locke and colleagues (2002) found that the SEP was the major source of any vigorous activity, but there was not the expected decline in activity in other parts of the day; in fact, the SEP seemed to motivate people to think about being more active outside of programs.

It may be the regular and structured nature of programs that stimulates adherence, and this structure is often maintained following the program. The SWEAT trial in Western Australia enrolled 126 sedentary women into center-based versus home-based exercise programs and observed better adherence and higher total energy expenditure 12 months later in women who had started with a center-based program (Cox et al. 2003). The goal setting that is possible in theoretically sound SEPs may improve long-term adherence even in socially disadvantaged populations.

Many of the evidence-based trials have been organized around SEPs as central parts of the intervention. For example, the evidence that "lifestyle" exercise prevents diabetes (the U.S. Diabetes Prevention Project) was based on an initial structured program with frequent contacts, "lifestyle coaches," and center-supervised physical activity sessions twice weekly (Knowler et al. 2002). Although the focus was on encouraging lifestyle physical activity, the setting for encouragement, training, and motivation was a center-based program led by trained experts. This was required to increase the probability of enough behavior change to examine whether it was then associated with reduced diabetes incidence. The diet and physical activity programs together lowered the risk of diabetes by 58% in those with impaired glucose tolerance (a group at very high risk of developing diabetes).

Unstructured (Lifestyle) Programs

Lifestyle activity involves integrating physical activity into daily life; as such, it is often described as **"active living."** It is closer to a population health strategy, and its recommendations are closely related to the moderate-intensity threshold for health proposed by the 1996 U.S. Surgeon General. The arguments for "active living" include the following:

- There is a selection bias in the type of people who enroll in structured exercise programs; currently these form a small proportion of the total sedentary population.

- Many of the opportunities for increasing total physical activity occur outside of structured programs, and the range of domains for active living offer more opportunities for a variety of activity to suit individual preferences.

- If activity is integrated into daily routines, there is a greater probability of increasing activity and maintaining it as part of a changed lifestyle.

- Active living can be supported by environmental and policy changes—building sidewalks and bike lanes, providing better public transport systems, and making communities more pleasant, interconnected, and walkable.

The intervention literature supports the benefits of lifestyle approaches to physical activity. Key examples of the evidence for lifestyle interventions (promoting "incidental" physical activity) was provided by Andersen and colleagues (1999) and Dunn and colleagues (1999). The first of these was a small RCT among obese women, with a lifestyle versus 16-week structured diet and aerobic exercise program. Weight loss was similar at 8 kg in both groups after 4 months, but weight regain was smaller in the lifestyle group (0.08 kg) than the structured group (1.6 kg) at 12 months (Andersen et al. 1999). A large study was Project Active, an RCT with around 120 women and men per arm allocated to traditional exercise prescriptions and individual supervised sessions, compared with participants in the lifestyle group, who were advised to choose activities of daily living they enjoyed. Maintained increases in fitness, energy expenditure, and heart disease risk factors, such as blood pressure, were comparable between the groups at 24 months (Dunn et al. 1999).

These well-designed studies indicate that life-style interventions may be as effective as structured programs, provided that home-based participants receive training and follow-up. There is evidence that the active living interventions are more cost-effective and deliver the ancillary benefits of physical activity in reducing other risk factors. Cost-effectiveness was shown in an analysis of Project Active, where lifestyle intervention costs at 24 months were around $17 per participant per month, which was a third of the costs of the structured program; similarly, costs per unit increase in energy expenditure in the lifestyle program were a third to a fifth of the costs in the structured program (Sevick et al. 2000). The convenience factor and potential population-wide generalizability of lifestyle programs are important. In these respects, lifestyle programs appear to have promise and should be encouraged where it is feasible to ensure that quality home-based programs are delivered. However, the adherence issue remains unresolved and it may be that structured programs show better adherence, possibly because of the selection effects of who enrolls in them but possibly because structure and supervision enhance maintained activity for some people.

In summary, there is evidence of success for both SEPs and lifestyle physical activity programs. They target different groups, the SEPs being desirable for those willing and able to attend structured programs. We should not view these approaches as competing or consider that one is better than the other; rather they are options for enhancing health through physical activity and exercise. Some individuals will probably prefer to use the structured approach to increasing activity and obtaining the health benefits of such a change, and others will prefer the lifestyle approach. The best approach for an individual is the one he or she will adopt and maintain.

Physical Activity and Exercise Programs for Subpopulations and Groups

Physical activity and exercise programs may be differently implemented and demonstrate differing effectiveness in diverse population groups and sub-samples. This section reviews interventions targeting minority populations, women, older adults, and youth and adolescents. There are too many studies to review them individually, so principles for best practice are described as the evidence allows and review papers are cited for further reference.

Physical Activity and Exercise Programs for Minority and Disadvantaged Populations

Minority populations are often the least physically active in the population and often have the added burden of social disadvantage and impoverished physical environments as barriers to physical activity levels (Brownson et al. 2000). There is limited good quality evidence for interventions with these population groups, because many studies have small samples, nonexperimental designs (Bauman 1998), and short-term follow-up (Taylor et al. 1998). Apart from these methodological issues, which indicate the need for better-designed studies to generate evidence, several issues appear to improve the relevance and likely effectiveness of minority-targeted interventions. These include ensuring cultural appropriateness, using social support and community supports, developing community partnerships and coalitions, and considering the physical environment, especially in socially disadvantaged neighborhoods (Taylor et al. 1998).

Physical Activity and Exercise Programs for Women

In most population surveys, women seem to report less leisure-time physical activity than men across the age groups (Brownson et al. 2000). The gender differential begins with declines in physical activity among midadolescent girls, so gender-focused interventions might consider this as a pivotal point for initial intervention. One concern is in the ways that physical activity is measured, because women

When pleasant and convenient infrastructure for physical activity is provided by public funds, the activity level among the population is likely to increase.

may report less leisure-time physical activity but carry out greater volumes of domestic-setting energy expenditure at low intensity (Brownson et al. 2000). The barriers to women being more physically active include responsibilities for young children, unsafe environments, and lack of social support. Effective programs might consider the time constraints for women, the need for social supportive interventions, and the need for urban safety such as provided by group walking programs. The population goal is to integrate physical activity into everyday life—where active living may be an appropriate strategy.

Physical Activity and Exercise for Older Adults

Aging is associated with declines in physical activity, strength, and fitness but is also associated with increased risks of noncommunicable diseases, decreased mobility, and increased injurious falls. The special intervention outcomes sought following physical activity and exercise programs among older adults include improving functional status and mobility, preventing falls, and improving mental

health and social functioning (Taylor et al. 2004). Different types of physical activity are required for elderly people; in particular, strength training and progressive resistance training reduce depression, enhance functional status, and improve glucose homeostasis. Programs such as walking may be most efficient for encouraging physical activity, but facilities-based programs for strength training are important (van der Bij et al. 2002). The best initiatives may be multilevel interventions, with community supports to reinforce supervised programs. Tailoring programs to individual needs and capacity and including balance, gait, and resistance training may best reduce the incidence of injurious falls.

Physical Activity and Exercise for Children and Adolescents

Little is known about the genesis of inactivity through childhood. Few interventions or even prevalence estimates are available for preschool-aged children, but declines in energy expenditure in this age group may result from increasing rates of working parents and increased childcare center attendance. For school-age children, interventions are divided into those at school and those outside of school (Timperio et al. 2004). The latter include active commuting interventions, such as walking to school programs.

There is little evidence that increasing in-school physical education classes will increase physical activity sufficiently to meet recommended thresholds. More comprehensive multiple-level interventions such as the Child and Adolescent Trial for Cardiovascular Health (CATCH) include both in-school and out-of-school components, with the latter focusing on family supports and parental modeling for physical activity behavior; results are encouraging even at 3 years of follow-up (Nader et al. 1999). Challenges to the evidence base include lack of objective measures in some research, although there are now established observational measures of playgrounds and school environments that can quantify physical activity levels. It may be that multisetting programs are required, with insufficient increases possible through active commuting, in-school programs, or family supports alone (Timperio et al. 2004).

Summary of the Effectiveness of Physical Activity Interventions

There are many ways of assessing the settings for physical activity interventions and for appraising the evidence of program effectiveness across settings. This poses challenges for researchers and practitioners and does not give clear guidance on recommended approaches for interventions to promote exercise and physical activity. This has been the subject of a recent systematic review carried out by the Centers for Disease Control and Prevention in the United States; this review summarized evidence for public health interventions to increase physical activity (Kahn et al. 2002). This systematic review identified different areas for physical activity interventions and summarized program effects against standardized criteria. For some areas there are insufficient studies to make clear recommendations, and for other areas the recommendations can be graded from not very strong to very strongly recommended, based on the evidence available. A summary of the major recommendations of this systematic review is presented in table 21.3. These were divided into programs emphasizing informa-

tion transfer, those emphasizing behavioral and social approaches, and those focusing on environmental and policy approaches. They are ranked in the table from largest population target to smaller group and individual approaches. Despite the clear recommendations, numerous cross-cutting research issues remain to be resolved in all of these areas. The broad strategies targeting environmental and policy interventions show promise, but it is too early to develop definitive policy and practice recommendations here. Other issues, such as some settings and population groups mentioned throughout this chapter, were not specifically considered by the review.

Levels of leisure-time physical activity and fitness have remained relatively stable in recent decades in many populations, and population levels of obesity have increased. The reasons for this increase in obesity include both increases in energy intake and also likely total decreases in energy expenditure, through increased inactive and sitting time and decreased incidental physical activity. Programs consisting of multiple strategies remain the best approach for changing social norms to facilitate population-level participation in more physical activity, but the evidence on the independent effects of each specific component remains to be elucidated. For example, we may know whether a program worked, but it may not be possible to determine how the successful ones worked; this will lead to research to identify effective components (intervening causal variables, or mediators) through which the program effects operate (figure 21.5 on page 333). Sometimes, effects will be different across population groups, and a program that works with Caucasian males might not work with Hispanic females. Such research will enable the development of more efficient targeted research, focusing programs on those most likely to change.

Figure 21.5 shows that the hypothetical program brings about substantial changes in self-efficacy and in motivational readiness to be more active (stage of change) but that the situation-specific confidence variable of "self-efficacy" is more strongly related to physical activity or fitness outcomes. This means that a program like this one can target increasing self-efficacy more than trying to influence motivational readiness; furthermore, structural issues such as program attendance are also related to long-term outcomes, so efforts should be put into maintaining weekly program attendance.

Table 21.3 Systematic Review by Centers for Disease Control and Prevention–Expert Group on the Evidence for Physical Activity Intervention Effectiveness

Intervention category and examples of the type, setting, and audience for the program	Recommendation of the expert group from a systematic review of PA program effectiveness
[Info] Community-wide education, multistrategy and multicomponent interventions targeting whole communities (or large population segments); paid media messages, PSAs with other ancillary programs and strategies	Strongly recommended (some of these were strongly led by mass media campaign elements)
[Info] Mass media campaigns alone (campaigns using paid or unpaid mass media to communicate and persuade people to be more active)	Insufficient evidence (or maybe achievable outcomes are proximal, such as increased understanding and community awareness raising)
[E&P] Creation or enhanced access to places for PA combined with informational outreach activities; improved physical activity environments for children; includes some worksite programs	Strongly recommended (see Sallis et al. 1998; intervention evidence still limited to few settings)
[E&P] Other environmental changes—transportation policy and infrastructure changes to promote nonmotorized transit; urban planning approaches and their impact on PA	Evidence not available; some cross-sectional associations shown, but data needed on the response following interventions in this area
[Info] Point-of-decision prompts—an environmental setting for providing information encouraging activity; for example, stair studies, with signage to encourage stair rather than escalator use (this is one kind of easily disseminated, low-cost environmental change that alters a single behavior such as choices about stair use, operated "on site and immediately")	Recommended
[Info] Classroom-based (curriculum-based) health education focused on providing information to children and youth about being active	Insufficient evidence on physical activity time as specific outcome; most showed knowledge improvements
[Beh/Soc] School-based physical education—modify school curricula and increase activity time in school programs, especially through PE classes being more active, or increased implementation	Strongly recommended (but small effects); impact on overall PA not clear
[Beh/Soc] Nonfamily social support—interventions to in-crease social support of others, friends, such as community walking groups, faith-based settings	Strongly recommended (increases in PA and in fitness generally seen)
[Beh/Soc] Classroom-based health education with TV and video game turnoff component; restricting television watching	Insufficient evidence (few studies) for PA outcomes; effects seen on reduced TV viewing
[Beh/Soc] College and university student-age interventions around physical education or health education courses—mostly lecture-based lessons	Insufficient evidence (based on only two published studies)
[Beh/Soc] Family-based social support programs include additional effects of family involvement in school-based programs	Insufficient evidence (but may be part of more comprehensive programs, e.g., CATCH intervention; Nader et al. 1999)
[PA] Individually adapted health behavior change—individually tailored programs (usually theoretically based) but often with participants who are self-selected; usually individual programs, sometimes small groups; often structured exercise programs	Strongly recommended (increases observed in PA and in fitness as outcomes); also some "good" cost-effectiveness seen

Info = information transfer; PSA = public service announcement; Beh/Soc = intervention emphasizing behavioral and social approaches; E&P = intervention focusing on environmental and policy approaches; PA = physical activity; PE = physical education; CATCH = Child and Adolescent Trial for Cardiovascular Health.

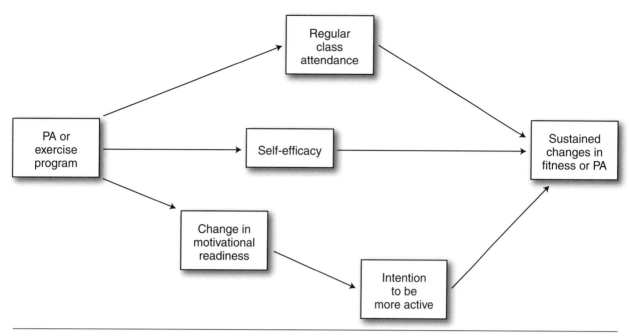

Figure 21.5   Mediators of change in physical activity habits. The width of the arrows indicates the strength of the association (in statistical models, these are usually indicated by path coefficients—analogous to standardized regression coefficients).

KEY TERMS

active living: Way of life in which physical activity is valued and integrated into daily life.

effectiveness: Effects of physical activity or exercise programs in real-world conditions, delivered by the real health care system, professionals, or relevant organizations. Effectiveness imposes less control on the delivery and implementation of programs than efficacy trials; hence results, if positive, may be more generalizable to the population.

efficacy: Effects of controlled evaluation of physical activity or exercise programs under laboratory or carefully controlled conditions.

generalizability: Confidence with which the effects of a study might apply to others like the research participants, or whether results might be expected to be replicated if the intervention were conducted in the whole population.

physical activity programs, exercise programs, or interventions: Specific purposive efforts to engage with individuals or populations to increase defined and measurable elements of physical activity or exercise and to attribute changes to participation in such programs.

structured exercise programs (SEPs): Planned exercise programs, consisting of continuous aerobic exercise or strength training, conducted under professional supervision, often in facilities or gyms.

STUDY QUESTIONS

1. What elements of research design provide confidence in the scientific quality of the evidence provided in a published intervention study?

2. What is the generalizability of a study's findings? How does it affect the interpretation of the findings of published studies?

3. What might be some of the elements of a lifestyle or incidental physical activity intervention compared with an exercise trial? What might participants be asked to do?

4. How does a population approach differ from an individual intervention? List three examples for the appropriate use of each method.

5. What caused health agencies and governments to change from recommending "exercise programs" to recommending "active living" programs? Do you agree with this shift in emphasis?

6. How would you judge a self-report measure of physical activity to be of good quality compared with poor quality?

7. What kinds of settings would you invest in if you ran a state health department and were asked to invest $1 million in physical activity and exercise programs?

8. What kinds of programs and groups would be more suitable for structured exercise programs, and who would benefit more from unstructured home-based programs?

9. Which other subpopulations (other than those already discussed—women, older adults, children, minorities) might warrant special attention to the types of exercise or physical activity programs that you might develop?

V

New Challenges and Opportunities

You have now been exposed to the broad area of physical activity and health, and you should have a better understanding of how our field has developed over the past several decades. In part II, five international experts presented current data and understanding of how physical activity affects the human organism and specific physiological systems. The 11 chapters in part III contain what is perhaps the core material of the book. A group of distinguished investigators presented material on what we know about how regular physical activity, or in some cases sedentary lifestyles, affect several different health outcomes. You have learned that a sedentary and unfit way of life has devastating effects on many health outcomes and decreases longevity. Part IV includes two chapters on what we know about specific amounts, types, and intensities of physical activity in relation to health outcomes and how this information can be integrated to produce recommendations for physical activity programs and approaches for implementing these programs in a variety of settings. Part V reviews current knowledge on how genetic factors regulate changes in various outcome variables in response to changes in physical activity. Finally, the editors present some concluding thoughts on the integration of data, thoughts, and concepts presented here and what this means for our field. The chapter concludes with a discussion of research areas on which we should focus our attention in the future.

Courtesy of Claude Bouchard

Genetic Differences in the Relationships Among Physical Activity, Fitness, and Health

▓ Tuomo Rankinen, PhD ▓ Claude Bouchard, PhD

This book provides ample evidence of the benefits of regular physical activity and the advantages of maintaining a reasonable level of health-related fitness. What has not been discussed in any detail thus far is the magnitude of individual differences in the relationship between physical activity or fitness and health outcomes and the role of genetic heterogeneity in accounting for human variation. These topics are the focus of this chapter. After a brief overview of the human genome, the unique features of human **genes** and the regulation of their expression or repression are described. The extent of variation in the genome of *Homo sapiens* is reviewed. Next we discuss how genetic variation affects sedentary people. The magnitude of the individual differences in response to regular exercise is then defined. We discuss the role of genes and DNA sequence variants in the response of blood pressure, lipids and lipoproteins, glucose and insulin, adiposity, and cardiorespiratory endurance. Finally the implications for public health and individualized preventive medicine approaches are highlighted.

Basics of Human Genetics

The first section of the chapter is devoted to defining the human genome and its key characteristics. Subsequently, the human gene, gene expression regulation, and the role of **alternative splicing** are briefly described.

Human Genome

The biology of the gene and the characteristics of the human genome are complex topics about which more is learned every day. The blueprint of the human body is contained in the genetic code specified in the deoxyribonucleic acid (DNA) sequence of chromosomes found in every nucleated cell. Human **genome** is a term that refers to the total genetic information in human cells. It consists of 22 pairs of autosomes and two sex chromosomes in somatic (non-reproductive) cells. Male and female gametes (germline cells) each contain a nucleus in which 23 chromosomes are normally present. The gametes are haploid, which means they contain a single set of chromosomes. The female gamete has a single copy of each of 22 autosomes and an X chromo-

some. The male gamete has a similar complement of autosomes and either an X or a Y chromosome. At fertilization, the nuclear contents of the female and male gametes fuse, and the diploid number of chromosomes (23 pairs) is restored with either two X chromosomes (XX, a female zygote) or an X and a Y chromosome (XY, a male zygote).

Chromosomes are composed of long chains of DNA and basic and acidic proteins packed with the DNA. The genetic material in each chromosome is a long string of the four DNA bases: adenine (A), cytosine (C), guanine (G), and thymine (T), joined together via phosphate bonds. The two complementary strands are precisely folded and twisted around one another to form a double helix, with the informative base on the inside. The strands of paired complementary sequences of nucleotides (bases) are held together with relatively weak hydrogen bonds (figure 22.1). C pairs with G, and A pairs with T. The 23 chromosomes contain about 2 m of linear DNA, or about 3 billion pairs of nucleotides. The linear structure of bases in the DNA strands is called the primary structure of the chromosome. The secondary structure of the chromosome arises when the two complementary strands of DNA twist to form a double helix. One turn of the helix is called a pitch and accommodates 10 nucleotides.

The chromosomes are arranged by size and position of the centromere; autosomes are numbered from 1 to 22, and the sex chromosomes are noted as X and Y. The short arm of a chromosome is denoted as p and the long arm as q. Each arm is subdivided into regions numbered consecutively from the centromere to the telomere (tip of the chromosome), and each band within a given region is identified by a number. With this nomenclature, it is possible to specify any chromosomal region by a "cytological address"; for example, 2p25 refers to chromosome 2, p arm, region 2, band 5 (figure 22.2 on page 340). However, the sequence of almost all the DNA bases of the entire human genome is now available (www.ncbi.nlm.nih.gov), and this is rapidly changing the way an address on a chromosome is defined. Indeed, it is now possible to specify a physical position on a given human chromosome in terms of the exact base number in a sequence ranging from one to millions. Thus, the availability of the DNA sequence for each chromosome makes it possible to go beyond the crude cytological address and specify an actual

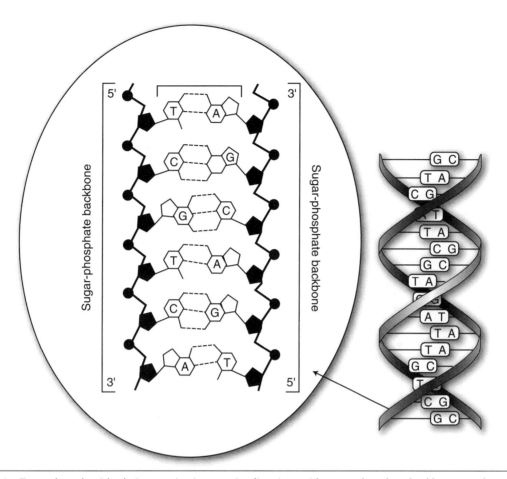

Figure 22.1   Two polynucleotide chains running in opposite directions with sugar-phosphate backbones on the outside and the nitrogenous bases inside paired to each other by hydrogen bonds. Adenine (A) pairs with thymine (T), whereas cytosine (C) pairs with guanine (G).

Reprinted from R. Roberts, J. Torobin, T. Parker and R. Bies, 1992, *A primer of molecular biology* (New York, NY: Elsevier), 22, with kind permission of Springer Science and Business Media.

physical position on a chromosome (in terms of base number).

Moreover, each mitochondrion of a cell contains several copies of circular, double-stranded DNA molecules composed of 16,569 base pairs. This is a small number of base pairs compared with nuclear DNA. Mitochondrial DNA (mtDNA) is able to replicate itself independently of nuclear DNA and has its own system of **transcription** and **translation.** Mitochondrial DNA is inherited from the mother through the cytoplasm of the ovum. This small DNA codes for 13 polypeptides associated with the regeneration of adenosine triphosphate (ATP) in the mitochondrion and for two ribosomal [RNA (rRNA)] and 22 transfer ribonucleic acid (tRNA) molecules. The remaining genetic information required for the synthesis of the proteins of the mitochondrion originates from nuclear DNA. The

importance of mtDNA cannot be overestimated because it has been implicated in a number of fundamental biological processes associated with the ability of the cell to meet changing energy needs and also with aging, a number of anomalies and diseases, and cell death.

Human Gene

A typical gene (figure 22.3) consists of coding sequences **(exons),** noncoding regions **(introns),** and regulatory sequences located both before (5' end; **promoter** region) and after (3' untranslated region [UTR]) the coding regions of the gene. The number of exons is quite heterogeneous, with a range from one (e.g., intronless G-protein coupled receptor genes) to several hundreds (e.g., titin gene with 363 exons). Introns were originally considered

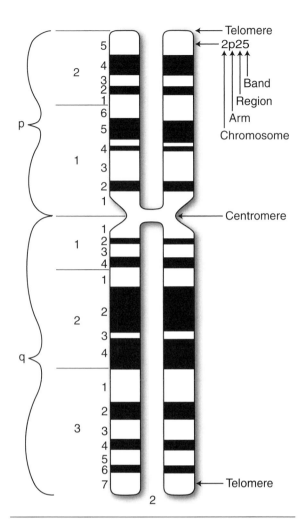

Figure 22.2   Numbering system used to specify a cytological address on a chromosome.

Reprinted, by permission, from R.M. Malina, C. Bouchard and O. Bar-Or, 2004, *Growth, maturation, and physical activity,* 2nd ed. (Champaign, IL: Human Kinetics), 371.

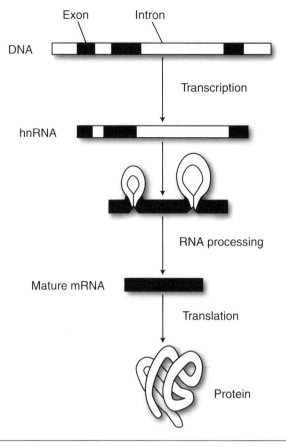

Figure 22.3   A gene includes introns and exons that are transcribed into messenger RNA (mRNA). The introns are spliced out as the transcript is processed into heterogeneous RNA (hnRNA) and then into mature mRNA that is subsequently translated into a polypeptide.

Reprinted, by permission, from R.M. Malina, C. Bouchard and O. Bar-Or, 2004, *Growth, maturation, and physical activity,* 2nd ed. (Champaign, IL: Human Kinetics), 373; adapted from Francomano and Kazazian (1986).

Important features of the human genome

- 3 billion base pairs of DNA
- About 20,000 genes
- Alternative promoters
- Alternative splicing for about 75% of multi-exonic genes
- About 2,000 **transcription factors**
- Some genes with only one exon and no intron
- Most genes with many exons

---

to be nonfunctional stretches of DNA between the coding regions, but they may harbor regulatory elements, such as alternative promoters, and splicing enhancers and suppressors.

The number of amino acids in proteins varies from very few to 1,000 and more, with an average of about 100. Thus, the average protein requires the coding information from about 300 DNA base pairs (see the next section). With at least 3 billion pairs of nucleotides in the haploid human genome, about 20 million genes could be encoded. In contrast, a much lower number of genes has been deduced from the completed human genome sequence. The current estimate is about 20,000 genes. However, a number of other lines of evidence have shown that the human genome encodes a much larger number of proteins than suggested by the simple sequence data.

Gene Expression

Gene expression can be simply defined as the process by which genes are activated or repressed in response to biological signals from the internal or external milieu. Gene expression is an area of active research because of its potential significance for understanding the processes of growth, maturation, and aging as well as adaptation to environmental challenges or abnormal cellular growth as in cancer.

Every nucleated cell contains a complete copy of the human genome. Some genes are expressed in most or all cells, because the gene product is essential for the function of the cell (housekeeping genes). However, the majority of the genes in the human genome are expressed only in specific organs, tissues, or cells and some are expressed only at certain stages of a cell cycle or during specific periods of the development of an organism. Moreover, the expression of a gene may be enhanced or depressed in response to external stimuli, such as changes in the metabolic milieu of the cell or in extracellular concentrations of certain ions and nutrients. To accommodate all these specific demands, gene expression must be closely controlled and coordinated. The adaptability and coordination of the gene expression process are achieved through several regulatory mechanisms, such as transcription factors, alternative splicing, alternative promoters, genomic imprinting, and gene silencing through epigenetic mechanisms.

Alternative Splicing

One of the most striking surprises generated by the production of the human genome sequence was the low number of genes. The previous estimates based on the number of expressed sequences ranged from 80,000 up to 150,000, whereas the human genome project revealed the presence of only about 20,000 annotated and predicted genes. This finding emphasized the fact that several genes must produce more than one transcript and that the phenomenon of one gene, several gene products, is more common than previously thought.

The main factor contributing to the disparity between the number of genes and gene transcripts is **alternative splicing.** Alternative splicing refers to a situation in which a single gene produces multiple mRNAs through different combinations of exons (figure 22.4). It is estimated that about 75% of the human multi-exon genes have alternative splice forms and that at least 70% of splice isoforms change the characteristics of the gene product (Modrek and Lee 2002). Alternative splicing may cause either inclusion or exclusion of one or several exons, or alternative 5' or 3' sites. These changes may induce an in-frame addition or deletion of functional units in the gene product (alternative exons), change the amino terminus of the polypeptide (alternative initiation), or modify the carboxyl terminus of the gene product because of frameshift or alternative termination. The regulation of alternative splicing is still poorly understood.

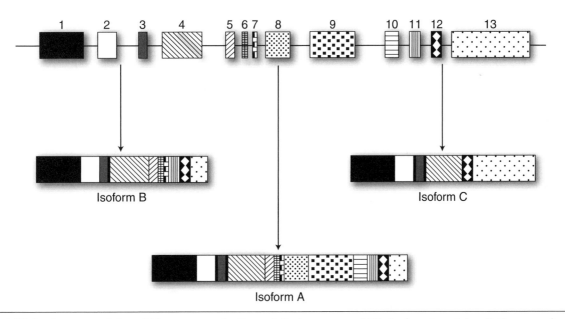

Figure 22.4 A gene consisting of 13 exons encodes three distinct messenger RNAs. Isoform A contains all 13 exons, whereas exons 8-10 and 5-11 are spliced off from isoforms B and C, respectively. Isoform C also uses an alternative stop codon.

Reprinted, by permission, from F.C. Mooren and K. Volker, 2005, *Molecular and cellular exercise physiology* (Champaign, IL: Human Kinetics), 42.

Events in Human Genes and Genomes

We know how genes determine the sequences of proteins. In this section the relationship between the encoded genetic information, proteins, and phenotypes is discussed.

From Genes to Proteins

The role of genetic information is to specify the sequence of amino acids that ultimately form all proteins synthesized by the cellular machinery. The set of all possible DNA triplets and the corresponding amino acids that they encode is called the genetic code. The process that allows the instructions contained in a given gene to be converted to a final gene product may appear to be simple but is in reality highly complex: the DNA sequence is first converted to a RNA sequence, which is then translated to a polypeptide giving rise to the final protein. The first step of the process, transcription, takes place in the nucleus of the cell. Subsequently, the mRNA specifies the primary sequence of a polypeptide, which may undergo posttranslational modifications, such as phosphorylation, methylation, acetylation, carboxylation, or glycosylation. The polypeptides may also be enzymatically cleaved to produce smaller functional products.

Transcription

The initiation of transcription requires the presence of transcription factors that bind to specific DNA sequence elements located in the immediate vicinity of a gene. These sequence elements are usually clustered upstream of the coding sequence and form the promoter region of the gene. Once the necessary transcription factors are bound to the promoter, an RNA polymerase binds to the transcription factor complex and is activated to start the synthesis of RNA. The common promoter elements recognized by several transcription factors include a TATA box (usually TATAAA) located about 25 base pairs (bp) before (–25 bp) the transcriptional start site, a CAAT box (–80 bp), and a GC box. In addition to the common promoter elements, enhancers, silencers, and response elements form a group of regulatory sequences that can enhance or inhibit the transcriptional activity of specific genes. These sequence elements are usually located quite far from the transcription initiation site (several thousands of base pairs). They bind gene regulatory proteins, and the DNA strand between the promoter and enhancer–silencer folds in a loop allowing the regulatory proteins to interact with the transcription factors bound to the promoter.

Once the synthesis of an RNA molecule from the DNA template is finished, the primary RNA transcript (i.e., the full copy of the original template DNA) undergoes various posttranscriptional modifications. These include removal of the unwanted internal segments (i.e., intronic sequences), rejoining of the remaining segments (exonic sequences), and capping at the 5' end of the transcript and polyadenylation at the 3' end. The removal of the intronic RNA segments is called RNA splicing. The process is directed by specific nucleotide sequences at the exon–intron boundaries (splice junctions).

Transcription Factors

Gene expression is acutely regulated in response to several external stimuli. These stimuli activate specific transcription factors, which bind to specific regulatory sequences in the promoter region (response elements) of the target genes, leading to their activation. The promoter region of a gene contains binding sites for various transcription factors, which are necessary for the initiation of gene transcription. Activation of RNA polymerases I and III, which transcribe housekeeping genes, requires a host of ubiquitous transcription factors. On the other hand, the transcription of polypeptide-encoding genes by RNA polymerase II uses complex sets of general and tissue-specific transcription factors. The complex of RNA polymerase II and general transcription factors are sufficient to initiate gene transcription at a minimum rate, which can be increased or turned off by additional positive or negative regulatory elements. However, the majority of the genes transcribed by RNA polymerase II show tissue-specific expression. The tissue specificity is achieved by interactions between special enhancer and silencer sequence and a variety of promoter sequence elements that are recognized only by tissue-specific transcription factors. The number of transcription factors, rather than the number of genes, has been suggested to be a key determinant of the biological complexity of an organism. For example, the human genome contains more than 2,000 genes for transcription factors, whereas they number about 500 and 700 in the worm (*C. elegans*) and fruit fly genomes, respectively (Szathmary et al. 2001).

Small RNAs

The latest breakthrough in our understanding of the regulation of gene expression is the discovery of the role played by small RNAs. Among other functions, small RNAs can suppress gene expression either by binding to target mRNA and repressing translation or by degrading the mRNA.

From Proteins to Phenotypes

Translation, or the synthesis of polypeptides from an mRNA template, takes place in the cytoplasm. After mRNA molecules migrate from the nucleus to the cytoplasm, they bind with the ribosomes where translation takes place. Ribosomes are large RNA–protein complexes providing a structural framework for polypeptide synthesis.

Proteins are ubiquitous throughout the body and constitute more than 50% of the dry weight of a typical cell. Proteins are best understood in the context of their functions, which are summarized in table 22.1. Nine types of proteins are indicated in the classification. Several subdivisions could be added, but this classification is sufficient to demonstrate the central role of proteins in mediating the chain of events between genetic specifications and the expression of phenotypic characteristics. For example, enzyme proteins are a very diverse class that have in common the capacity to increase the rate of biochemical reactions in cells. An enzyme typically has the property of accelerating a specific chemical reaction, although there are several exceptions.

As an example of the concepts presented here, consider an enzyme relevant to physical activity and health, glycogen synthase, which is a regulatory enzyme involved in the glycogen synthesis pathway of the liver and skeletal muscle. The enzyme exists in cells of both tissues in two forms—a less active, phosphorylated form and a more active, nonphosphorylated form. Phosphorylation is the process by which a phosphate molecule is added to a protein to alter its activation state. The phosphorylation of the enzyme is achieved by enzymes from the kinase family; dephosphorylation is brought about by the action of a phosphatase enzyme. This example shows that the effectiveness of a gene product (e.g., glycogen synthase in one tissue) is modulated in part by molecules that are themselves products of other genes. The resulting phenotype (i.e., measurable liver or muscle glycogen concentrations) is thus dependent on several genes. It is also dependent

Table 22.1   A Classification of Proteins by Function

Protein	Example
Structural	Collagen, the human body's most abundant protein, is found in various kinds of connective tissues.
Storage	Ovalbumin is a major source of material and energy during embryonic development.
Transport	Hemoglobin transports oxygen from areas of high concentration in the lungs to areas of lower concentration in the tissues.
Receptor	Insulin receptors are proteins found embedded in the cell membrane and exposed on the surface of the cell. When insulin and the receptor combine, a complex signaling cascade is activated that leads to glucose uptake.
Hormone	Growth hormone, released by the pituitary gland, stimulates growth of most body tissues and has widespread metabolic effects.
Protective	Antibodies are produced in response to the presence of foreign substances, organisms, or tissues in the body.
Contractile	Actin, myosin, and other contractile proteins arranged in orderly arrays in muscle fibers reduce length by sliding past each other in a controlled manner.
Regulatory	Regulatory proteins influence which genes are expressed and when they are expressed. Transcription factors are proteins that bind to DNA sequence and control the expression of specific genes.
Enzymes	Enzymes are the largest and most diverse class of proteins. Creatine kinase, which allows the phosphorylation of ADP into ATP using creatine phosphate as the substrate, is an example.

ADP = adenosine monophosphate; ATP = adenosine triphosphate.

Adapted, by permission, from S. Singer, 1985, *Human genetics: An introduction to the principles of heredity*, 2nd ed. (New York, NY: W.H. Freeman and Company), 60.

on other mechanisms, such as those related to the availability of glucose precursors and their entry into the glycolytic or glycogenic pathway, which may or may not be dependent on immediate genetic influences as well.

Acute exercise and regular exercise training induce several physiological responses that necessitate increased protein synthesis. The latter is necessary to meet the needs for higher levels of specific gene products (e.g., enzymes of energy production pathways), to replenish proteins that are catabolized during exercise, or to support the adaptive changes associated with the improved capacity to perform physical activity (e.g., enhanced blood flow in working muscles). It has been argued that the human genome evolved over a long period of time during which high levels of physical activity were necessary for survival. This would support the view that the link between exercise and the regulation of gene expression is deeply engrained in our biology.

An example of the effect of exercise on gene expression is muscle contraction–induced skeletal muscle hypertrophy. Muscle contraction has been shown to increase both transcription and translation of myofibrillar proteins, such as α-actin. Data from animal models suggest that a transcription factor called serum response factor and its response element in the promoter of the α-actin gene are involved in muscle contraction–induced α-actin expression and consequently muscular hypertrophy. Muscle overload also increases expression levels of several growth factors, such as insulin-like growth factor 1 (IGF-1) and signaling molecules (e.g., calcineurin).

This example illustrates the fact that the acute or chronic adaptation to exercise depends on the expression level of specific genes. However, the net effect of exercise-induced gene expression is determined by the integrated action of multiple genes. Recent advances in microarray technology have opened new opportunities to investigate the expression levels of thousands of genes simultaneously in a single experiment. This allows exploration of the effects of a specific stimulus, such as exercise, on the expression patterns of several gene families, such as transcription factors or genes involved in specific metabolic or physiologic pathways.

For instance, Roth and colleagues (2002) investigated the effects of a 9-week strength training program on the vastus lateralis gene expression profile in 20 sedentary participants. The strength

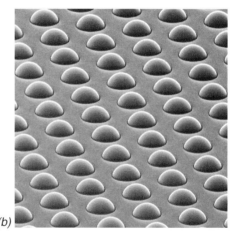

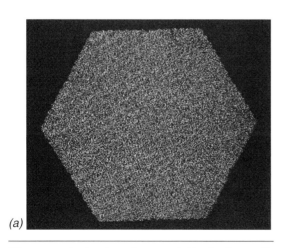

(a)

(b)

Microarray technology has revolutionized genetic research. High-density microarrays allow scientists to measure genome-wide expression levels or DNA sequence variations of a few hundred genes to hundreds of thousands of SNP loci in a single experiment. Image *(a)* shows an example of an array consisting of a fiber optic bundle with about 50,000 silica beads embedded in wells etched at the end of each fiber strand in the bundle. The beads are 3 micron in diameter *(b)* and can be coated with oligos *(c)* to identify a specific product (e.g., a single-nucleotide polymorphism).

Printed with permission from Illumina, Inc.

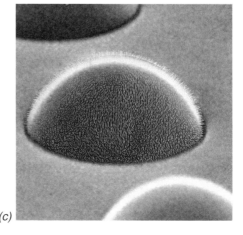

(c)

Key sources of variation among people in the human genome

- Independent assortment of chromosomes at gametogenesis
- Recombination events during meiosis (crossing over)
- Base substitution
- Base deletions
- Base mutations

training program consisted of unilateral knee extension exercises of the dominant leg. The participants exercised three times per week and each training session consisted of four sets of high-volume, heavy resistance knee extensions. A total of 69 genes showed >1.7-fold difference in expression levels after the training period in the pooled data. Fourteen of these genes were identified in all age-by-sex subgroups, 12 of them showing decreased and 2 showing increased expression levels after the training program (table 22.2).

Sequence Variation in the Human Genome

Two important events of meiosis contribute to the extraordinary amount of genetic diversity characteristic of humans and other sexually reproducing species. The first event is the independent assortment of chromosome pairs during their migration to daughter cells in gametogenesis. With 23 pairs of chromosomes in humans, there are 2^{23} or 8,388,608 different combinations of paternal and maternal chromosomes that can occur in a haploid gamete. The second event that enhances genetic variation is recombination. Before the migration of chromosomes to daughter cells, when homologous chromosomes are paired, crossing-over occurs. Crossing-over refers to the exchange of chromosomal segments between homologous chromosomes and results in the recombination of **alleles** (alternative forms of genes) between the homologous chromosomes of maternal and paternal origins. This is illustrated in figure 22.5. For example, if a pair of chromosomes carries three genes, each existing in two different forms in the population (e.g., A and a, B and b, C and c), a crossing-over taking place

Table 22.2 Summary of the Genes Showing >1.7-Fold Change in Expression Levels in Response to a 9-Week Strength Training Program

Gene name	Gene symbol	Fold change
Down-regulated genes		
Four-and-a-half LIM domains 1	FHL1	0.248
Myosin, light polypeptide 2	MYL2	0.262
Cold shock domain protein A	CSDA	0.265
Glyceraldehyde-3-phosphate dehydrogenase	GAPD	0.297
Actin, α 2	ACTA2	0.405
Myosin, light polypeptide 3	MYL3	0.442
Dynactin	ACTB	0.446
Eukaryotic translation elongation factor 1 γ	EEF1G	0.484
ATP synthase, mitochondrial F1 complex, β polypeptide	ATP5B	0.505
Troponin I	TNNI1	0.508
Actin-related protein 1, centractin α	ACTR1A	0.513
Topoisomerase (DNA) I	TOP1	0.547
Up-regulated genes		
Tetraspan 5	TM4SF9	2.131
TNF receptor–associated factor 6	N/A	1.852

LIM = comes from the three transcription factors, Lin-11, Isl-1, and Mec-3, in which the double zinc finger motif (LIM domain) was first identified (see Freyd, Kim, & Horvitz, 1990, *Nature* 344:876-879); ATP = adenosine triphosphate; TNF = tumor necrosis factor.

Data from S.M. Roth, R.E. Ferrell, D.G. Peters, E.J. Metter, B.F. Hurley and M.A. Rogers, 2002, "Influence of age, sex, and strength training on human muscle gene expression determined by microarray," *Physiol Genom* 10: 181-190.

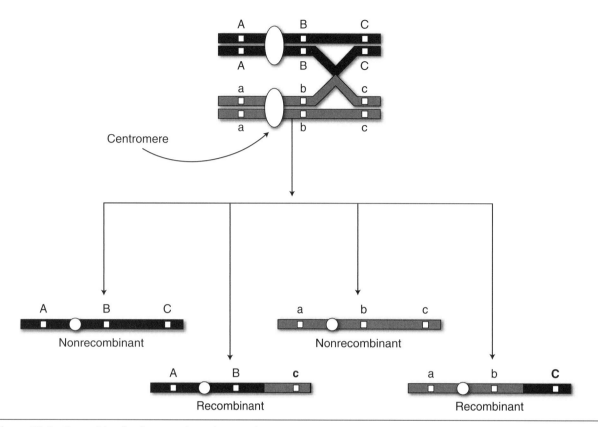

Figure 22.5 Recombination between homologous chromosomes.

Reprinted, by permission, from R.M. Malina, C. Bouchard and O. Bar-Or, 2004, *Growth, maturation, and physical activity,* 2nd ed. (Champaign, IL: Human Kinetics), 372.

between loci B and C will result in two recombinant chromosomes with new gene combinations and two nonrecombinant chromosomes carrying the parental gene combination. It is estimated that about two to three recombination events take place between each pair of homologous chromosomes (i.e., between pairs of chromosomes of maternal and paternal descent) during meiosis. The existing genetic variability in a population is thus amplified by "independent assortment" and "recombination" during meiosis to yield an almost infinite number of different unique gametes, ensuring the genetic uniqueness of each individual (except monozygotic [MZ] twins).

In addition, a major source of genetic variation is the variety of heritable changes (mutations) in the nucleotide sequence of the DNA. Mutations can be grouped into three classes:

- Base substitutions
- Deletions
- Insertions (figure 22.6)

Base substitutions usually involve the replacement of a single base. Synonymous or silent substitutions, which do not change an amino acid in the final gene product, are the most frequently observed in coding DNA. Nonsynonymous substitutions result in an altered codon that specifies either a different amino acid (a missense mutation) or a termination codon (a nonsense mutation). A missense mutation can induce either a conservative or nonconservative amino acid substitution. A conservative substitution refers to a situation in which the new amino acid is chemically similar to the old amino acid, whereas the amino acid introduced by a nonconservative substitution has different chemical characteristics. Thus, nonconservative substitutions are more likely to change the properties of the gene product than conservative substitutions. Deletions and insertions refer to the removal or addition, respectively, of one or a few nucleotides from the DNA sequence. These variations are relatively common in noncoding DNA. They are less frequent in exons where they may introduce frameshifts, that is, alter the normal translational reading frame of the gene and thereby change the final gene product.

Traditionally, mutations have been considered to be functionally significant only if they alter the amino acid sequence. However, it is now acknowledged that silent substitutions in exons as well as

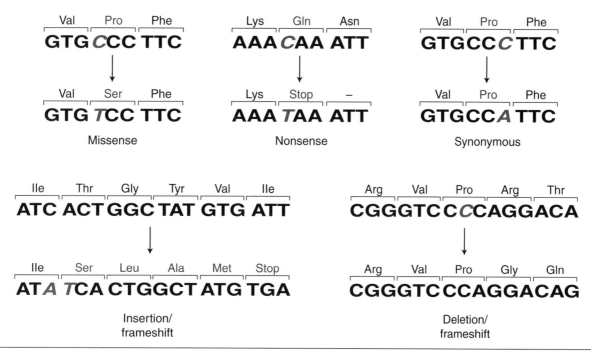

Figure 22.6 The top two lines present single base substitutions, where a change of a single nucleotide induces either a change in amino acid (missense) or a premature stop codon (nonsense) or has no effect on the gene product (synonymous—silent). The lower two lines show examples of small-scale insertion (left) and deletion (right) mutations that alter the translational reading frame of the gene (frameshift). Polymorphic nucleotides and resulting changes in amino acids are indicated with gray font.

Reprinted, by permission, from F.C. Mooren and K. Volker, 2005, *Molecular and cellular exercise physiology* (Champaign, IL: Human Kinetics), 45.

mutations in the noncoding sequence may also have strong effects on gene transcription and on the final gene product. For instance, it has been estimated that as much as 15% of point mutations that cause human diseases affect the splicing process. Mutations in the 5' regulatory region may disrupt a transcription factor binding site, a response element, or an enhancer or silencer sequence and thereby affect the rate at which a gene is transcribed. The 3' untranslated region harbors several sequence elements that affect nuclear transport, polyadenylation, subcellular targeting, and stability of mRNA. Mutations in these sequences could also potentially influence gene transcription and translation. Both synonymous and nonsynonymous substitutions in the coding sequence may alter splicing sites as well as splicing enhancers and silencers and thereby influence the properties of the mature polypeptide.

A question of concern is the extent of variation in the DNA sequences of human chromosomes. The concept of polymorphism is important when considering genetic variation. A polymorphism is defined as an alteration in DNA sequence that is present in the population with a frequency of at least 1%.

Genetic variation is quite extensive at the DNA sequence level. It is estimated that there is a base variation about every 600 to 1,000 bp in the human genome. A single change in a nucleotide base is known as a **single nucleotide polymorphism (SNP).** Because the genome has about 3 billion base pairs of DNA, this would imply that any given individual would carry 3 to 5 million variants when compared with the consensus *Homo sapiens* sequence of DNA. Already more than 10 million SNPs have been uncovered, and the number is growing. Polymorphic sequences are ubiquitous in introns, flanking regions of genes, and throughout the genome. Variable numbers of nucleotide sequences (e.g., CG or CAG repeats) are found throughout the genome and are highly polymorphic in populations. It is not uncommon to find 10 alleles or more of a given length polymorphism in such tandem repeats. They have become very useful markers of human diversity. In addition, insertions and deletions of from one to many hundreds of nucleotides are also found throughout the genome.

The proportion of genetic variation common to all humans and that specific to a particular population has been the object of discussion for several decades. It is now generally recognized that most

Extent of human DNA sequence variation

• Each individual carries at least 3 million DNA variants compared with the common *Homo sapiens* sequence.

• Most DNA sequence variants are seen in all ethnic groups.

• Only a small fraction (3-5%) of variants are specific to ethnic groups.

genetic variants are shared by the human species and that only about 3% to 5% are unique to specific populations. Genetic differences between populations or ethnic groups are thus relatively small compared with the overall genetic diversity observed in *Homo sapiens*.

Genetic Variation Among Sedentary People

The focus of this chapter is on human variation in the response to regular exercise and on the role of genetic differences. However, all the traits of interest to the physical activity and health paradigm are characterized by a significant genetic component even among totally sedentary people. A few examples will suffice to illustrate this point.

The heritability of cardiorespiratory endurance phenotypes in the sedentary state has been estimated from twin and family studies, the most comprehensive of these being the HERITAGE Family Study (Bouchard et al. 1998). An analysis of variance revealed a clear familial aggregation of $\dot{V}O_2max$ in the sedentary state. The variance in $\dot{V}O_2max$ (adjusted for age, sex, body mass, and body composition) was 2.7 times greater between families than within families, and about 40% of the variance in $\dot{V}O_2max$ was accounted for by family lines (figure 22.7). Maximum likelihood estimation of familial correlations (spouse, four parent–offspring, and three sibling correlations) revealed a maximal heritability of 51% for $\dot{V}O_2max$. However, the significant spouse correlation suggested that the genetic heritability was likely less than 50% (Bouchard et al. 1998). For additional information on the HERITAGE Family Study, visit www.pbrc.edu/HERITAGE.

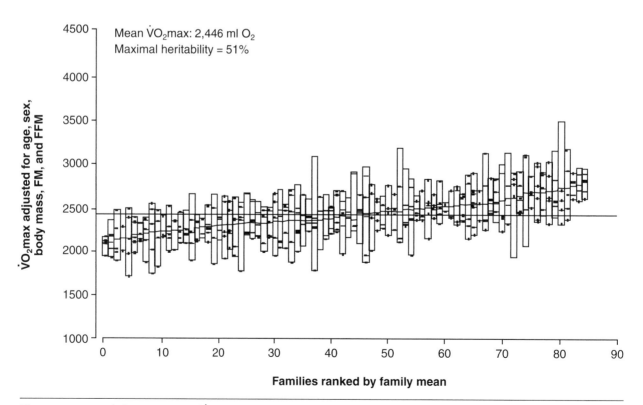

Figure 22.7 Familial aggregation of $\dot{V}O_2max$ in sedentary participants.

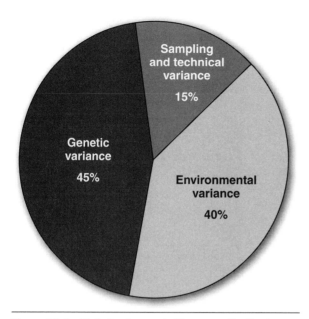

Figure 22.8 Variation in muscle fiber type distribution.

Reprinted, by permission, from J.A. Simoneau and C. Bouchard, 1995, "Genetic determinism of fiber type proportion in human skeletal muscle," *FASEB J.* 9(11): 1091-1095.

Research with MZ and dizygotic (DZ) twins, biological brothers, and unrelated individuals suggest that the fiber type composition of a mixed muscle (vastus lateralis), although mediated by the genes, is not completely regulated by genetic mechanisms and may also be influenced by regular exercise. A summary of the genetic, environmental, and methodological sources of variation in the proportion of type I fibers in human skeletal muscle is illustrated in figure 22.8. The genetic component accounts for about 45% of the variation in the proportion of type I muscle fibers in humans (Simoneau and Bouchard 1995).

The genotype also plays a role in the quantity of key enzymes in skeletal muscle. For example, phosphofructokinase and oxoglutarate dehydrogenase are often considered regulatory enzymes of the glycolytic and citric acid cycle pathways, respectively. The quantity of these two enzymes in muscle fibers is critical for their activities and also central to the flow of substrates through the glycolytic and citric acid pathways and, in turn, to the replenishment of ATP for the energy needs of the fiber. A study of young adult DZ and MZ twins as well as nontwin brothers suggested that at least 25% and perhaps more of the variation in the muscle content of these two key enzymes is associated with a genetic effect (Bouchard et al. 1986).

Twin and family studies indicate a significant genetic contribution to fat-free mass (FFM). In a cohort of 706 postmenopausal women, including 227 pairs of MZ twins and 126 pairs of DZ twins, FFM was estimated by dual-energy X-ray (DEXA). The results yielded a heritability estimate of 0.52 for FFM (Arden and Spector 1997). In the Quebec Family Study, path analysis of familial correlations computed among various pairs of relatives by descent or adoption indicated a genetic effect accounting for about 30% of the variance for FFM, as assessed by underwater weighing (Bouchard et al. 1988).

Individual Differences in Response to Regular Exercise

There are marked interindividual differences in the adaptation to exercise training. For example, in the HERITAGE Family Study, 742 healthy but sedentary participants followed an identical, well-controlled endurance-training program for 20 weeks. Despite the identical training program, increases in $\dot{V}O_2$max varied from no change to increases of more than 1 L/min (figure 22.9). This high degree of heterogeneity in responsiveness to a fully standardized exercise program in the HERITAGE Family Study was not accounted for by age, gender, or ethnic differences. A similar pattern of variation in training responses was observed for several other phenotypes, such as plasma lipid levels and submaximal exercise heart rate and blood pressure (Bouchard and Rankinen 2001). These data underline the notion that the effects of endurance training on cardiovascular and other relevant traits should be evaluated not only in terms of mean changes but also in terms of response heterogeneity.

Following are major areas of interest concerning the genetics of fitness and adaptation to exercise:

- Familial aggregation and genetic contribution to human variation in the sedentary state
- Individual differences in the response to regular exercise
- Familial aggregation of the response variation to regular exercise
- Genes and alleles contributing to human variation in trainability

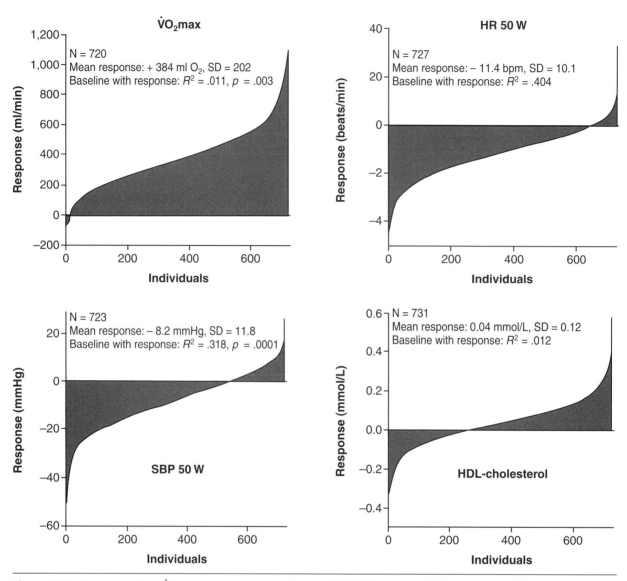

Figure 22.9 Heterogeneity of V̇O₂max, submaximal exercise heart (HR 50 W) and systolic blood pressure (SBP 50 W), and plasma HDL-cholesterol training responses in the HERITAGE Family Study.

Reprinted, by permission, from C. Bouchard and T. Rankinen, 2001, "Individual differences in response to regular physical activity," *Med Sci Sports Exerc* 33(6): S446-451.

Two questions come to mind as a result of the observations such as those depicted in figure 22.9. Are the high and low responses to regular exercise characterized by significant familial aggregation; that is, are there families with mainly low responders and others in which all family members show significant improvements? Is individual variability a normal biological phenomenon reflecting genetic diversity?

Genes and Responses to Exercise

We now turn our attention to the evidence for a role of specific gene and sequence variants in the range of responses to regular exercise. Blood pressure, lipids and lipoproteins, glucose and insulin, and cardiorespiratory endurance response phenotypes are discussed. For editorial considerations, all studies reviewed from this point onward in this chapter cannot be referenced individually. However, the interested student can find all these references in the fitness and performance gene map paper, version 2003 (Rankinen et al. 2004).

Genes and Blood Pressure Response to Regular Exercise

The 2003 update of the Human Performance and Health-Related Fitness Gene Map included 10 genes from 14 studies that have been investigated in relation to exercise training-induced changes in hemodynamic phenotypes (Rankinen et al. 2004). Findings on seven **candidate genes** (AMPD1, TTN, LPL, GNB3, BDKRB2, APOE, PPARA) were based on a single study. However, with three candidate genes, the positive associations were replicated in at least two studies. For example, in both the HERITAGE Family Study and the DNASCO Study cohorts, the angiotensinogen Met235Thr polymorphism (in which threonine is substituted for methionine) was associated with endurance training-induced changes in diastolic blood pressure in men.

Similarly, an association between the angiotensin-converting enzyme (ACE) gene I/D (insertion or deletion of a sequence) polymorphism and training-induced left ventricular (LV) growth has been reported in two studies (figure 22.10) (Montgomery et al. 1997; Myerson et al. 2001). In 1997, Montgomery and coworkers reported that the ACE D-allele was associated with greater increases in LV mass and with septal and posterior wall thickness after 10 weeks of physical training in British Army recruits (figure 22.10a). A similar training paradigm was repeated a few years later, and the training-induced increase in LV mass was 2.7 times greater in the D/D genotype compared with the I/I **homozygotes** (figure 22.10b).

A third candidate gene with positive evidence of associations from multiple studies is endothelial nitric oxide synthase (NOS3). In the HERITAGE Family Study, homozygotes for the glutamine allele at codon 298 (Glu298) had a more than three times greater reduction in submaximal exercise diastolic blood pressure after the training program than the homozygotes for the asparagine allele (Asp298Asp). A similar pattern was evident with the systolic blood pressure and rate–pressure product training responses (Rankinen, Rice, et al. 2000). In coronary artery disease patients, exercise training significantly improved acetylcholine-induced change in average peak velocity of coronary arteries. However, the training response was significantly blunted in the carriers of the *NOS3* –786C allele of a polymorphism located in the 5'-UTR of the NOS3 gene compared with the patients who were homozygotes for the –786T allele (Erbs et al. 2003).

Genes and Lipids and Lipoprotein Response to Regular Exercise

Although the first observations of the beneficial effects of regular physical activity on plasma lipid levels date back to the early 1970s, there are surprisingly few studies available on the associations between DNA sequence variation in candidate genes and training-induced changes in plasma lipids, lipoproteins, and apolipoproteins. In the HERITAGE Family Study, apolipoprotein E (APOE) genotypes

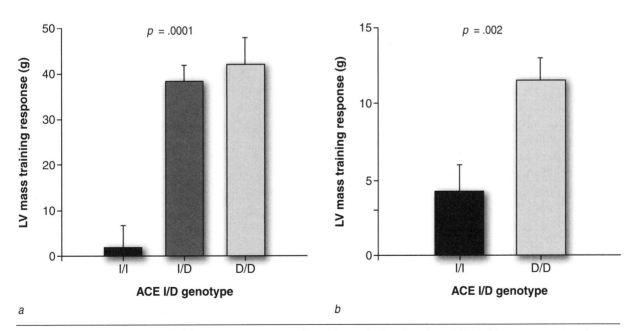

a

b

Figure 22.10 Panel *a* is modified from (Montgomery et al. 1997) and depicts data from 140 healthy army recruits who participated in a 10-week basic training program. Panel *b* summarizes data from a replication study (Myerson et al. 2001) using a similar training program with 141 healthy recruits.

Adapted from H.E. Montgomery et al., 1997, "Association of angiotensin-converting enzyme gene I/D polymorphism with change in left ventricular mass in response to physical training," *Circulation* 95: 741-747.

were associated with training-induced changes in plasma low-density lipoprotein (LDL) and high-density lipoprotein (HDL) cholesterol and plasma triglyceride levels (Leon et al. 2004). In both black and white participants, regular exercise lowered the LDL cholesterol levels in the E2/E3 and E2/E4 genotypes but not in the E3/E3 homozygotes and the E3/E4 **heterozygotes.** The LDL cholesterol training response in the E4/E4 homozygotes showed an ethnic difference: in black participants there was a significant reduction, whereas in white participants the LDL cholesterol levels tended to increase. HDL cholesterol and triglyceride training responses were associated with APOE genotypes only in white participants: the E2/E2, E2/E3, and E3/E3 genotypes showed more favorable changes in HDL cholesterol and triglyceride levels than the E2/E4 and E4/E4 genotypes.

In a small cohort of elderly men, the APOE E2 allele carriers showed greater increase in HDL cholesterol levels after a 9-month exercise training program than the E3 and E4 allele carriers (Hagberg et al. 1999). Other studies have suggested that there may be associations between HDL training responses and the lipoprotein lipase and endothelial lipase genotypes.

Genes and Glucose and Insulin Response to Regular Exercise

Only a few candidate gene studies are available regarding exercise training-induced changes in glucose and insulin metabolism phenotypes. In obese, type 2 diabetic Japanese women, carriers of the arginine allele at the Trp64Arg polymorphism in the β$_3$-adrenergic receptor (ADRB3) gene showed smaller reductions in fasting glucose and HbA$_{1c}$ levels after a combined low-calorie diet and exercise training program compared with noncarriers. In healthy Japanese males, a 3-month endurance-training program resulted in a significant reduction in fasting plasma glucose levels in the Trp64Trp homozygotes and Trp64Arg heterozygotes but not in the Arg64Arg homozygotes. Another Japanese study reported suggestive associations between Pro-12Ala (alanine is substituted for proline) genotypes of the peroxisome proliferative activated receptor γ (PPARG) gene and training-induced changes in fasting insulin and insulin resistance indexes in healthy males. The Pro12Ala heterozygotes showed reduction in both trait values, whereas no changes were observed in the Pro12Pro homozygotes. In a

small cohort of older hypertensive men, the ACE I/I homozygotes had the greatest improvements in insulin sensitivity and the greatest decline in the acute insulin response to glucose compared with those who were homozygous for the D allele.

In the HERITAGE Family Study, a genome-wide **linkage** scan revealed a quantitative trait locus (QTL) on chromosome 7q31 for the fasting insulin response to endurance training in white families (Lakka et al. 2003). A single-nucleotide polymorphism in the 5' untranslated region of the leptin (LEP) gene, which is located within the QTL region, was not associated with the insulin training response. However, there was a significant interaction between the LEP genotype and a Lys109Leu polymorphism (lysine is replaced by leucine) of the leptin receptor (LEPR) gene. Changes in fasting insulin levels did not differ among the LEP genotypes in those participants who were homozygotes for the LEPR Lys109 allele. However, among the LEPR Leu109 allele carriers, the LEP A/A homozygotes showed significantly greater reduction in fasting insulin levels than the heterozygotes and the LEP G/G homozygotes (figure 22.11).

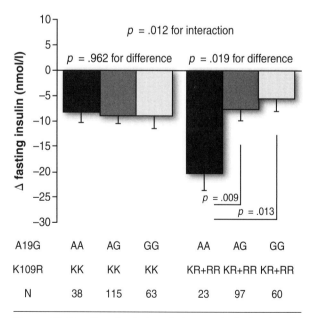

Figure 22.11 Exercise-induced changes in fasting insulin according to the leptin (LEP) A19G polymorphism and the leptin receptor (LEPR) K109R polymorphism.

Copyright © 2004 American Diabetes Association. From *Diabetes*, Vol. 53, 2004; 1603-1608. Reprinted with permission from *The American Diabetes Association*.

Genes and Cardiorespiratory Endurance in Response to Regular Exercise

In addition to examining the role of genetic heterogeneity in the ability to improve the diabetes and cardiovascular disease risk factor profile in response to a fully standardized program, one can ask whether the same concept applies to the gains in cardiorespiratory endurance or exercise tolerance. A number of studies indicate that such is the case, and the genetic dissection of the trainability phenotype is underway.

In pairs of MZ twins, the $\dot{V}O_2$max response to standardized training programs showed six to nine times more variance between genotypes (between pairs of twins) than within genotypes (within pairs of twins) based on the findings of three independent studies (Bouchard et al. 1992). Thus, gains in absolute $\dot{V}O_2$max were much more heterogeneous between pairs of twins than within pairs of twins. The results of one such study are summarized in figure 22.12. The MZ twins exercised for 20 weeks using a standardized and demanding endurance training program (Bouchard et al. 1992). In the HERITAGE Family Study, the increase in $\dot{V}O_2$max

in 481 individuals from 99 two-generation families of Caucasian descent showed 2.6 times more variance between families than within families, and the model-fitting analytical procedure yielded a maximal heritability estimate of 47% (Bouchard et al. 1999). Thus, the extraordinary heterogeneity observed for the gains in $\dot{V}O_2$ among adults is not random and is characterized by a strong familial aggregation (figure 22.13).

We are dealing here with extremely complex phenotypes, and the genetic architecture of their responsiveness to regular exercise is likely to depend on a large number of genes. A favorite candidate gene among exercise scientists has been the ACE gene, which was investigated in 28 reports as reviewed in a recent publication of the Human Gene Map for Performance and Health-Related Fitness (Rankinen et al. 2004).

In exercise-related studies, the I allele has been reported to be more frequent in Australian elite rowers and in Spanish endurance athletes than in sedentary controls. Likewise, British long-distance runners tended to have a greater frequency of the I allele than sprinters. In a group of postmenopausal women who were selected on the basis of their physical activity levels, the I/I genotype was associated with greater $\dot{V}O_2$max and maximal arterial–venous oxygen difference compared with the D/D homozygotes. The I allele was also associated with greater increase in muscular endurance and efficiency after 10 weeks of physical training in British Army recruits.

In contrast, the frequency of the D allele was found to be higher in elite swimmers than in sedentary controls. Moreover, in almost 300 sedentary but healthy white offspring from the HERITAGE Family Study, the D/D homozygotes showed the greatest improvements in $\dot{V}O_2$max and maximal power output following a controlled and supervised 20-week endurance training program (Rankinen, Perusse, et al. 2000). Furthermore, some studies suggest that the D allele is associated with greater muscular strength gains in response to resistance training. However, several reports showed no associations between the ACE genotype and performance phenotypes.

The observations on the ACE gene need to be put in perspective. The statistical evidence for an association or against an association is not very strong in the published reports. The most likely explanation for seemingly inconsistent results from association studies in different populations

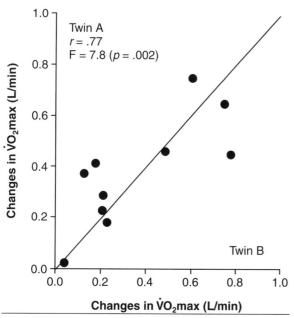

Figure 22.12 Monozygotic (MZ) twin response to training.

Adapted from D. Prud'homme, C. Bouchard, C. Leblanc, F. Landry and E. Fontaine, 1984, "Sensitivity of maximal aerobic power to training is genotype dependent," *Medicine and Science in Sports and Exercise* 16:489-493, and C. Bouchard, F.T. Dionne, J.-A. Simoneau and M.R. Boulay, 1992, "Genetics of aerobic and anaerobic performances," *Exercise and Sport Sciences Reviews* 20:27-58.

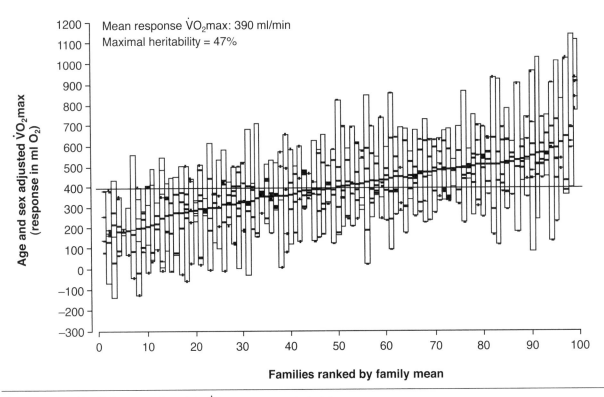

Figure 22.13 Familial aggregation of the V̇O₂max response to training.

Reprinted from C. Bouchard, P. An, T. Rice, J.S. Skinner, J.H. Wilmore, J. Gagnon, L. Perusse, A.S. Leon and D.C. Rao, 1999, "Familial aggregation of VO(2max) response to exercise training: Results from the HERITAGE Family Study," *J Appl Physiol* 87(3): 1003-1008. Used with permission.

is typically the overinterpretation of marginal statistical evidence (Cardon and Palmer 2003). Thus, the available data are still too fragmented to fully evaluate the role played by the ACE gene in the variation of human physical performance level. The same conclusion can be reached about all the other genes studied to date.

Genome-wide linkage scans for V̇O₂max and maximal power output responses to regular exercise did not produce any strong QTLs. However, several suggestive linkage signals were found, such as those on 4q26, 5q23, and 13q12 (Bouchard et al. 2000; Rico-Sanz et al. 2004). The lack of particularly strong QTLs most likely reflects the polygenic nature of these traits. The V̇O₂max capacity is influenced by several intermediate phenotypes (cardiac output, oxygen transport capacity, oxidative capacity of the working muscles), and multiple genes contribute to the variation in each of these subphenotypes. Consequently, the detection of these gene effects requires denser microsatellite marker sets for linkage analyses or a dense panel of SNPs for genome-wide association scans.

Although exercise-related traits are mainly polygenic and multifactorial in nature, much can be

learned from some monogenic disorders characterized by compromised exercise capacity or exercise intolerance. These disorders affect only a few individuals, but they provide interesting examples of genetic defects that have profound effects on the ability to perform physical activity, usually attributable to compromised energy metabolism. Although these genetic defects compromise exercise capacity, there is no evidence that overexpression of these genes leads to improved physical performance. However, it is important to understand the molecular mechanisms contributing to both ends of the distribution of cardiorespiratory endurance and its trainability. Table 22.3 lists some of the genes that have been associated with a decreased exercise capacity (Rankinen et al. 2004).

Trait-Specific Response to Exercise

This chapter has provided ample evidence that the response to regular exercise varies considerably among individuals. One of the lessons from these studies is that there are people who do not enjoy

Table 22.3   Genes Encoded by Nuclear and Mitochondrial DNA in Which Mutations Have Been Reported in Patients With Exercise Intolerance

Gene	OMIM number	Location
Nuclear DNA		
CPT2	255110	1p32
PGAM2	261670	7p13-p12
LDHA	150000	11p15.4
PYGM	232600	11q12-q13.2
PFKM	232800	12q13.3
ENO3	131370	17pter-p11
ACADVL	201475	17p13-p11
SGCA	600119	17q21
PHKA1	311870	Xq12-q13
PGK1	311800	Xq13
Mitochondrial DNA		
MTTL1	590050	3,230-3,304
MTND1	51600	3,307-4,262
MTTI	590045	4,263-4,331
MTTM	590065	4,402-4,469
MTTY	590100	5,826-5,891
MTCO1	516030	5,904-7,445
MTTS1	590080	7,445-7,516
MTTK	590060	8,295-8,364
MTCO3	516050	9,207-9,990
MTND4	516003	10,760-12,137
MTTL2	590055	12,266-12,336
MTTE	590025	14,674-14,742
MTCYB	516020	14,747-15,887

OMIM = Online Medelian Inheritance in Man.

Reprinted, by permission, from Rankien et al. 2004.

any improvement in a given risk factor or biological indicator of health or fitness. It is unfortunate that some people are less responsive than others.

Fortunately, the heterogeneity in responsiveness to regular exercise is a trait-specific phenomenon. In other words, an individual who is a low responder for one phenotype may be an average or even a high responder for another trait. This phenomenon can be illustrated by calculating correlation coefficients between the training responses among various traits. A strong correlation would indicate a uniform response pattern across traits, whereas the lack of correlation would support the independence of the responses to regular

Important Key Messages From Recent Research

- Individual differences exist in response of risk factors to regular exercise.
- Risk factor responses are not correlated with the gains in $\dot{V}O_2$max.
- Risk factor response changes are poorly correlated among themselves.
- There is no indication that there are individuals who are nonresponders across all fitness and risk factors.

exercise. Table 22.4 summarizes the correlation coefficients between $\dot{V}O_2max$ and several cardiovascular disease risk factor training responses in the HERITAGE Family Study cohort. The table clearly shows that endurance training-induced changes in risk factors do not correlate well with changes in cardiorespiratory fitness level. On the other hand, the correlations summarized in table 22.5 indicate that it is impossible to predict the effects of regular exercise on one risk factor by observing the training response of another risk factor. Thus far, there is no indication that there are people who are nonresponders across all the markers of risk for diseases.

These observations emphasize that even if a person cannot improve cardiorespiratory fitness with regular exercise, it is likely that he or she will achieve other significant health benefits from a physically active lifestyle. This has important public health implications.

Table 22.4   Correlation Coefficients Between Endurance Training-Induced Changes in Cardiorespiratory Fitness ($\dot{V}O_2max$) and Risk Factors in the HERITAGE Family Study

Response phenotype	Black participants		White participants	
	Males	Females	Males	Females
Body weight	.20	.11	−.09	−.08
Fat mass	−.03	.05	−.19	−.18
Insulin	.05	.10	−.04	.03
Cholesterol	.13	−.05	.00	.01
Triglycerides	.05	−.01	.12	−.10
Systolic blood pressure	.22	.01	−.11	−.01

Table 22.5   Correlation Coefficients Between Risk Factor Training Responses in the HERITAGE Family Study Cohort

	Black participants					White participants				
	FM	Ins	Chol	TG	SBP	FM	Ins	Chol	TG	SBP
Females										
Body weight	.92	.24	.34	.23	−.06	.84	.17	.22	.06	.10
Fat mass		.16	.31	.23	.04		.14	.21	.09	.06
Insulin			.19	.07	−.04			−.02	.04	.11
Cholesterol				.25	−.06				.31	.16
Triglycerides					−.01					.14
Males										
Body weight	.84	.20	.29	.33	.24	.83	−.03	.13	.13	.06
Fat mass		.24	.14	.18	.16		.03	.07	.08	.04
Insulin			.23	.24	.19			.09	.03	−.05
Cholesterol				.33	.18				.31	−.11
Triglycerides					.15					.14

FM = fat mass; Ins = plasma insulin; Chol = plasma cholesterol; TG = plasma triglycerides; SBP = systolic blood pressure.

Summary

The past decade has witnessed remarkable progress in genomics and human genetics. The availability of the DNA sequence of the human genome has changed our ability to study the genetic basis of complex multifactorial traits and to develop novel treatments for several chronic diseases. The recent advances in molecular genetics are starting to affect exercise science. The availability of powerful methods such as microarray technology to measure gene expression and targeting specific genes in knockout and transgenic animal models will greatly add to our research capability in the investigation of basic and applied exercise science issues.

Although the research on molecular genetics of physical activity, health-related fitness, and health-related outcomes is still in its infancy, we recognize that understanding the effects of DNA sequence variation on interindividual differences in responsiveness to acute exercise and regular exercise holds great promise. Such data not only would help to develop more concrete public health measures regarding the role of physical activity in the prevention and treatment of chronic diseases but also would provide an opportunity to individualize preventive medicine.

KEY TERMS

allele: Alternative form of a genetic locus; a single allele for each locus is inherited from each parent.

alternative splicing: Different ways of combining a gene's exons to make variants of the complete protein.

candidate gene: A gene suspected of being involved in a trait or a disease.

exon: An amino-acid coding DNA sequence of a gene.

gene: Fundamental physical and functional unit of heredity. A gene is an ordered sequence of nucleotides located in a particular position on a particular chromosome that encodes a specific functional product (i.e., a protein or RNA molecule).

gene expression: Process by which a gene's coded information is converted into a functional product. Expressed genes include those that are transcribed into mRNA and then translated into protein and those that are transcribed into RNA but not translated into protein (e.g., transfer and ribosomal RNAs and small RNAs).

genome: All the DNA bases and sequences of a particular organism; the size of a genome is generally given as its total number of bases or base pairs.

heterozygote: Presence of different alleles at a given genetic locus.

homozygote: Two identical alleles at a given genetic locus.

intron: DNA sequence that interrupts the amino-acid coding sequence in a gene; an intron is transcribed into RNA but is spliced out of the message before it is translated into protein.

linkage: Proximity of two or more markers on a chromosome; the closer the markers, the lower the probability that they will be separated during DNA recombination at meiosis and hence the greater the probability that they will be inherited together.

promoter: A DNA site to which RNA polymerase will bind and initiate transcription.

single nucleotide polymorphism (SNP): DNA sequence variation that occurs when a single nucleotide (A, T, C, or G) in a genome sequence is altered.

transcription: Synthesis of an RNA copy from a sequence of DNA.

transcription factor: Protein that binds to the regulatory region of a gene and participates in the regulation of gene expression.

translation: Process by which the genetic code carried by mRNA directs the incorporation of amino acids and the synthesis of a protein.

STUDY QUESTIONS

1. How can we account for the fact that human tissues produce more proteins than there are genes?

2. Describe how differences in gene expression can influence a phenotype affected by regular exercise.

3. What are the major sources of human genetic variation?

4. How extensive is human genetic variation?

5. What is a quantitative trait locus?

6. How extensive are individual differences in response to regular exercise?

7. Is the ACE I/D polymorphism a good predictor of blood pressure response to regular exercise?

8. What are the implications of the differences in the responsiveness to regular exercise for the promotion of physical activity?

An Integrated View of Physical Activity, Fitness, and Health

William L. Haskell, PhD ■ Steven N. Blair, PED ■ Claude Bouchard, PhD

CHAPTER OUTLINE

Regardless of a person's socioeconomic status, ethnicity, or culture, a common goal for most people is to live a healthy, enjoyable, peaceful, and productive life. As has been documented in a number of the chapters in this book, having a physically active lifestyle can contribute significantly to this goal. A major challenge for many people of all ages is building back into their otherwise sedentary lives sufficient activity to promote good health. Although regular activity is necessary, it is not sufficient in itself to ensure optimal health. Thus, a person's health-oriented physical activity plan needs to be integrated into a general wellness plan that includes a variety of other health behaviors, such as eating a healthy diet, avoiding tobacco, not abusing alcohol or drugs, and managing stress. The activity plan will vary as life conditions change—being in school, changing jobs, moving from one community to another, getting married, having children, recovering from an illness, and retirement. As people move through the life span, their life goals tend to change and so will some of the specific features of their activity plans. In this chapter, we present key challenges and issues that a person will encounter when attempting to integrate a health-oriented plan of physical activity into a technology-dominated environment that has reduced dramatically the need for human **energy expenditure.** We also present some strategies on how to successfully meet these challenges.

Chapter Overview

In most technologically advanced cultures where there has been a large decrease in the daily physical activity required of many adults to survive, there is a general understanding that being sedentary throughout the day contributes to a variety of health problems. However, many people are unsuccessful in acting on this knowledge and remain sedentary or, at best, sporadically active. This inability to implement and maintain an effective activity plan can result from a number of countervailing factors that vary from time to time for a person as well as from person to person.

Factors frequently reported to prevent an active lifestyle include lack of time, laziness, uncertainty about how to begin, existing illness or injury, concerns about the risks of exercise, lack of access to convenient or safe areas to exercise, and confusion about the difference between being busy and being

physically active. The challenge to the public and health professionals to overcome these barriers has been only partly met, with ultimate success requiring the following:

- More research on human behavior change
- Government policies that support an active lifestyle throughout the life span
- **Built environments** that make physical activity convenient and safe wherever people live, work, and go to school
- Medical and health care systems that give priority to lifestyle change and disease prevention
- An educational system that teaches all students about the health benefits of activity, shows students how to initiate and sustain a successful activity program, and provides opportunities for participation in activities that contribute to a lifelong active lifestyle

> To reverse the population trend toward a more sedentary lifestyle, individual behavior change needs to be supported by public policy that promotes health and physical education, preventive medical services, convenient access to exercise facilities and services, and construction of an activity-friendly built environment.

Physical Activity Versus Inactivity: Universal Value Versus Damaging Consequences

The failure to maintain a physically active lifestyle has major negative health consequences regardless of a person's age, gender, race, or ethnicity. Despite this knowledge, for an increasing portion of the world's population less and less physical activity is required to get where they need to go, to earn a living, to take care of their home and family, and to enjoy recreation. A multifactor approach will be necessary to reverse this trend.

Universal Value of a Physically Active Lifestyle

The promotion of increased physical activity is an excellent public health intervention in that a physically active lifestyle

- has a positive impact in the prevention and treatment of a wide variety of chronic diseases,

- has unique and independent positive effects on physical and mental functioning and quality of life,

- acts synergistically with other behaviors to improve health, and

- contributes to the health and well-being of children and adults of all cultures and ethnicities throughout their life span.

This is not to say that physical activity is a panacea for maintaining good health, but rather that given the human constitution, populations with a physically active lifestyle will have a health and wellness advantage over those who remain sedentary. The results from the INTERHEART study indicate the population-based effects of a physically active lifestyle, in this case the prevention of myocardial infarction in people living in 52 countries representing every inhabited continent on earth (Yusuf et al. 2004). This was a case-control study where major behavioral and biological risk factors were determined for 15,152 men and women who had a myocardial infarction and 14,820 age- and sex-matched control participants free of cardiovascular disease. Participants were considered "active" if they reported usually performing ≥4 hr per week of mod-

erate- or vigorous-intensity exercise or "inactive" if they were routinely less active than that. Regardless of other risk factors and independent of ethnicity and country of origin, fewer participants with myocardial infarction were classified as active compared with the matched controls. After adjustment for other risk factors, active persons had a significantly lower risk of myocardial infarction than inactive persons (odds ratio = 0.86; 95% confidence interval [CI] = 0.76-0.97), and the contribution of being inactive to having a myocardial infarction (population attributable risk) was 12.2% (p < .0001). That a combination of lifestyle risk factors has a major impact on risk of myocardial infarction in this highly diverse population is demonstrated in figure 23.1. When being active is added to nonsmoking and daily intake of fruit and vegetables, the odds ratio for developing a myocardial infarction decreases to 0.21 (95% CI = 0.17-0.25).

> Physical activity promotion is an excellent public health intervention because of its positive effects on various chronic medical conditions, physical and mental function, and quality of life in children and adults regardless of culture or ethnicity throughout their life span.

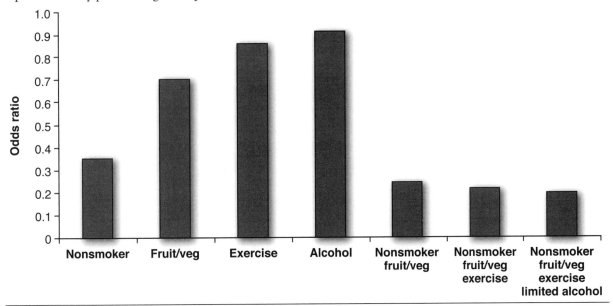

Figure 23.1   Odds ratios for nonsmokers versus current smokers, participants who consumed fruit and vegetables daily versus infrequent consumption, frequent exercisers (≥4 hr/week) versus infrequent exercisers, participants who consumed alcohol regularly (three or more times per week) versus those who consumed alcohol infrequently. Odds ratios are adjusted for age, country, and all other risk factors (Yusuf et al. 2004).

Other studies have documented that people who are physically active, remain relatively lean, frequently eat fruit and vegetables (and less processed foods), don't smoke cigarettes, and drink limited amounts of alcohol have very favorable health outcomes. For example, in 84,129 nurses followed for 14 years, those who were nonsmokers, performed ≥30 min of moderate- or vigorous-intensity exercise per day, and ate a diet high in fruit and vegetables had a risk of heart attack or stroke that was 57% less (relative risk 0.43, 95% CI = 0.33-0.55) than women without these three characteristics (Stampfer et al. 2000). When women who had these three characteristics and also were of normal body weight (body mass index [BMI] ≤25) and consumed ≤5 g of alcohol per day, their risk of a heart attack or stroke was 83% lower (relative risk = 0.17, 95% CI = 0.07-0.41) than women without these characteristics (figure 23.2). A separate analysis of data from this large group of nurses indicated that physical inactivity (<3.5 hr/week) and excess weight (BMI ≥25.0) together could account for 31% of all premature deaths, 59% of deaths from cardiovascular disease, and 21% of deaths from cancer among nonsmoking women (Hu et al. 2004). These and other data strongly support the substantial value of integrating activity with other established health habits to prevent disease and promote health.

Challenge of Labor-Saving Technology

For some people, just performing their **required activities of daily living** provides sufficient physical activity to promote good health. For example, there still are farmers, construction workers, professional gardeners, and others who perform physical labor much of the day, as well as those who walk or ride bicycles an hour or more per day for transportation. These individuals do not need to seek other activity for health outcomes. However, in technologically advanced societies, fewer and fewer people achieve an adequate level of physical activity by just the activities they typically perform for self- and home care, transportation, and occupation. Over

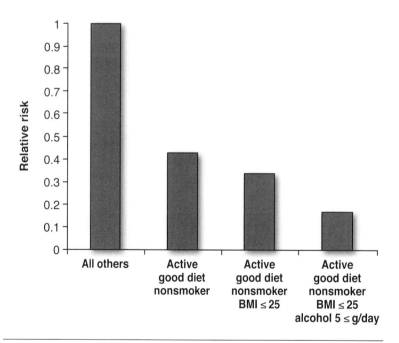

Figure 23.2 Relative risk for heart attack or stroke in nurses based on their health-related habits, including physical activity, diet, cigarette smoking, obesity as indicated by a body mass index ≥25, and excessive alcohol consumption (≥5 g/day). For example, compared with nurses without these habits, nurses who were regularly active, had a diet high in fruits and vegetables and low in animal fat and processed foods, and were not cigarette smokers had a risk of heart attack or stroke in the future that was 60% lower (see text for additional information).

Data from M.J. Stampfer, F.B. Hu, J.E. Manson, E.B. Rimm and W.C. Willett, 2000, "Primary prevention of coronary heart disease in women through diet and exercise," *New England Journal of Medicine* 343: 16-22.

a relatively short time—since the beginning of the industrial revolution circa 1800—a significant amount of the physical activity required for most people to survive has been eliminated and replaced by machines. The list of technological advances that reduce the daily **energy expenditure** of an increasing number of people throughout the world is large and continues to grow—electricity, steam and gas engines, telephones, automobiles, trains, elevators and escalators, tractors and cranes, television, computers, and the Internet.

Although few actual measurements have been made that systematically document the decline in energy expenditure attributable to all of these advances in technology, a number of measurements have been made of the energy cost associated with the physical activity required by various occupations. Also, several investigators have measured the physical activity or energy expenditure of people who live without the use of modern technologies. Montgomery (1978) collected energy expenditure data on members of a primitive Indian tribe of hunters, gatherers, and farmers living in Peru and documented that their

primitive way of living required an average daily energy expenditure of about 60 kcal per kilogram of body weight per day for men and about 44 for women. This is in comparison with about 34 and 32 kcal per kilogram of body weight per day for current-day healthy but sedentary middle-aged men and women, respectively (Simons-Morton et al. 2000).

Old Order Amish living in Ontario, Canada, refrain from driving automobiles and from using electrical appliances and other modern conveniences. Manual labor farming is their primary occupation. The amount of physical activity performed daily by 98 Old Order Amish men and women was assessed using standard questionnaires, pedometers, and a log for 7 consecutive days (Bassett et al. 2004). The average number of steps per day for the Amish men was 18,425 and for women 14,196, and the men reported walking 12.0 hr/week and the women 5.7 hr/week. None of the men and only 9% of the women were obese (BMI >30), whereas 25% of the men and 26% of the women were overweight or obese (BMI >25) (see figure 23.3). By contrast, a majority of U.S. adults (20-74

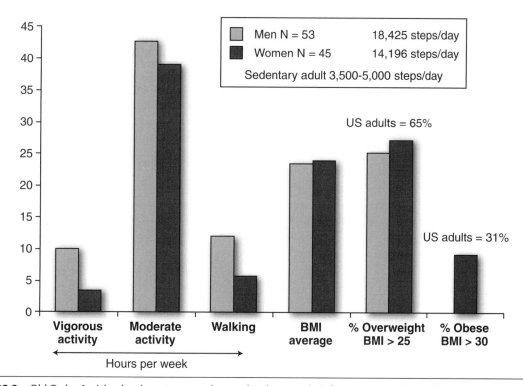

Figure 23.3   Old Order Amish who do not use modern technology on their farms are very physically active and exhibit much less obesity compared with men and women who use such technology in many aspects of their lives (see text for detailed explanation of data).

Data from D.R. Bassett, P.L. Schneider and G.E. Huntington, 2004, "Physical activity in an Old Order Amish community," *Medicine and Science in Sports and Exercise* 36: 79-85, and A.A. Hedley, C.L. Ogden, C.L. Johnson, M.D. Carroll, L.R. Curtin and K.M. Flegal, 2004, "Prevalence of overweight and obesity among US children, adolescents, and adults; 1999-2002," *Journal of the American Medical Association* 292: 2847-2850.

years old) walk less than 2 to 3 hr/week and accumulate less than 5,000 steps per day, approximately 65% have BMI >25, and 31% are considered obese (BMI 30) (Hedley et al. 2004).

Typical Occupational Activities

For most people, the opportunities to build increased physical activity back into their occupation are very limited. An increasing number of jobs require that for people to be most productive, they need to stay in one place, very frequently sitting, for much of the work day. Many office workers can now sit at their desks and through the use of advanced computer and communication technology perform most of their job tasks without even standing—they can communicate electronically with a coworker down the hall or a customer on the other side of the world. In the 1950s, most major office and manufacturing buildings employed night patrolmen who walked the perimeter of the building for much of their shift. By the 1990s, most of these "patrolman" jobs were replaced by "watchmen" who, instead of walking to provide security, sat and watched a bank of video screens that displayed various locations in the building (one watchman can provide security to a greater area than one patrolman, thus increasing worker productivity). A number of similar situations can be cited demonstrating that individual worker productivity tends to be increased using new technologies by reducing workers' need to move about. Thus, for these types of jobs, there is an economic disincentive to increase opportunities for employees to be physically active during work hours. This situation is compounded when these employees frequently are required to work extended hours. It is less expensive for companies to have fewer employees work longer hours than to hire additional employees, partly attributable to cost of employee benefits other than salary, especially health insurance. As the use of technologies spreads to more occupations, especially in developing countries, there will be a further increase in the proportion of the population who become sedentary.

Typical Modes of Transportation

Daily physical activity has decreased substantially because of advancing technology regarding personal transportation. The rise in the use of the automobile during the 20th century significantly reduced average daily physical activity of many children and adults. Few programs have been successful in convincing people to substitute walking or bicycling for the use of their cars. Such a change would have a number of beneficial consequences in addition to the increased physical activity, including reductions in environmental pollution, traffic congestion, the need for new roads and parking structures, and—in some countries—reliance on foreign oil. The use of the automobile for transportation has been both facilitated and made necessary by the automobile-friendly design of many communities and by government subsidization of road and parking lot construction (partly in response to vigorous lobbying by automobile manufacturers and dealers, tire manufacturers, and the petroleum industry).

In many communities, safe sidewalks and bike paths are not available or do not lead from one functional destination to another (e.g., from a residential area to a shopping center, school, or office complex). The Institute of Medicine and Transportation Research Board of the U.S. National Academies of Science has released a report titled "Does the Built Environment Influence Physical Activity? Examining the Evidence." Although the research on this topic is at an early stage, preliminary results indicate that there are associations between the built environment and physical activity. These findings need to be confirmed with prospective observations and ideally with experiments, but at this point it seems reasonable to assume that some environments are more favorable than others when it comes to encouraging individuals to be physically active (Committee of Physical Activity, Health, Transportation, and Land Use 2005).

Typical Domestic Activities

As with the reduction in the need for physical activity on the job, technology has substantially reduced the amount of human energy required for domestic activities and self-care. Devices such as automatic clothes washers and dryers, dishwashers, self-propelled vacuum cleaners, and power lawn mowers make life much easier for many people but reduce their physical activity on most days (Lanningham-Foster et al. 2003). Although the amount of energy saved by using these devices is quite small on any one day (<50 kcal/day), a reduction of only 25 kcal/day over the year would equal about 9,125 kcal, which is equivalent to the amount of energy in 2.6 lb or 1.2 kg of body mass (figure 23.4). To put this in context, the average weight gain for U.S. adults during the 1990s, when major increases in the prevalence of obesity were observed, was about 1 lb

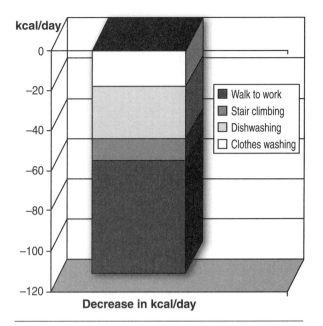

Figure 23.4 The participants (122 healthy adults) completed the indicated tasks with and without the aid of equipment or machines while the energy expenditure was measured. The distance walked or driven was 0.8 miles (1.28 km), the number of stairs climbed was 219, and the time spent washing clothes and dishes each day was 18.7 and 17.1 min, respectively.

Reprinted, by permission, from L. Lanningham-Foster, L.J. Nysse and J.A. Levine, 2003, "Labor saved, calories lost: The energetic impact of domestic labor-saving devices," *Obes Res* 11: 1178-1181.

per year (United States Department of Health and Human Services 2004). A decline in energy expenditure of this magnitude is sufficient to account for the obesity epidemic that has been observed in the United States over the past 25 years.

The automobile also has significantly reduced domestic-related physical activity by its frequent use for short trips for shopping, going to the cleaners and other errands, and taking children to school. It is highly unlikely that individuals will give up their dishwasher, clothes washer, or power lawn mower in order to increase their activity, but an improved system of safe sidewalks and bike paths throughout the community and financial disincentives rather than incentives to use the automobile might contribute to more physical activity to complete many short trips. Another example of small changes in activity having a substantial effect on energy expenditure is the observation that frequent "fidgeting" (can't sit still) can amount to 350 kcal/day. Obese men and women were documented as sitting 164 min/day longer than lean individuals (Levine et al. 2005).

Typical Leisure Time

In the 1960s and 1970s, the prediction by experts regarding the impact of technology on U.S. society over the following 50 years was that many adults would spend much less time at their occupations and have a great deal of time to pursue "leisure" activities. This prediction has not been realized, with many adults working more than 40 hr per week, spending more time commuting to work via automobile or mass transit, or working more than one job. Also, the increasing percentage of households where both partners are gainfully employed has further increased the overall amount of time spent working, performing domestic chores, or commuting and has reduced time available for recreation or leisure activities.

Compounding the problem of reduced leisure time has been the widespread pursuit of sedentary leisure-time activities, especially watching television, video monitors, and computer screens. How most people spend their leisure time depends on interests, priorities, and opportunities. It is in the area of leisure time that most people have the greatest opportunity to build physical activity back into their lives. There is a paradox regarding leisure time and amount of physical activity. Surveys show that the most frequently given reason for not being more physically active is lack of time, but this is in a population that reports at least several hours of TV watching each day. The problem is one of assigning priorities to leisure-time activities, not an absolute lack of leisure time itself. This issue appears especially important for children, because a national survey of television and video viewing and computer use found that for boys and girls ages 8 to 18 years, parents reported an average of 5.7 hr/day of screen or monitoring viewing, and for children as young as 2 to 7 they reported 2.8 hr/day (Henry J. Kaiser Family Foundation 1999) (figure 23.5). With every passing year, new technologies are providing more and more options for sedentary activities by children and youth using computers, video players, and other electronic devices.

Developing and Implementing Physical Activity Plans

Developing and implementing strategies that will assist generally inactive people to become more physically active need to be a high priority for

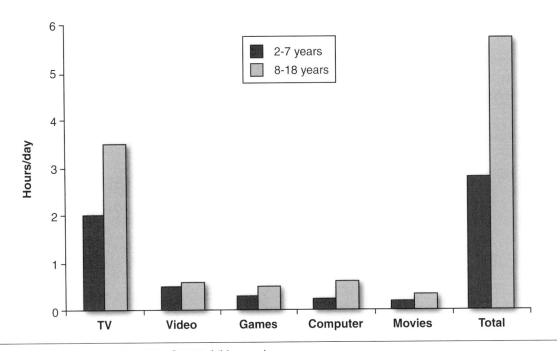

Figure 23.5 Survey conducted in 1999 of 3,158 children and parents.

Data from Henry J. Kaiser Family Foundation, 1999, *Kids and media at the millennium: A comprehensive national survey of children's media use* (Menlo Park, CA: Author).

government agencies and private organizations. There are no simple solutions to reversing the downward trend in activity for many people, but research and experience have provided a number of good leads on what needs to be done.

Developing Good Physical Activity Knowledge, Attitudes, and Skills

Given the tendency for current society to engineer a great deal of physical activity out of most people's lives, there is a public health need to develop system approaches to reintroduce activity so that most people can achieve active lifestyles throughout their life span. Because of the complexity of physical activity as a behavior and the multitude of factors that influence the process of a person becoming and remaining physically active, there is no one set formula for achieving this goal (chapter 21). However, adequate and accurate information, a positive attitude, and good skills are valuable tools in achieving an active lifestyle.

Knowledge

For most people, understanding why an active lifestyle is important for their health and what options they have to become and stay active is a good first step to an active lifestyle. Ideally, this knowledge is acquired as a youth from parents as well as teachers,

but it is never too late to obtain such information from teachers, health care providers, exercise specialists, books, videos, and the Internet. The more relevant or personalized this information is, the more likely it will be retained and acted upon. As in other areas of health behavior, just having the facts about what needs to be done and how to do it is not sufficient to ensure that the behavior is performed.

The primary reason many people are physically active is not because they want to prevent disease or maintain their physical independence. Some are active because they enjoy participating in sports and other active recreational activities. The goal of many hikers or backpackers is to enjoy the outdoors; promoting good health is a secondary issue for them. Other people, primarily for environmental or cost reasons, walk or use bicycles as their major means of transportation. For these people, it is important to encourage and possibly reward their behavior and find ways to support their active lifestyle by making the activity as convenient as possible. Also, there still is a segment of the population who obtain adequate physical activity during their occupation or domestic activities. These people need to know if their "required" activity meets the dose of activity that promotes good health and how to supplement this required activity if necessary.

Attitudes

One goal of any physical activity intervention should be to promote positive attitudes about the experience and seek ways to maintain or even enhance these attitudes over time. Although positive attitudes don't automatically lead to the desired behavior, negative attitudes, especially when derived or reinforced by negative experiences associated with activity, are a major deterrent to achieving an active lifestyle. A variety of factors can go into the formation of attitudes about physical activity, including past personal experiences with exercise; attitudes about exercise by parents, teachers, physicians, mentors, and other role models; and the importance of the potential benefits of being active or the negative consequences of being inactive. People who have been sedentary for a number of years often view themselves as inactive and have a difficult time viewing themselves as ever being physically active or fit. This negative attitude about activity frequently is made worse when the person is obese or has a chronic illness that contributes to a decrease in his or her functional capacity. A change in this attitude may only occur when the person has actually been successful in increasing activity and not as a precursor to an increase in activity.

Skills

A major reason many people continue to be physically active during leisure time is the enjoyment it provides. For activities that require a special skill, such as most sports, having a certain level of skill significantly adds to the enjoyment of participation. Beginning to develop such skills as a child is thought to facilitate participation as an adult. Thus, including programs for "lifetime activities" in schools for all students may be a way to facilitate sport participation in adults as well as youth. These programs would include activities such as tennis, volleyball, soccer, basketball, cycling, jogging and running, weight training, and dance. The focus here is not on competition or winning but on attaining basic skills, understanding the health and performance benefits, and enjoying participation. This type of program is in sharp contrast to a majority of school programs that focus on competitive football, basketball, and baseball for only the highly skilled and motivated students. For some basic activities that can be a major component of a health-oriented activity program, no or few special skills are needed. The best example here is walking, which for many

adults is, or should be, a major component of their activity plan. We do not have extensive data showing that teaching "lifetime activities" in schools will increase activity levels, but it seems clear that highly competitive sports are not the answer to mass sedentary behavior. It would be hard to emphasize competitive sports more than has been done in the United States over the past several decades (consider the number of pages of the typical daily newspaper that are devoted to sports coverage, in contrast to international events), yet this interest in competitive sports has not resulted in an active society.

Two Approaches to Physical Activity Plans

As part of a comprehensive program designed to promote good health, all individuals should have a physical activity plan. Ideally this plan should be individualized to meet each person's current needs (goals), health status, physical capacity, interests, skills, schedule, and resources, at the same time taking into account current required daily activity. This plan should include activities to be proscribed because of risk of injury or illness as well as activities prescribed for their health benefits. In its simplest form, this plan is no more than the current public health recommendations of "moderate intensity activity for at least 30 minutes per day on 5 or more days of the week." However, although good advice, this very general statement is primarily directed at the average sedentary middle-aged person and does not address any number of specific issues that are deterrents to maintaining an active lifestyle. Many inactive individuals would benefit from a more detailed and personalized plan.

Class Approach

The traditional approach for many people to build more activity into their life is to join an exercise

---

Although general physical activity and health guidelines (moderate-intensity activity for $\geq$30 min on 5 or more days of the week) are valuable for informing the public what they need to do, a personalized and detailed activity plan should be implemented that considers goals, current activity, health status, and family, work, and social obligations.

facility or exercise class. This approach can be the major part of a successful plan for some people, because it can provide a safe place with proper facilities (exercise equipment, showers), good exercise instruction and leadership, and a positive social environment. Some classes are designed for people with special interests or needs and can provide a very personalized program or meet special needs, such as the medical supervision provided in many classes for people with cardiovascular disease (CVD), diabetes, or chronic pulmonary disease.

Some of the limitations to this approach include the logistics of time and location. For many busy people with work and family responsibilities, it is very difficult to visit an exercise facility or class at least several times per week. It is not only the time in class but also the time required to commute to such a facility that can be a major deterrent. For people with limited incomes, the financial costs to join a facility or class may make participation difficult or impossible. Also, participating in a typical exercise class two times per week might not fully replace the decrease in energy expenditure experienced by many people with otherwise very sedentary lifestyles.

For example, an administrative assistant who, with access to improved computer and communication technology, reduces the amount of time walking around the office building 5 min/hr (40 min per 8 hr day or 200 min per 5-day work week) and now drives a car to work rather than taking a bus, which required 20 min more walking per workday ($20 \times 5 = 100$ min/week), needs to replace the energy expended walking versus sitting for 300 min or 5 hr per week. If the net decrease in energy expenditure between walking and sitting at a desk or driving averages 2.5 **metabolic equivalents (METs)** (3.5 METs walking – 1 MET resting = 2.5 METs), then the person needs to replace about 750 MET-minutes of activity per week (300 min $\times$ 2.5 METs). Thus, to replace this amount of physical activity during two 45 min classes per week, it would be necessary to exercise at an average intensity of approximately 8 METs (vigorous or hard intensity). At three 45 min exercise sessions per week, the average intensity would need to be approximately 5.5 METs (moderate intensity).

Active Living Concept

A second approach to building physical activity back into a person's life that has received much more attention of late has focused on making or taking opportunities throughout the day to be active. So, instead of going to a class or a gym to work out several times a week, a person makes time to be active while going on with daily life. Although such recommendations in a somewhat isolated way have been made by many people and organizations for a long time (e.g., take the stairs, park in the back of the lot), starting around 1990, this approach began to be formalized, discussed, evaluated, debated, and promoted. In Canada, this idea of integrating physical activity into daily life to create an active lifestyle was labeled as **active living** (Fitness Canada 1991). Although active living has been given a variety of meanings and definitions, it was defined by Fitness Canada (1991) as "a way of life in which physical activity is valued and integrated into daily life" (p. 4) or by Makosky (1994) as "a way of life in which individuals make useful, pleasurable and satisfying physical activities as an integral part of their daily routine" (p. 272).

A key to success is to help sedentary individuals learn to apply cognitive and behavioral strategies such as self-monitoring, goal setting, and problem solving to increase physical activity (Blair et al. 2001). Some individuals may use these approaches to increase their lifestyle physical activities and others may use them to adopt and maintain more traditional forms of physical activity such as walking, jogging, or sports. Behavioral intervention approaches have been found to be comparable to more traditional and structured approaches in helping sedentary persons to become and stay more physically active up for up to 24 months (Dunn et al. 1999).

This concept, be it called active living, lifestyle activity, or some other name, is in part supported by the research that has evaluated the health and performance benefits of shorter versus longer bouts of activity, but the two ideas are not the same. Most of the experimental studies evaluating short bouts of activity have used activity bouts in the 8 to 10 min range, with few data on the value of discreet bouts of moderate-intensity exercise of 5 min or less. Short bouts of vigorous exercise, such as stair climbing, spread throughout the day have produced improvements in health-related fitness. For example, in a study of young sedentary women, those who performed stair climbing 5 days per week for 7 weeks (climbed 199 steps in 135 s, one time per day in week 1 progressing to six times per day in weeks 6 and 7) had a reduction in heart rate and blood lactate concentration during submaximal exercise and an increase in high-density lipoprotein cholesterol compared with women who remained sedentary (Boreham et al. 2000). Some epidemiological studies relating habitual physical activity to chronic disease or all-cause mortality have collected data supporting the concept that activity accumulated throughout the days in bouts of various lengths and intensities confers health benefits (Lee et al. 2000).

Overcoming Barriers to Achieving a Physically Active Lifestyle

As noted earlier in this chapter, there are a number of barriers to initiating and especially maintaining a physically active lifestyle.

Barriers

Frequently these barriers relate to the need for sedentary people to make time to build activity back into their daily life. Often people feel too busy with work, family, commuting, and social engagements to take time to be active. As individuals age, a major barrier to being physically active is the concern about the health or injury risks of increased activity, especially if the person has led a sedentary life or has one or more medical conditions.

Yet even some very busy women who have a full-time job and a family to care for make time to be physically active—their answer is that their health and appearance are a priority and they have good organizational and time management skills. It is a major challenge for health professionals to help people give regular activity a sufficiently high priority to overcome time demands and other barriers. Somehow the negative health consequences of being inactive as well as the benefits of being active need to be made salient to the person as a first step in an activity plan.

Strategies for Increasing Physical Activity

Various strategies for helping people achieve an active lifestyle work differently for different people. For example, some people are loners and prefer to exercise on their own, whereas others benefit from having access to a highly social environment. Some employees may find it very difficult to make time for a 20 min walk alone during lunch but will readily commit to walking and talking with a coworker on a regular basis. Appointments to go for a walk or perform other exercise with a friend, colleague, or family member should carry the same obligation as any other type of appointment. For people who benefit from positive social interactions, well-led exercise classes of numerous types can facilitate frequent and long-term participation.

Those who believe they are too busy to make time for physical activity need to closely review their schedules and look for even small opportunities to begin to build activity back—climbing the stairs instead of using the escalator, taking a 10 min walk during at least one coffee break each work day, spending some of the extra time between plane connections walking in the terminal. For some people, performing these short bouts of activity is the first step in building more activity into their lives and thinking of themselves as active rather than sedentary.

For some people, interaction with an authority figure or role model may help make an active lifestyle a higher priority. The results of numerous surveys have indicated that if a person's physician or other health care provider makes a strong recommendation regarding an active lifestyle and monitors this recommendation even to a small extent, the likelihood that the person will become or stay active is increased. People who are respected for how they lead their lives can be positive role models for people of all ages in developing an active lifestyle—this seems to be true throughout the life span.

Older individuals who are afraid that physical activity may be harmful because of their age or physical condition should receive specific personalized instructions from their health care provider or staff regarding their activity plan, including an opportunity to get questions answered and a schedule for follow-up. There are many options for even people with quite low exercise capacities attributable to injury, illness, or age to perform health-promoting physical activity. At the very low end of the activity–fitness range, a person should at least get out of a bed or chair and move about periodically throughout the day to help prevent thrombosis, orthostatic hypotension, insulin resistance, muscle weakness, and loss of balance. To provide additional assurance for at-risk patients that activity can be increased with relative safety, supervised exercise class or individual instruction by a well-trained exercise leader should be considered.

One procedure that appears to help many people improve adherence to their activity plan is "self-monitoring" (Dunn et al. 1999). This procedure can take different forms but basically consists of developing a logging or tracking system that allows individuals to compare their activity performance over time with their activity plan. For example, a

It is easy to become so overcommitted to work and other obligations that the instant we have a moment to ourselves, sedentary options seem irresistibly attractive. If we are to remain healthy, we must either make leisure-time choices that include getting the exercise we need in ways we enjoy, or learn how to incorporate adequate physical activity into our daily tasks.

person could record her activity plan on a calendar or electronic organizer for the next 3 months (e.g., walk briskly 30 min at noon Monday, Wednesday, and Friday and in the morning on Saturday and Sunday) and then check off or document each time she performs that activity. This process provides a visual history of adherence to the plan and acts as a frequent reminder when an activity session is missed.

Integrating Physical Activity With Other Aspects of Life

Much of how we spend our daily lives is based on a number of trade-offs among those things we have to accomplish daily to survive and meet our obligations to others and those things we select to do during our free or leisure time. In attempting to build more physical activity into existing busy schedules, we must recognize that some balance needs to be achieved among these various interests and time commitments but that physical activity needs to be given sufficient priority to ensure a physically active lifestyle. All too often physical activity gets dropped from the schedule because of the immediate need to get other things done, with the notion that the activity will get done later. Is it more important to go to a movie, watch a television show, read a

book, or have coffee with a friend than to go for a 30 min walk, bicycle ride, or swim? The individual has to make this decision, but he or she should do so fully aware of the long-term consequences of being physically inactive.

Need to Modify the Activity Plan As Life Evolves

As a person progresses through life, his or her activity plan should be considered a dynamic and not a static tool for helping to maintain good health. Although the overall plan of maintaining an active lifestyle remains the same, some specific features of the plan will need to change in response to changes in the person's age, living environment, physical fitness, and health.

Frequently, it is during major life transition periods, such as leaving high school or college, getting a full-time job, changing jobs, moving to a new home, getting married, having children, and retirement, that a number of health-related behaviors, including physical activity, encounter pressure to change. With each of these life changes, a person takes on new responsibilities, the daily schedule is changed, and frequently the opportunities for activity are reduced. When a person is young and in school, her activity plan should include not only the activity performed at recess, in physical education classes, or as part of sports teams but also non-school-related activity, such as supervised and nonsupervised play or recreation, household chores, and personal transportation. Once out of high school, people typically have less structured time for activity and their plan will substantially depend on college or work obligations. When a person who has been in college takes a full-time job, frequently moving to a new community and meeting new friends, his activity plan will be influenced by the need to identify new space or facilities, find time for activity, and find new persons to exercise with.

> The basic components of an effective activity plan will remain relatively constant over the life span, but its implementation will change with age, other life commitments (school, employment, family), health status, and personal goals.

With more and more people living into their 70s, 80s, and 90s, becoming or remaining physically active takes on increased public health importance. In addition to helping to reduce the incidence or severity of various chronic diseases, frequent activity is critical to maintain physical functioning, independent living, and adequate quality of life. Many people who reach their late 60s or older realize that if they had known they were going to live so long they would have taken better care of themselves. Of particular importance is the need for older persons to retain sufficient muscle strength, endurance, balance, and flexibility to independently perform a wide range of activities of daily living. A comprehensive retirement plan (or nonretirement plan if a person continues to work) should include an activity plan that considers the special needs and goals of the older person. This activity plan should focus on maintaining the capacity to independently climb stairs, get out of a bathtub, dress and undress, walk several blocks without stopping, carry groceries home from the market, make the bed, wash dishes and clothes, and perform other self-care and household chores. The results of numerous research projects along with increasing experience from community-based exercise programs for older persons have documented the importance of a personal physician's endorsement of an activity

plan, personalized instruction on how to conduct an effective and safe plan, and convenient and safe facilities to carry out the plan.

Integration of Physical Activity With Other Health Behaviors for Optimal Health

Whereas physical activity has a number of unique and independent effects on a person's health, it also interacts in a variety of important ways with other health behaviors or conditions. To obtain the greatest benefit from a physically active lifestyle, a person should consider how activity interacts with such behaviors as sleep, stress management, and nutrition. For example, if the activity being performed to enhance health significantly interferes with a person's ability to get adequate sleep, then health status and quality of life will be negatively affected. Similarly, physical activity and other strategies for managing stress will interact. If the activity is viewed by the person as another burden on his or her limited time and adds to feelings of stress or anxiety, the activity plan should be revised or other adjustments made to reduce the person's overall time burden.

Because food is the source of all the energy and nutrients required for physical activity, eating and being physically active should be closely linked for optimal health. The most obvious link here is the relationship among the number of calories consumed, the number of calories expended during physical activity, and body mass, especially adiposity. To achieve optimal long-term health, calories consumed need to be balanced against calories expended at an optimal body mass—generally recommended to be a BMI between 18.5 and 25. Over weeks or months if the calories expended during activity significantly increase, then calorie intake will need to be increased to maintain a specific body weight, with the opposite being true if there is a sustained decrease in activity. Maintaining this calorie balance at an optimal body weight is not an easy task for many people but appears to be easier to achieve at higher levels of activity (Hill et al. 2003). At lower levels of activity where total daily energy expenditure is less than 30 kcal/kg, calorie intake can easily exceed calorie expenditure resulting in an increase in body weight. Also, to maintain body weight with a sedentary lifestyle and low calorie intake, very careful food selection is needed to obtain the diversity of micronutrients required for optimal health.

> Moderate-intensity physical activity has numerous health benefits that occur independently of other health behaviors, but such activity appears to be even more effective when well integrated with other health-promoting actions.

Maximizing Benefits and Minimizing Risks

Physical activity can aggravate existing disease or cause a variety of injuries as well as provide substantial health and performance benefits. (See chapter 19 for more a detailed discussion of risks associated with increased activity.) The activity plan should be individualized to minimize these risks while maximizing benefits as defined by the primary goals of the participant. To obtain this balance of benefits versus risks, the plan should be individualized based on age, past medical history, current health status, other health-related behaviors, and physical fitness. Attention to risk should increase if the person has cardiorespiratory, metabolic, or musculoskeletal disease; has had joint or bone injuries; is obese; is at high risk of CVD because of smoking, hypertension, or dyslipidemia; or has a very low exercise capacity attributable to age or disability. Other factors that will increase the risk of exercise-induced injury include shoes that don't provide appropriate support for walking, jogging, or running; slick or uneven exercise surfaces that contribute to falls (especially by older persons); and high environmental temperatures and humidity, which can cause heat injuries. For a comprehensive discussion of how to develop a personalized activity plan that considers all of these issues, refer to *ACSM's Guidelines for Exercise Testing and Prescription* (ACSM 2005).

> Because the promotion of physical activity can aggravate existing disease or cause injury as well as provide health benefits, people should consider personal characteristics (illness, prior injury), types of activity (high intensity or impact), and environmental conditions (slippery path, high temperature) that can substantially increase risk when designing a personalized activity plan.

Research Questions and Issues

This book has cited scientific documentation that a physically active lifestyle leads to a higher state of health and wellness across the life span. The results of this wide range of research support current recommendations by a number of government agencies, health organizations, and medical and exercise associations. However, a number of important questions need to be answered through scientific investigation to provide a more comprehensive understanding of the biological, clinical, and behavioral issues involved in maintaining an active lifestyle throughout the life span. Following is a summary of key research issues that should be investigated.

Most of the data supporting a favorable relationship between higher levels of physical activity or physical fitness and morbidity or mortality attributable to coronary heart disease (CHD), stroke, type 2 diabetes, and various cancers are from observational studies that are limited in their ability to demonstrate causality. Well-designed randomized clinical trials are needed in persons at increased risk for specific clinical events to test whether an increase in activity significantly reduces specific clinical events.

A low rate of clinical events such as heart attacks, strokes, and type 2 diabetes, if attributable to high levels of physical activity, must be mediated by biological changes. For example, it appears that the reduced risk of coronary artery disease may be attributable to improvements in a number of biological processes including lipid metabolism, blood pressure, fibrinolysis, insulin-mediated glucose uptake, and coronary artery dilation. However, which of these or other biological changes are necessary to reduce CHD clinical events needs to be established, and the situation is similar for other clinical outcomes. Knowing what biological changes are necessary to improve clinical status will be valuable in establishing who is most likely to receive benefit for that outcome and to more accurately develop the dose component of their activity plan.

A small body of research has demonstrated that the tissues and organs challenged by regular exercise adapt by activating the transcription of genes in key pathways and attenuating the transcription of others. Different types of exercise modalities are associated with different profiles of changes in gene expression and thus protein synthesis. Further research of this type should be conducted because it has great potential for establishing how exercise affects basic biological functions and how adaptation to exercise occurs at the molecular level.

A number of issues regarding the dose of activity required for specific health outcomes still need be to be resolved using innovative scientific methods. For example, experimental data are needed on what health outcomes will be produced by accumulating 30 to 60 min of activity throughout the day in bouts of just several minutes duration each as well as on the specific benefits derived from just two long bouts of activity per week, like 90 min hikes on each day of the weekend.

The role of physical activity in preventing unhealthy weight gain, contributing to weight loss, and preventing weight regain after weight loss is not well understood and requires substantial scientific investigation. Controversy surrounds how much activity is needed to prevent undesired weight gain or significantly contribute to weight loss. Also, the relative benefits of endurance versus resistance exercise in the management of body mass and composition is not well understood. Anecdotes and nonscientific observations suggest that some individuals appear to be quite sedentary yet maintain weight over time. Other individuals are relatively active but gain weight. These individual differences are probably attributable to genetic and familial factors, but specific information on these points is needed. To more effectively conduct research on physical activity and obesity, measurement tools are needed that accurately assess energy intake and energy expenditure in real-life situations over extended periods of time.

Studies conducted over the past 2 decades have demonstrated substantial interindividual differences in the response to a change in **physical activity level,** many of which are attributable to genetic variation. This genetic research needs to continue using advanced techniques to reveal the true basis of the heterogeneity in responsiveness to regular exercise and to progressively move the field toward individualized exercise recommendations and individualized preventive medicine.

One of the most challenging issues facing exercise and health professionals is how to help a large proportion of the population achieve a physically active lifestyle throughout their life span. Innovative research is needed to evaluate theory-based activity interventions that are targeted at specific subsets of the population. Priority needs to be given to persons in medically underserved populations, including those with low incomes and limited formal education and ethnic minorities. Much additional research is needed on the effect of the built environment on physical activity patterns. We also need to evaluate how environmental changes accompanied by behavioral and educational interventions might interact to affect activity levels.

Summary and Conclusions

The primary objective for having an active lifestyle throughout the life span is to contribute to a healthy, enjoyable, productive, and long life. It is possible for some minority of people to be quite sedentary throughout most of their life and still have success and satisfaction, remain reasonably free of major diseases, and avoid weight gain during a long life. In the future, we may be able to identify the genetic and biological profile that makes this situation possible, but for the present we do not have the knowledge or procedures to determine whom these people might be. Thus, the public health goal should be to facilitate an active lifestyle throughout the life span for the entire population by means of the following:

- Improved education to increase the public's knowledge about why physical activity is important and how to go about developing a physically active lifestyle.

- Changes in the built environment that make safe activity available to all.

- Adoption of polices that encourage activity during all aspects of life (occupation, education, transportation, retirement).

- A health care system that aggressively promotes disease prevention, health promotion, and quality of life by means of improved health behaviors, including a physically active lifestyle.

KEY CONCEPTS

active living: For definition, see page 333.

built environment: Environment as modified by the construction of any structure or place including, but not limited to, homes, commercial buildings, schools, churches, roads, sidewalks, parks, and recreation centers in the context of its potential effects on daily physical activity.

energy expenditure: The sum of three factors:

- Resting energy expenditure to maintain basic body functions (approximately 60% of total energy requirements)

- Processing of food eaten throughout the day, which includes digestion, absorption,

transport, and disposition of nutrients (about 10% of total energy requirements)

- Nonresting energy expenditure, primarily in the form of physical activity (about 30% of total energy requirements)

metabolic equivalent (MET): For definition, see page 19.

required activities of daily living: That activity required by a person to survive and lead a productive life.

physical activity level: For definition, see page 189.

STUDY QUESTIONS

1. Briefly describe three major changes that have occurred in the United States that have significantly contributed to the reduction in "daily required physical activity" for a large proportion of the population.

2. What information would you want to have before helping a 43-year-old healthy but very sedentary woman, who is a senior administrative assistant in a large corporation and lives with her husband and two teenage daughters, develop her physical activity plan?

3. Contrast the advantages and disadvantages of a class-based activity program versus an "active living" approach for a 55-year-old business executive who frequently travels to other cities.

4. Provide four examples of how the built environment has reduced the opportunities for many people to perform physical activity throughout the day.

5. In helping to design a physical activity plan for an obese (BMI = 36) 17-year-old girl, what information would you want to know about her activity history, medical status, school activities, family arrangements, and social activities?

6. Describe the role that positive or negative attitudes about physical activity have in maintaining an active lifestyle throughout the life span. What conditions or experiences lead to negative attitudes? Propose strategies to improve these attitudes.

7. Briefly discuss three topics where more research is needed in the area of physical activity and health and suggest specific research projects that would help address these issues.

References

Chapter 1

Aboderin, I., A. Kalache, Y. Ben-Shlomo, J.W. Lynch, C.S. Yajnik, D. Kuh, and D. Yach. 2001. *Life course perspectives on coronary heart disease, stroke, and diabetes: Key issues and implications for policy and research.* Geneva: World Health Organization.

Bouchard, C. 1994. Physical activity, fitness, and health: Overview of the consensus symposium. In *Toward active living. Proceedings of the International Conference on Physical Activity, Fitness, and Health*, edited by H.A. Quinney, L. Gauvin, and A.E.T. Wall. Champaign, IL: Human Kinetics, pp. 7-14.

Bouchard, C., and R.J. Shephard. 1994. Physical activity, fitness, and health: The model and key concepts. In *Physical activity, fitness, and health*, edited by C. Bouchard, R.J. Shephard, and T. Stephens. Champaign, IL: Human Kinetics, pp. 77-88.

Brown, J.R., and G.P. Crowden. 1963. Energy expenditure ranges and muscular work grades. *British Journal of Industrial Medicine* 20: 227.

Dishman, R.K. 1988. Determinants of participation in physical activity. In *Exercise, fitness and health: A consensus of current knowledge*, edited by C. Bouchard, R.J. Shephard, T. Stephens, J.R. Sutton, and B.D. McPherson. Champaign, IL: Human Kinetics, pp. 33-48.

Flegal, K.M., B.I. Graubard, D.F. Williamson, and M.H. Gail. 2005. Excess deaths associated with underweight, overweight, and obesity. *Journal of the American Medical Association* 293(15): 1861-1867.

Krauss, H., and W. Raab. 1961. *Hypokinetic diseases.* Springfield IL: Charles C Thomas.

Malina R.M. 1991. Darwinian fitness, physical fitness and physical activity. In *Applications of biological anthropology*, edited by C.G.N. Mascie-Taylor and G.W. Lasker. Cambridge, UK: Cambridge University Press, pp. 143-184.

McIntosh, P.C. 1980. *"Sport for All" programmes throughout the world.* Paris: UNESCO.

Mokdad, A.H., J.S. Marks, D.F. Stroup, and J.L. Gerberding. 2004. Actual causes of death in the United States, 2000. *Journal of the American Medical Association* 291: 1238-1245.

Paffenbarger, R., R.T. Hyde, and A.L. Wing. 1990. Physical activity and physical fitness as determinants of health and longevity. In *Exercise, fitness and health: A consensus of current knowledge*, edited by C. Bouchard, R.J. Shephard, T. Stephens, J.R. Sutton, and B.D. McPherson. Champaign, IL: Human Kinetics, pp. 33-48.

Pate, R.R. 1988. The evolving definition of fitness. *Quest* 40: 174-179.

Pucher, J. 1997. Bicycling boom in Germany: A revival engineered by public policy. *Transportation Quarterly* 51: 31-36.

Reddy, K.S. 2004. Cardiovascular disease in non-Western countries. *New England Journal of Medicine* 350: 2438-2440.

World Health Organization. 1948. *Constitution of the World Health Organization. Basic documents.* Geneva: World Health Organization.

World Health Organization. 1968. *Meeting of investigators on exercise tests in relation to cardiovascular function.* Geneva: World Health Organization.

World Health Organization. 1998. *The world health report 1998—Life in the 21st century: A vision for all.* Geneva: World Health Organization.

World Health Organization. 2002. *The world health report 2002—Reducing risks, promoting healthy life.* Geneva: World Health Organization.

Yusuf, S., S. Hawken, S. Ounpuu, T. Dans, A. Avzeum, F. Lanas, M. McQueen, A. Budaj, P. Pais, J. Varigos, and L. Lisheng. 2004. Effect of potentially modifiable risk factors associated with myocardial infarction in 52 countries (the INTERHEART study) case-control study. *Lancet* 364: 937-952.

Chapter 2

American College of Sports Medicine. 1975. *Guidelines for graded exercise testing and prescription.* Philadelphia: Lea & Febiger.

American College of Sports Medicine. 1978. Position statement—The recommended quantity and quality of exercise for developing and maintaining fitness in healthy adults. *Medicine and Science in Sports and Exercise* 10: vii-x.

American Heart Association. 1975. *Exercise testing and training of individuals with heart disease or at high risk for its development.* Dallas: American Heart Association.

Bouchard, C. 1991. Heredity and the path to overweight and obesity. *Medicine and Science in Sports and Exercise* 23: 285-291.

Bouchard, C., P. An, T. Rice, J.S. Skinner, J.H. Wilmore, J. Gagnon, L. Perusse, A.S. Leon, and D.C. Rao. 1999. Familial aggregation of VO_2max response to exercise training: Results from the HERITAGE Family Study. *Journal of Applied Physiology* 87: 1003-1008.

Bowerman, W.J., and W.E. Harris. 1967. *Jogging.* New York: Grosset and Dunlap.

Cooper, K.H. 1968. *Aerobics.* New York: Bantam Books.

Crespo, C.J., S.J. Keteyian, G.W. Heath, and C.T. Sempos. 1996. Leisure-time physical activity among US adults. Results from the Third National Health and Nutrition Examination Survey. *Archives of Internal Medicine* 156(1): 93-98.

Duncan, J.J., N.F. Gordon, and C.B. Scott. 1991. Women walking for health and fitness. How much is enough? *Journal of the American Medical Association* 266: 3295-3299.

Fletcher, G.F., S.N. Blair, J. Blumenthal, C. Caspersen, B. Chaitman, S. Epstein, H. Falls, E.S. Froelicher, V.F. Froelicher, and I.L. Pina. 1992. Statement on exercise. Benefits and recommendations for physical activity programs for all Americans. A statement for health professionals by the Committee on Exercise and Cardiac Rehabilitation of the Council on Clinical Cardiology, American Heart Association. *Circulation* 86: 340-344.

Haskell, W.L. 1984. The influence of exercise on the concentrations of triglyceride and cholesterol in human plasma. *Exercise and Sport Sciences Reviews* 12: 205-244.

Karvonen, M.J., E. Kentala, and O. Mustala. 1957. The effects of training on heart rate: A longitudinal study. *Annals of Medicine Experimental Biology Fennica* 35: 307-315.

King, A.C. 1991. Community intervention for promotion of physical activity and fitness. *Exercise and Sport Sciences Reviews* 19: 211-259.

King, A.C. 1994. Community and public health approaches to the promotion of physical activity. *Medicine and Science in Sports and Exercise* 26: 1405-1412.

Marcus, B.H., V.C. Selby, R.S. Niaura, and J.S. Rossi. 1992. Self-efficacy and the stages of exercise behavior change. *Research Quarterly for Exercise and Sport* 63: 60-66.

Marcus, B.H., and L.R. Simkin. 1993. The stages of exercise behavior. *Journal of Sports Medicine and Physical Fitness* 33: 83-88.

Morris, J.N., and M.D. Crawford. 1958. Coronary heart disease and physical activity of work: Evidence of a national necropsy survey. *British Medical Journal* 30: 1485-1496.

Morris, J.N., J.A. Heady, P.A. Raffle, C.G. Roberts, and J.N. Parks. 1953. Coronary heart disease and physical activity at work. *Lancet* 265: 1053-1057, 1111-1120.

NIH Consensus Development Panel on Physical Activity and Cardiovascular Health. 1996. Physical activity and cardiovascular health. *Journal of the American Medical Association* 276: 241-246.

Paffenbarger, R.S., Jr., S.N. Blair, and I.M. Lee. 2001. A history of physical activity, cardiovascular health and longevity: The scientific contributions of Jeremy N Morris, DSc, DPH, FRCP. *International Journal of Epidemiology* 30: 1184-1192.

Paffenbarger, R.S., Jr., A.L. Wing, and R.T. Hyde. 1978. Physical activity as an index of heart attack risk in college alumni. *Journal of Epidemiology* 108: 161.

Pate, R.R., M. Pratt, S.N. Blair, W.L. Haskell, C.A. Macera, C. Bouchard, D. Buchner, W. Ettinger, G.W. Heath, A.C. King, A. Kriska, A.S. Leon, S.E. Marcus, J. Morris, R.S. Paffenbarger, Jr., K. Patrick, M.L. Pollock, J.M. Rippe, J.F. Sallis, and J.H. Wilmore.

1995. A recommendation from the Centers for Disease Control and Prevention and the American College of Sports Medicine. *Journal of the American Medical Association* 273: 402-407.

Tipton, C.M. 1984. Exercise, training, and hypertension. *Exercise and Sport Sciences Reviews* 12: 245-306.

U.S. Department of Agriculture. 2004. Nutrition and Your Health: Dietary Guidelines for Americans, 2005 Dietary Guidelines Advisory Committee Report. Washington, DC: U.S. Department of Agriculture.

U.S. Department of Health and Human Services. 1996. *Physical activity and health: A report of the Surgeon General*. Atlanta: USDHSS–CDC.

WHO–FIMS Committee on Physical Activity for Health. 1995. Exercise for health. *Bulletin of the World Health Organization* 73: 135-136.

Chapter 3

Australian Bureau of Statistics. 2003. *Australian Social Trends 2003*. Canberra: Australian Bureau of Statistics.

Bonen, A., and S.M. Shaw. 1995. Recreational exercise participation and aerobic fitness in men and women: Analysis of data from a national survey. *Journal of Sports Sciences* 13: 297-303.

Carnethon, M.R., S.S. Gidding, R. Nehgme, S. Sidney, D.R. Jacobs, and K. Liu. 2003. Cardiorespiratory fitness in young adulthood and the development of cardiovascular disease risk factors. *Journal of the American Medical Association* 290: 3092-3100.

Comstock, R.D., E.M. Castillo, and P. Lindsay. 2004. Four-year review of the use of race and ethnicity in epidemiologic and public health research. *American Journal of Epidemiology* 159: 611-619.

Fitness Canada. 1983. *Fitness and lifestyle in Canada*. Ottawa: Government of Canada.

Grunbaum, J.A., L. Kann, S.A. Kinchen, B. Williams, J.G. Ross, R. Lowry, and L. Kolbe. 2002. Youth risk behavior surveillance—United States, 2001. Surveillance summaries, June 28, 2002. *Morbidity and Mortality Weekly Reports* 51 (No. SS-4): 1-64.

Ham, S.A., M.M. Yore, J.E. Fulton, and H.W. Kohl, III. 2004. Prevalence of no leisure-time physical activity—35 states and the District of Columbia, 1988-2002. *Morbidity and Mortality Weekly Reports* 53: 82-86.

Helakorpi, S., K. Patja, R. Prattala, A.R. Aro, and A. Uutela. 2002. *Health behaviour and health among Finnish adult population, Spring 2002*. Helsinki: Finnish National Public Health Institute.

Janicke, B., D. Coper, and U.-A. Janicke. 1986. Motor activity of different-aged Cercopithecidae: Silver-leafed monkey (Presbytis cristatus Esch.), lion-tailed monkey (Macaca silenus L.), Moor Macaque (Macaca maura Cuv.) as observed in the zoological garden, Berlin. *Gerontology* 32: 133-140.

Rantanen, T., K. Masaki, D. Foley, G. Izmirlian, L. White, and J.M. Guralnik. 1998. Grip strength changes over 27 yr in Japanese-American men. *Journal of Applied Physiology* 85: 2047-2053.

Skinner, J.S., A. Jaskolski, A. Jaskolska, J. Krasnoff, J. Gagnon, A.S. Leon, D.C. Rao, J.H. Wilmore, and C. Bouchard. 2001. Age, sex, race, initial fitness, and response to training: The HERITAGE Family Study. *Journal of Applied Physiology* 90: 1770-1776.

Sports Council and Health Education Authority. 1992. *Allied Dunbar national fitness survey: Main findings.* London: The Sports Council and Health Education Authority.

Statistics Canada. 2002. *Health indicators*, May 2002. Catalogue No. 82-221-XIE. Retrieved February 10, 2006, from http://www.statcan.ca/Daily/English/020508/td020508.htm.

Talbot, L.A., E.J. Metter, and J.L. Fleg. 2000. Leisure-time physical activities and their relationship to cardiorespiratory fitness in healthy men and women 18-95 years old. *Medicine and Science in Sports and Exercise* 32: 417-425.

Teers, R. 2001. Physical activity. In *Health survey for England—The health of minority ethnic groups '99.* Edited by B. Erens, P. Primatesta, and G. Prior. London: The Stationary Office.

Trappe, S.W., D.L. Costill, M.D. Vukovich, J. Jones, and T. Melham. 1996. Aging among elite distance runners: A 22-yr longitudinal study. *Journal of Applied Physiology* 80: 285-290.

World Health Organization. 2000. *Health and health behaviour among young people.* WHO Policy Series: Health Policy for Children and Adolescents Issue 1: International Report. Geneva: World Health Organization.

Chapter 4

American College of Sports Medicine. 2000. *ACSM's guidelines for exercise testing and prescription.* Baltimore: Lippincott Williams & Wilkins.

Åstrand, P-O., T.E. Cuddy, B. Saltin, and J. Stenberg. 1964. Cardiac output during submaximal and maximal work. *Journal of Applied Physiology* 19: 268-274.

Åstrand, P-O., K. Rodahl, H.A. Dahl, and S.B. Strømme. 2003. *Textbook of work physiology.* 4th ed. Champaign, IL: Human Kinetics, p. 373.

Bouchard, C., T.P. An, T. Rice, J.S. Skinner, J.H. Wilmore, J. Gagnon, L. Pérusse, A.S. Leon, and D.C. Rao. 1999. Family aggregation of VO₂max response to exercise: Results from the HERITAGE Family Study. *Journal of Applied Physiology* 87: 1003-1008.

Bouchard, C., E.W. Daw, T. Rice, L. Pérusse, J. Gagnon, M.A. Province, A.S. Leon, D.C. Rao, J.S. Skinner, and J.H. Wilmore. 1998. Family resemblance for VO₂max in the sedentary state: The HERITAGE Family Study. *Medicine and Science in Sports and Exercise* 30: 252-258.

Brooks, G.A. 1985. Anaerobic threshold: Review of the concept and directions for future research. *Medicine and Science in Sports and Exercise* 17: 22-34.

Dempsey, J.A., L. Adams, D.M. Ainsworth, R.F. Fregosi, C.G. Gallagher, A. Guz, B.D. Johnson, and S.K. Powers. 1996. Airway, lung, and respiratory muscle function during exercise. In *Handbook of physiology.*

Section 12: Exercise: Regulation and integration of multiple systems, edited by L.B. Rowell and J.T. Shepherd. New York: Oxford University Press, pp. 448-514.

Ekblom, B., P-O. Åstrand, B. Saltin, J. Stenberg, and B. Wallström. 1968. Effect of training on circulatory response to exercise. *Journal of Applied Physiology* 24: 518-528.

Gaesser, G.A., and G.A. Brooks. 1984. Metabolic bases of excess post-exercise oxygen consumption: A review. *Medicine and Science in Sports and Exercise* 19: 29-43.

Holloszy, J.O., and W.M. Kohrt. 1995. Exercise. In *Handbook of physiology. Section 11: Aging*, edited by E.J. Masoro. New York: Oxford University Press, pp. 633-666.

Howley, E.T., and B.D. Franks. 2003. *Health fitness instructor's handbook.* 4th ed. Champaign, IL: Human Kinetics.

Powers, S.K., and E.T. Howley. 2004. *Exercise physiology: Theory and applications to fitness and performance.* New York: McGraw-Hill.

Powers, S., W. Riley, and E. Howley. 1980. A comparison of fat metabolism in trained men and women during prolonged aerobic work. *Research Quarterly for Exercise and Sport* 52: 427-431.

Rowell, L.B. 1993. *Human cardiovascular control.* New York: Oxford University Press.

Rowell, L.B., D.S. O'Leary, and D.L. Kellogg, Jr. 1996. Integration of cardiovascular control systems in dynamic exercise. In *Handbook of physiology. Section 12: Exercise: Regulation and integration of multiple systems*, edited by L.B. Rowell and J.T. Shepherd. New York: Oxford University Press, pp. 770-840.

Chapter 5

Braun, B., M.B. Zimmermann, and N. Kretchmer. 1995. Effects of exercise intensity on insulin sensitivity in women with non-insulin-dependent diabetes mellitus. *Journal of Applied Physiology* 78: 300-306.

Cadroy, Y., F. Pillard, K.S. Sakariassen, C. Thalamas, B. Boneu, and D. Riviere. 2002. Strenuous but not moderate exercise increases the thrombotic tendency in healthy sedentary male volunteers. *Journal of Applied Physiology* 93: 829-833.

Febbraio, M.A., and B. Klarlund Pedersen. 2002. Muscle-derived interleukin-6: Mechanisms for activation and possible biological roles. *FASEB Journal* 16: 1335-1347.

Haskell, W.L. 1994. Health consequences of physical activity: Understanding and challenges regarding dose-response. *Medicine and Science in Sports and Exercise* 26: 649-660.

King, D.S., R.J. Baldus, R.L. Sharp, L.D. Kesl, T.L. Feltmeyer, and M.S. Riddle. 1995. Time course for exercise-induced alterations in insulin action and glucose tolerance in middle-aged people. *Journal of Applied Physiology* 78: 17-22.

Klarlund Pedersen, B.K., and L. Hoffman-Goetz. 2000. Exercise and the immune system: Regulation, integration, and adaptation. *Physiological Reviews* 80: 1055-1081.

MacDonald, J.R. 2002. Potential causes, mechanisms, and implications of post exercise hypotension. *Journal of Human Hypertension* 16: 225-236.

Malm, C., Ö. Ekblom, and B. Ekblom. 2004. Immune system alteration in response to two consecutive soccer games. *Acta Physiologica Scandinavica* 180: 143-155.

Murphy, M.H., A.M. Nevill, and A.E. Hardman. 2000. Different patterns of brisk walking are equally effective in decreasing postprandial lipaemia. *International Journal of Obesity* 24: 1303-1309.

Pate, R., M. Pratt, S.N. Blair, W.L. Haskell, C.A. Macera, C. Bouchard, D. Buchner, W. Ettinger, G.W. Heath, A.C. King, A. Kriska, A.S. Leon, B.H. Marcus, R. Paffenbarger, S.K. Patrick, M.L. Pollock, J.M. Rippe, J. Sallis, and J.H. Wilmore. 1995. Physical activity and public health: A recommendation from the Centers for Disease Control and Prevention and the American College of Sports Medicine. *Journal of the American Medical Association* 273: 402-407.

Stubbs, J.R., D.A. Hughes, A.M. Johnstone, G.W. Horgan, N. King, and J.E. Blundell. 2004. A decrease in physical activity affects appetite, energy, and nutrient balance in lean men feeding ad libitum. *International Journal of Obesity* 79: 62-69.

Stubbs, R.J., A. Sepp, D.A. Hughes, A.M. Johnstone, N. King, G. Horgan, and J.E. Blundell. 2002. The effect of graded levels of exercise on energy intake and balance in free-living women. *International Journal of Obesity* 26: 866-869.

Taylor-Tolbert, N., D. Dengel, M.D. Brown, S.D. McCole, R.E. Pratley, M.I. Ferrell, and J.M. Hagberg. 2000. Ambulatory blood pressure after acute exercise in older men with essential hypertension. *American Journal of Hypertension* 13: 44-51.

Tsetsonis, N.V., and A.E. Hardman. 1996. Reduction in postprandial lipidemia after walking: Influence of exercise intensity. *Medicine and Science in Sports and Exercise* 28: 1235-1242.

Tsetsonis, N.V., A.E. Hardman, and S.S. Mastana. 1997. Acute effects of exercise on postprandial lipidemia: A comparative study in trained and untrained middle-aged women. *American Journal of Clinical Nutrition* 65: 525-533.

Wojtakzewski, J.F.P., S.B. Jørgensen, C. Frøsig, C. MacDonald, J.B. Birk, and E.A. Richter. 2003. Insulin signalling: Effects of prior exercise. *Acta Physiologica Scandinavica* 178: 321-328.

Womack, C.J., P.R. Nagelkirk, and A.M. Coughlin. 2003. Exercise-induced changes in coagulation and fibrinolysis in healthy populations and patients with cardiovascular disease. *Sports Medicine* 33: 795-807.

Young, J.C., J. Enslin, and B. Kuca. 1989. Exercise intensity and glucose tolerance in trained and untrained subjects. *Journal of Applied Physiology* 67: 39-43.

Chapter 6

Bonadonna, R.C., L.C. Groop, K. Zych, M. Shank, and R.A. DeFronzo. 1990. Dose dependent effect insulin on plasma free fatty acid turnover and oxidation in humans. *American Journal of Physiology (Endocrinology Metabolism)* 259: E736-E750.

Borer, K.T. 2003. *Exercise endocrinology.* Champaign, IL: Human Kinetics.

Engdahl, J.H., J.D. Veldhuis, and P.A. Farrell. 1995. Altered pulsatile insulin secretion associated with endurance training. *Journal of Applied Physiology* 79: 1977-1985.

Farrell, P.A. 1992. Exercise effects on regulation of energy metabolism by pancreatic and gut hormones. In *Perspectives in exercise science and sports medicine* edited by D.K.R. Lamb, R. Murray, and C. Gisolfi. Indianapolis, IN: Brown and Benchmark, pp. 383-434.

Galbo, H. 1983. *Hormonal and metabolic adaptation to exercise.* Stuttgart: Georg Thieme Ferlag.

Gosselink, K., R.R. Roy, H. Zhong, R.E. Grindeland, A.J. Bigbee, and V.R. Edgerton. 2004. Vibration-induced activation of muscle afferents modulates bioassayable growth hormone release. *Journal of Applied Physiology* 96: 2097-2102.

Gustafson, A.B., P.A. Farrell, and R.K. Kalkhoff. 1990. Impaired plasma catecholamine response to submaximal treadmill exercise in obese women. *Metabolism* 39: 410-417.

Hansen, A.P. 1973. Serum growth hormone response to exercise in non-obese and obese normal subjects. *Scandinavian Journal Clinical Laboratory Investigation* 31: 175-178.

Kjaer M, P.A. Farrell, N.J. Christensen, and H. Galbo. 1986. Increased epinephrine response and inaccurate glucoregulation in exercising athletes. *Journal of Applied Physiology* 61: 1693-1700.

Kjaer, M., and H. Galbo 1988. Effect of physical training on the capacity to secrete epinephrine. *Journal of Applied Physiology* 64: 11-16.

Krotkiewski, M., K. Mandroukas, L. Morgan, T. William-Olsson, G.E. Feurle, H. vonSchenck, P. Bjorntorp, L. Sjostrom, and U. Smith. 1983. Effect of physical training on adrenergic sensitivity in obesity. *Journal of Applied Physiology* 55: 1811-1817.

Loucks, A.B. 2001. Physical health of the female athlete: Observations, effects, and causes of reproductive disorders. *Canadian Journal of Applied Physiology* 26(Suppl): S176-S185.

Mikines, K.J., F. Dela, B. Sonne, P.A. Farrell, E.A. Richter, and H. Galbo. 1987. Insulin action and insulin secretion: Effects of different levels of physical activity. *Canadian Journal of Sports Medicine* 12: 113-116.

Mikines, K.J., F. Dela, B. Tronier, and H. Galbo. 1989. Effect of 7 days of bed rest on dose-response relation between plasma glucose and insulin secretion. *American Journal of Physiology (Endocrinology Metabolism)* 257: E43-E48.

Mikines, K.J., B. Sonne, P.A. Farrell, B. Tronier, and H. Galbo. 1988a. The effect of training on the dose response relationship for insulin action in man. *Journal of Applied Physiology* 66: 695-703.

Mikines, K.J., B. Sonne, P.A. Farrell, B. Tronier, and H. Galbo. 1988b. Effect of physical exercise on sensitivity and responsiveness to insulin in man. *American Journal of Physiology (Endocrinology Metabolism)* 254: E248-E259.

Viru, A. 1985a. *Hormones in muscular activity—Adaptive effects of hormones in exercise.* Boca Raton, FL: CRC Press.

Viru, A. 1985b. *Hormones in muscular activity—Hormonal ensemble in exercise.* Boca Raton, FL: CRC Press.

Warren, M.P., and N.W. Constantini, editors. 2000. *Sports endocrinology.* Totowa, NJ: Humana Press.

Chapter 7

Allen, D.G., A.A. Kabbara, and H. Westerblad. 2002. Muscle fatigue: The role of intracellular calcium stores. *Canadian Journal of Applied Physiology* 27: 83-96.

Baldwin, K.M., and F. Haddad. 2002. Skeletal muscle plasticity: Cellular and molecular responses to altered physical activity paradigms. *American Journal of Physical Medicine and Rehabilitation* 81(Suppl): S40-S51.

Berchtold, M.W., H. Brinkmeier, and M. Müntener. 2000. Calcium ion in skeletal muscle: Its crucial role for muscle function, plasticity and disease. *Physiological Reviews* 80: 1216-1265.

Booth, F.W., M.V. Chakravarthy, and E.E. Sprangenburg. 2002. Exercise and gene expression: Physiological regulation of the human genome through physical activity. *Journal of Physiology* 543: 399-411.

Brooks, G.A. 1998. Mammalian fuel utilization during sustained exercise. *Comparative Biochemistry and Physiology: Part B* 120: 89-107.

Carmelli, E., R. Coleman, and A.Z. Reznick. 2002. The biochemistry of aging muscle. *Experimental Gerontology* 37: 477-489.

Clausen, T. 2003. Na$^+$-K$^+$ pump regulation and skeletal muscle contractility. *Physiological Reviews* 83: 1269-1324.

Connett, R.J., C.R. Honig, T.E.J. Gayeski, and G.A. Brooks. 1990. Defining hypoxia: A systems view of VO$_2$, glycolysis, energetics and intracellular PO$_2$. *Journal of Applied Physiology* 68: 833-842.

Doherty, J.J. 2003. Invited review: Aging and sarcopenia. *Journal of Applied Physiology* 95: 1717-1727.

Fitts, R.H. 1994. Cellular mechanisms of muscle fatigue. *Physiological Reviews* 74: 49-94.

Green, H.J. 2000. Muscular factors in endurance. In: *Encyclopedia of sports medicine. Endurance in sports,* edited by R. Shephard and P-O. Åstrand. Oxford, UK: Blackwell Science, pp. 156-163.

Hochachka, P.W. 1994. *Muscles as molecular and metabolic machines.* Boca Raton, FL: CRC Press.

Holloszy, J.O. 2003. A forty-year memoir of research on the regulation of glucose transport into muscle. *American Journal of Physiology* 284: E453-E467.

Holloszy, J.O., and F.W. Booth. 1976. Biochemical adaptations to endurance training in muscle. *Annual Review of Physiology* 38: 273-291.

Hood, D.A. 2001. Invited review: Contractile activity-induced mitochondrial biogenesis in skeletal muscle. *Journal of Applied Physiology* 90: 1137-1157.

Juel, C., and A.P. Halestrap. 1999. Lactate transport in skeletal muscle-role of regulation of monocarboxylate transporter. *Journal of Applied Physiology* 517: 633-642.

MacLean, P.S., D. Zheng, and G.L. Dohm. 2000. Muscle glucose transporter (GLUT 4) gene expression during exercise. *Exercise and Sport Sciences Reviews* 28: 148-152.

Margreth, A., E. Damiani, and E. Bortaloso. 1999. Sarcoplasmic reticulum in aged skeletal muscle. *Acta Physiologica Scandinavica* 167: 331-338.

Pette, D. 2002. The adaptive potential of skeletal muscle fibers. *Canadian Journal of Applied Physiology* 27: 423-448.

Reggiani, C., R. Bottinelli, and G.J.M. Stienen. 2000. Sarcomeric myosin isoforms: Fine tuning a molecular motor. *News in Physiological Sciences* 15: 26-33.

Sahlin, K., M. Tonkonogi, and K. Söderlund. 1998. Energy supply and muscle fatigue in humans. *Acta Physiologica Scandinavica* 162: 261-266.

Schiaffino, S., and C. Reggiani. 1996. Molecular diversity of myofibrillar proteins: Gene regulation and functional significance. *Physiological Reviews* 76: 371-423.

Short, K.R., and K.S. Nair. 2001. Does aging adversely affect muscle mitochondrial function? *Exercise and Sport Sciences Reviews* 29: 118-123.

Turcotte, L.P. 2000. Muscle fatty acid uptake during exercise. Possible mechanisms. *Exercise and Sport Sciences Reviews* 28: 4-9.

Vandervoort, A.A. 2002. Aging in the human neuromuscular system. *Muscle & Nerve* 25: 17-25.

Chapter 8

Churchill, J.D., R. Galvez, S. Colcombe, R.A. Swain, A.F. Kramer, and W.T. Greenough. 2002. Exercise, experience and the aging brain. *Neurobiology of Aging* 23: 941-955.

Gleeson, M. 2000. Mucosal immune responses and risk of respiratory illnesses in elite athletes. *Exercise Immunology Review* 6: 5-42.

Landers, D.M., and S.J. Petruzello. 1994. Physical activity, fitness and anxiety. In *Physical activity, fitness and health,* edited by C. Bouchard, R.J. Shephard, and T. Stephens. Champaign, IL: Human Kinetics, pp. 868-882.

Mackinnon, L.T. 2000. Chronic exercise training effects on immune function. *Medicine and Science in Sports and Exercise* 32(Suppl. 7): S369-S376.

Marshall, J.C. 1998. The gut as a potential trigger of exercise-induced inflammatory responses. *Canadian Journal of Physiology and Pharmacology* 76: 479-484.

Moses, F.M. 1994. Physical activity and the digestive processes. In *Physical activity, fitness, and health,* edited by C. Bouchard, R.J. Shephard, and T. Stephens. Champaign, IL: Human Kinetics, pp. 383-400.

Nieman, D.C. 2000. Special feature for the Olympics: Effects of exercise on the immune system: Exercise effects on systemic immunity. *Immunology and Cell Biology* 78: 496-501.

North, T.C., P. McCullagh, and Z.V. Tran. 1990. Effect of exercise on depression. *Exercise and Sport Sciences Reviews* 18: 379-416.

Poortmans, J.R., and J. Vanderstraeten. 1994. Kidney function during exercise in healthy and diseased humans. *Sports Medicine* 18: 419-437.

Rowell, L.B. 1971. Visceral blood flow and metabolism during exercise. In *Frontiers of fitness*, edited by R.J. Shephard. Springfield, IL: Charles C Thomas, pp. 210-229.

Shephard, R.J. 1997a. Curricular physical activity and academic performance. *Pediatric Exercise Science* 9: 113-126.

Shephard, R.J. 1997b. *Physical activity, training and the immune response*. Carmel, IN: Cooper.

Shephard, R.J., and R. Futcher. 1997. Physical activity and cancer: How may protection be maximized? *Critical Reviews in Oncogenesis* 8: 219-272.

van Nieuwenhoven, M.A., F. Brouns, and R.J. Brummer. 1999. The effect of physical activity on parameters of gastrointestinal function. *Neurogasteroenterology & Motility* 11: 431-439.

Wade, O.L., and J.M. Bishop. 1962. *Cardiac output and regional blood flow*. Oxford, UK: Blackwell Scientific.

Chapter 9

Arraiz, G.A., D.T. Wigle, and Y. Mao. 1992. Risk assessment of physical activity and physical fitness in the Canada Health Survey Mortality Follow-up Study. *Journal of Clinical Epidemiology* 45: 419-428.

Blair, S.N., J.B. Kampert, H.W. Kohl, C.E. Barlow, C.A. Macera, R.S. Paffenbarger, Jr., and L.W. Gibbons. 1996. Influences of cardiorespiratory fitness and other precursors on cardiovascular disease and all-cause mortality in men and women. *Journal of the American Medical Association* 276: 205-210.

Blair, S.N., and Wei, M. Sedentary habits, health, and function in older women and men. 2000. *American Journal of Health Promotion* 15: 1-8.

Church, T.S., Y.J. Cheng, C.P. Earnest, C.E. Barlow, L.W. Gibbons, E.L. Priest, and S.N. Blair. 2004. Exercise capacity and body composition as predictors of mortality among men with diabetes. *Diabetes Care* 27: 83-88.

Church, T.S., J.B. Kampert, L.W. Gibbons, C.E. Barlow, and S.N. Blair. 2001. Usefulness of cardiorespiratory fitness as a predictor of all-cause and cardiovascular disease mortality in men with systemic hypertension. *American Journal of Cardiology* 88: 651-656.

FitzGerald, S.J., C.E. Barlow, J. Kampert, J.R. Morrow, A.W. Jackson, and S.N. Blair. 2004. Muscular fitness and all-cause mortality: Prospective observations. *Journal of Physical Activity and Health* 1: 7-18.

Kampert, J.B., S.N. Blair, C.E. Barlow, and H.W. Kohl. 1996. Physical activity, physical fitness, and all-cause and cancer mortality: A prospective study of men and women. *Annals of Epidemiology* 6: 452-457.

Kohl, H.W., III. 2001. Physical activity and cardiovascular disease: Evidence for a dose response. *Medicine and Science in Sports and Exercise* 33: S472-S483.

Kohl, H.W., III, M.Z. Nichaman, R.F. Frankowski, and S.N. Blair. 1996. Maximal exercise hemodynamics and risk of mortality in apparently healthy men and women. *Medicine and Science in Sports and Exercise* 28: 601-609.

Lee, C.D., S.N. Blair, and A.S. Jackson. 1999. Cardiorespiratory fitness, body composition, and all-cause and cardiovascular disease mortality in men. *American Journal of Clinical Nutrition* 69: 373-380.

Lee, I.M., and P.J. Skerrett. 2001. Physical activity and all-cause mortality: What is the dose-response relation? *Medicine and Science in Sports and Exercise* 33: S459-S471.

Morris, J.N., D.G. Clayton, M.G. Everitt, A.M. Semmence, and E.H. Burgess. 1990. Exercise in leisure time: Coronary attack and death rates. *British Heart Journal* 63: 325-334.

Morris, J.N., R. Pollard, M.G. Everitt, and S.P.W. Chave. 1980. Vigorous exercise in leisure-time: Protection against coronary heart disease. *Lancet* 11: 1207-1210.

Oguma, Y., H.D. Sesso, R.S. Paffenbarger, Jr., and I.M. Lee. 2002. Physical activity and all cause mortality in women: A review of the evidence. *British Journal of Sports Medicine* 36: 162-172.

Paffenbarger, R.S., Jr. 1988. Contributions of epidemiology to exercise science and cardiovascular health. *Medicine and Science in Sports and Exercise* 20: 426-438.

Paffenbarger, R.S., Jr., S.N. Blair, and I.M. Lee. 2001. A history of physical activity, cardiovascular health and longevity: The scientific contributions of Jeremy N Morris, DSc, DPH, FRCP. *International Journal of Epidemiology* 30: 1184-1192.

Paffenbarger, R.S., Jr., R.T. Hyde, A.L. Wing, and C.-C. Hsieh. 1986. Physical activity, all-cause mortality, and longevity of college alumni. *New England Journal of Medicine* 314: 605-613.

Paffenbarger, R.S., Jr., R.T. Hyde, A.L. Wing, I-M. Lee, D.L. Jung, and J.B. Kampert. 1993. The association of changes in physical-activity level and other lifestyle characteristics with mortality among men. *New England Journal of Medicine* 328: 538-545.

Paffenbarger, R.S., Jr., and I-M. Lee. 1998. A natural history of athleticism, health and longevity. *Journal of Sports Sciences* 16: S31-S45.

Rastogi, T., M. Vaz, D. Spiegelman, K.S. Reddy, A.V. Bharathi, M.J. Stampfer, W.C. Willett, and A. Ascherio. 2004. Physical activity and risk of coronary heart disease in India. *International Journal of Epidemiology* 33: 1-9.

Sawada, S.S., T. Muto, H. Tanaka, I.M. Lee, R.S. Paffenbarger, Jr., M. Shindo, and S.N. Blair. 2003. Cardiorespiratory fitness and cancer mortality in Japanese men: A prospective study. *Medicine and Science in Sports and Exercise* 35: 1546-1550.

Chapter 10

American College of Sports Medicine. 1994. American College of Sports Medicine position stand: Exercise for patients with coronary artery disease. *Medicine and Science in Sports and Exercise* 26: i-v.

American Thoracic Society. 1999. Pulmonary rehabilitation—1999. *American Journal of Respiratory and Critical Care Medicine* 159(5 Pt 1): 1666-1682.

Blair, S.N., Y. Cheng, and J.S. Holder. 2001. Is physical activity or physical fitness more important in defining health benefits? *Medicine and Science in Sports and Exercise* 33(6 Suppl): S379-S399; discussion S419-S420.

Blair, S.N., H.W. Kohl, III, C.E. Barlow, R.S. Paffenbarger, Jr., L.W. Gibbons, and C.A. Macera. 1995. Changes in physical fitness and all-cause mortality. A prospective study of healthy and unhealthy men. *Journal of the American Medical Association* 273: 1093-1098.

Gordon, N.F., M. Gulanick, F. Costa, G. Fletcher, B.A. Franklin, E.J. Roth, and T. Shephard. 2004. Physical activity and exercise recommendations for stroke survivors: An American Heart Association scientific statement from the Council on Clinical Cardiology, Subcommittee on Exercise, Cardiac Rehabilitation, and Prevention; the Council on Cardiovascular Nursing; the Council on Nutrition, Physical Activity, and Metabolism; and the Stroke Council. *Circulation* 109: 2031-2041.

Hooi, J.D., A.D. Kester, H.E. Stoffers, M.M. Overdijk, J.W. van Ree, and J.A. Knottnerus. 2001. Incidence of and risk factors for asymptomatic peripheral arterial occlusive disease: A longitudinal study. *American Journal of Epidemiology* 153: 666-672.

Katzmarzyk, P.T., and I. Janssen. 2004. The economic costs associated with physical inactivity and obesity in Canada: An update. *Canadian Journal of Applied Physiology* 29: 90-115.

Kelley, G.A., and K.S. Kelley. 2000. Progressive resistance exercise and resting blood pressure: A meta-analysis of randomized controlled trials. *Hypertension* 35: 838-843.

Kohl, H.W., III. 2001. Physical activity and cardiovascular disease: Evidence for a dose response. *Medicine and Science in Sports and Exercise* 33(6 Suppl): S472-S483; discussion S493-S494.

Lakka, T.A., J.M. Venalainen, R. Rauramaa, R. Salonen, J. Tuomilehto, and J.T. Salonen. 1994. Relation of leisure-time physical activity and cardiorespiratory fitness to the risk of acute myocardial infarction. *New England Journal of Medicine* 330: 1549-1554.

Lee, C.D., A.R. Folsom, and S.N. Blair. 2003. Physical activity and stroke risk: A meta-analysis. *Stroke* 34: 2475-2481.

Oga, T., K. Nishimura, M. Tsukino, S. Sato, and T. Hajiro. 2003. Analysis of the factors related to mortality in chronic obstructive pulmonary disease: Role of exercise capacity and health status. *American Journal of Respiratory and Critical Care Medicine* 167: 544-549.

Paffenbarger, R.S., Jr., R.J. Brand, R.I. Sholtz, and D.L. Jung. 1978. Energy expenditure, cigarette smoking, and blood pressure level as related to death from specific diseases. *American Journal of Epidemiology* 108: 12-18.

Paffenbarger, R.S., Jr., A.L. Wing, and R.T. Hyde. 1978. Physical activity as an index of heart attack risk in college alumni. *American Journal of Epidemiology* 108: 161-175.

Pate, R.R., M. Pratt, S.N. Blair, W.L. Haskell, C.A. Macera, C. Bouchard, D. Buchner, W. Ettinger, G.W. Heath, A.C. King, et al. 1995. Physical activity and public health. A recommendation from the Centers for Disease Control and Prevention and the American College of Sports Medicine. *Journal of the American Medical Association* 273: 402-407.

Pereira, M.A., A.R. Folsom, P.G. McGovern, M. Carpenter, D.K. Arnett, D. Liao, M. Szklo, and R.G. Hutchinson. 1999. Physical activity and incident hypertension in black and white adults: The Atherosclerosis Risk in Communities Study. *Preventive Medicine* 28: 304-312.

Pescatello, L.S., B.A. Franklin, R. Fagard, W.B. Farquhar, G.A. Kelley, and C.A. Ray. 2004. American College of Sports Medicine position stand. Exercise and hypertension. *Medicine and Science in Sports and Exercise* 36: 533-553.

Pollock, M.L., B.A. Franklin, G.J. Balady, B.L. Chaitman, J.L. Fleg, B. Fletcher, M. Limacher, I.L. Pina, R.A. Stein, M. Williams, and T. Bazzarre. 2000. AHA Science Advisory. Resistance exercise in individuals with and without cardiovascular disease: Benefits, rationale, safety, and prescription: An advisory from the Committee on Exercise, Rehabilitation, and Prevention, Council on Clinical Cardiology, American Heart Association; Position paper endorsed by the American College of Sports Medicine. *Circulation* 101: 828-833.

Rasmussen, F., J. Lambrechtsen, H.C. Siersted, H.S. Hansen, and N.C. Hansen. 2000. Low physical fitness in childhood is associated with the development of asthma in young adulthood: The Odense schoolchild study. *European Respiratory Journal* 16: 866-870.

Regensteiner, J.G., and W.R. Hiatt. 2002. Current medical therapies for patients with peripheral arterial disease: A critical review. *American Journal of Medicine* 112: 49-57.

Satta, A. 2000. Exercise training in asthma. *Journal of Sports Medicine and Physical Fitness* 40: 277-283.

Tanasescu, M., M.F. Leitzmann, E.B. Rimm, W.C. Willett, M.J. Stampfer, and F.B. Hu. 2002. Exercise type and intensity in relation to coronary heart disease in men. *Journal of the American Medical Association* 288: 1994-2000.

Wannamethee, S.G., A.G. Shaper, and M. Walker. 2000. Physical activity and mortality in older men with diagnosed coronary heart disease. *Circulation* 102: 1358-1363.

Chapter 11

American College of Sports Medicine. 2001. Appropriate intervention strategies for weight loss and prevention of weight regain for adults. *Medicine and Science in Sports and Exercise* 33: 2145-2156.

Arner, P., E. Kriegholm, P. Engfeldt, and J. Bolinder. 1990. Adrenergic regulation of lipolysis in situ at rest and during exercise. *Journal of Clinical Investigation* 85: 893-898.

Björntorp, P. 1990. "Portal" adipose tissue as a generator of risk factors for cardiovascular disease and diabetes. *Arteriosclerosis* 10: 493-496.

Blundell, J.E., R.J. Stubbs, D.A. Hughes, S. Whybrow, and N.A. King. 2003. Cross talk between physical activity and appetite control: Does physical activity stimulate appetite? *Proceedings of the Nutrition Society* 62: 651-661.

Cole, T.J., M.C. Bellizzi, K.M. Flegal, and W.H. Dietz. 2000. Establishing a standard definition for child overweight and obesity worldwide: International survey. *British Medical Journal* 320: 1240-1243.

Deurenberg-Yap, M., and P. Deurenberg. 2003. Is a re-evaluation of WHO body mass index cut-off values needed? The case of Asians in Singapore. *Nutrition Reviews* 61: S80-S87.

Donnelly, J.E., J.O. Hill, D.J. Jacobsen, J. Potteiger, D.K. Sullivan, S.L. Johnson, K. Heelan, M. Hise, P.V. Fennessey, B. Sonko, T. Sharp, J.M. Jakicic, S.N. Blair, Z.V. Tran, M. Mayo, C. Gibson, and R.A. Washburn. 2003. Effects of a 16-month randomized controlled exercise trial on body weight and composition in young, overweight men and women: The Midwest Exercise Trial. *Archives of Internal Medicine* 163: 1343-1350.

Erlichman, J., A.L. Kerbey, and W.P. James. 2002. Physical activity and its impact on health outcomes. Paper 2: Prevention of unhealthy weight gain and obesity by physical activity: An analysis of the evidence. *Obesity Reviews* 3: 273-287.

Ford, E.S., A.H. Mokdad, and W.H. Giles. 2003. Trends in waist circumference among U.S. adults. *Obesity Research* 11: 1223-1231.

Goodpaster, B.H., J. He, S. Watkins, and D.E. Kelley. 2001. Skeletal muscle lipid content and insulin resistance: Evidence for a paradox in endurance-trained athletes. *Journal of Clinical Endocrinology and Metabolism* 86: 5755-5761.

Institute of Medicine. 2002. *Dietary reference intake for energy, carbohydrate, fiber, fat, fatty acids, cholesterol, protein and amino acids.* Washington, DC: National Academy Press.

Janssen, I., S.B. Heymsfield, and R. Ross. 2002. Application of simple anthropometry in the assessment of health risk: Implications for the Canadian Physical Activity, Fitness and Lifestyle Appraisal. *Canadian Journal of Applied Physiology* 27: 396-414.

Janssen, I., P.T. Katzmarzyk, R. Ross, A.S. Leon, J.S. Skinner, D.C. Rao, J.H. Wilmore, T. Rankinen, and C. Bouchard. 2004. Fitness alters the associations of BMI and waist circumference with total and abdominal fat. *Obesity Research* 12: 525-537.

Jeffery, R.W., and J. Utter. 2003. The changing environment and population obesity in the United States. *Obesity Research* 11(Suppl): 12S-22S.

Katzmarzyk, P.T. 2002. The Canadian obesity epidemic, 1985-1998. *Canadian Medical Association Journal* 166: 1039-1040.

Katzmarzyk, P.T., and I. Janssen. 2004. The economic costs associated with physical inactivity and obesity in Canada: An update. *Canadian Journal of Applied Physiology* 29: 90-115.

Kelley, D.E., B.H. Goodpaster, and L. Storlien. 2002. Muscle triglyceride and insulin resistance. *Annual Reviews in Nutrition* 22: 325-346.

Malnick, S.D., M. Beergabel, and H. Knobler. 2003. Non-alcoholic fatty liver: A common manifestation of a metabolic disorder. *Quarterly Journal of Medicine* 96: 699-709.

Matsuzawa, Y. 2002. Importance of adipocytokines in obesity-related diseases. *International Journal of Obesity* 26(Suppl 1): S63.

Mayer, J. 1955. The physiological basis of obesity and leanness. I. *Nutrition Abstracts and Reviews* 25: 597-611.

Mokdad, A.H., B.A. Bowman, E.S. Ford, F. Vinicor, J.S. Marks, and J.P. Kaplan. 2001. The continuing epidemics of obesity and diabetes in the United States. *Journal of the American Medical Association* 286: 1195-2000.

National Institutes of Health National Heart Lung and Blood Institute. 1998. Clinical guidelines on the identification, evaluation, and treatment of overweight and obesity in adults: The evidence report. *Obesity Research* 6: S51-S210.

Rexrode, K.M., V.J. Carey, C.H. Hennekens, E.E. Walters, G.A. Colditz, M.J. Stampfer, W.C. Willett, and J.E. Manson. 1998. Abdominal adiposity and coronary heart disease in women. *Journal of the American Medical Association* 280: 1843-1848.

Ross, R., D. Dagnone, P.J. Jones, H. Smith, A. Paddags, R. Hudson, and I. Janssen. 2000. Reduction in obesity and related comorbid conditions after diet-induced weight loss or exercise-induced weight loss in men. A randomized, controlled trial. *Annals of Internal Medicine* 133: 92-103.

Ross, R., and I. Janssen. 2001. Physical activity, total and regional obesity: Dose-response considerations. *Medicine and Science in Sports and Exercise* 33(6 Suppl): S521-S527.

Ross, R., I. Janssen, J. Dawson, A.M. Kugl, J. Kuk, S. Wong, T.B. Nguyen-Duy, S.J. Lee, K. Kilpatrick, and R. Hudson. 2004. Exercise with or without weight loss is associated with reduction in abdominal and visceral obesity in women. A randomized controlled trial. *Obesity Research* 12: 789-798.

Saris, W.H., S.N. Blair, M.A. van Baak, S.B. Eaton, P.S. Davies, L. Di Pietro, M. Fogelholm, A. Rissanen, D. Schoeller, B. Swinburn, A. Tremblay, K.R. Westerterp, and H. Wyatt. 2003. How much physical activity is enough to prevent unhealthy weight gain? Outcome of the IASO 1st Stock Conference and consensus statement. *Obesity Reviews* 4: 101-114.

U.S. Surgeon General. 2001. *Call to action to prevent and decrease overweight and obesity.* Washington, DC: Department of Health and Human Services.

Williamson, D.F., J. Madans, R.F. Anda, J.C. Kleinman, H.S. Kahn, and T. Byers. 1993. Recreational physical activity and ten-year weight change in a US national cohort. *International Journal of Obesity and Related Metabolic Disorders* 17: 279-286.

World Health Organization. 1998. *Obesity: Preventing and managing the global epidemic* (WHO/NUT/NCD/98.1.1998). Geneva: World Health Organization.

World Health Organization Expert Consultation. 2004. Appropriate body-mass index for Asian populations and its implications for policy intervention. *Lancet* 363: 157-163.

Chapter 12

Boyle, J.P., A.A. Honeycutt, K.M. Narayan, M.T.J. Hoerger, L.S. Geiss, H. Chen, and T.J. Thompson. 2001. Projection of diabetes burden through 2050: Impact of changing demography and disease prevalence in the U.S. *Diabetes Care* 24: 1936-1940.

Eriksson, K.F., and F. Lindgarde. 1991. Prevention of type 2 (non-insulin-dependent) diabetes mellitus by diet and physical exercise. *Diabetologia* 34: 891-898.

Goodyear, L.J., and B.B. Kahn. 1998. Exercise, glucose transport, and insulin sensitivity. *Annual Reviews of Medicine* 49: 235-261.

Helmrich, S.P., D.R. Ragland, R.W. Leung, and R.S. Paffenbarger. 1991. Physical activity and reduced occurrence of non-insulin-dependant diabetes mellitus. *New England Journal of Medicine* 325: 147-152.

Joslin, E.P., H.F. Root, P. White, and A. Marble. 1935. *The treatment of diabetes mellitus.* 5th ed. Philadelphia: Lea & Febiger.

Knowler, W.C., E. Barrett-Connor, S.E. Fowler, R.F. Hamman, J.M. Lachin, E.A. Walker, D.M. Nathan (the Diabetes Prevention Program Research Group). 2002. Reduction in the incidence of type 2 diabetes with lifestyle intervention or metformin. *New England Journal of Medicine* 346: 393-403.

Manson, J.E., D.M. Nathan, A.S. Krolewski, M.J. Stampfer, W.C. Willett, and C.H. Hennekens. 1992. A prospective study of exercise and incidence of diabetes among US male physicians. *Journal of the American Medical Association* 268: 63-67.

Manson, J.E., E.B. Rimm, M.J. Stampfer, G.A. Colditz, W.C. Willett, A.S. Krolewski, B. Rosner, C.H. Hennekens, F.E. Speizer. 1991. Physical activity and incidence of non-insulin dependent diabetes mellitus in women. *Lancet* 338: 774-778.

Mokdad, A.H., E.S. Ford, B.A. Bowman, W.H. Dietz, F. Vinicor, V.S. Bales, and J.S Marks. 2003. Prevalence of obesity, diabetes, and obesity-related health risk factors. *Journal of the American Medical Association* 289: 76-79.

Mokdad, A.H., E.S. Ford, B.A. Bowman, D.E. Nelson, M.M. Engelgau, F. Vinicor, and J.S. Marks. 2000. Diabetes trends in the U.S.: 1990-1998. *Diabetes Care* 23: 1278-1283.

Pan, X., G. Li, and Y. Hu. 1995. Effect of dietary and/or exercise intervention on incidence of diabetes in 530 subjects with impaired glucose tolerance from 1986-1992. *Chinese Journal of Internal Medicine* 34: 108-112.

Pan, X.R., G.W. Li, Y.H. Hu, J.X. Wang, W.Y. Yang, Z.X. An, Z.X. Hu, J. Lin, J.Z. Xiao, H.B. Cao, P.A. Liu, X.G. Jiang, Y.Y. Jiang, J.P. Wang, H. Zheng, H. Zhang, P.H. Bennett, B.V. Howard. 1997. Effects of diet and exercise in preventing NIDDM in people with impaired glucose tolerance. The Da Qing IGT and Diabetes Study. *Diabetes Care* 20: 537-544.

Pate, R.R., M. Pratt, S.N. Blair, W.L. Haskell, C.A. Macera, C. Bouchard, D. Buchner, W. Ettinger, G.W. Heath, A.C. King, A. Kriska, A.S. Leon, B.H. Marcus, J. Morris, R.S. Paffenbarger, K. Patrick, M.L. Pollock, J.M. Rippe, J. Sallis, and J.H. Wilmore. 1995. Physical activity and public health. A recommendation from the Centers for Disease Control and Prevention and the American College of Sports Medicine. *Journal of the American Medical Association* 273: 402-407.

Tomas, E., A. Zorzano, and N.B. Ruderman. 2002. Exercise and insulin signaling: A historical perspective. *Journal of Applied Physiology* 93: 765-772.

Torjesen, P.A., K.I. Birkeland, S.A. Anderssen, I. Hjermann, I. Holme, and P. Urdal. 1997. Lifestyle changes may reverse development of the insulin resistance syndrome. The Oslo Diet and Exercise Study: A randomized trial. *Diabetes Care* 20: 26-31.

Tuomilehto, J., J. Lindstrom, J.G. Eriksson, T.T. Valle, H. Hamalainen, P. Ilanne-Parikka, S. Keinanen-Kiukaanniemi, M. Laakso, A. Louheranta, M. Rastas, V. Salminen, and M. Uusitupa. 2001. Prevention of type 2 diabetes mellitus by changes in lifestyle among subjects with impaired glucose tolerance. *New England Journal of Medicine* 344: 1343-1350.

Chapter 13

Byers, T., M. Nestle, A. McTiernan, C. Doyle, A. Currie-Williams, T. Gansler, and M. Thun. 2002. American Cancer Society guidelines on nutrition and physical activity for cancer prevention: Reducing the risk of cancer with healthy food choices and physical activity. *CA: A Cancer Journal for Clinicians* 52: 92-119.

Giovannucci, E., M. Leitzmann, D. Spiegelman, E.B. Rimm, G.A. Colditz, M.J. Stampfer, and W.C. Willett. 1998. A prospective study of physical activity and prostate cancer in male health professionals. *Cancer Research* 58: 5117-5122.

International Agency for Research on Cancer (IARC). 2002. *Weight control and physical activity. IARC Working Group on the Evaluation of Cancer-Preventive Agents.* Lyon: IARC.

Irwin, M.L., Y. Yasui, C.M. Ulrich, D. Bowen, R.E. Rudolph, R.S. Schwartz, M. Yukawa, E. Aiello, J.D. Potter, and A. McTiernan. 2003. Effect of exercise on total and intra-abdominal body fat in postmenopausal women: A randomized controlled trial. *Journal of the American Medical Association* 289: 323-330.

Jemal, A., R.C. Tiwari, T. Murray, A. Ghafoor, A. Samuels, E. Ward, E.J. Feuer, and M.J. Thun. 2004. Cancer statistics, 2004. *CA: A Cancer Journal for Clinicians* 54: 8-29.

Lee, I.-M., and Y. Oguma. In press. Physical activity. In *Cancer epidemiology and prevention*, edited by D. Schottenfeld and J.F.J. Fraumeni. San Francisco: Oxford University Press.

Lee, I.-M., R.S. Paffenbarger, Jr., and C.C. Hsieh. 1991. Physical activity and risk of developing colorectal cancer among college alumni. *Journal of the National Cancer Institute* 83: 1324-1329.

Lee, I.-M., R.S. Paffenbarger, Jr., and C.C. Hsieh. 1992. Physical activity and risk of prostatic cancer among college alumni. *American Journal of Epidemiology* 135: 169-179.

Lee, I.-M., K.M. Rexrode, N.R. Cook, C.H. Hennekens, and J.E. Buring. 2001. Physical activity and breast cancer risk: The Women's Health Study (United States). *Cancer Causes and Control* 12: 137-145.

Lee, I.-M., H.D. Sesso, and R.S. Paffenbarger, Jr. 1999. Physical activity and risk of lung cancer. *International Journal of Epidemiology* 28: 620-625.

Lee, I. M., H.D. Sesso, and R.S. Paffenbarger, Jr. 2001. A prospective cohort study of physical activity and body size in relation to prostate cancer risk (United States). *Cancer Causes and Control* 12: 187-193.

Leung, P.S., W.J. Aronson, T.H. Ngo, L.A. Golding, and R.J. Barnard. 2004. Exercise alters the IGF axis in vivo and increases p53 protein in prostate tumor cells in vitro. *Journal of Applied Physiology* 96: 450-454.

Mao, Y., S. Pan, S.W. Wen, and K.C. Johnson. 2003. Physical activity and the risk of lung cancer in Canada. *American Journal of Epidemiology* 158: 564-575.

Martinez, M.E., E. Giovannucci, D. Spiegelman, D.J. Hunter, W.C. Willett, and G.A. Colditz. 1997. Leisure-time physical activity, body size, and colon cancer in women. Nurses' Health Study Research Group. *Journal of the National Cancer Institute* 89: 948-955.

McTiernan, A., C. Kooperberg, E. White, S. Wilcox, R. Coates, L.L. Adams-Campbell, N. Woods, and J. Ockene. 2003. Recreational physical activity and the risk of breast cancer in postmenopausal women: The Women's Health Initiative Cohort Study. *Journal of the American Medical Association* 290: 1331-1336.

Strong, K., C. Mathers, S. Leeder, and R. Beaglehole. 2005. Preventing chronic diseases: How many lives can we save? *Lancet* 366: 1578-1582.

Thune, I., T. Brenn, E. Lund, and M. Gaard. 1997. Physical activity and the risk of breast cancer. *New England Journal of Medicine* 336: 1269-1275.

Chapter 14

Arthritis Foundation. 2001. *Primer on the rheumatic diseases*, edited by J.H. Klippel. 12th ed. Atlanta: Arthritis Foundation.

Buckwalter, J.A. 2003. Sports, joint injury, and post-traumatic osteoarthritis. *Journal of Orthopaedic and Sports Physical Therapy* 33: 578-588.

Centers for Disease Control and Prevention. 2001. Prevalence of disabilities and associated health conditions among adults—United States, 1999. *Morbidity and Mortality Weekly Report* 50: 120-125.

Felson D.T., R.C. Lawrence, P.A. Diepp, et al. 2000. Osteoarthritis: New insights part I: The disease and its risk factors. *Annals of Internal Medicine* 133: 635-646.

Finkelstein E.A., I.C. Fiebelkorn, P.S. Corso, and S.C. Binder. 2004. Medical expenditures attributable to injuries—United States, 2000. *Morbidity and Mortality Weekly Report* 53: 1-4.

Hootman, J.M., C.A. Macera, B.A. Ainsworth, et al. 2001. Association among physical activity level, cardiorespiratory fitness and risk of musculoskeletal injury. *American Journal of Epidemiology* 154: 251-258.

Khan, K., H. McKay, P. Kannus, et al. 2001. *Physical activity and bone health*. Champaign, IL: Human Kinetics.

Loud, K.J., C.M. Gordon, L.J. Micheli, and A.E. Field. 2005. Correlates of stress fractures among preadolescent and adolescent girls. *Pediatrics* 115: 399-406.

National Center for Injury Prevention and Control. 2002. *Activity report 2001 CDC's unintentional injury prevention program*. Atlanta: Centers for Disease Control and Prevention, National Center for Injury Prevention and Control.

Powell, K.E., G.W. Heath, M.J. Kresnow, et al. 1998. Injury rates from walking, gardening, weightlifting, outdoor bicycling and aerobics. *Medicine and Science in Sport and Exercise* 30: 1246-1249.

Praemer A., S. Furner, and D.P. Rice. 1999. *Musculoskeletal conditions in the United States*. Rosemont, IL: American Academy of Orthopaedic Surgeons.

Singh, M.A.F. 2002. Exercise to prevent and treat functional disability. *Clinics in Geriatric Medicine* 18: 431-462.

Spirduso, W.W., and D.L. Cronin. 2001. Exercise dose-response effects on quality of life and independent living in older adults. *Medicine and Science in Sport and Exercise* 33(6 Suppl): S598-S608.

Stuck, A.E., J.M. Walthert, T. Nikolaaus, et al. 1999. Risk factors for functional status decline in community-living elderly people: A systematic literature review. *Social Science and Medicine* 48: 445-469.

van Mechelen, W. 1992. Running injuries: A review of the epidemiological literature. *Sports Medicine* 14: 320-335.

Vuori, I.M. 2001. Dose-response of physical activity and low pack pain, osteoarthritis and osteoporosis. *Medicine and Science in Sports and Exercise* 33(6 Suppl): S551-S586.

Wolf, S.L., H.X. Barnhart, N.G. Kutner, et al. 2003. Selected as the best paper in the 1990's: Reducing frailty and falls in older persons: An investigation of tai chi and computerized balance training. *Journal of the American Geriatrics Society* 51: 1794-1803.

World Health Organization. 2004. Global Burden of Disease Project. Revised 2002 Global burden of disease estimates by regions. Retrieved October 27, 2006, from www3.who.int/whosis/menu.cfm?path=whosis,burden ,burden_estimates,burden_estimates_2002N.

Chapter 15

Braith, R.W. 1998. Exercise training in patients with CHF and heart transplant recipients. *Medicine and Science in Sports and Exercise* 30: S367-S378.

Campbell, W.W., M.L. Barton, Jr., D. Cyr-Campbell, S.L. Davey, J.L. Beard, G. Parise, and W.J. Evans. 1999. Effects of an omnivorous diet compared with a lactoovovegetarian diet on resistance-training-induced changes in body composition and skeletal muscle in older men. *American Journal of Clinical Nutrition* 70: 1032-1039.

Cooper, K.H. 1968. *Aerobics.* New York: M. Evans.

De Lorme, T., and A.L. Watkins. 1945. Restoration of power by heavy resistance exercises. *Journal of Bone Joint Surgery* 27: 645-647.

Feigenbaum, M.S., and M.L. Pollock. 1999. Prescription of resistance training for health and disease. *Medicine and Science in Sports and Exercise* 31: 38-45.

Fiatarone Singh, M.A. 2001. Elderly patients and frailty. In *Resistance training for health and rehabilitation*, edited by J.E. Graves and B.A. Franklin. Champaign, IL: Human Kinetics, pp. 181-213.

Hellekson, K.L. 2002. NIH releases statement on osteoporosis prevention, diagnosis, and therapy. *American Family Physician* 66: 161-162.

McCartney, N. 1998. Role of resistance training in heart disease. *Medicine and Science in Sports and Exercise* 30: S396-S402.

McCartney, N. 1999. Acute responses to resistance training and safety. *Medicine and Science in Sports and Exercise* 31: 31-37.

McKelvie, R.S., K.K. Teo, R. Roberts, N. McCartney, D. Humen, T. Montague, K. Hendrican, and S. Yusuf. 2002. Effects of exercise training in patients with heart failure: The Exercise Rehabilitation Trial (EXERT). *American Heart Journal* 14: 423-430.

Pescatello, L.S., B.A. Franklin, R. Fagard, W.B. Farquhar, G.A. Kelley, and C.A. Ray. 2004. American College of Sports Medicine position stand. Exercise and hypertension. *Medicine and Science in Sports and Exercise* 36: 533-553.

Phillips, S.M. 2004. Protein requirements and supplementation in strength sports. *Nutrition* 20: 689-695.

Pollock, M.L., B.A. Franklin, G.J. Balady, B.L. Chaitman, J.L. Fleg, B. Fletcher, M. Limacher, I.L. Pina, R.A. Stein, M. Williams, and T. Bazzarre. 2000. AHA Science Advisory. Resistance exercise in individuals with and without cardiovascular disease: Benefits, rationale, safety, and prescription: An advisory from the Committee on Exercise, Rehabilitation, and Prevention, Council on Clinical Cardiology, American Heart Association; Position paper endorsed by the American College of Sports Medicine. *Circulation* 10: 1828-1833.

Reinsch, S., P. MacRae, P.A. Lachenbruch, and J.S. Tobis. 1992. Attempts to prevent falls and injury: A prospective community study. *Gerontologist* 32: 450-456.

Rice, B., I. Janssen, R. Hudson, and R. Ross. 1999. Effects of aerobic or resistance exercise and/or diet on glucose tolerance and plasma insulin levels in obese men. *Diabetes Care* 22: 684-691.

Tseng, B.S., D.R. Marsh, M.T. Hamilton, and F.W. Booth. 1995. Strength and aerobic training attenuate muscle wasting and improve resistance to the development of disability with aging. *Journals of Gerontology. Series A. Biological Sciences and Medical Sciences* 50: 113-119.

van Baar, M.E., W.J. Assendelft, J. Dekker, R.A. Oostendorp, and J.W. Bijlsma. 1999. Effectiveness of exercise therapy in patients with osteoarthritis of the hip or knee: A systematic review of randomized clinical trials. *Arthritis and Rheumatology* 42: 1361-1369.

World Health Organization. 1997. *Obesity: Preventing and managing the global epidemic.* Report of a WHO Consultation on Obesity, June 3-5, 1997. Geneva: WHO.

Chapter 16

American Psychiatric Association. 1994. *Diagnostic and statistical manual of mental disorders.* 4th ed. Washington, DC: American Psychiatric Association.

Bahrke, M.S., and W.P. Morgan. 1978. Anxiety reduction following exercise and meditation. *Cognitive Therapy and Research* 2: 323-333.

Blumenthal, J.A., M.A. Babyak, K.A. Moore, W.E. Craighead, S. Herman, P. Khatri, R. Waugh, M.A. Napolitano, L.M. Forman, M. Appelbaum, P.M. Doraiswamy, and K.R. Krishnan. 1999. Effects of exercise training on older patients with major depression. *Archives of Internal Medicine* 159: 2349-2356.

Broocks A., B. Bandelow, G. Pekrun, A. George, T. Meyer, U. Bartmann, U. Hillmer-Vogel, and E. Rüther. 1998. Comparison of aerobic exercise, clomipramine, and placebo in the treatment of panic disorder. *American Journal of Psychiatry* 155: 603-609.

Buckworth, J., and R.K. Dishman. 2002. *Exercise psychology.* Champaign IL: Human Kinetics.

Dishman, R.K. 1985. Medical psychology in exercise and sport. *Medical Clinics of North America* 9: 123-143.

Dunn, A.L., M.H. Trivedi, J.B. Kampert, C.G. Clark, and H.O. Chambliss. 2005. Exercise treatment for depression: Efficacy and dose response. *American Journal of Preventive Medicine* 28: 1-8.

Martinsen, E.W., J.S. Raglin, A. Hoffart, and S. Friis. 1998. Tolerance to intensive exercise and high levels of lactate in panic disorder. *Journal of Anxiety Disorders* 12: 333-342.

McNair, D.N., M. Lorr, and L.F. Droppleman. 1992. *Profile of Mood State manual.* San Diego: Educational and Industrial Testing Service.

Morgan, W.P. 1979. Negative addiction in runners. *The Physician and Sportsmedicine* 7:57-70.

Morgan, W.P., editor. 1997. *Physical activity and mental health.* Washington, DC: Taylor and Francis.

Morgan, W.P., J.A. Roberts, F.R. Brand, and A.D. Feinerman. 1970. Psychological effects of chronic physical activity. *Medicine and Science in Sports and Exercise* 2: 213-217.

Motl, R.W., A.S. Birnbaum, M.Y. Kubik, and R.K. Dishman. 2004. Naturally occurring changes in physical activity are inversely related to depressive symptoms during early adolescence. *Psychosomatic Medicine* 66: 336-342.

O'Connor, P.J., J.S. Raglin, and E.W. Martinsen. 2000. Physical activity, anxiety, and anxiety disorders. *International Journal of Sport Psychology* 31: 136-155.

Petruzzello, S.J., D.M. Landers, B.D. Hatfield, K.A. Kubitz, and W. Salazar. 1991. A meta-analysis on the anxiety reducing effects of acute and chronic exercise: Outcomes and mechanisms. *Sports Medicine* 11: 143-182.

Phillips, K.A., S.L. McElroy, P. Keck, H.G. Pope, and J. Hudson. 1993. Body dysmorphic disorder: Thirty cases of imagined ugliness. *American Journal of Psychiatry* 150: 302-308.

Pitts, F.J, and J.J. McClure. 1967. Lactate metabolism in anxiety neurosis. *New England Journal of Medicine* 277: 1329-1336.

Raglin, J.S. 1997. Anxiolytic effects of physical activity. In *Physical activity and mental health*, edited by W.P. Morgan. Washington, DC: Taylor and Francis, pp. 107-126.

Raglin, J.S., and G.S. Wilson. 2000. Overtraining in athletes. In *Emotion in sports*, edited by Y.L. Hanin. Champaign, IL: Human Kinetics, pp. 191-207.

Spielberger, C.D., R.L. Gorsuch, R.E. Lushene, P.R. Vagg, and G.A. Jacobs. 1983. *Manual for the State–Trait Anxiety Inventory (form Y)*. Palo Alto, CA: Consulting Psychologists Press.

Chapter 17

Alpert, B.S., and J.H. Wilmore. 1994. Physical activity and blood pressure in adolescents. *Pediatric Exercise Science* 6: 361-380.

Armstrong, N., and B. Simons-Morton. 1994. Physical activity and blood lipids in adolescents. *Pediatric Exercise Science* 6: 381-405.

Armstrong, N., and W. van Mechelen. 2000. *Paediatric exercise science and medicine*. Oxford, UK: Oxford University Press.

Bailey, D.A., and A.D. Martin. 1994. Physical activity and skeletal health in adolescents. *Pediatric Exercise Science* 6: 330-347.

Bar-Or, O., and T. Baranowski. 1994. Physical activity, adiposity, and obesity among adolescents. *Pediatric Exercise Science* 6: 348-361.

Bar-Or, O., and T.W. Rowland. 2004. *Pediatric exercise medicine*. Champaign, IL: Human Kinetics.

Blair, S.N., D.G. Clark, K.J. Cureton, and K.E. Powell. 1989. Exercise and fitness in childhood: Implications for a lifetime of health. In *Youth, exercise, and sport. Perspectives in exercise science and sports medicine*, edited by C.V. Gisolfi and D.R. Lamb. Indianapolis: Benchmark Press, pp. 401-430.

Burke, G., L. Webber, S. Srinivasan, B. Radhakrishnamurthy, D. Freedman, and G. Beenson. 1986. Fasting plasma glucose and insulin levels and their relationship to cardiovascular risk factors in children: Bogalusa Heart Study. *Metabolism* 35: 441-446.

Cavill, N., S. Biddle, and J.F. Sallis. 2001. Health enhancing physical activity for young people: Statement of the United Kingdom Expert Consensus Conference. *Pediatric Exercise Science* 13: 12-25.

Corbin, C.B., and R.P. Pangrazi. 2004. *Physical activity for children: A statement of guidelines for children ages 5-12*. Reston, VA: National Association for Sport and Physical Education.

Gutin, B., S. Owens, F. Treiber, S. Islam, W. Karp, and G. Slavens. 1997. Weight independent cardiovascular fitness and coronary risk factors. *Archives of Pediatric and Adolescent Medicine* 151: 462-465.

Hager, R.L., L.A. Tucker, and G.T. Seljaas. 1995. Aerobic fitness, blood lipids, and body fat in children. *American Journal of Public Health* 85: 1702-1706.

Health Canada. 2004. *Canada's physical activity guide for children*. Retrieved November 1, 2004, from www.healthcanada.ca/paguide.

Kahle, E.B., W.B. Zipfe, D.R. Lamb, C.A. Horswill, and K.M. Ward. 1996. Association between mild, routine exercise and improved insulin dynamics and glucose control in obese adolescents. *International Journal of Sports Medicine* 17: 1-6.

Kemper, H. 2000. Skeletal development during childhood and adolescence. *Pediatric Exercise Science* 12: 198-216.

Malina, R.M. 1996. Tracking of physical activity and physical fitness across the lifespan. *Research Quarterly for Exercise and Sport* 67(Suppl. 3): S48-S57.

Morrow, J.R., and P.S. Freedson. 1994. Relationship between habitual physical activity and aerobic fitness in adolescents. *Pediatric Exercise Science* 6: 315-329.

Pinhas-Hamiel, O., L.M. Dolan, S.R. Daniels, D. Standiford, P.R. Khoury, and P. Zeitler. 1996. Increasing incidence of non-insulin-dependent diabetes mellitus among adolescents. *Journal of Pediatrics* 128: 608-615.

Riddoch, C. 1998. Relationships between physical activity and physical health in young people. In *Young and active? Young people and health-enhancing physical activity—Evidence and implications*, edited by S. Biddle, J. Sallis, and N. Cavill. London: Health Education Authority, pp. 17-39.

Rowland, T.W. 1990. *Exercise and children's health*. Champaign, IL: Human Kinetics.

Rowland, T.W. 1998. The biological basis of physical activity. *Medicine and Science in Sports and Exercise* 30: 392-399.

Sallis, J.F., and K.P. Patrick. 1994. Physical activity guidelines for adolescents: Consensus statement. *Pediatric Exercise Science* 6: 302-314.

Steinberger, J., and A.P. Rocchini. 1991. Is insulin resistance responsible for the lipid abnormalities seen in obesity? *Circulation* 84(Suppl II): II-5.

Tolfrey, K., A.M. Jones, and I.G. Campbell. 2000. The effect of aerobic exercise training on the lipid-lipoprotein profile of children and adolescents. *Sports Medicine* 29: 99-112.

Tomkinson, G.R., L.A. Leger, T.S. Olds, and G. Cazorla. 2003. Secular trends in the performance of children and adolescents (1980-2000). *Sports Medicine* 33: 285-300.

Chapter 18

Administration on Aging, U.S. Department of Health and Human Services. 2004. *A profile of older Americans.* Washington, DC: Administration on Aging, U.S. Department of Health and Human Services.

American College of Sports Medicine. 1998. Exercise and physical activity for older adults. *Medicine and Science in Sports and Exercise* 30: 992-1008.

Bijnen, F.C.H., E.J.M. Feskens, C.J. Caspersen, W.L. Mosterd, and D. Kromhout. 1998. Age, period, and cohort effects on physical activity among elderly men during 10 years of follow-up: The Zutphen Elderly Study. *Journal of Gerontology: Medical Sciences* 53A: M235-M241.

Bortz, W.M. 1982. Disuse and aging. *Journal of the American Medical Association* 248: 1203-1208.

Caspersen, C.J., M.A. Pereira, and K.M, Curran. 2000. Changes in physical activity patterns in the United States, by sex and cross-sectional age. *Medicine and Science in Sports and Exercise* 32: 1601-1609.

Centers for Disease Control. 1995. Prevalence of recommended levels of physical activity among women—Behavioral Risk Factor Surveillance System, 1992. *Morbidity and Mortality Weekly Report* 44: 105-113.

Colditz, G.A. 1999. Economic costs of obesity. *Medicine and Science in Sports and Exercise* 31: S663-S667.

Dempsey, J.A., and D.R. Seals. 1995. Aging, exercise, and cardiopulmonary function. In *Perspectives in exercise science and sports medicine. Volume 8. Exercise in older adults*, edited by D.R. Lamb, C.V. Gisolfi, and E.R. Nadel. Traverse City, MI: Cooper, pp. 237-304.

Dice, J.F. 1993. Cellular and molecular mechanisms of aging. *Physiological Reviews* 73: 149-159.

DiPietro, L. 2001. Physical activity in aging: Changes in patterns and their relations to health and function. *Journal of Gerontology: Medical Sciences* 56A: 13-22.

DiPietro, L., D.F. Williamson, C.J. Caspersen, and E. Eaker. 1993. The descriptive epidemiology of selected physical activities and body weight among adults trying to lose weight: The Behavioral Risk Factor Surveillance System Survey, 1989. *International Journal of Obesity* 17: 69-76.

Evenson, K.R., W.D. Rosamond, J. Cai, A.V. Diez-Roux, and F.L. Brancati. 2002. Influence of retirement on leisure-time physical activity. The Atherosclerosis Risk in Communities Study. *American Journal of Epidemiology* 155: 692-699.

Federal Interagency Forum on Aging Related Statistics. 2004. *Older Americans 2004: Key indicators of well-being.* Washington, DC: Federal Interagency Forum on Aging Related Statistics.

Heath, G.W., J.M. Hagberg, A.A. Ehsani, and J.O. Holloszy. 1981. A physiological comparison of young and older endurance athletes. *Journal of Applied Physiology* 51: 634-640.

King, A.C., M. Kiernan, R.F. Oman, H.C. Kraemer, M. Hull, and D. Ahn. 1997. Can we identify who will adhere to long-term physical activity? Signal detection methodology as a potential aid to clinical decision making. *Health Psychology* 16: 380-389.

Lakatta, E.G. 1993. Cardiovascular regulatory mechanisms in advanced age. *Physiological Reviews* 78: 413-467.

National Center for Health Statistics. 2003. Reported physical activity among older persons living in the United States between 1985 and 1995, retrieved August 15, 2004, from www.agingstats.gov.

National Health Interview Survey–Health Promotion Disease Prevention supplement, 1985-1995. Participation in activities among older persons, 1985-1995, retrieved August 15, 2004, from www.NCHS.com.

U.S. Department of Health and Human Services. 1996. *Physical activity and health: A report of the Surgeon General.* Washington, DC: U.S. Government Printing Office.

Chapter 19

Bouter, L.M. 1986. Spanningsbehoefte en ongevalsrisico bij sportbeoefening. *Geneeskunde en Sport* 19: 205-208.

Burke, A.P., A. Farb, G.T. Malcom, Y-h. Liang, J.E. Smialek, and R. Virmani. 1999. Plaque rupture and sudden death related to exercise in men with coronary artery disease. *Journal of the American Medical Association* 281: 921-926.

Conn, J.M., J.L. Annest, and J. Gilchrist. 2003. Sports and recreation related injury episodes in the US population, 1997-99. *Injury Prevention* 9: 117-123.

Langdeau, J.B., and L.P. Boulet. 2001. Prevalence and mechanisms of development of asthma and airway hyperresponsiveness in athletes. *Sports Medicine* 31: 601-616.

Loucks, A.B., and S.M. Horvath. 1985. Athletic amenorrhoea: A review. *Medicine and Science in Sports and Exercise* 17: 56-72.

Mosterd, W.L. 1999. Plotse dood bij sport in Nederland. *Bijblijven* 15 okt: 68-74.

Otis, C.L., B. Drinkwater, M. Johnson, A. Loucks, and J. Wilmore, J. 1997. The female athlete triad. *Medicine and Science in Sports and Exercise* 29: i-ix.

Parkkari, J., U.M. Kujala, and P. Kannus. 2001. Is it possible to prevent sports injuries? Review of controlled clinical trials and recommendations for future work. *Sports Medicine* 31: 985-995.

Pope, R.P., R.D. Herbert, J.D. Kirwan, and B.J. Graham. 2000. A randomized trial of preexercise stretching for prevention of lower-limb injury. *Medicine and Science in Sports and Exercise* 32: 271-277.

Renström, P.A.F.H., editor. 1993. *Sports injuries: Basic principles of prevention and care. Encyclopaedia of Sports Medicine Vol. IV.* Oxford, UK: Blackwell Scientific.

Renström, P.A.F.H., editor. 1994. *Clinical practice of sports injuries: Prevention and care. Encyclopaedia of Sports Medicine Vol. V.* Oxford, UK: Blackwell Scientific.

Schmikli, S.L. 2002. 97/98 survey van sportblessures. In *Trendrapport Bewegen en Gezondheid 2000/2001*, edited by W.T.M. Ooijendijk, V.H. Hildebrandt, and

M. Stiggelbout. TNO Arbeid, TNO PG. Hoofddorp: Netherlands (chapter 7, pp. 79-88).

Shrier, I. 2000. Stretching before exercise: An evidence based approach. *British Journal of Sports Medicine* 34: 324-325.

Siscovick, D.S., N.S. Weiss, R.H. Fletcher, and T. Lasky. 1984. The incidence of primary cardiac arrest during vigorous exercise. *New England Journal of Medicine* 311: 874-877.

Thacker, S.B., J. Gilchrist, D.F. Stroup, and C.D. Kimsey. 2004. The impact of stretching on sports injury risk: A systematic review of the literature. *Medicine and Science in Sports and Exercise* 36: 371-378.

van Mechelen, W. 1992. *Aetiology and prevention of running injuries* (dissertation). Amsterdam: Free University of Amsterdam.

Vuori, I. 1995. Reducing the number of sudden deaths in exercise. *Scandinavian Journal of Medicine and Science in Sports* 5: 267-268.

Vuori, I. 1997. *Perspectives on health and exercise*, edited by J. McKenna and C. Riddoch. New York: Palgrave Macmillan.

Chapter 20

Ainsworth, B.E., W.L. Haskell, M.C. Whitt, M.I. Irvin, A.S. Schwartz, S.J. Straith, W.L. O'Brian, D.R. Basset, K.H. Schmitz, P.O. Emplaincourt, D.R. Jacobs, and A.S. Leon. 2000. Compendium of physical activities: an update of activity codes and Met intensities. *Medicine and Science in Sports and Exercise* 32: S498-S519.

Blair, S.N., H.W. Kohl, III, R.S. Paffenbarger, D.G. Clark, K.H. Cooper, and L.W. Gibbons. 1989. Physical fitness and all-cause mortality: A prospective study of healthy men and women. *Journal of the American Medical Association* 262: 2395-2401.

Gyntelberg, F., R. Brennan, J.O. Holloszy, G. Schonfeld, M. Rennie, and S. Weidman. 1977. Plasma triglyceride lowering by exercise despite increased food intake in patients with Type IV hyperlipoproteinemia. *American Journal of Clinical Nutrition* 30: 716-720.

Hardman, A.E. 2001. Issues of fractionalization of exercise (short vs long bouts). *Medicine and Science in Sports and Exercise* 33: S421-S427.

Howley, E.T. 2001. Type of activity: resistance, aerobic and leisure versus occupational physical activity. *Medicine and Science in Sports and Exercise* 33: S364-S369.

Kujala, U.M., J. Kaprio, S. Sarna, and M. Koskenvuo. 1998. Relationship of leisure-time physical activity and mortality: The Finnish twin cohort. *Journal of the American Medical Association* 279: 440-444.

Laukkanen, J.A., T.A. Lakka, R. Rauramaa, R. Kuhanen, J.M. Venalainen, R. Salonen, and J.T. Salonen. 2001. Cardiovascular fitness as a predictor of mortality in men. *Archives of Internal Medicine* 161: 825-831.

Leon, A.S., J. Connett, D.R. Jacobs, and R. Rauramaa. 1987. Leisure-time physical activity levels and risk of dose response and death: The Multiple Risk Factor Intervention Trial. *Journal of the American Medical Association* 258: 2388-2395.

Manson, J.E., P. Greenland, A.Z. LaCroix, M.L. Stefanick, C.P. Mouton, A. Oberman, M.G. Perri, D.S. Sheps, M.B. Pettinger, and D.S. Siscovick. 2002. Walking compared with vigorous exercise for the prevention of cardiovascular events in women. *New England Journal of Medicine* 347: 716-726.

Myers, J., M. Prakash, V. Froelicher, D. Do, S. Partington, and J.E. Atwood. 2002. Exercise capacity and mortality among men referred for exercise testing. *New England Journal of Medicine* 346: 793-801.

Paffenbarger, R.S., R.T. Hyde, A.L. Wing, and C.C. Hsieth. 1986. Physical activity, all-cause mortality, and longevity of college alumni. *New England Journal of Medicine* 314: 605-613.

Pate, R., M. Pratt, S.N. Blair, W.L. Haskell, C.A. Macera, C. Bouchard, D. Buchner, W. Ettinger, G.W. Heath, A.C. King, A. Kriska, A.S. Leon, B.H. Marcus, R. Paffenbarger, S.K. Patrick, M.L. Pollock, J.M. Rippe, J. Sallis, and J.H. Wilmore. 1995. Physical activity and public health: A recommendation from the Centers for Disease Control and Prevention and the American College of Sports Medicine. *Journal of the American Medical Association* 273: 402-407.

Pescatello, L.S., A.E. Fargo, C.N. Leach, and A.E. Scherzer. 1991. Short-term effect of dynamic exercise on arterial blood pressure. *Circulation* 83: 1557-1561.

Pollock, M.L., J. Ayers, and A. Ward. 1977. Cardiorespiratory fitness response to differing intensities and durations of training. *Archive of Physical Medicine and Rehabilitation* 58: 467-473.

Saris, W.H.M., S.N. Blair, M.A. van Back, E.A. Eaton, P.S.W. Davies, L. Di Pietro, M. Fogelholm, A. Rissanen, D. Schoeller, B. Swinburn, A. Tremblay, K.R. Westerpert, and H. Wyatt. 2003. How much physical activity is enough to prevent unhealthy weight gain? Outcome of the IASO 1st Stock Conference and consensus statement. *Obesity Reviews* 4: 101-114.

United States Department of Health and Human Services and United States Department of Agriculture. 2005. *Dietary guidelines for Americans 2005*. Washington, DC: United States Department of Health and Human Services and United States Department of Agriculture.

Williams, P. 1997. Relationship of distance run per week to coronary heart disease risk factors in 8283 male runners: The National Runners' Health Study. *Archives of Internal Medicine* 157: 191-198.

Wilmore, J.H., J. Royce, R.N. Girandola, F.I. Katch, and V.L. Katch. 1970. Physiological alterations resulting from a 10-week jogging program. *Medicine and Science in Sports* 2: 7-14.

Chapter 21

Andersen R.E., T.A. Wadden, S.J. Bartlett, B. Zemel, T.J. Verde, and S.C. Franckowiak. 1999. Effects of lifestyle activity vs structured aerobic exercise in obese women: A randomized trial. *Journal of the American Medical Association* 281: 335-340.

Bauman, A. 1998. Quasi-experimental designs in public health research, Part II, Unit 21. In *Public Health*

Research Methods, edited by C.B. Kerr and R. Taylor. Sydney: McGraw-Hill, pp. 137-144.

Bauman, A., and B. Bellew. 1999. Environmental and policy approaches to promoting physical activity. In *Health in the Commonwealth—sharing solutions 1999/2000*. London: Commonwealth Secretariat, pp. 38-41.

Boule, N.G., G.P. Kenny, E. Haddad, G.A. Wells, and R.J. Sigal. 2003. Meta-analysis of the effect of structured exercise training on cardiorespiratory fitness in type 2 diabetes mellitus. *Diabetologia* 46: 1071-1081.

Brownson, R.C., A.A. Eyler, A.C. King, D.R. Brown, Y.L. Shyu, and J.F. Sallis. 2000. Patterns and correlates of physical activity among US women 40 years and older. *American Journal of Public Health* 90: 264-270.

Cox, K.L., V. Burke, T.J. Gorely, L.J. Beilin, and I.B. Puddey. 2003. Controlled comparison of retention and adherence in home- vs center-initiated exercise interventions in women ages 40-65 years: The SWEAT Study (Sedentary Women Exercise Adherence Trial). *Previews in Medicine* 36: 17-29.

Dishman, R.K., B. Oldenburg, H. O'Neal, and R.J. Shephard. 1998. Worksite physical activity interventions. *American Journal of Previews in Medicine* 15: 344-361.

Dunn, A.L., B.H. Marcus, J.B. Kampert, M.E. Garcia, H.W. Kohl, III, and S.N. Blair. 1999. Comparison of lifestyle and structured interventions to increase physical activity and cardiorespiratory fitness: A randomized trial. *Journal of the American Medical Association* 281(4): 327-334.

Eden, K.B., T.C. Orleans, C.D. Mulrow, N.J. Pender, and S.M. Teutsch. 2002. Counseling by clinicians: Does it improve physical activity: A summary of the evidence for the U.S. Preventive Services Task Force. *Annals of Internal Medicine* 137: 208-215.

Glanz, K., B.K. Rimer, and F.M. Lewis. 2002. *Health behavior and health education: Theory, research, and practice*. 3rd ed. San Francisco: Jossey-Bass.

Kahn, E.B., L.T. Ramsey, R.C. Brownson, G.W. Heath, E.H. Howze, K.E. Powell, E.J. Stone, M.W. Rajab, and P. Corso. 2002. The effectiveness of interventions to increase physical activity. A systematic review. *American Journal of Previews in Medicine* 22(4 Suppl): 73-107.

Knowler, W.C., E. Barrett-Connor, S.E. Fowler, R.F. Hamman, J.M. Lachin, E.A. Walker, D.M. Nathan, and the Diabetes Prevention Program Research Group. 2002. Reduction in the incidence of type 2 diabetes with lifestyle intervention or metformin. *New England Journal of Medicine* 346: 393-403.

Nader, P.R., E.J. Stone, L.A. Lytle, C.L. Perry, S.K. Osganian, S. Kelder, L.S. Webber, J.P. Elder, D. Montgomery, H.A. Feldman, M. Wu, C. Johnson, G.S. Parcel, and R.V. Luepker. 1999. Three-year maintenance of improved diet and physical activity: The CATCH cohort. Child and Adolescent Trial for Cardiovascular Health. *Archives of Pediatrics and Adolescent Medicine* 153: 695-704.

Pate, R.R., M. Pratt, S.N. Blair, W.L. Haskell, C.A. Macera, C. Bouchard, D. Buchner, W. Ettinger, G.W. Heath, A.C. King, et al. 1995. Physical activity and public health: A recommendation from the Centers for Disease Control and Prevention and the American College of Sports Medicine. *Journal of the American Medical Association* 273: 402-407.

Sallis, J.F., A. Bauman, and M. Pratt. 1998. Environmental and policy interventions to promote physical activity. *American Journal of Previews in Medicine* 15: 379-397.

Sevick, M.A., A.L. Dunn, M.S. Morrow, B.H. Marcus, G.J. Chen, and S.N. Blair. 2000. Cost-effectiveness of lifestyle and structured exercise interventions in sedentary adults: Results of project ACTIVE. *American Journal of Preventive Medicine* 19: 1-8.

Taylor, A.H., N.T. Cable, G. Faulkner, M. Hillsdon, M. Narici, and A.K. ven der Bij. 2004. Physical activity and older adults: A review of health benefits and the effectiveness of interventions. *Journal of Sports Sciences* 22(8): 703-725.

Taylor, W.C., T. Baranowski, and D.R. Young. 1998. Physical activity interventions in low-income, ethnic minority, and populations with disability. *American Journal of Previews in Medicine* 15: 334-343.

Timperio, A., J. Salmon, and K. Ball. 2004. Evidence based strategies to promote physical activity among children, adolescents and young adults: Review and update. *Journal of Science and Medicine in Sport* 7: S20-S29.

Tudor-Locke, C., R. Jones, A.M. Myers, D.H. Paterson, and N.A. Ecclestone. 2002. Contribution of structured exercise class participation and informal walking for exercise to daily physical activity in community-dwelling older adults. *Research Quarterly in Exercise and Sport* 73: 350-356.

van der Bij, A.K., M.G. Laurant, and M. Wensing. 2002. Effectiveness of physical activity interventions for older adults: A review. *American Journal of Previews in Medicine* 22: 120-133.

U.S. Department of Health and Human Services. 1996. *Physical activity and health: A report of the Surgeon General*. Pittsburgh: U.S. Department of Health and Human Services, Centers for Disease Control and Prevention, National Center for Chronic Disease Prevention and Health Promotion.

Chapter 22

Arden, N.K., and T.D. Spector. 1997. Genetic influences on muscle strength, lean body mass, and bone mineral density: A twin study. *Journal of Bone and Mineral Research* 12: 2076-2081.

Bouchard, C., P. An, T. Rice, J.S. Skinner, J.H. Wilmore, J. Gagnon, L. Perusse, A.S. Leon, and D.C. Rao. 1999. Familial aggregation of VO$_2$max response to exercise training: Results from the HERITAGE Family Study. *Journal of Applied Physiology* 87: 1003-1008.

Bouchard, C., E.W. Daw, T. Rice, L. Perusse, J. Gagnon, M.A. Province, A.S. Leon, D.C. Rao, J.S. Skinner, and J.H. Wilmore. 1998. Familial resemblance for VO$_2$max in the sedentary state: The HERITAGE Family Study. *Medicine and Science in Sports and Exercise* 30: 252-258.

Bouchard, C., F.T. Dionne, J.A. Simoneau, and M.R. Boulay. 1992. Genetics of aerobic and anaerobic performances. *Exercise and Sport Sciences Reviews* 20: 27-58.

Bouchard, C., A.S. Leon, D.C. Rao, J.S. Skinner, J.H. Wilmore, and J. Gagnon. 1995. The HERITAGE Family Study. Aims, design, and measurement protocol. *Medicine and Science in Sports and Exercise* 27: 721-729.

Bouchard, C., L. Perusse, C. Leblanc, A. Tremblay, and G. Theriault. 1988. Inheritance of the amount and distribution of human body fat. *International Journal of Obesity* 12: 205-215.

Bouchard, C., and T. Rankinen. 2001. Individual differences in response to regular physical activity. *Medicine and Science in Sports and Exercise* 33: S446-S451.

Bouchard, C., T. Rankinen, Y.C. Chagnon, T. Rice, L. Perusse, J. Gagnon, I. Borecki, P. An, A.S. Leon, J.S. Skinner, J.H. Wilmore, M. Province, and D.C. Rao. 2000. Genomic scan for maximal oxygen uptake and its response to training in the HERITAGE Family Study. *Journal of Applied Physiology* 88: 551-559.

Bouchard, C., J.A. Simoneau, G. Lortie, M.R. Boulay, M. Marcotte, and M.C. Thibault. 1986. Genetic effects in human skeletal muscle fiber type distribution and enzyme activities. *Canadian Journal of Physiology and Pharmacology* 64: 1245-1251.

Cardon, L.R., and L.J. Palmer. 2003. Population stratification and spurious allelic association. *Lancet* 361: 598-604.

Erbs, S., Y. Baither, A. Linke, V. Adams, Y. Shu, K. Lenk, S. Gielen, R. Dilz, G. Schuler, and R. Hambrecht. 2003. Promoter but not exon 7 polymorphism of endothelial nitric oxide synthase affects training-induced correction of endothelial dysfunction. *Arteriosclerosis, Thrombosis, and Vascular Biology* 23: 1814-1819.

Hagberg, J.M, R.E. Ferrell, L.I. Katzel, D.R. Dengel, J.D. Sorkin, and A.P. Goldberg. 1999. Apolipoprotein E genotype and exercise training-induced increases in plasma high-density lipoprotein (HDL)- and HDL2-cholesterol levels in overweight men. *Metabolism* 48: 943-945.

Lakka, T.A., T. Rankinen, S.J. Weisnagel, Y.C. Chagnon, T. Rice, A.S. Leon, J.S. Skinner, J.H. Wilmore, D.C. Rao, and C. Bouchard. 2003. A quantitative trait locus on 7q31 for the changes in plasma insulin in response to exercise training: The HERITAGE Family Study. *Diabetes* 52: 1583-1587.

Leon, A.S., K. Togashi, T. Rankinen, J.P. Despres, D.C. Rao, J.S. Skinner, J.H. Wilmore, and C. Bouchard. 2004. Association of apolipoprotein E polymorphism with blood lipids and maximal oxygen uptake in the sedentary state and after exercise training in the HERITAGE Family Study. *Metabolism* 53: 108-116.

Modrek, B., and C. Lee. 2002. A genomic view of alternative splicing. *Nature Genetics* 30: 13-19.

Montgomery, H.E., P. Clarkson, C.M. Dollery, K. Prasad, M.A. Losi, H. Hemingway, D. Statters, M. Jubb, M. Girvain, A. Varnava, M. World, J. Deanfield, P. Talmud, J.R. McEwan, W.J. McKenna, and S. Humphries. 1997. Association of angiotensin-converting enzyme gene I/D polymorphism with change in left ventricular mass in response to physical training. *Circulation* 96: 741-747.

Myerson, S.G., H.E. Montgomery, M. Whittingham, M. Jubb, M.J. World, S.E. Humphries, and D.J. Pennell. 2001. Left ventricular hypertrophy with exercise and ACE gene insertion/deletion polymorphism: A randomized controlled trial with Losartan. *Circulation* 103: 226-230.

Rankinen, T., L. Perusse, J. Gagnon, Y.C. Chagnon, A.S. Leon, J.S. Skinner, J.H. Wilmore, D.C. Rao, and C. Bouchard. 2000. Angiotensin-converting enzyme I/D polymorphism and trainability of the fitness phenotypes. The HERITAGE Family Study. *Journal of Applied Physiology* 88: 1029-1035.

Rankinen, T., L. Perusse, R. Rauramaa, M.A. Rivera, B. Wolfarth, and C. Bouchard. 2004. The human gene map for performance and health-related fitness phenotypes: The 2003 update. *Medicine and Science in Sports and Exercise* 36: 1451-1469.

Rankinen, T., T. Rice, L. Perusse, Y.C. Chagnon, J. Gagnon, A.S. Leon, J.S. Skinner, J.H. Wilmore, D.C. Rao, and C. Bouchard. 2000. NOS3 Glu298Asp genotype and blood pressure response to endurance training: The HERITAGE Family Study. *Hypertension* 36: 885-889.

Rico-Sanz, J., T. Rankinen, T. Rice, A.S. Leon, J.S. Skinner, J.H. Wilmore, D.C. Rao, and C. Bouchard. 2004. Quantitative trait loci for maximal exercise capacity phenotypes and their responses to training in the HERITAGE Family Study. *Physiological Genomics* 16: 256-260.

Roth, S.M., R.E. Ferrell, D.G. Peters, E.J. Metter, B.F. Hurley, and M.A. Rogers. 2002. Influence of age, sex, and strength training on human muscle gene expression determined by microarray. *Physiological Genomics* 10: 181-190.

Simoneau, J.A., and C. Bouchard. 1995. Genetic determinism of fiber type proportion in human skeletal muscle. *FASEB Journal* 9: 1091-1095.

Szathmary, E., F. Jordan, and C. Pal. 2001. Molecular biology and evolution. Can genes explain biological complexity? *Science* 292: 1315-1316.

Chapter 23

American College of Sports Medicine. 2005. *ACSM's guidelines for exercise testing and prescription.* 7th ed. Philadelphia: Lippincott Williams & Wilkins.

Bassett, D.R., P.L. Schneider, and G.E. Huntington. 2004. Physical activity in an Old Order Amish community. *Medicine and Science in Sports and Exercise* 36: 79-85.

Blair, S.N., A.L. Dunn, B.H. Marcus, R.A. Carpenter, and P. Jaret. 2001. *Active living every day—20 weeks to lifelong vitality.* Champaign, IL: Human Kinetics.

Boreham, C.A., F.M. Wallace, and A. Nevill. 2000. Training effects of accumulated daily stair-climbing exercise in previously sedentary young women. *Preventive Medicine* 30: 277-281.

Committee of Physical Activity, Health, Transportation, and Land Use. 2005. *Does the built environment influence physical activity? Examining the evidence.* Special Report 282. Washington, DC: Transportation Research Board, Institute of Medicine, National Academies of Science.

Dunn, A.L., B.H. Marcus, J.B. Kampert, M.E. Garcia, H.W. Kohl, III, and S.N. Blair. 1999. Comparison of lifestyle and structured interventions to increase physical activity and cardiorespiratory fitness: A randomized trial. *Journal of the American Medical Association* 281: 327-334.

Fitness Canada. 1991. *Active living: A conceptual overview.* Ottawa: Government of Canada.

Hedley, A.A., C.L. Ogden, C.L. Johnson, M.D. Carroll, L.R. Curtin, and K.M. Fegal. 2004. Prevalence of overweight and obesity among US children, adolescents, and adults; 1999-2002. *Journal of the American Medical Association* 292: 2847-2850.

Henry J. Kaiser Family Foundation. 1999. *Kids and media at the millennium: A comprehensive national survey of children's media use.* Menlo Park, CA: Henry J. Kaiser Family Foundation.

Hill, J.O., H.R. Wyatt, G.W. Reed, and J.C. Peters. 2003. Obesity and the environment: Where do we go from here? *Science* 299: 853-855.

Hu, F.B., W.C. Willett, T. Li, M.J. Stampfer, G.A. Colditz, and J.E. Manson. 2004. Adiposity compared to physical activity in predicting mortality among women. *New England Journal of Medicine* 351: 2694-2703.

Lanningham-Foster, L., L.J. Nysse, and J.A. Levine. 2003. Labor saved, calories lost: The energetic impact of domestic labor-saving devices. *Obesity Research* 11: 1178-1181.

Lee, I-M., H.D. Sesso, and R.S. Paffenbarger. 2000. Physical activity and coronary heart disease risk in men: Does the duration of exercise episodes predict risk? *Circulation* 102: 981-986.

Levine, J.A., L.M. Lanningham-Foster, S.K. McCrady, A.C. Krizan, L.R. Olson, P.H. Kane, M.D. Jensen, and M.M. Clark. 2005. Interindividual variation in posture allocation: Possible role in human obesity. *Science* 307: 584-586.

Makosky, L. 1994. The active living concept. In *Toward active living: Proceedings of the International Conference on Physical Activity, Fitness and Health*, edited by H.A. Quinney, L. Gauvin, and A.E. Wall. Champaign, IL: Human Kinetics, pp. 272-276.

Montgomery, E. 1978. Toward representative energy expenditure data: The Machiguenga Study. *Federation Proceedings* 37: 61-64.

Simons-Morton, D.G., P. Hogan, A.L. Dunn, L. Pruitt, A.C. King, B.D. Levine, and S.T. Miller. 2000. Characteristics of inactive primary care patients: Baseline data from the Activity Counseling Trial. *Preventive Medicine* 31: 513-521.

Stampfer, M.J., F.B. Hu, J.E. Manson, E.B. Rimm, and W.C. Willett. 2000. Primary prevention of coronary heart disease in women through diet and exercise. *New England Journal of Medicine* 343: 16-22.

U.S. Department of Health and Human Services. 2004. *Advance data from vital and health statistics: 347, Mean body weight, height and body mass index, United States 1960-2002.* Washington, DC: U.S. Department of Health and Human Services.

Dietary guidelines for Americans 2005. Yusuf, S., S. Hawken, S. Ounpuu, T. Dans, A. Avzeum, F. Lanas, M. McQueen, A. Budaj, P. Pais, J. Varigos, and L. Lisheng. 2004. Effect of potentially modifiable risk factors associated with myocardial infarction in 52 countries (the INTERHEART study) case-control study. *Lancet* 364: 937-952.

Index

About the Editors

Claude Bouchard, PhD, is the executive director of the Pennington Biomedical Research Center, a campus of the Louisiana State University System, and holds the George A. Bray Jr. chair in nutrition. He was director of the Physical Activity Sciences Laboratory at Laval University, Quebec City, Canada, for over 20 years. Dr. Bouchard holds a BPed from Laval University, an MSc in exercise physiology from the University of Oregon at Eugene, and a PhD in population genetics from the University of Texas at Austin.

For four decades, his research has dealt with the role of physical activity, and the lack thereof, on physiology, metabolism, and indicators of health, taking into account genetic uniqueness. He has performed research on the contributions of gene sequence variation and the benefits to be expected from regular activity in terms of changes in cardiovascular and diabetes risk factors.

Dr. Bouchard has served as program leader for four consensus conferences and symposia pertaining to various aspects of physical activity and health. He has published more than 850 scientific papers and has edited four books and several monographs dealing with physical activity and health.

Dr. Bouchard is former president of the Canadian Society for Applied Physiology and the North American Association for the Study of Obesity. He is president of the International Association for the Study of Obesity (2002-2006), a fellow of the American College of Sports Medicine (ACSM), and has been a member of the Scientific Advisory Board of the Cooper Research Institute for the last decade.

Steven N. Blair, PED, is the president and CEO of the Cooper Institute and is one of the world's most eminent epidemiologists in the area of physical activity and health. Dr. Blair has three honorary doctoral degrees in the United States, England, and Belgium.

For nearly 40 years, he has researched and done public health work in the areas of physical activity and health. He has published more than 350 scientific articles, including one on fitness and mortality that has been cited over 1,100 times.

He is past president of the American College of Sports Medicine and a fellow in many organizations. He has received numerous honors, including the ACSM Honor Award. He was the senior scientific editor of the Surgeon General's Report on Physical Activity and Health.

William Haskell, PhD, is emeritus professor of medicine at Stanford Prevention Research Center, Stanford School of Medicine. He holds an honorary MD degree from Linkoping University in Sweden.

For 40 years, his research has investigated the relationships between physical activity and health. He has been involved at the national and international levels in the development of physical activity and fitness guidelines and recommendations for physical activity in health promotion and disease prevention.

Dr. Haskell has served as principal investigator on major NIH-funded research projects demonstrating the health benefits of physical activity. For 11 years, he was a member of the planning committee and faculty for the CDC-sponsored research course on physical activity and public health. From 1968 to 1970, he was program director for the President's Council on Physical Fitness and Sports.

He is past president of the American College of Sports Medicine and founder and past president of the American College of Sports Medicine Foundation. He is a fellow with the Exercise and Rehabilitation Council, American Heart Association, and American Association of Cardiovascular and Pulmonary Rehabilitation.

About the Contributors

Adrian Bauman, MB, BS, MPH, PhD, FAFPHM, is a professor of public health and director of the Centre for Physical Activity and Health in the School of Public Health at the University of Sydney. Dr. Bauman obtained his MPH and PhD in public health with a focus on health promotion program evaluation and epidemiology. He has more than 20 years of research experience in this area and has received multiple research grants from national funding agencies on physical activity and public health. Bauman chaired the NSW Premiers Task Force on Physical Activity from 1996 to 2004 and was an expert member of the WHO Reference Group on Diet and Physical Activity from 2002 to 2004. He established the Physical Activity and Health and the Obesity and Overweight research centers at Sydney University. Dr. Bauman is a member of the American College of Sport Medicine and the International Society of Behavioral Nutrition and Physical Activity.

Loretta DiPietro, PhD, MPH, is associate fellow and associate professor of epidemiology and public health at the John B. Pierce Laboratory in New Haven, Connecticut. Dr. DiPietro's background includes field work and laboratory- and clinic-based studies of physical activity, and she has received research funding from NIA in this area. DiPietro is a fellow of the American College of Sports Medicine and a recipient of the R. Tait McKenzie Prize (AAPHERD), and she served as the epidemic intelligence service officer for the Centers for Disease Control and Prevention from 1990 to 1991.

Peter A. Farrell, PhD, is professor and chair in the department of exercise and sport science at East Carolina University. Dr. Farrell has studied hormonal adaptations to both acute and chronic physical activity since 1980. He has authored more than 90 peer-reviewed publications and chapters with a special emphasis on how physical activity alters insulin secretion, insulin action, and hormonal regulation of muscle protein synthesis. Dr. Farrell conducted some of the original research investigating the role of endorphins during exercise. He has been supported by the National Institutes of Health for his work centering on the regulation of skeletal muscle protein synthesis after resistance exercise. He is a leader in the promotion of regular exercise for people with type 1 and type 2 diabetes mellitus. Farrell is a member of both the American College of Sports Medicine and the American Physiological Society. He has also been a Citation Award winner from the American College of Spots Medicine (2004).

Daniel I. Galper, PhD, is a licensed clinical psychologist with a specialization in health psychology and behavioral medicine. Dr. Galper has been actively involved in physical activity and exercise research for more than a decade. As a graduate student at Virginia Tech, Dr. Galper worked with Richard Winett on studies to promote the adoption and maintenance of physical activity in sedentary adults and with David Altman at Wake Forest University School of Medicine to develop and pilot-test a home-based physical activity program for older adults. Since 1995 Dr. Galper has collaborated with the Cooper Institute in Dallas, Texas, which has certified him as an advanced physical activity specialist. Dr. Galper is currently a senior research associate at the Mood Disorders Research Program and Clinic, department of psychiatry, University of Texas Southwestern Medical Center at Dallas, where he is project director for a four-year NIMH-funded clinical trial, Treatment with Exercise Augmentation for Depression (TrEAD), evaluating the benefits of exercise for people with clinical depression.

Laurie J. Goodyear, PhD, is the senior investigator and section head at the Joslin Diabetes Center in Boston and an associate professor of medicine at Harvard Medical School. Dr. Goodyear has an MS degree in exercise physiology and a PhD in cell biology. She has authored and coauthored numerous publications and obtained several NIH grants. Dr. Goodyear is a member of the American Physiological Society, the American Diabetes Association, and the American College of Sports Medicine. She is a recipient of several awards, including the Mary K. Iacocca Fellow while at the Joslin Research Laboratory (1992); the New Investigator Award from the American College of Sports Medicine (1993); the Career Development Award from the American

Diabetes Association (1993); and the Alumna of the Year Award from the School of Public Health at the University of South Carolina (2002).

Howard J. Green, PhD, is a distinguished professor emeritus in the department of kinesiology at the University of Waterloo in Ontario. Dr. Green has more than 40 years of teaching and research experience in muscle physiology and has published more than 240 research articles and reviews. He has done extensive researched on muscle, which has addressed exercise and training, altitude, and disease (e.g., COPD, CHF), including investigating the cellular basis of weakness and fatigue that accompany those disease states. Dr. Green is a member of the Canadian Society for Exercise Physiology and the American College of Sports Medicine. Among other awards, he has received the 1987 Outstanding Research Award from the Canadian Association of Sport Sciences, the 2001 Award of Excellence in Research from the University of Waterloo, and the 2002 Citation Award (Research) from the American College of Sports Medicine.

Adrianne E. Hardman, MSc, PhD, is professor emeritus of human exercise metabolism at the school of sport and exercise sciences at the University of Loughborough, UK. Adrianne is known internationally for research on the acute effects of exercise bouts on lipoprotein metabolism—particularly in the postprandial period. She was a member of the Scientific Advisory Board for the 1992 United Kingdom National Fitness Survey; a speaker on issues of fractionalization (intermittent versus continuous exercise) at the 2000 Consensus Symposium organized by Health Canada and the Centers for Disease Control and Prevention; and the recipient of the first research grants funded by the British Heart Foundation for studies on the acute effects of exercise. She has received many invitations to speak on the acute response to physical activity and exercise at scientific meetings in the United Kingdom, Europe, and beyond. Dr. Hardman is a member of the American College of Sport Medicine and the British Association of Sport and Exercise Sciences. She received her PhD from Loughborough University and became an honorary fellow of the British Association of Sport and Exercise Sciences in 2001.

Jennifer M. Hootman, PhD, ATC, FACSM, is an epidemiologist in the Arthritis Program at the Centers for Disease Control and Prevention in Atlanta. She trained clinically in sports medicine, practiced as a certified athletic trainer for 10 years,

and taught anatomy and joint evaluation at the university level. Dr. Hootman's doctoral work in epidemiology focused on injury and physical activity epidemiology methods, and she completed a postdoctoral fellowship in arthritis epidemiology at the Centers for Disease Control and Prevention. Dr. Hootman served as an associate editor and on the editorial board of the *Journal of Athletic Training* and the *Journal of Physical Activity and Health* and is a member of the American College of Sports Medicine and the Association of Rheumatology Health Professionals. She has won many awards, including the Student Research Award from the Southeast American College of Sports Medicine in 1998; the Student Travel Award from the American Public Health Association in 2000; and the Journal of Athletic Training, Clint Thompson Award for Outstanding Non-Research Manuscript in 2004.

Edward T. Howley, PhD, is a professor of exercise science at the University of Tennessee. Dr. Howley is the coauthor of an exercise physiology textbook and coeditor of a fitness testing and prescription textbook. He taught exercise physiology for more than 30 years, during which time he received numerous awards for outstanding teaching. Other awards include the University of Wisconsin's School of Education Alumni Achievement Award (2002); the Gatorade Sports Science Institute Excellence in Education Award (2002); the Manhattan College Physical Education Service Award (1988); the SE Chapter of the American College of Sports Medicine Service Award (1994); the SE Chapter of the American College of Sports Medicine Scholar Award (1998); and induction into the American Academy of Kinesiology and Physical Education (1999). From 2002 to 2003 Dr. Howley served as president of American College of Sports Medicine, and he is also a member of the American Physiological Society.

Ian Janssen, PhD, is an assistant professor in the School of Physical and Health Education and the department of community health and epidemiology at Queen's University in Ontario. Dr. Janssen has published more than 50 peer-reviewed publications in the area of obesity, physical activity, and cardiovascular disease. He also has expertise and training in exercise physiology and physical activity epidemiology. Janssen is a member of the Canadian Society for Exercise Physiology (CSEP) and the North American Association for the Study of Obesity (NAASO). Awards include the Queen's National Scholar Award from Queen's University

(2004) and the Governor General's Academic Gold Medal Award from Queen's University (2002).

Peter T. Katzmarzyk, PhD, is an associate professor in the School of Physical and Health Education at Queen's University in Ontario. Dr. Katzmarzyk received his PhD in exercise science from Michigan State University under the supervision of Robert Malina with an MSc in physical anthropology, and he completed a postdoctoral fellowship at Laval University under the direction of Claude Bouchard. A member of the Physical Activity Task Force of the International Association for the Study of Obesity (IASO), the Canadian Society for Exercise Physiology, and fellow of the American College of Sports Medicine, Dr. Katzmarzyk has won the American College of Sports Medicine New Investigator Award (2003) and the 2002 Canadian Society for Exercise Physiology Young Investigator Award (2002) and was recognized as a Queen's National Scholar, Queen's University (2002-2007).

Michael J. LaMonte, PhD, is director of the epidemiology division at the Cooper Institute, which studies physical activity, fitness, and health outcomes with a focus on mortality. Prior to his current position, Dr. LaMonte directed the exercise testing laboratory and research program at LDS Hospital at the University of Utah in Salt Lake City. He completed a postdoctoral fellowship in physical activity epidemiology at the University of South Carolina. He is a member of the American College of Sports Medicine and the American Heart Association.

I-Min Lee, MBBS, ScD, is trained in medicine and epidemiology. She currently serves as associate professor of medicine at Harvard Medical School and associate professor of epidemiology at the Harvard School of Public Health. Her research interests focus on the role of physical activity in promoting health and enhancing longevity and on women's health. She has authored more than 150 scientific publications, most of which relate to physical activity. Dr. Lee is widely recognized for her expertise in physical activity and health. She has been invited to serve on several expert panels and to speak at numerous national and international meetings. She was part of the team that authored the landmark Surgeon General's Report on Physical Activity and Health in 1996. Dr. Lee was awarded the 1999 Young Epidemiologist Award by the Royal Society of Medicine (UK) and the 2002 Golden Shoe award from WalkBoston, a pedestrian advocacy group, for her work on physical activity and women's health. She is an elected member of the American Epidemiological Society and a fellow of the American College of Sports Medicine. Dr. Lee serves on the editorial boards of *Medicine and Science in Sports and Exercise* and *Harvard Women's Health Watch*.

Neil McCartney, PhD, is professor and chair of the department of kinesiology at McMaster University in Ontario. He has served as director of the McMaster University Centre for Health Promotion and Rehabilitation as well as the director of the McMaster University Cardiac Rehabilitation and Seniors' Exercise and Wellness Programs. Dr. McCartney is internationally recognized for his work on strength training in seniors and other special populations such as those with coronary artery disease and neuromuscular disorders. He is a member of the American College of Sport Medicine and the Canadian Association of Cardiac Rehabilitation. His research has demonstrated the safety of resistance training in people with heart disease, the effectiveness of long-term resistance training in healthy seniors, and that resistance training can promote positive changes in the muscle phenotype of individuals with neuromuscular disorders such as limb girdle and fascioscapulohumeral muscular dystrophy.

Russell R. Pate, PhD, is a professor in the department of exercise science in the Arnold School of Public Health at the University of South Carolina. He has taught exercise science for more than 25 years. He is a nationally recognized expert on physical activity and physical fitness. He was the lead author of the landmark CDC-ACSM position statement on physical activity and public health, published in the *Journal of the American Medical Association*. Dr. Pate is a member of the American College of Sports Medicine (having served as its president) and the North American Society for Pediatric Exercise Medicine. He has been the recipient of Citation Awards from the American College of Sports Medicine and the Alliance Scholar Award from the American Alliance for Health, Physical Education, Recreation and Dance.

Stuart M. Phillips, PhD, is an associate professor in the department of kinesiology at McMaster University in Ontario. Dr. Phillips received his BSc in biochemistry, his MSc in nutritional biochemistry, and his PhD in physiology (exercise). He has published more than 40 peer-reviewed publications in human exercise physiology with a focus on muscle physiology. A member of the Canadian Society for

Exercise Physiology and the American College of Sports Medicine, Dr. Phillips has received the New Investigator Award from the Canadian Institutes for Health Research, the Ontario Premier's Research Excellence Award, and the CSEP New Investigator Award.

John S. Raglin, PhD, is a professor in the department of kinesiology at Indiana University in Bloomington. He is a fellow of the American College of Sports Medicine, the American Psychological Association, and the American Academy of Kinesiology and Physical Activity. Dr. Raglin's research focuses on how psychological and physiological variables interact to influence sport performance and exercise behavior.

Tuomo Rankinen, PhD, is an assistant professor in the Human Genomics Laboratory at Pennington Biomedical Research Center in Louisiana. Dr. Rankinen has 10 years of research experience on gene–physical activity interactions on health-related outcomes. He has authored more than 140 original scientific and professional publications, most of them on the genetics of responsiveness to exercise training. He is also the lead author of the Human Gene Map For Performance and Health-Related Fitness Phenotypes review, an annual review published in *Medicine and Science in Sports and Exercise* since 2001. Rankinen has been the project director of the HERITAGE Family Study and has received independent research funding from the National Institutes of Health and the National Heart, Lung, and Blood Institute. He is a member of the American College of Sports Medicine (fellow), the American Physiological Society, and the American Heart Association. He is the recipient of the International Olympic Committee Sport Science Award for the HERITAGE Family Study (1999) and the American College of Sports Medicine New Investigator Award (2001).

Robert Ross, PhD, received his doctoral degree in exercise physiology from the Université de Montréal in 1992. He is currently a professor in the School of Physical and Health Education and the department of medicine, division of endocrinology and metabolism, at Queen's University in Ontario. His research focuses on the characterization of obesity and the development of lifestyle-based strategies designed to prevent and reduce obesity and related comorbid conditions, in particular, insulin resistance. Dr. Ross is recognized internationally as a leader in the area of obesity, physical activity, and metabolism and has published more than 100

manuscripts and book chapters on these and related subjects. He has received a Premiere's Research Excellence Award and the Young Investigator Award from the Canadian Society for Clinical Nutrition, and he was recently awarded a Queen's University research chair.

Thomas Rowland, MD, is a pediatric cardiologist. He has extensive research experience in exercise in children. He has served as the director of pediatric exercise physiology laboratory at the Baystate Medical Center in Springfield, MA, the president of the North American Society for Pediatric Exercise Medicine, editor of *Pediatric Exercise Science*, and director of pediatric cardiology at Baystate Medical Center from 1975 to 2005. He published the textbook *Children's Exercise Physiology* (2005), providing state-of-the-art information in the field. Dr. Rowland is a member of the North American Society for Pediatric Exercise Medicine and the American College of Sports Medicine. He was the winner of the Honor Award from the New England chapter of the American College of Sports Medicine in 1988.

Roy J. Shephard, MB, BS, MD (London), PhD, DPE, is a professor emeritus of applied physiology at the University of Toronto. For more than 50 years he has written, conducted research, and taught at the university level in all areas of exercise physiology. Dr. Shephard has won Honour Awards and served as president of both the American College of Sport Medicine and the Canadian Association of Sport Sciences. He has also received honorary doctorates from Ghent University and Université de Montréal. Dr. Shephard has written more than 100 books on many aspects of exercise physiology. He has worked as a professor of applied physiology, director of the School of Physical Education and Health and professor of preventive medicine, faculty of medicine, University of Toronto.

Willem van Mechelen, MD, PhD, FACSM, FECSS, is a professor of occupational and sports medicine in the department of public and occupational health at the VU University Medical Center in Amsterdam. Dr. van Mechelen has practical and research experience in occupational medicine, work and health, work-related musculoskeletal disorders, sports medicine, physical fitness and physical activity epidemiology, sports injury epidemiology, and applied exercise physiology. He is a fellow of the ACSM and the European College of Sport Sciences. Dr. van Mechelen has been an editorial board member of nine sport medicine journals, a board member of the Dutch College of Sports Medicine, president of the Dutch Society for Human Movement Sciences, and a board member of the Dutch Epidemiological Society.

Esther van Sluijs, PhD, is a career development fellow at MRC Epidemiology Unit at Strangeways Research Laboratory in the United Kingdom. Dr. van Sluijs received her MSc in human movement sciences and her PhD in public health and epidemiology with a focus on physical activity promotion in general practice. She is a member of the EMGO Institute/Department of Public and Occupational Health at the VU University Medical Centre in Amsterdam.

Evert A.L.M. Verhagen, PhD, is a researcher at the Intermunicipal Health Department for the region of the Zuid-Hollandse eilanden in the Netherlands. He received his MSc in human movement sciences and his PhD in sports medicine with a focus on ankle sprain prevention. After his PhD he was involved in several projects on methodology, ankle sprains, pediatrics, and women's sports injuries. Dr. Verhagen is a member of several medical committees, including the committee Winter Sports and Health of the Dutch Skiing Federation, and the Dutch Platform for Injury Registration.

Gregory S. Wilson, PED, is chair and associate professor in the department of exercise and sport science at the University of Evansville. Dr. Wilson has cowritten seven chapters for various edited textbooks dealing with topics in exercise and sport psychology, and he has coedited a textbook titled *Applying Sport Psychology: Four Perspectives.* This text brought together more than 60 contributors; each chapter presented the views of a prominent researcher, applied consultant, coach, and athlete on various topics. He has also published articles on topics including the effects of anxiety and overtraining on sport performance and the effects of coping strategies on physical health. Dr. Wilson is a member of the American College of Sports Medicine and the American Alliance for Health, Physical Education, Recreation and Dance. He has been the recipient of both the Exemplary Teacher of the Year Award (2004) and the Dean's Teaching Award from the University of Evansville (2001).